WA 1259851 8
D0301802

INTRODUCTION TO LIMNOLOGY

LIMNOLOGY

Stanley I. Dodson

INTRODUCTION TO LIMNOLOGY

Stanley I. Dodson

Boston Burr Ridge, IL Dubuque, IA Madison, WI New York San Francisco St. Louis
Bangkok Bogotá Caracas Kuala Lumpur Lisbon London Madrid Mexico City
Milan Montreal New Delhi Santiago Seoul Singapore Sydney Taipei Toronto

INTRODUCTION TO LIMNOLOGY

Published by McGraw-Hill, a business unit of The McGraw-Hill Companies, Inc., 1221 Avenue of the Americas, New York, NY 10020.

 This book is printed on recycled, acid-free paper containing 10% postconsumer waste.

1 2 3 4 5 6 7 8 9 0 DOW/DOW 0 9 8 7 6 5 4 3

ISBN 0-07-287935-1

Publisher: *Margaret J. Kemp*
Senior developmental editor: *Kathleen R. Loewenberg*
Executive marketing manager: *Lisa L. Gottschalk*
Senior project manager: *Mary E. Powers*
Production supervisor: *Kara Kudronowicz*
Lead media project manager: *Judi David*
Media technology producer: *Renee Russian*
Senior coordinator of freelance design: *Michelle D. Whitaker*
Cover/interior designer: *Jamie E. O'Neal*
Cover image: *Daryl Benson/Masterfile*
Interior design image: *water lily image ©Corbis*
Senior photo research coordinator: *Lori Hancock*
Compositor: *Carlisle Communications, Ltd.*
Typeface: *10/12 Times Roman*
Printer: *R. R. Donnelley Willard, OH*

The credits section for this book begins on page 377 and is considered an extension of the copyright page.

Library of Congress Cataloging-in-Publication Data

Dodson, Stanley I.
Introduction to limnology / Stanley Dodson. — 1st ed.
p. cm.
Includes index.
ISBN 0-07-287935-1
1. Limnology. II. Title.

GB1603.2.D63 2005
551.48—dc22 2003060397
CIP

www.mhhe.com

dedication

My family—parents and two sisters—moved to western Colorado when I was quite young, so I got to "develop" in an arid environment, where I learned about irrigation, cottonwood trees, trout, and the value of water. My sisters and I ran outside and danced to celebrate the rare rainstorms. My parents taught me to appreciate nature and showed me how to be a naturalist long before I was a limnologist.

Ginny Dodson, my dear wife, gave the sustained encouragement and support that kept this project on track. She is a skilled teacher and a limnologist who was always willing to contribute her time and expertise in discussions about ways to improve the book. She offered excellent suggestions at all stages of the development of this book and she generously read more revisions than one would think possible.

G. Evelyn Hutchinson taught an ecology class at Yale that introduced me to aquatic communities. He was a naturalist who had an immense love for aquatic organisms and their environment. He also had a great desire to apply population ecology and genetics to limnology. He showed me what it meant to be a creative scholar.

John L. Brooks was also a Yale professor when I was an undergraduate. He welcomed me into his research group to visit Connecticut lakes, collect and identify zooplankton, and be a co-author on one of his excellent papers. This is why I still believe that both field trips and undergraduate research participation are critical to ecological education.

W. Thomas (Tommy) Edmondson was my graduate advisor at the University of Washington, Seattle. He is best known for his study of Lake Washington, as it became polluted with phosphate-rich sewage, and during its subsequent improvement after sewage was diverted from the lake. When his students complained about the Seattle weather, he liked to remind us that "rain makes lakes." Tommy made it possible for me to study effects of predation on the zooplankton communities of small ponds high in the mountains of Colorado.

Edward Asahel Birge was an early pioneer of limnology at the University of Wisconsin, starting in 1875. He served as a Professor of Natural History, Dean of Letters and Sciences, and for three years, Acting President of the University. He also served as an irritant and perhaps a stimulus to G. Evelyn Hutchinson and his student Ray Lindeman (who helped invent "ecosystems" in Minnesota). I never met Professor Birge, but his influence lies heavily over Wisconsin—this book was written on the fourth floor of Birge Hall.

Arthur Hasler spent 41 years at the University of Wisconsin, where he, like Franz Ruttner in Austria (one of the first limnologists), understood and taught fundamental concepts of aquatic ecology. He extended the Wisconsin Idea ("The Boundaries of the University are the Boundaries of the State") to include the idea of applying limnological knowledge to practical problems of fish management and water quality. Art, influenced by the great conservationist Aldo Leopold, is primarily responsible for the vision of a large general limnology course with something of importance to offer any educated person.

John Magnuson hired me into the zoology faculty of the University of Wisconsin. As director of our Center for Limnology, John has consistently brought the cutting edge of limnology to Wisconsin. His people skills and love of limnology have maintained an environment in which new ideas are constantly forming and inspiring research. It took John about 30 years to teach me the value of collaboration, but, to his credit, I finally got the message.

about the author

Stanley I. Dodson received his Ph.D. from the Zoology Department of the University of Washington in 1970, on the interaction of size-selective predators and zooplankton community structure. He teaches courses in general ecology, plankton ecology, summer limnology, and toxicology, and coordinates an ecology internship program, all at the University of Wisconsin, Madison. He also directs undergraduate research projects and serves as major advisor for several graduate programs. Stanley is a freshwater ecologist, focusing on community ecology of zooplankton and population ecology of *Daphnia* (the water flea). He has studied size-selective predation on zooplankton, *Daphnia* and copepod relationships, and effects of primary productivity on zooplankton biological diversity. Stanley has developed a whole-animal bioassay using *Daphnia* sex ratio and morphology, characters sensitive to environmental contaminants, including common agricultural and industrial chemicals. He is exploring the possibility that some herbicides disrupt a hormone-like system in *Daphnia.* His field research is focused on questions about the relationship between land-use practices and zooplankton community structure.

contents

Chapter 3 Diversity of Aquatic Organisms: The Single-Celled and Colonial Organisms 64

Chapter 4 Diversity of Aquatic Organisms: Rotifers, Annelids, and Arthropods 84

Chapter 5 Diversity of Aquatic Organisms: Larger Organisms 120

Chapter 6 Population Dynamics in Limnology: Population Size Changing with Time 142

Chapter 7 Community Ecology: Species Interactions and Community Structure 160

Chapter 8 Community Ecology: Freshwater Communities Changing Through Time 188

Chapter 9 Aquatic Ecosystems and Physiology: Energy Flow 208

Chapter 10 Aquatic Ecosystems: Chemical Cycles 230

Chapter 11 Water in Landscapes 264

Chapter 12 The Citizen Limnologist 298

Chapter 13 Field and Laboratory Exercises 316

Appendix 1 Spearman Rank Correlation Test 353

Appendix 2 The X^2 Test of Independence, with Emphasis on the X^2 Median Test 355

preface

WHY I WROTE THIS BOOK

Introduction to Limnology presents limnology the way I have learned it from my teachers, and especially the way I have experienced it at the University of Wisconsin. For the past several decades, the UW limnology course has included about 150 undergraduate students each fall, and an additional 50 students in the summer. The emphasis of the course is on the broad diversity of topics that make up limnology: the standard chemistry, physics, and biology of water, as well as environmental, management, and conservation topics. There is even room for the occasional poem or story about a beautiful lake or stream. Limnology is taught here as a general science course with both lecture and laboratory experiences. *Introduction to Limnology* differs from other texts on the subject by having a strong emphasis on ecology, with a distinct liberal arts orientation.

FOR THE BEGINNER

This text is written at an introductory level that will appeal to the undergraduate instructor, teaching assistant, and student. If the teacher happens to be a specialist in limnology, the introductory material can easily be supplemented with results of specific research.

Reflecting my own experiences, there are more lake than stream examples, and I have given biology, especially ecology, more emphasis than is traditional in a limnology text. Examples in the text are mostly from North America, but I have made an effort to include selected international examples. (I have not used too many Wisconsin examples. Really.)

The information and ideas in this book will provide the foundation to allow any college student or other citizen limnologist to make informed decisions in specific situations concerning water resource management and protection. This book serves as a step in the education of scientists, environmentalists, naturalists, water resource managers, educators, and conservationists.

LOGICAL ORGANIZATION

Chapter 1 begins with a general introduction to the subject matter and the history of limnology. Chapter 2 is a broad introduction to the physics and chemistry of water, an environment that humans find alien. In the next three chapters (3, 4, and5) we meet the major groups of organisms that live in fresh water, from microscopic bacteria to the charismatic megafauna. Chapter 6 focuses on populations and population dynamics of individual species, applying concepts of theoretical and population ecology to limnology. Freshwater communities are treated in two chapters: Chapter 7 describes examples of freshwater communities and the biological interactions that allow species to sometimes live together in the same habitat, and chapter 8 focuses on how communities change. The next two chapters (9 and 10) are concerned with ecosystem topics. Chapter 9 discusses the flow of energy in lakes and streams, both in the habitat (ecosystem) and in individual organisms (physiological ecology). Chapter 10 deals with major aquatic chemical cycles. Material in chapter 11 describes the landscape setting and morphometry of lakes and streams, flow patterns of water through the landscape components of the hydrological cycle, and the birth, development, and extinction of lakes and streams. And, finally, chapter 12 explores how the citizen limnologist participates in human–water interactions. The additional laboratory and field exercises in chapter 13 provide examples of ways to experience limnology hands-on.

PEDOGOGY DEVELOPED FOR THE STUDENT

Each chapter begins with a thought-provoking quote and an outline that lets the reader see the structure of the chapter at a glance. Important terms are bolded, and referenced quotations are set off by indentations and shading. To assist in teaching and learning, each chapter ends with a **unique Study Guide,** comprised of practice questions, lists of major concepts, vocabulary words, specific examples, reminders of special contributions of individual limnologists, additional readings, and links to relevant websites. There are even crossword puzzles. These study guides can be used to generate discussions, to write or prepare for exams, and as a starting place for further study.

24 CHAPTER 1

Study Guide

Chapter 1 Introduction to Limnology

Questions

1. What is limnology?
2. Why is limnology difficult to define?
3. How long has limnology been a science?
4. What is a lake?
5. What is a model?
6. What is a field station?
7. What are different ways of looking at limnology? Which aspect of limnology most attracts you?
8. After reading chapter 1, and taking into account your earlier experiences with water and even limnology, answer the question: "What does it mean to think like a limnologist?"
9. Since you are taking a limnology course, you have probably had experiences with lakes or streams. Describe your first memorable, limnological experience.
10. Hutchinson (1965) titled a collection of essays on ecology *The Ecological Theater and the Evolutionary Play.* Close your eyes and imagine the limnological theater.
11. Analyze the following statement according to your understanding of science and environmental science: *The global climate is warming steadily, and this process, which will result in changes in the distribution and abundance of terrestrial and aquatic organisms, is caused by human activity that is putting gases such as carbon dioxide and chlorofluorocarbons into the atmosphere.*
12. Go to a library or go online and look for articles on limnology in one of the journals listed in this chapter. How would you describe the limnological articles in the journal?
13. Use a database to search for more information on (1) limnology, (2) the location of the field station nearest you, and (3) your favorite lake or stream.
14. Check out the website for the Organization of Biological Field Stations (address is at the end of the chapter). If you had the necessary time and funds, what would be your choice for a dream course at a summer biological field station?
15. Draw and label all parts of the vertical temperature profile of a lake, as it is currently drawn.
16. What are the similarities and differences for descriptive and predictive models? Describe at least one actual example for each model.
17. What is the basic process used in hypothesis testing?
18. How do you classify yourself—as a scientist or an environmentalist?

Words Related to General Limnology

aquatic backwater biogeochemistry brackish
ASLO bathymetric bog canal

FIELD AND LABORATORY EXERCISES

Several cohorts of summer limnology students have tested the field and laboratory exercises presented in chapter 13. Although the exercises are designed for 8-hour lab periods, they can easily be broken into shorter time segments. These exercises emphasize a hypothesis-testing approach in a natural history context, and include examples of simple experimental design, low-tech sampling and observation techniques, and simple methods for making, interpreting, and reporting results and observations.

CUSTOM WEBSITE

Accompanying this textbook is a website with valuable resources for both instructors (PowerPoint presentations, animations, active art) and students (practice, quizzes, career information, tips on writing papers, hotlinks).

ACKNOWLEDGEMENTS

It is traditional to say that there are too many people to thank, and that it would be too embarrassing to leave out someone. This is true, especially for this book! Thanks for helpful conversations about limnology are due to hundreds of limnologists, aquatic ecologists, teachers, students, naturalists, and citizens interested in lakes.

Many people assisted me in putting this book together. Bob Rogers, Sheri Snavely, and Kirk Jensen encouraged the development of this book in its formative stages and provided reviews of early drafts. Over the last 10 years, Bill Feeny has prepared drawings and charts for teaching summer limnology, some of which served as the basis for figures in the text. Kandis Elliot gave excellent advice on how to write, which I have been partly able to follow. She also did a masterful job of turning my primitive drawings into beautiful and helpful art for this book. What a delight to be able to work with an artist who is also a limnologist! Editors at McGraw-Hill have provided outstanding assistance in the final stages of publication. Special thanks are due Kathy Loewenberg and Mary Powers for their organizational skills, excellent advice, and untiring patience. This book is based on several drafts, which were improved by thoughtful reviews by many limnologists, but especially Matt Brewer, Robert Bohanan, Ginny Dodson, John Havel, Carla Cáceres, Scott McNaught, Donald Roeder, Alan Covich, several anonymous readers—the kind professors who took the time to respond to e-mail questions about content and format—and the following careful reviewers:

Reviewers

A. Ross Black
Eastern Washington University

William R. DeMott
Indiana University-Purdue University at Fort Wayne

Dennis Englin
The Master's College

Alexander Karatayev
Stephen F. Austin State University

Bill Perry
Illinois State University

Andrew P. Wold
Fond du Lac Tribal and Community College

Val H. Smith
University of Kansas

Lloyd Wright
Hocking College

Grace Wyngaard
James Madison University

INTRODUCTION TO LIMNOLOGY

Stanley I. Dodson

1

"Water is the beginning of all things."

Thales of Miletus, 600 B.C.

Introduction to Limnology

outline

LIMNOLOGY DEFINED

Water is a critical resource for all of life on Earth. Fresh or salty water is the environment for most organisms on Earth, and living organisms, including limnologists, are typically about 70% water. Terrestrial organisms often find that the availability of water constrains growth, activity, and reproduction. Except for ice deep in glaciers, human activity has impacted all water on Earth (Vitousek et al., 1997). In many parts of the world, water is a life-or-death issue for satisfying thirst and supplying agricultural needs. Water is in short supply in many parts of the world, and water conservation is of prime importance in many cultures (figure 1.1). As the human population grows exponentially, requiring ever more water, it is critical that as many people as possible are familiar with fundamental limnological concepts. Citizen limnologists play an important role in conservation and the on-going development of policy for management and use of water and related resources in lakes and streams.

Although the emphasis of this limnology text is on lakes and related still-water bodies (**lentic** systems), it also uses examples from streams and related moving-water bodies (**lotic** systems). Lake and stream examples are selected to demonstrate specific principles of limnology. To aid the reader, all of the lakes and streams mentioned in the text are located on a global map (figures 1.2 and 1.3) or on a map of North America (figure 1.3).

While the surface impression of a lake is often beautiful and interesting, limnologists, like most people, are also interested in what lies beneath the surface. Looking into a lake is the essence of limnology—limnologists are scientists who look beneath the surface. In 1854, Thoreau wrote:

> "The water [of Walden Pond] is so transparent that the bottom can easily be discerned at the depth of twenty-five feet or thirty feet. Paddling over it, you may see, many feet beneath the surface, the schools of perch and shiners, perhaps an inch long, yet the former easily distinguished by their transverse bars, and you think that they must be ascetic fish that find a subsistence there."

FIGURE 1.1 Tjanjundinya, and elder of the Pitjantjatjara people, with traditionally dressed hair. He is covering a gnamma (rock pool) to conserve water from animals and evaporation. The gnamma is in a granite outcrop at North Oparinna in the Musgrave Ranges, North-Western Australia, in 1940. **Source:** Photograph PRG 1218 Series 3, Part 1, 1031D from the Mountford-Sheard Collection, State Library of South Australia. Reproduced by kind permission of his daughter Nganyinytja. Photograph courtesy of the State Library of South Australia.

In 1892, the French-speaking Swiss scientist, François-Alphonse Forel (1841–1912), coined the word **limnology** (from the Greek *limne,* meaning "pool" or "marsh"). Forel (figure 1.4) was born and raised in a small town on the shores of Lake Geneva (Lac Léman in French; figure 1.5), a large, deep, and clear lake located in the Alps on the border between southwestern Switzerland and France. At the age of 50, Forel, by then a professor of medicine at the University of Lausanne, Switzerland (also on Lake Geneva) undertook a description of the lake, primarily because he was interested in archeological artifacts hidden below the surface. His monograph, "*Le Léman: monographie limnologique,*" a three-volume work published between 1892 and 1902, established limnology as a science. Forel took a multidisciplinary approach to his studies, focusing on geology, physics, chemistry, biology, and archaeology. For his new science, Forel invented the word "limnology" because other suitable names were already being used by oceanographers:

> "This book is a **limnological** monograph. I have to explain this new word and apologize, if necessary. . . . The object of my description is a part of the Earth, which is geography. *The geography of the ocean is called oceanography.* But a lake, as big as it is, is by no means an ocean; its limited area gives it a unique quality, which is very different from that of the endless ocean. I had to find a more modest word to describe my work, such as the word 'limnography.' But, because a limnograph is a tool to measure the water level of lakes, I had to fabricate the new word limnology. **Limnology is therefore the oceanography of lakes**" (Forel, 1892).

Forel's definition, which constrained limnology to the study of lakes, persisted until the founding conference of the International Association of Theoretical and Applied Limnology in Kiel, Germany in 1922. At this conference, it was acknowledged that limnologists, in practice, studied wetlands and streams as well. To reflect this range of environments, the definition of limnology was broadened from "the oceanography of lakes" to "the science of inland waters."

Modern limnologists often define limnology to reflect their own ideas and interests. For example, Edmondson (1991) studied complex lake systems; his definition stresses multidisciplinarity (figure 1.6). Horne and Goldman (1994), from the arid western United States, remind us that inland **saline** lakes are also a part of limnology. Wetzel's (2001) perspective of aquatic energy flow emphasizes the chemical and physical aspects of limnology. At its essence, the definition of limnology is no different from that adopted in 1922; the study of inland waters, including lakes, streams, and wetlands (these are **aquatic** habitats).

EARLY LANDMARKS IN LIMNOLOGY

Limnology as a distinct scientific discipline is relatively young, having its roots in the late 18th century when the systematic collection of limnological data began in western Europe. The birth of limnology followed and benefited from the "scientific revolution" of the 15th through 18th centuries, which brought about great advances in both scientific knowledge and in the way science was conducted.

Limnology is an eclectic, multidisciplinary science, drawing upon many different areas of science including geography, geology, chemistry, physics, mathematics and statistics, and various biological disciplines, including taxonomy and systematics and all branches of ecology. The history of limnology parallels the development of the other natural sciences (see Edmondson, 1991).

In its infancy, between about 1850 and 1920, limnology was principally a European science. In addition to the work of Forel, studies by early limnological pioneers

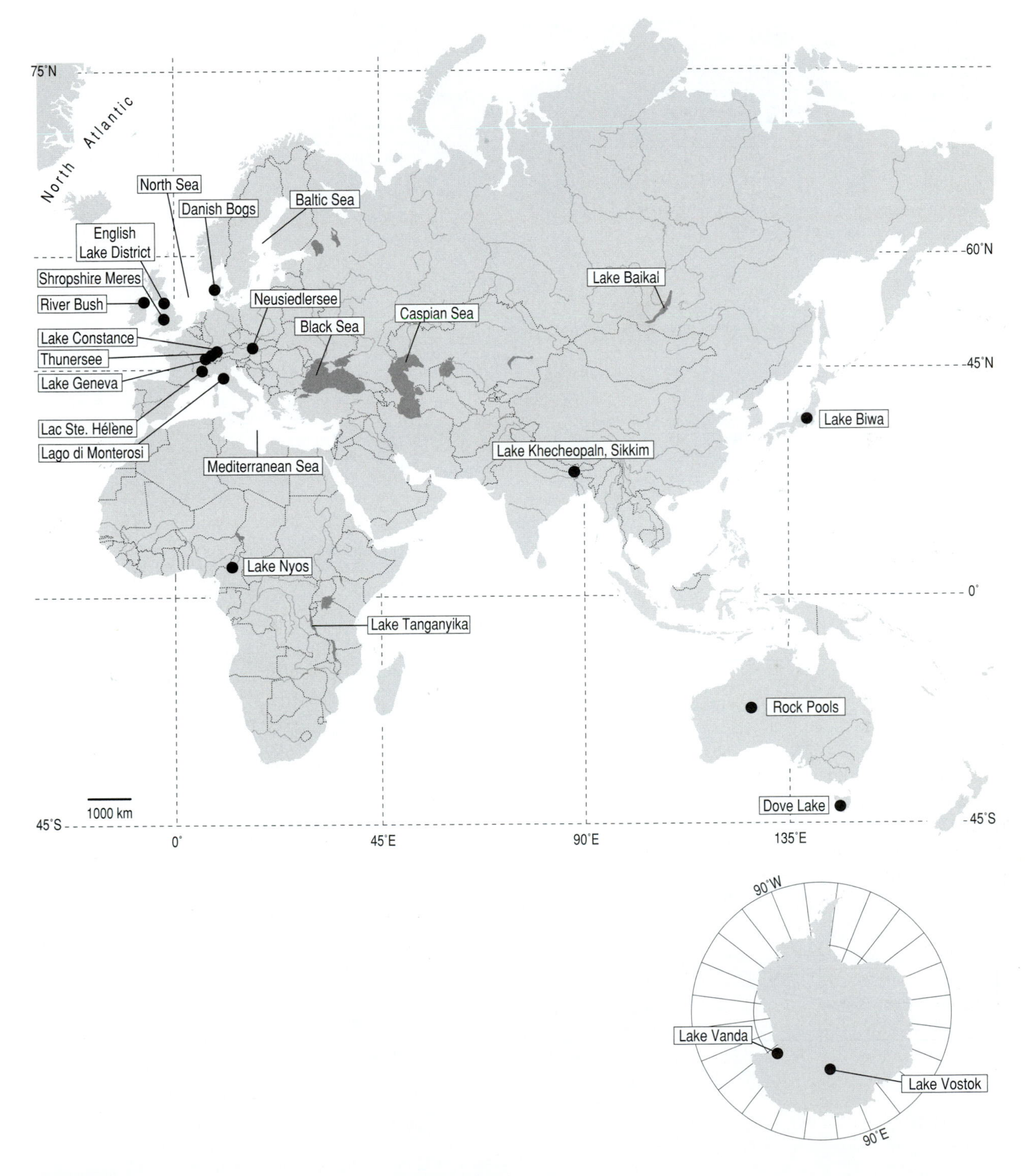

FIGURE 1.2 Locations of lakes and streams mentioned in this text.

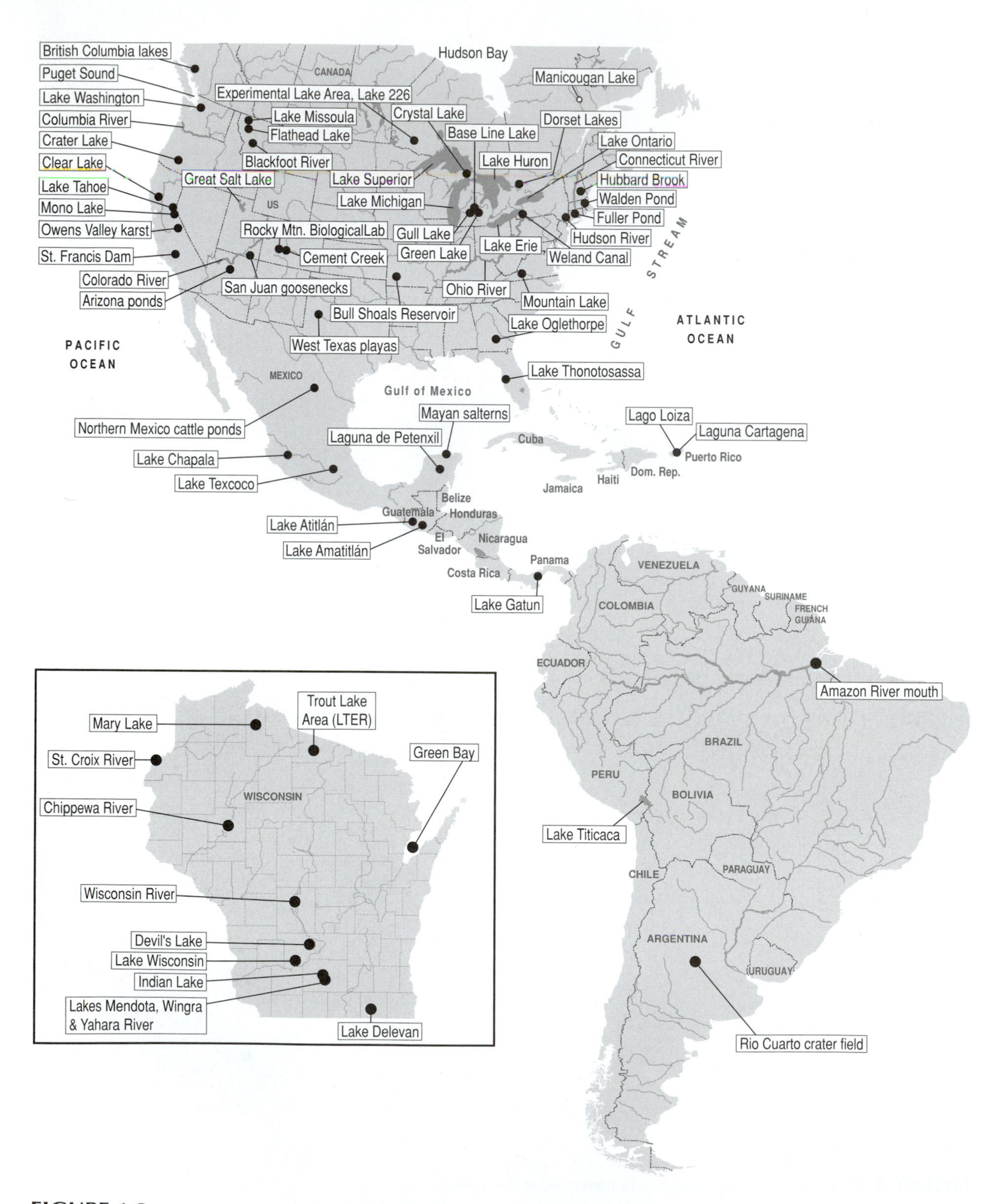

FIGURE 1.3 A map of the North American locations of lakes and streams mentioned in this text.

FIGURE 1.4 François-Alphonse Forel. **Source:** Photograph complements of the Institut F.-A. Forel, University of Geneva, Switzerland.

who set the stage for later work include, in England, Baird's (1850) report on freshwater microcrustaceans and Phillips's (1884) charming description of the biology of nuisance algae blooms. August Thienemann (1905) published a paper on stream insects to begin a 60-year-long career studying the biology and chemical limnology of German lakes and streams. Other outstanding early European limnological studies include Naumann's (1917) study of algae in Swedish lakes and a series of studies of temperature and light absorbance of German lakes beginning with Halbfass (1903). Early North American limnologists were very much in the shadow of European scientists. For example, Birge (1897, 1906), in his early studies of lake biology, physics, and chemistry in Wisconsin, interpreted his results in the context of European limnology.

Key Early Inventions and Techniques

The following are a sampling of important inventions and techniques that made limnology possible.

FIGURE 1.5 Map of Lac Léman (Lake Geneva), Switzerland. **Source:** From Forel, 1892.

FIGURE 1.6 Professor W. T. Edmondson on Lake Washington, Seattle, collecting aquatic organisms using a fine-mesh net, with the assistance of Diane Crosetto and David Allison. **Source:** Robert W. Kelley, courtesy of Arni Litt, University of Washington, Seattle WA, USA.

FIGURE 1.7 *Daphnia,* the water flea. Tommy Edmondson admired this image for two reasons. The *Daphnia* resemble the logo of a cleaning product "Old Dutch Cleanser," whose motto was "Chases Dirt." The image and motto resonate with the ability of *Daphnia* to efficiently reduce nuisance algae in lakes. Also, the multiple claws reminded Edmondson of the multiple images used by Duchamp to indicate movement in his 1912 "Nude Descending a Staircase"—it looks as if Swammerdam's artistic convention was 240 years ahead of its time. **Source:** From Swammerdam (1669).

Aquatic Organisms

Many of the most important aquatic organisms are too small to be seen with the unaided eye. The systematic observation and description of small, aquatic organisms gained momentum with the invention of the microscope and the explorations of the Dutch microscopists in the middle of the 17th century. Swammerdam (1669) published the first drawings of the water flea *Daphnia* (figure 1.7) and Leeuwenhoek (1674) observed the green, filamentous alga *Spirogyra.* These Dutch scientists were the first to describe in detail the minute members of the aquatic world that are such an important part of limnology. Yet, despite these early discoveries, the microscope was not routinely focused on aquatic organisms until the late 18th century.

Plankton Nets

In 1845, the German biologist Johannes Müller first towed a conical net of fine-meshed silk behind a boat cruising the North Sea (Koller, 1958). Müller reported that the net efficiently filtered out of the water mostly microscopic, floating organisms—both **producers** (photosynthetic organisms such as algae) and **consumers** (animals that eat algae and their micropredators). These midwater organisms were subsequently given the name **plankton** (from the Greek *planktos,* meaning "wanderer") by yet another German biologist, Victor Hensen (Henson, 1887). Hensen chose the name "plankton" to emphasize the fact that these organisms, because of their small size, are more or less at the mercy of water currents. The animals are free to wander through a relatively homogeneous, three-dimensional environment:

> "A dark illimitable ocean without bound, without dimension, where length, breadth, and height and time and place are lost."
>
> *John Milton,* Paradise Lost

Vertical Temperature Profile

De Saussure (1779) measured water temperatures at the bottom of several Swiss lakes, including Lake Geneva. He achieved this by lowering an alcohol thermometer encased in dense wood of high thermal capacity into the deep water for several hours and quickly hauling it up to the surface and reading its temperature. As technology for measuring temperature improved, more was learned about the thermal properties of lakes in space and time. An important limnological milestone occurred in 1849, when de Fischer-Foster and Brunner (1849) published the first **vertical profile** of temperature for a lake, the Thunersee in Switzerland (figure 1.8a). (Note that **see** means "lake" in the German

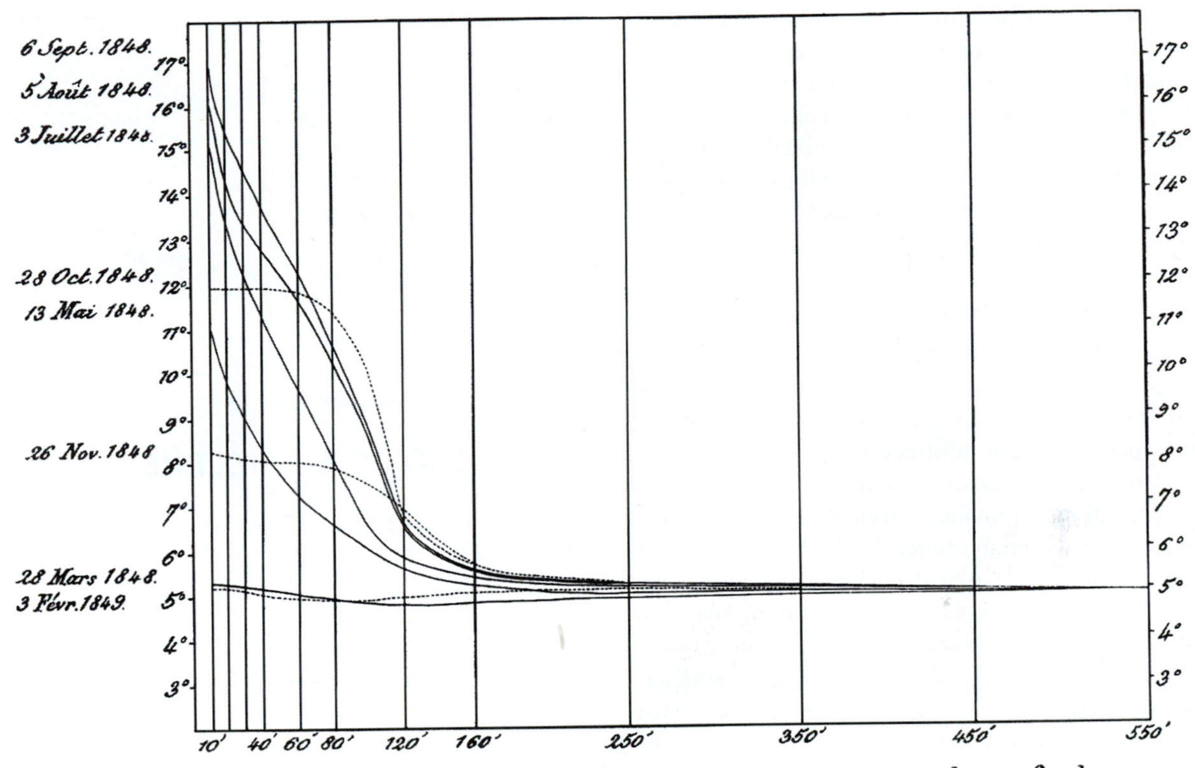

(a)

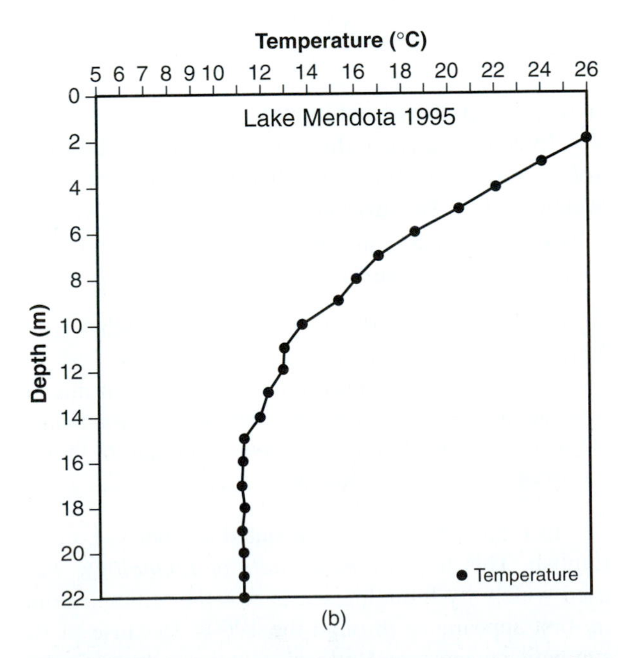

(b)

FIGURE 1.8 (a) The first graph of temperature-versus-lake depth. The several lines are the relationship between depth and temperature for different times of year. (b) The current format for a graph of the relationship between temperature and lake depth (the line connecting the solid black dots). Notice that depth is along the right axis, and goes from zero at the top to the greatest depth at the bottom. Temperature is across the top of the graph. **Sources:** (a) De Fischer-Foster and Brunner, 1849; (b) University of Wisconsin Limnology Class data, summer 1995.

language.) The shape of the vertical profile graph is the same today as in the 1849 figure, except that to make the relationship with depth seem more intuitive, we now typically put depth on the vertical axis, with zero depth at the top (figure 1.8b). The vertical temperature profile (figure 1.8b) is so fundamental to limnological understanding that it is ubiquitous and can be regarded as a **logo** (graphical symbol) for limnology.

Oxygen Concentration

Oxygen was one of the earliest and most important chemicals that limnologists measured in water. Hutchinson (1957) wrote that oxygen concentrations were the single most important information a limnologist could have in order to understand the nature of a lake. Oxygen concentration provides information about the history of water, its suitability for life, and an overview of water chemistry. Analytical chemical techniques for dissolved chemicals were developed during the 19th century, including a method for determining oxygen concentration in water (Winkler, 1889). Recent advances in electronic measurement of oxygen have greatly improved accuracy and efficiency of determinations of oxygen and many other important aquatic chemicals.

Contour Maps

Lines on a **contour** map indicate equivalent elevation, or equivalent depth, in the case of an aquatic habitat. Mariners first developed maps with lines of equal depth (the lines are called **isobaths;** the maps are **bathymetric**) for harbors in the mid-16th century (Dent, 1999). The contour map of Lac de Ste. Hélène, France (figure 1.9), is one of the first contour maps of a lake published by a limnologist (Delebecque, 1898). Contour maps allow limnologists to calculate lake volumes, which lead to descriptions of whole-lake chemical budgets and physical models of water movements.

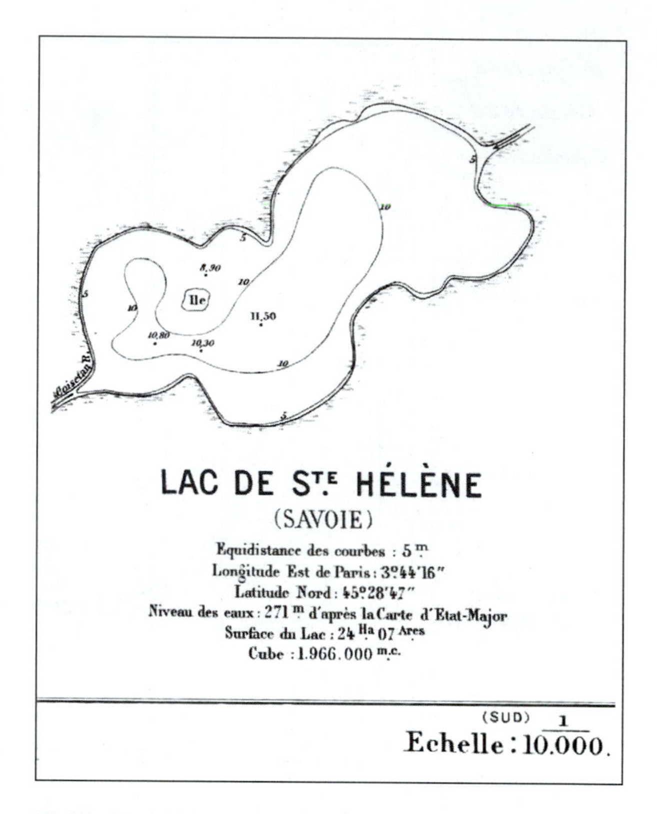

FIGURE 1.9 An early bathymetric map of a French lake, Lac de Ste. Hélène. The lines of equal depth contours (= courbes) are given for 5 and 10 m in this rather shallow lake; the maximum depth is 11.5 m and there is a small, steep-sided island (Ile) in the lake. **Source:** Delebecque, 1898.

Limnological Textbooks

Textbooks summarize the state of knowledge in a field. The first limnology textbook, *Handbuch der Seenkunde: Allgemeine Limnologie,* was published in German (Forel, 1901). Limnology became more accessible to English speakers with the publication of the first American (and English language) textbook, *The Life of Inland Waters: An Elementary Text Book of Freshwater Biology for American Students,* by Needham and Lloyd (1916), two professors of limnology at Cornell University. In North America, this text was followed by *Limnology* by a University of Michigan professor, Paul Welch (1935). An Austrian limnologist, Franz Ruttner (1940), provided a synthesis of the rapidly growing science, whose objective was to:

> "draw from the vast assemblage of facts what is necessary to make the principles clear and to illustrate the causal relationships, and to present this material in a precise and clear form so that even the beginner can gain a comprehensive picture of the whole science of limnology."

In 1953, Frey and Fry translated Ruttner's text into English. This text, *Fundamentals of Limnology,* was used widely by limnologists, in several editions, from its first appearance through the 1960s. Because of its synthetic perspective, Ruttner's text was a dramatic departure from traditional, data-rich limnological education (e.g., Welch, 1935).

FIGURE 1.10 Professor G. Evelyn Hutchinson at Yale. The tortoise and the piece of coral are related, respectively, to Professor Hutchinson's enthusiasm for evolution and biogeochemistry. **Source:** Phyllis Crowley © 2003.

G. Evelyn Hutchinson (figure 1.10) had the goal of bringing together all limnological knowledge in his four-volume *A Treatise on Limnology.* Professor Hutchinson made a good start with Volume I (*Geography, Physics, and Chemistry*) and Volume II (*Introduction to Lake Biology and the Limnoplankton*), published in 1957 and 1967, respectively. His masterful synthesis of lake geography, physics, chemistry, and biology established him as the father of modern limnology. These two initial volumes were followed by a third volume (Hutchinson, 1975) on the ecology of aquatic macrophytes (or water plants) and algae that grow on surfaces (inanimate substrate, plants and animals). Hutchinson began his fourth volume of the *Treatise* (Hutchinson, 1993) in the 1970s; it was published posthumously with the assistance of dozens of his students and colleagues. At the time of its publication, Hutchinson's treatise was an outstanding review and synthesis of limnology. The knowledge explosion since the 1950s has discouraged other limnologists from repeating Hutchinson's feat.

During his career, Professor Hutchinson pioneered or refined the fields of **paleolimnology** (studies based on material from lake sediments, including **microfossils** and chemicals such as heavy metals, nutrients, and algal pigments), **biogeochemistry** (global chemical cycles), and population biology. His students developed much of what we now know as the modern science of ecology.

Since the *Treatise,* limnological information has continued to grow exponentially. It is probably no longer possible, or even desirable, to collect all limnological knowledge into a single set of books. Even with computer databases and Internet search engines, it is impossible to surf the entire limnological information tsunami. It is useful to keep in mind, however, that using the Internet, it takes only about 5 minutes to answer virtually any limnological question or to locate an image of an aquatic organism. As we gain experience by taking courses, doing research, and writing papers, we refine the filter we need to separate Internet gold from dross. For the person who can think like a limnologist, the Internet is a vast storehouse of information.

The goal of the text in your hands is to provide an introduction—an overview—of limnology that emphasizes aquatic ecology and makes connections between limnology and people. In order to think like a limnologist, it is necessary to speak the language. This means understanding conventions such as the metric measuring system; recognizing limnological categories such as different kinds of lakes, streams, and wetlands; getting hands-on experience working in aquatic systems; using limnological journals and Internet databases; and having a background in physics, chemistry, math, biology, and ecology, which are all relevant to limnology.

NAMES OF BODIES OF INLAND WATER

Liquid water on the surface of Earth comes in a variety of shapes that are given a variety of names, depending on the water body's size, shape, duration, geography, origin, and whether it is moving or still. The same kind of inland water body will have different (English) names in different parts of the world because of regional differences in language.

Perhaps the question most frequently asked of limnologists is, "What is the difference between a lake and a pond?" This is a difficult question to answer because there is not a clear, rigorous distinction between different kinds of inland water bodies, which are typically categorized according to historical and cultural whims. Table 1.1 lists some of the many names given

Table 1.1 Common Names of Water Bodies

Bodies of liquid waters on the surface of the planet are called by a multitude of names, depending on the water body's size, geography, whether it is moving or still, and the culture of the people in the neighborhood.

Name of Water Body	Example—See Color Plates	Brief Description
Lentic		
Lake	1, 2, 3, 4, 5, 20, 22, 26, 28, 30, 31, 33, 40	Relative large body of water, typically deeper than 3 meters, with an area of greater than about 1–10 hectares (ha). Often shows thermal stratification. Shallow water bodies can be called *shallow lakes* if they are large. Large lakes are larger than about 500 km^2, the largest being the **brackish** (salty) Caspian Sea (436,400 km^2), Lake Superior (83,300 km^2), and Lake Baikal (31,500 km^2), the deepest lake in the world.
Pond	6, 7, 8, 9, 10, 11	A relatively small body of water, with an area of 1–10 ha or less, and often shallow enough (less than 3 m) to be easily mixed from top to bottom by light wind. Ephemeral ponds are also called *pools, pans,* or *playas.* Sinkholes are ponds (often steep-sided) in **karst** (limestone) regions. Small lakes or ponds are sometimes called *meres;* **salterns** are ponds used in salt production. Includes artificial farm ponds.
Reservoir	10, 12, 28	An artificial pond or lake, created by placing a dam in a valley or depression—small reservoirs include cattle ponds, stock tanks, and farm ponds. The morphology, hydrology, and ecology of reservoirs are distinct from those of natural ponds or lakes.
Wetland	13, 37	Characterized by soil saturated with water, but with standing water less than about 1 m deep, often with extensive areas of floating and emergent vegetation. Wetlands are also called *marshes* or *carrs.* Swamps have trees standing in water. A *morass* is a dangerous wetland. Wetlands can be lentic (associated with a lake) or lotic (e.g., the Everglades, Florida).
Rock Pool	14	Very small body of water, usually less than 10 m^2 surface area and less than 2 m deep; often a depression in rock. Also called **tinajas, gnammas,** or *pot holes.*
Tree Hole	38	Small pools of water inside trees but open to the air, caused by decay of the wood.
Bog	15	Usually a small lake or pond with acidic, tea-colored water and accumulated organic matter (peat) and often with a floating mat of plants and peat.

to water bodies. (Note that the sizes of lakes and streams are expressed in metric units—see table 1.2 for a translation to English units.)

There is more to a water body than a simple, physical definition. Lakes have been described in a variety of ways by different limnologists. The prominent Wisconsin limnologist E. A. Birge (1907) (figure 1.11) created a physiological metaphor to describe lakes. In this description, the lake sounds as if it has a body, just like an individual person. While this romantic concept of "lake as an individual organism" is not generally accepted by modern limnologists, it is the forerunner of a widely-accepted limnological perspective that includes energy and chemical budgets:

Table 1.1 Common Names of Water Bodies *Continued*

Bodies of liquid waters on the surface of the planet are called by a multitude of names, depending on the water body's size, geography, whether it is moving or still, and the culture of the people in the neighborhood.

Name of Water Body	Example— See Color Plates	Brief Description
Lotic		
Rivers and **streams**	16,17,27	Moving water can be called a *river, stream, creek, crick, branch, rivulet, trace, brook . . .* depending on size and regional preferences. Rivers are the largest. Relatively still areas in streams are called *pools;* fastest (turbulent) water is called a *riffle* or *white water; runs* are sections of smooth flow. A *reach* is any section of a stream under observation. Waterfalls can create lake-like (or pond-like) plunge basins.
Spring		A depression receiving a perceptible flow of ground water. Alkaline springs (and an associated marsh) are called *fens.* Small springs without an obvious single source are called *seeps.*
Caves pools and streams	36	Caves (usually in limestone) often contain pools and streams of running water.
Intermittent streams		Streams that flow only part of the year. Intermittent streams are the flowing analogues of temporary ponds.
Canal		An artificial waterway, connecting streams or lakes, used for boat traffic. Canals are a major corridor for dispersal of aquatic organisms.
Backwater		Still water in the flood plain of a river, but not necessarily connected to the main channel during times of low flow. These are called by a variety of regional names, including *slough, bayou, billabong,* or *oxbow lake.*
Estuary		A wetland influenced by a river and the sea—the water is brackish (less salty than seawater).

"An inland lake has often been compared to a living being, and this has always seemed to me one of the happiest of the attempts to find resemblances between animate and inanimate objects. Unlike many such comparisons, which turn on a single point of resemblance and whose fitness disappears as soon as the objects are viewed from a different position, the appropriateness of this increases rather than diminishes as our knowledge both of lakes and living beings is enlarged. The lake, like the organism, has its birth and its periods of growth, maturity, old age, and death . . . the rhythm of seasonal activity . . . but in no respect does [a lake] resemble an organism more closely than in...its *respiration.*

Another early ecological definition of a lake is credited to Stephen A. Forbes (1925) in a lecture titled *The Lake as a Microcosm,* which he originally presented to the Peoria (Illinois) Scientific Association in 1887. In this paper, Forbes laid out the foundation for a modern limnology that focused on the interactions of the living organisms with the environment and with one another:

"A lake is to the naturalist a chapter out of the history of a primeval time, for the conditions of life are primitive, the forms of life are, as a whole, relatively low and ancient, and the system of organic interactions by which they influence and

FIGURE 1.11 E. A. Birge and C. Juday on Lake Mendota, Wisconsin, collecting zooplankton. **Source:** WHI-3176 Wisconsin State Historical Society.

control each other has remained substantially unchanged from a remote geological period.

The animals of such a body of water are, as a whole, remarkably isolated—closely related among themselves in all their interests, but so far independent of the land about them that if every terrestrial animal were suddenly annihilated it would doubtless be long before the general multitude of the inhabitants of the lake would feel the effects of this event in any important way. It is an islet of older, lower life in the midst of the higher, more recent life of the surrounding region. It forms a little world within itself—a **microcosm** within which all the elemental forces are at work and the play of life goes on in full, but on so small a scale as to bring it easily within the mental grasp. . . . If one wishes to become acquainted with the black bass, for example, he will learn but little if he limits himself to that species. He must evidently study also the species upon which it depends for its existence, and the various conditions upon which these depend. He must likewise study the species with which it comes in competition, and the entire system of conditions affecting their prosperity; and by the time he has studied all these sufficiently he will find that he has run through the whole complicated *mechanism* of the aquatic life of the locality, both animal and vegetable, of which his species forms but a single element."

ENVIRONMENTALISM AND/OR TRADITIONAL SCIENCE IN LIMNOLOGY

Limnology is a complex science that draws upon many other scientific disciplines, such as chemistry, physics, and ecology. Different fields of science are in themselves complex. For example, ecology is a large field that included several different perspectives. The public understanding of ecology is based on the media (TV, newspapers, magazines, movies) and public education. This is the **environmental science** perspective, which is distinct in several ways from the perspective of **ecological science.** Students learning to think like limnologists need to be able to distinguish the two perspectives.

Environmental science grows out of traditional science, in the context of an ever increasing human population and the perception of increasingly limited natural resources—the result of growing public knowledge and concern for the environment. Environmentalists depend on traditional scientists for information. The perspective of the environmentalist includes:

Characteristic Activities: The Environmentalist

- Engages in making changes in the political arena.
- Uses scientific information to support a social agenda and develop resource management policy.
- Has little use for quantitative analysis of data and hypothesis testing.
- Communicates using government reports and the public media.

Distinguishing Characteristics

- Embraces controversy in society.
- Values social skills such as persuasion and negotiation.
- Usually has a definite opinion or position on any given ecological topic.
- May enjoy nature, but typically does not participate in research.

Teaching

- Lectures, but tends not to do formal teaching.

The science of ecology takes a different approach to ecology (and ecological limnology) than does the environmentalist. The scientific perspective includes:

Characteristic Activities: The Scientist

- Loves systematic classification of things or ideas.
- Looks for and often finds pattern in nature.
- Uses quantitative tools—quantitative observation, experiments, models, statistics, and hypothesis testing to generate and interpret data.
- Communicates new results using the scientific literature.

Distinguishing Characteristics: The Scientist

- Avoids controversy in society—wants to be objective.
- Values creativity.
- Has alternative hypotheses—won't give a straight answer.
- Is obtrusive and often manipulates nature.

Teaching

- Gives formal lectures, mentors students, with the teaching theme focused on understanding nature.

There are a number of other perspectives, in addition to the environmental and scientific perspectives, that fall within the broad field of ecology. There are conservationists, managers, teachers in public education, and even recreational ecologists, all with distinct ways of looking at nature.

It is possible to be a limnologist as either a scientist or an environmental scientist, or both. G. Evelyn Hutchinson is an example of a scientist who focused on the quantitative search for pattern in nature. He had a profound influence on the development of the sciences of ecology and limnology, but had little to do directly with limnology as an environmental science. Jack Vallentyne, who trained as a limnological chemist, embraces environmentalism, using his knowledge to influence human ecology and politics (Vallentyne, 1974). One of Hutchinson's students, Tommy Edmondson, was both a scientist and an environmentalist. He was an authority on aquatic organisms and on the chemistry of lakes, and he put his career as a scientist at risk by taking a stand on the importance of diverting sewage (and its phosphate content) to improve water quality in Lake Washington (Edmondson, 1991). The limnological studies of Birge, Forbes, and other North American limnologists, along with their European counterparts (e.g., Thienemann, 1925), contributed significantly to the early development of ecology as a science and an environmental science. Forbes, in particular, was one of the first scientists to combine the empirical knowledge of anglers and fisheries managers with academic knowledge of fish biology, to produce the field of ecology, which has since been both pure and applied, local and global (Schneider, 2000).

This limnology textbook presents limnology as a multidisciplinary science and as environmental science. Chapters 2 through 11 are in the scientific mode, while chapter 12 focuses on the environmentalism of the citizen limnologist. The chapter 13 laboratory and field exercises give an introduction to how science and environmental science are actually conducted.

THE ROLE OF MODELING AND STATISTICAL THINKING IN LIMNOLOGY

Scientists, whether ecologists, chemists, or physicists, look for pattern in nature. They use a variety of quantitative tools such as modeling and hypothesis testing using statistics.

Modeling

Levins (1968) calls the scientific model the basic unit of theoretical investigation, a reconstruction of nature for the purpose of study. A **model** is a simplified representation of something. Limnologists often model water bodies to better understand how they work or how they can be managed. A model can be **descriptive** (a narrative story) or **predictive,** and the format is often *graphical* (as in figures 1.5 and 1.13), *physical* (a 3-D model of a lake), or *mathematical* (using equations to represent interactions).

Descriptive Models

A descriptive model names things and shows relationships. Charles Elton (1927) published an ecology text with "*box and arrow*" models (figure 1.12). The figure, which he calls variously "food and enemies" and the "nitrogen cycle," is one of the earliest such visual models of an ecological system, or **ecosystem.** Note that a system is made up of **components** (which Elton represented as boxes containing species names and parts of the environment) and **interactions** among the components (which Elton represented as arrows). An ecosystem is a system defined by biological interactions measured by energy or chemicals.

Elton's (1927) descriptive graphical model was the progenitor of two distinct kinds of modern graphical

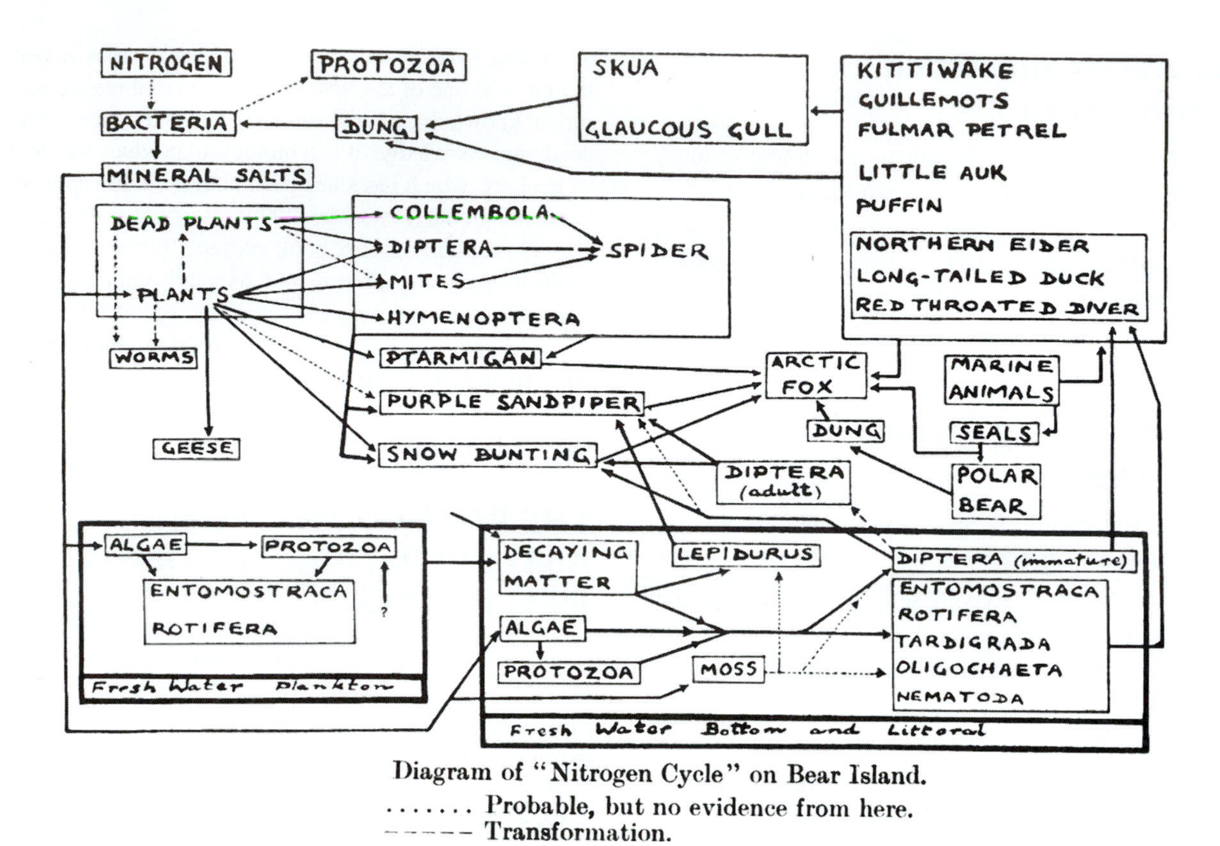

FIGURE 1.12 Diagram of "Nitrogen Cycle." **Source:** JOURNAL OF ECOLOGY (Sommerhayes & Elton, 1923, Fig. 2 in "Contributions to the ecology of Spitsbergen and Bear Island. Journal of Ecology, Blackwell Publishers).

(descriptive) models of limnological systems: the "**food web** model" and the "**trophic** model" (*trophic* refers to food, and by extension, energy or chemical nutrients). The salient difference is that the trophic model deals with energy or nutrient flow and incorporates living and nonliving things into an ecosystem, while the food web deals with organisms (only living things) in a community.

The food web model (of a group of species) shows who eats whom, and thereby indicates relationships of **competition** (struggle for a limited resource such as food) and **predation** (one organism using another for food). This food web model is like a menu (figure 1.13) and is characteristic of community ecology (see chapter 7). The boxes in a food web model contain species or groups of species that have the same diets and the same predators. The food web model is used by **community** ecologists—ecologists who study groups of species. In the model, arrows point from food to consumer.

The trophic model (figure 1.14) employs the concept of storage compartments of energy or of a specific nutrient (such as nitrogen) in a lake, and uses arrows to show how the energy or nutrient moves between storage compartments. Storage compartments could include the land surrounding a lake or stream, open water, the sediments, wetlands, and living compartments such as the plankton or aquatic vegetation. Ecologists and oceanographers have adopted the ecosystem perspective to describe and measure diets of organisms and transfers of chemicals (especially carbon and nitrogen) among species or among parts of the habitat. The ecosystem concept is currently one of the most important ecological concepts (see chapter 9), and limnologists are proud that the concept emerged from a lake study (Lindeman, 1942).

Predictive Models

Ecologists love to predict the future by using mathematical models to project what we know and understand about ecology. Early predictive models were mathematical, based on simple equations, and tended to focus on making predictions about population dynamics. (For more information on the development of

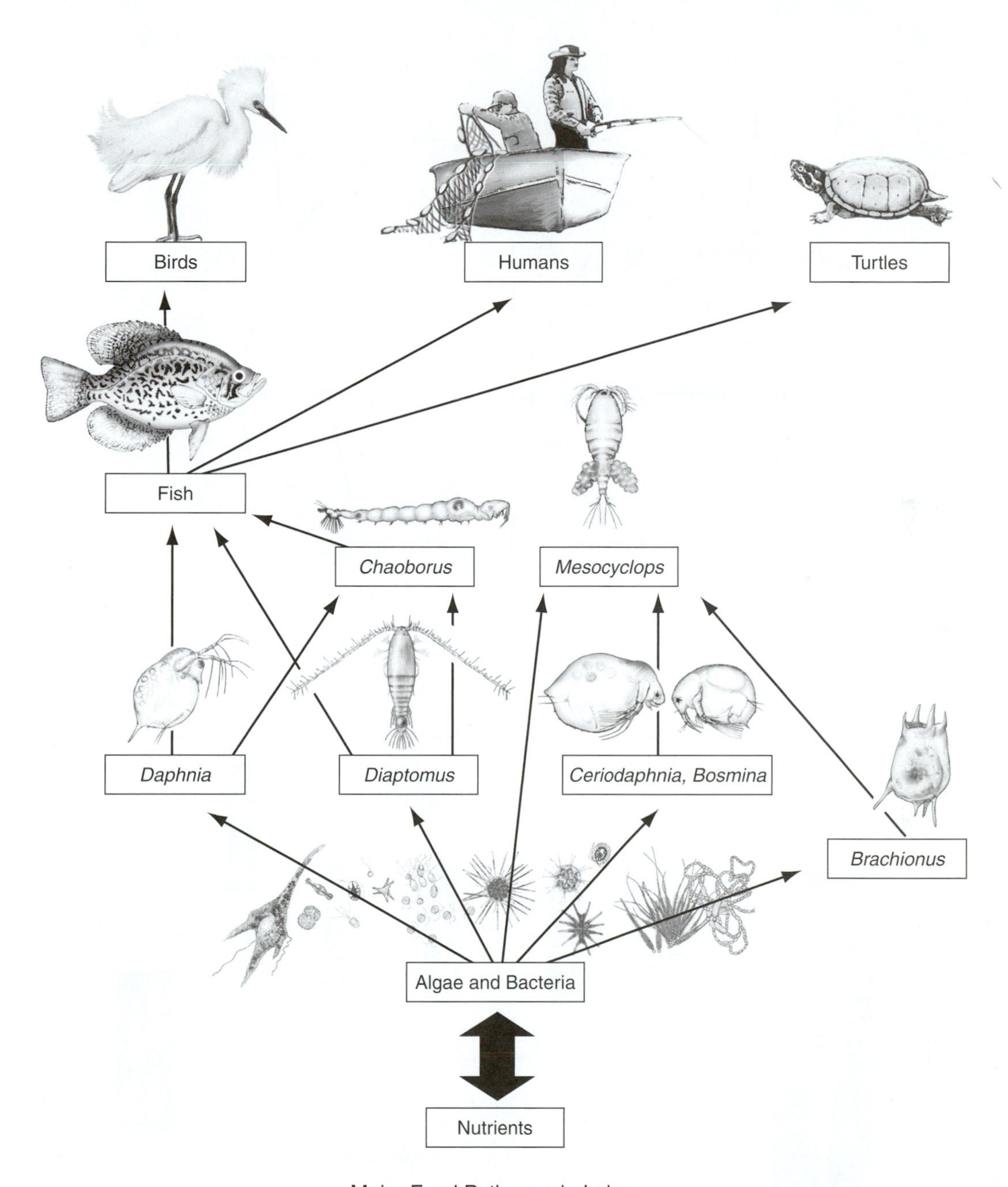

FIGURE 1.13 An example of a generalized food web model of a lake. The boxes are labeled with names of common aquatic taxa; the arrows point from prey to predator.

these early models, see Hutchinson, 1978 and Stearns, 1992.) These mathematical models were first used in the 18th century to predict changes in human population sizes and to predict the chance of giving birth or dying at a given age. Insurance companies still use this approach to estimate premiums. Limnologists use the same kind of models to predict future population sizes of aquatic organisms (as described in chapter 6), given specific environmental conditions such as food availability or predator abundance.

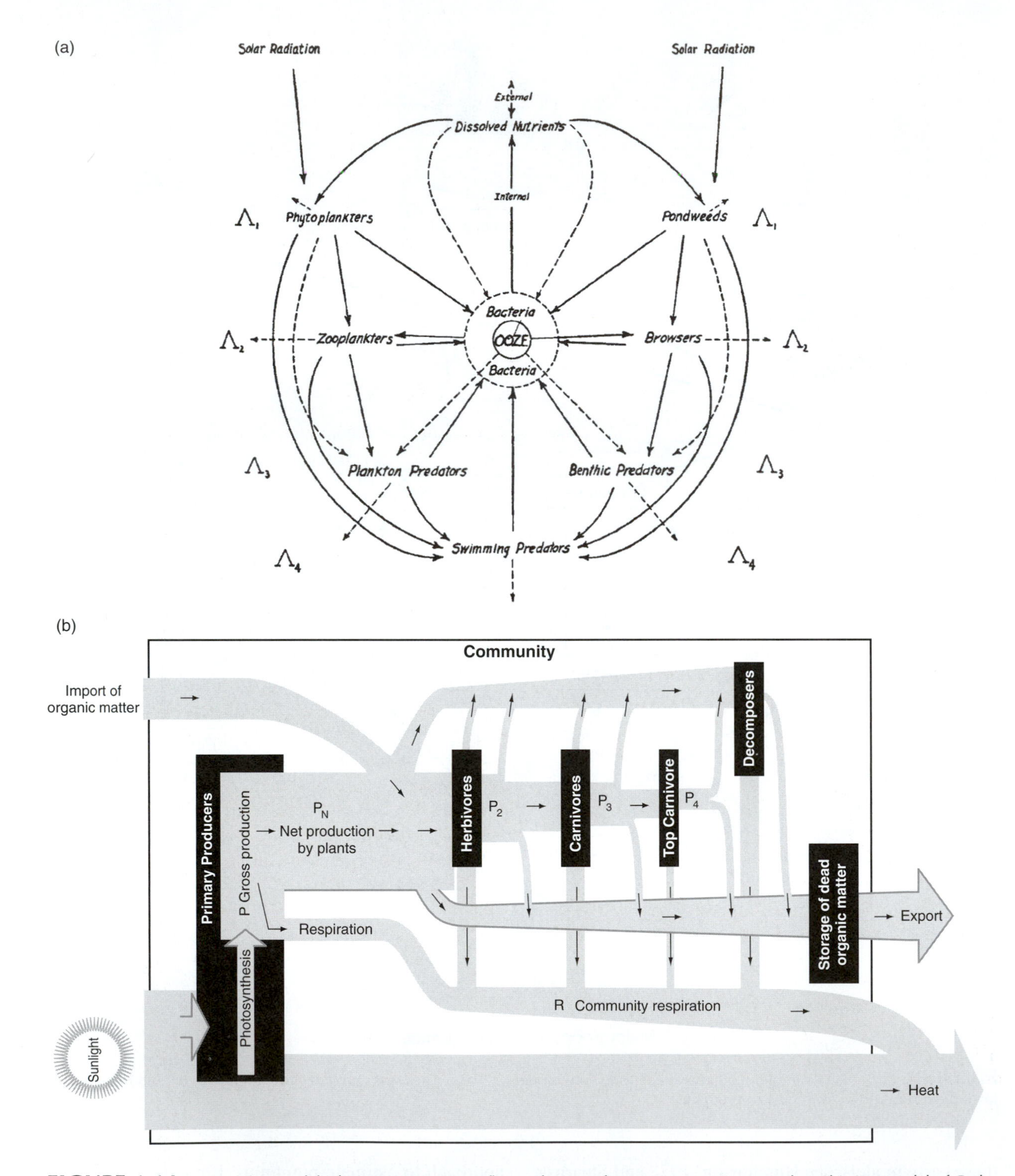

FIGURE 1.14 (a) Trophic model. The arrows represent flows of energy from one reservoir to another. This is a model of Cedar Bog Lake, Minnesota. (b) Trophic model. The arrows represent flows of energy from one reservoir to another. This is a model of Silver Spring, Florida. Besides sunlight, a major input of energy into this system was from bread thrown in by tourists ("import of organic matter"). **Source:** Figure courtesy of the Ecological Society of America. (a) From Lindeman, 1942; (b) inspired by Odum, 1971.

Predictive models are used in two ways. Environmental and resource managers and policy makers want models that make accurate predictions. They might ask questions about how much to harvest an aquatic resource or how to set regulation limits for nutrients or contaminants in water. On the other hand, research scientists often match the predictions of models to observations of nature. Scientists learn the fastest when their models fail to predict nature.

Modelers strive to optimize generality, reality, and precision (Levins, 1968). The earliest predictive models were based on very simple assumptions because of the difficulty in calculating predictions by hand using paper and pencil. However, computers have made predictive models easy to use and have allowed for some increase in generality, reality, and precision. Because the computer does calculations rapidly, we can make the model very **complex** (having many components and interactions). It is possible to write equations to **simulate** (mimic) the interactions we think are important in limnological systems. For example, to predict the amount of algae on a specific day of the year in a specific lake, we can use what we know about influences of algal nutrients, climate, and multiple species of consumers. This approach was used to predict the effect of an algae-eating mussel on water clarity in Lake Mendota (Reed-Andersen et al., 2000). The model predicted that if the mussel invades the lake, water clarity will increase. Luecke et al. (1992) used a computer simulation model to predict the amount (as grams per liter) of the algae-eating *Daphnia* (a small crustacean; also called the *water flea*) in the open water of lakes. The model predicts the daily amount of *Daphnia* based on how fast they reproduce in response to the food supply and how fast they are eaten by predators. This is a valuable model because *Daphnia* often have a large effect on the amount and kinds of algae living in a lake (i.e., water quality) and *Daphnia* are important fish food (an aspect of fish management).

Riessen (1999) simulates the life history of *Daphnia,* using an elegant computer model. This simulation model allows limnologists to test theories about effects of food and predation on the population dynamics and life history of this important freshwater species.

Whether descriptive or predictive, simulation models run the risk of the Scylla of simplicity and the Charybdis of complexity. (In ancient Greek stories, the Argonauts have to carefully navigate between the reef Scylla and the whirlpool Charybdis—two dangers, either of which was difficult to avoid without encountering the other.) Early mathematical models were general and could be applied to any system, but were too simple to make accurate predictions. Modern, complex simulation models tend to be as difficult to understand as the nature they emulate. Simulation models are simpler than nature, but they are often so complex (have enough components and flows) that they can be as difficult to understand as nature herself:

> "If our models become as complex as nature itself they will be as difficult to understand as nature is. This is not a joke: the point of modeling is to increase our understanding of nature, not merely [!] to reproduce it on the computer" (Mangel & Clark, 1988).

Hypothesis Testing

Scientific understanding of pattern or process in aquatic systems is refined (made more accurate in terms of prediction) via hypothesis testing. A **hypothesis** is a carefully worded statement about pattern or process—sometimes called a "best guess." The hypothesis often is stated in terms of cause and effect. Statistical thinking is employed to gauge the probability that hypotheses are correct, in light of experimental results or observations of nature (natural experiments). An **experiment** is a test of a hypothesis, often with multiple (**replicate**) parts for statistical purposes, and often involving a comparison of the reactions of two sets of replicates, with one set exposed to a causal factor (experimental **treatment**) and one set similar in all respects to the experimental treatment, except not exposed to the causal factor (**control** treatment). Small-scale experiments are done in the laboratory or in containers in lakes and streams, and are often carefully designed to have enough replicates and the right conditions to test a hypothesis. Sometimes hypotheses concern whole lakes or rivers, for which replication is an issue. So-called **natural experiments** are still possible by taking advantage of natural variation among lakes and variation from year to year (Edmondson, 1993). For example, careful statistical analysis of several years of data from Lake Washington suggested that the abundance of *Daphnia* depended on both the quality of algal nutrients in the water and the abundance of predators.

The one factor in an experiment that differs between control and treatment is hypothesized to be the cause of the predicted result in the experimental replicates. Experiments are most informative when they are carefully *designed* to have sufficient replicates and just the right treatments to allow a clear interpretation of the results. For example, we might hypothesize that adding phosphate to lake water will increase algal density. To do the experiment, we would enclose lake water in several containers (the replicates). The number of replicates would be chosen based on previous knowledge of variation in response among replicates. To half the containers we would add phosphate (the experimental treatment). Control containers would receive no additional phosphate. All containers would be incubated for the same amount of time under the same conditions, and algal concentration would be analyzed at the same time at the end of the experiment.

The process of hypothesis testing includes:

1. *Building* a hypothesis based on experience or existing data;
2. *Using* the hypothesis to make a prediction in a system, such as a lake—if we do an experiment by making a change in a system, then we use the hypothesis to predict the result of the change we make;
3. *Comparing* (often using statistical techniques) what actually happens in the system to what we predicted;
4. *Changing* the hypothesis so it would have predicted the observed result; then
5. *Starting* over again at #2, by continuing to observe the system, and possibly by further manipulating something in the system.

Hypothesis testing is the essence of research science, and it is used in several of the exercises in chapter 13. The purpose of this introductory text is to familiarize the student with the concepts, range of knowledge, and language of limnology, so as to provide a basis for future hypothesis testing.

Examples of hypotheses that can be tested with statistical analysis of appropriately collected data include:

- The average size of bluegill sunfish is greater in lakes that have larger populations of plankton, their favorite food.
- In lakes, water clarity decreases with an increase in phosphate concentration.

MESOCOSMS AND WHOLE-LAKE EXPERIMENTS

Limnological models are often tested in relatively natural settings. Small experiments set in a lake or stream are termed **mesocosms** and are either **enclosures** (to keep organisms inside) or **exclosures** (to keep organisms out). This technique was pioneered by a marine ecologist interested in measuring the effect of ecological factors on the distribution and abundance of intertidal organisms (Connell, 1961). The technique was enthusiastically adopted by limnologists, who now consider mesocosm experiments an indispensable part of their toolbox.

The whole-lake experiment is another characteristic limnological tool. One of the first whole-lake experiments was done by Johnson and Hasler (1954). They added lime to one of two small lakes in the Upper Peninsula of Michigan (United States) to explore the effects of pH on primary production and trout production. Whole-lake experiments are now commonplace, and much of modern limnology is based on studies at sites such as the Experimental Lake Area (western Ontario, Canada) or the North Temperate Lake program of the National Science Foundation (NSF)–Long Term Ecological Research Program (LTER) centered on lakes in northern Wisconsin, United States.

Whole-lake experiments can either be done using one lake over many years (Edmondson, 1991) or (exchanging space for time) using many lakes observed for a short amount of time (Stemberger et al., 2001). Although whole-lake experiments are the most realistic conditions a limnologist can use, they often suffer from low or nonexistent replication—the large scale of lakes inhibits replication, given the typical constraints of low budgets and limited time.

THE ROLE OF FIELD STATIONS

The physical setting of limnology is often a research or summer school **field station** located on the shores of a lake. Limnological field stations typically provide both instruction (undergraduate and graduate courses) and opportunities for research at all levels. For many limnologists, courses and associated research experiences at field stations have been wonderful personal and career-defining experiences. Accumulated research, often by multiple collaborators at well-established field stations (e.g., at Plön in Germany, Windermere in the UK, Pallanza in Italy, and Vilas County in the United States), has provided long-term limnological data sets that are invaluable for observing phenomena that change over decades or even centuries (Magnuson, 1990).

Limnology students can choose field stations from a variety of styles. Use the Internet to find the home page of the Organization of Biological Field stations. Some of the earliest field stations occupied elegant, converted villas or country homes. The station at Pallanza, Italy includes a serene, walled-in garden. The old main building at the Freshwater Institute on Lake Windermere, in the

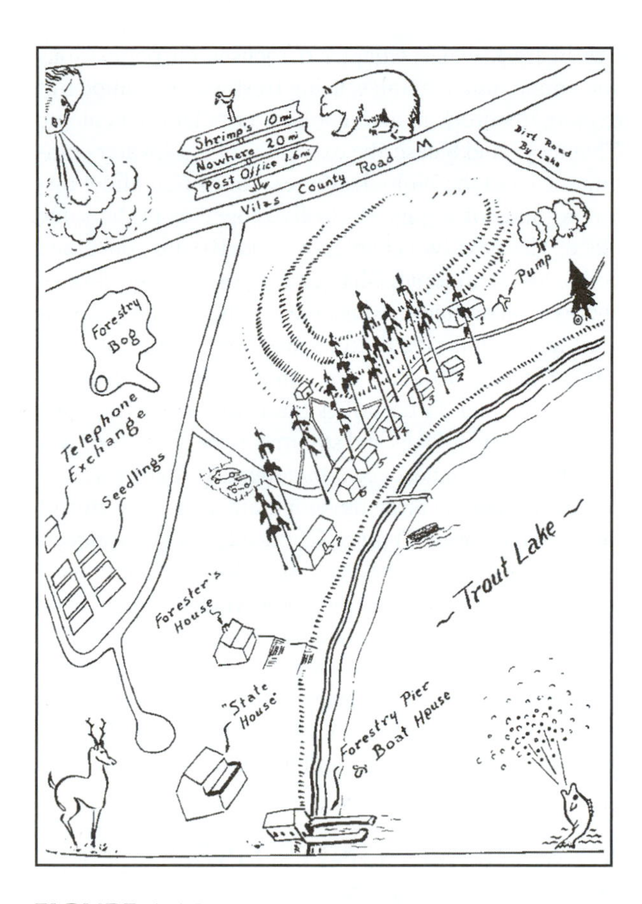

FIGURE 1.15 A sketch by Lowell E. Noland of the Trout Lake, Wisconsin, field station used by Birge and Juday. **Source:** Drawing by Lowell E. Noland, courtesy of the University of Wisconsin Center for Limnology, Madison WI, USA.

English Lake District, is a large, no-nonsense, rectangular country house built of local stone. A large house on the shore of Grosser Plönersee, Plön, Germany was converted into the famous Hydrobiologische Anstalt (1892–1959). It is now a hotel, but was used as a research facility by August Thienemann from 1917 to 1957.

Other early field stations were rustic; for example, the original field station at Trout Lake (Vilas County), Wisconsin (figure 1.15), the University of Montana field station on Yellow Bay of Flathead Lake, and the field station at the Rocky Mountain Biological Laboratory in Gothic, Colorado. When these camps were first started, students lived in tents or rustic cabins and most of the teaching was done outside. Recently, these camps have been modernized and now offer indoor plumbing, heated rooms, clean laboratories, and even telephone and Internet access.

An expansion of limnological facilities began in the 1960s and 1970s, during the "post-war" economic boom and the emphasis on science generated by the Russian satellite, Sputnik. These newer facilities are generally well-equipped and attractive. The post-1970 buildings at Gull Lake (Hickory Corners, Michigan) are elegant, modern, institutional buildings, and the 1963 Center for Limnology is an exceptionally beautiful structure hugging the shore of Lake Mendota (Madison, Wisconsin).

THE ROLE OF JOURNALS AND ASSOCIATIONS

The major storehouse of limnological information rests in libraries of large research institutions. The information is contained in monographs and books, and primarily in journal articles. Much of the information is being transferred to the Internet.

There are about a dozen journals that together publish the bulk of limnological reports, amounting to several thousand papers each year. Journals that focus on general limnology and are available and read globally include (with the country of origin):

Advances in Limnology (the title changed in 1995 from *Ergebnisse der Limnologie*) (Germany)

Archiv für Limnologie (Germany)

Canadian Journal of Fisheries and Aquatic Science (Canada)

Freshwater Biology (publication of the Freshwater Institute) (Great Britain)

Hydrobiologia (Germany, Belgium, and The Netherlands)

Journal of Limnology (until recently called the *Memoire dell'Istituto Italiano de Idrobiologia Dott. Marco de Marchi* and distributed as part of a publication-swapping arrangement) (Italy)

Journal of Plankton Research (Great Britain)

Journal of the North American Benthological Society (the publication of the North American Benthological Society, or **NABS**) (Canada and United States)

Limnology and Oceanography (the publication of the American Society of Limnology & Oceanography, or **ASLO**) (United States)

Proceedings (Verhandlungen) of the International Association of Theoretical and Applied Limnology (called **SIL** from the Latin abbreviation for "International Limnology Society") (Germany)

Transactions of the American Fisheries Society (United States)

In addition, many regional journals are sponsored by research institutes and field stations. Limnological papers also appear in well-known scientific journals such as *The American Naturalist, Ecology, Nature, Oecologia, Oikos,* and *Science.*

Professional scientific societies (e.g., ASLO, NABS, SIL) have been very important in promoting communication of limnological research. Several of these societies sponsor large, annual meetings to allow limnologists to present new work and talk to colleagues. These meetings are essential to limnology because they generate excitement and new ideas, allow rapid communication of new ideas, and provide role models for new limnologists. Participation in a national meeting can be an unforgettable experience—a milestone in the education of a new limnologist.

Many journals are available electronically, depending on subscription and licensing agreements, which are specific to individual libraries. Where the resources are available, journal articles can be read via the Internet, using journal aggregators and archives such as "ProQuest Research Library," "BioOne," "Academic Search," or "JSTOR." Consult your local librarian to learn which resources are available.

THE ORGANIZATIONAL SCHEME OF THE PRESENTATION OF LIMNOLOGY IN THIS TEXT

This book starts with a general introduction to the subject matter and the history of limnology. The second chapter is a broad introduction to the physics and chemistry of water—an environment that humans find alien. Basic concepts in this chapter are a necessary foundation for the rest of the text, which explores limnology, and especially ecology, in greater depth. Chapter 2 provides an important overview of basic environmental considerations of limnology. In the next three chapters (3 to 5), we meet the major groups of organisms that live in fresh water. Chapter 6 focuses on populations and population dynamics, using freshwater examples to explore theoretical and applied population ecology. Chapters 7 and 8 describe examples of freshwater communities and the biological interactions that allow different kinds of organisms to live together in the same place. The next two chapters (9 and 10) are concerned with ecosystem topic. Chapter 9 focuses on the flow of energy in lakes and streams, both in the habitat (ecosystem) and in individual organisms (physiological ecology). Chapter 10 deals with major aquatic chemical cycles. Chapter 11 describes the landscape setting and morphometry of lakes and streams, flow patterns of water through the landscape components of the hydrological cycle, and the birth, development, and extinction of lakes and streams. Chapter 12 introduces environmental science, using the theme of the "citizen limnologist," or application of limnological knowledge in the social context of conservation and management of aquatic resources. Chapter 13, an assortment of limnological exercises for the laboratory or field, introduces limnological data collection, water chemistry observation, physics and biology, and hypothesis testing.

MEASURING UNITS USED IN LIMNOLOGY

Limnologists primarily use the metric measuring system. The advantage of the metric system is that it is based on powers of 10, so different units are easily interconverted (table 1.2). The metric system is widely used in science and is commonly used by most of the world outside the United States. However, many people in the United States are unfamiliar with the metric system, and many U.S. water-related fields rarely use it. U.S. engineers, water and fish managers, and conservationists typically use the English measuring system. Commonly used units from the metric system are explained in terms of English-based, U.S. units in table 1.2.

Table 1.2 Approximate English Equivalents of Metric Units

In the metric system, *kilo* means 1000, *centi* means 1/100 or 10^{-2}, *milli* means 1/1000 or 10^{-3}, and *micro* means 1/1000000 or 10^{-6}.

What Is Being Measured	Metric Measurement Unit	Abbreviation	Approximate English Equivalent
Length	Meter	m	39.37 inches, or a little over a yard (which is 36 inches); a fathom is 2 yards, or about 2 meters
Length	Centimeter	cm	About 1/3 of an inch; 1/100 of a meter
Length	Millimeter	mm	About 1/32 of an inch; 1/10 of a cm
Length	Micrometer	µm	One-millionth of a meter, one 1/1000 of a mm, and 0.00003937 inches
Area	Hectare	ha	2.47 acres
Area	Square meter	m^2	About 1.2 square U.S. yards; there are 10,000 square centimeters (cm^2) in a m^2
Area	Square centimeter	cm^2	There are about 6.5 cm^2 in a square inch
Volume	Cubic meter	m^3	About 1.3 cubic U.S. yards; about 35 cubic feet; and about 28 bushels; there are about 811 cubic meters in 1 acre-foot
Volume	Liter	l	1.06 U.S. liquid quart; there are 1000 liters in a m^3
Volume	Milliliter	ml, cm^3 (=cc)	There are about 28 ml in a fluid ounce; 1 cm^3 is the same as 1 ml
Mass	Gram	g	There are about 28 grams in an avoirdupois ounce; 1 cm^3 of water weighs very close to 1 gram depending on temperature
Mass	Kilogram	kg	About 2.2 pounds; a liter of water weighs essentially 1 kg depending on temperature
Mass	Metric ton	t	Very similar to a long (U.S.) ton of 2000 pounds; a m^3 (1000 liters) of water weighs 1 metric ton, or about 2200 pounds
Rate	Km per hour	Km hr^{-1}	0.62 miles per hour; 28 cm sec^{-1}
Rate	Cubic meter per second	m^3 sec^{-1}	35 cubic feet sec^{-1}; 15,850 gallons min^{-1}; or 22.8 million gallons day^{-1}

Study Guide

Chapter 1 Introduction to Limnology

Questions

1. What is limnology?
2. Why is limnology difficult to define?
3. How long has limnology been a science?
4. What is a lake?
5. What is a model?
6. What is a field station?
7. What are different ways of looking at limnology? Which aspect of limnology most attracts you?
8. After reading chapter 1, and taking into account your earlier experiences with water and even limnology, answer the question: "What does it mean to think like a limnologist?"
9. Since you are taking a limnology course, you have probably had experiences with lakes or streams. Describe your first memorable, limnological experience.
10. Hutchinson (1965) titled a collection of essays on ecology *The Ecological Theater and the Evolutionary Play.* Close your eyes and imagine the limnological theater.
11. Analyze the following statement according to the perspectives of science and environmentalism: *The global climate is warming steadily, and this process, which will result in changes in the distribution and abundance of terrestrial and aquatic organisms, is caused by human activity that is putting gases such as carbon dioxide and chlorofluorocarbons into the atmosphere.*
12. Go to a library or go online and look for articles on limnology in one of the journals listed in this chapter. How would you describe the limnological articles in the journal?
13. Use a database to search for more information on (1) limnology, (2) the location of the field station nearest you, and (3) your favorite lake or stream.
14. Check out the website for the Organization of Biological Field Stations. If you had the necessary time and funds, what would be your choice for a dream course at a summer biological field station?
15. Draw and label all parts of the vertical temperature profile of a lake, using the modern format.
16. What are the similarities and differences for descriptive and predictive models? Describe at least one actual example for each model.
17. What is the basic process used in hypothesis testing?
18. How do you classify yourself—as a scientist or an environmentalist?

Words Related to General Limnology

aquatic	backwater	biogeochemistry	brackish
ASLO	bathymetric	bog	canal

community
competition
complex
component
consumer
contour
control
descriptive
ecosystem
enclosure
environmental science
estuary
exclosure
experiment
field station
food web
gnamma
hypothesis
interactions
intermittent stream
isobath
karst
lake
lentic
limnology
logo
lotic
mesocosm
microcosm
microfossil
model
NABS
natural experiment
paleolimnology
plankton
pond
predation
predictive
producer
replicate
reservoir
river
rock pool
saline
saltern
science
see (German)
SIL
simulate
spring
stream
tinaja
treatment
tree hole
trophic
vertical profile
wetland

Major Examples and Species Names to Know

Find lakes on a map, such as figures 1.2 and 1.3

Lake Geneva, Switzerland

Lake Washington, Washington State, United States

Thunersee, Switzerland and the first vertical temperature profile for a lake

Lac Ste. Hélène, France, an early bathymetric map

Daphnia
black bass
Spirogyra
bluegill sunfish
Scylla and Charybdis
water flea

What Was a Limnological Contribution of These People?

W. T. Edmondson
Stephen A. Forbes
G. Evelyn Hutchinson
Franz Ruttner
Jack Vallentyne
Fischer-Foster and Brunner
François-Alphonse Forel
Ray Lindeman (use the Internet)
August Thienemann

Additional Resources

Further Reading

The student, researcher, or manager wishing to find the final word on a limnological question should turn to the limnology texts by Kalff (2002) and Wetzel (2001):

Kalff, J. 2002. *Limnology.* Upper Saddle River, NJ: Prentice Hall. This text is in the same tradition as Wetzel's—a little less encyclopedic, but perhaps easier to read.

Wetzel, R. G. 2001. *Limnology.* 3rd ed. Academic Press. This encyclopedic limnology text is the single, most important limnological reference source for technical limnological details. Physical and chemical topics are especially well covered.

Other supplementary books of general interest include:

Bayly, I.A.E., J. A. Bishop, and I. D. Hiscock. 1967. *An illustrated key to the genera of the Crustacea of Australian inland waters.* Melbourne: Australian Society for Limnology. This small book provides a survey of the many species of small crustaceans that can be found in Australian rock pools, and gives the reader an impression of the ecological richness of these seemingly severe environments.

Beckel, A. L. 1987. *Breaking new waters: A century of limnology at the University of Wisconsin.* Transactions of the Wisconsin Academy of Sciences, Arts and Letters. Special Issue. 122 pp. This essay takes the reader back to the early days of limnology in Wisconsin, with Birge and Juday, and then recounts the Art Hasler story.

Brönmark, C., and L.–A. Hansson. 1998. *The biology of lakes and ponds.* Oxford: Oxford University Press. This small book focuses on the ecological aspects of modern limnology.

Edmondson, W. T. 1991. *The uses of ecology: Lake Washington and beyond.* Seattle: University of Washington Press.

Elster, H. J. 1974. History of Limnology. *Mitt. int. Ver. Limnol* 20: 7–13. This scientific paper reviews the history of limnology from the German perspective.

Hutchinson, G. E. 1979. *The kindly fruits of the Earth: Recollections of an embryo ecologist.* New Haven, CT: Yale University Press. 264 pp. This book includes several thought-provoking essays that have influenced many ecologists—aquatic and terrestrial.

Lampert, W., and U. Sommer. 1997. *Limnoecology: The ecology of lakes and streams.* English ed. New York: Oxford University Press.

Thienemann, A. 1925. *Die Binnengewässer: Einzeldarstellungen aus der limnologie und ihren nachbargebieten.* Stuttgart, Germany: E. Schweizerbart. This is a series of publications established by Thienemann, covering many aspects of limnology from 1925 to 1992.

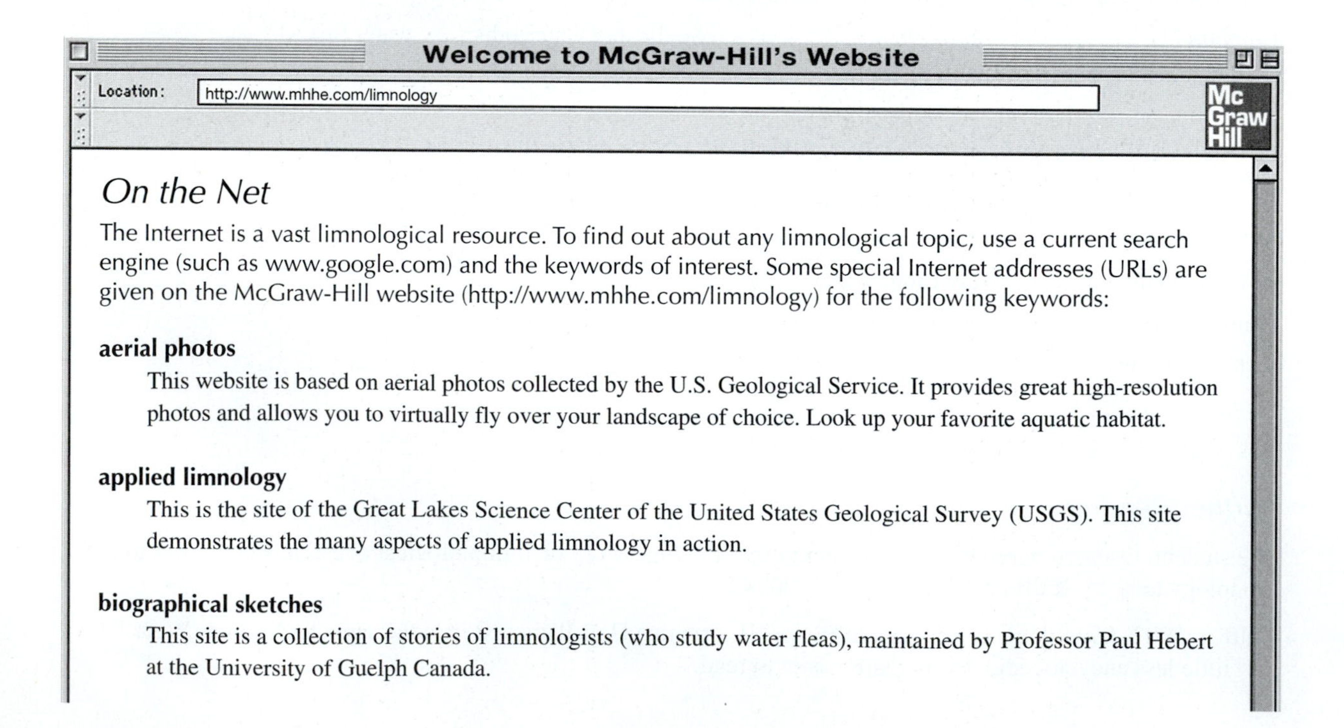

Welcome to McGraw-Hill's Website

Location: http://www.mhhe.com/limnology

Mc Graw Hill

On the Net

The Internet is a vast limnological resource. To find out about any limnological topic, use a current search engine (such as www.google.com) and the keywords of interest. Some special Internet addresses (URLs) are given on the McGraw-Hill website (http://www.mhhe.com/limnology) for the following keywords:

aerial photos

This website is based on aerial photos collected by the U.S. Geological Service. It provides great high-resolution photos and allows you to virtually fly over your landscape of choice. Look up your favorite aquatic habitat.

applied limnology

This is the site of the Great Lakes Science Center of the United States Geological Survey (USGS). This site demonstrates the many aspects of applied limnology in action.

biographical sketches

This site is a collection of stories of limnologists (who study water fleas), maintained by Professor Paul Hebert at the University of Guelph Canada.

equipment, books, maps

This site provides names of limnological organizations and companies that you can use with a search engine (such as www.google.com) to find images of limnological equipment, books, and maps.

field stations

These sites provide access to information about a large number of field stations—good places to gain limnological experience during the summer! Check the websites of the American Society of Limnology and Oceanography, North American Benthological Society, International Society of Limnology, and Ecological Society of America for a nearby meeting.

Forel Institute

This is the site of the F.-A. Forel Institute on the shores of Lake Geneva, Switzerland.

Great Lakes

This site is maintained by the Great Lakes Research Laboratory (National Oceanic and Atmospheric Administration). This particular page provides lake levels for the North American Great Lakes.

historical instruments

A site for great pictures of historical instruments used in oceanography that are similar to those used in limnology. See The National Oceanographic and Atmospheric Authority.

Jack Vallentyne

This site provides a biographical sketch of this influential Canadian limnologist who has taken the environmental science path.

oceanography

This marine site has good links to pictures of equipment used to sample water and pictures of aquatic organisms; also has links to marine scientists and pictures of zooplankton, and links to freshwater sites.

Water Research Network

Visit this site for an incredible international limnological clearinghouse—the Water Research Network—a "multidisciplinary database of research, researchers and institutions dealing with freshwater issues all over the world. The network includes natural and social sciences as well as the humanities."

2

"... fish live in an environment alien to us."

Allen & Hoekstra, 1992, p. 111

Setting the Stage: Water as an Environment

Small-Scale Characteristics of Water as a Substance

outline

THE DIVERSITY OF LIMNOLOGICAL ORGANISMS

Aquatic organisms are involved in an evolutionary play in an ecological theater (Hutchinson, 1965). This metaphor organizes the complex science of limnology. Hutchinson spoke of an ecological theater in the most general terms, but for this textbook of limnology, the metaphor focuses on a play that takes place on (and especially in) the freshwater theater.

The stage of this freshwater ecological theater is set according to a variety of physical and chemical characteristics. To understand limnology, it is crucial to understand the settings of the freshwater stage. We will begin our understanding of the limnological environment with an exploration of characteristics of water as a substance—first at a small, molecular level and then at larger, more familiar scales, where water has vertical structure and has significant mass moving as currents and waves at the scale of the entire lake.

Water, the **milieu** (environment) of lakes and streams, is an unfamiliar, alien environment for people. Water differs from air in being denser and holding more heat and less oxygen per unit volume. Water commonly exists in all three physical **phases** (a physically distinct form of matter). The solid phase is ice, the liquid phase is water, and the gas phase is water vapor. The gaseous atmosphere in which we live has a **density** of about 1.2 grams per liter, while water weighs 1000 grams per liter.

Like air, water is a **fluid.** Fluids are different from solids in that fluids have low internal (e.g., crystalline) structure and fluids distort, but do not break, when they receive a force. By this definition, water, ice, and even glass can act like fluids. These "solids" actually distort slowly when stressed.

Liquid water density has a subtle but critical relation to temperature that is an exaggeration of the thermal air currents we experience on land. Just as the physics and chemistry of air structure all aspects of life on land, so do the physics and chemistry of water constrain possibilities of life in lakes and streams.

FUNDAMENTAL CHARACTERISTICS OF WATER AS A SUBSTANCE

This section looks at water with a microscope. The small-scale physical and chemical characteristics of water are of ecological importance, especially to small organisms that live in a microscopic world. Also, in some cases, knowing about the small scale will assist us in understanding larger-scale phenomena. The reference text *Standard Methods for the Examination of Water and Wastewater* is a useful resource for water physics and chemistry (APHA, 1998).

Molecular Shape

Many special characteristics of water are a result of its molecular shape. The water molecule has two hydrogen atoms jutting out from one side of the oxygen. The hydrogen atoms are separated by an angle of about 104.5°. This angle is the result of the way electrons are arranged in the molecule, especially around the oxygen atom. The water molecule resembles a wedge of pie (figure 2.1), and this asymmetrical shape has a number of consequences for the chemistry and physics of water.

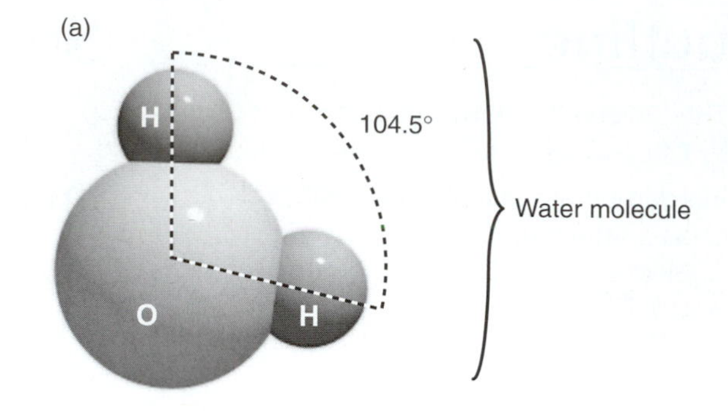

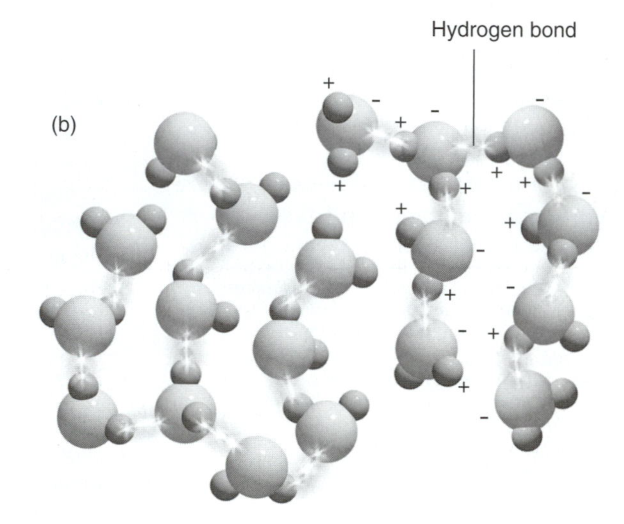

FIGURE 2.1 The water molecule is *asymmetrical,* with a separation of charges. (a) The polar water molecule, with the oxygen atom relatively negative, and the hydrogen atoms relatively positive. (b) The dotted lines represent hydrogen bonds, due to *electrostatic* attraction between negative and positive regions of water molecules.

Polarity

Because the two hydrogen atoms are on one side of the oxygen, the molecule is **polar,** with a negative pole on the side away from the hydrogen atoms and a positive pole between the hydrogen atoms. The polar water molecule readily forms weak **hydrogen bonds** with other polar molecules, including other water molecules, clay particles, and many organic molecules. Hydrogen bonds are the result of attraction between the negative pole of one molecule and the positive pole of another molecule. The tendency to form hydrogen bonds between water molecules has several important consequences: Water is viscous and has a surface tension, it is a strong solvent, its density changes with temperature, and it acts as a coolant.

Viscosity

Because water molecules tend to bond together, energy is needed to change their relative positions. We perceive the molecular (hydrogen) bonding as internal stickiness, or **viscosity.** Viscosity significantly slows the movement of particles in water only over small distances, at a scale of millimeters (mm) or less. When water is observed at small spatial and temporal scales, it looks like a thick, sticky fluid (figure 2.2). This stickiness is not really perceived by large humans, but it is a major environmental constraint for small organisms and small particles, less than a millimeter or so in diameter (Vogel, 1988). Under most conditions, the outer surface of animals in water has a layer of water a fraction of a millimeter thick. This "boundary" layer is the result of viscosity. Any fluid flowing past a surface flows more slowly the closer the fluid is to the surface. The flow is zero at the surface. This sub-millimeter-thick boundary layer of reduced flow (of water sticking to the surface) is of no consequence to large animals (whales, people, fish), but is of great significance to organisms less than a millimeter long (such as plankton). The biological importance of the boundary layer is **scale dependent.** The

FIGURE 2.2 A high-speed photo of the splash produced by a drop of water falling into milk. A crater has formed, and the central drop and the droplets around the rim are moving upward. The tails of the crater droplets give an impression of the viscosity of the fluid.

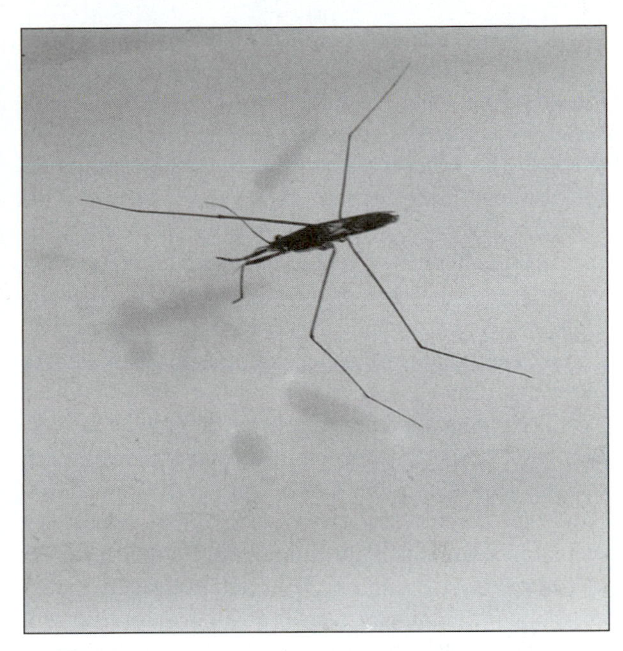

FIGURE 2.3 A water strider and its shadow in shallow water. The dimples made by the feet pressing on surface film produce shadows around the body shadow.

intensity of a scale dependent phenomenon, such as viscosity, depends on the size scale. Viscosity is only significant at the millimeter scale or smaller. At larger scales, viscosity is not as important as other forces (discussed in the Laminar Flow section).

The boundary layer of water has important implications for our understanding of how small organisms move and feed in water. The viscous boundary layer clogs filters (Cheer & Koehl, 1987), adds mass, and produces a force that impedes movement. On the bright side, the viscosity of water slows sinking rate. Also, on the one hand, small organisms expend energy to overcome viscosity as they swim, but on the other hand, they don't need brakes—viscosity stops movement almost instantly (Purcell, 1977).

At the micrometer scale, water in nature is variable in its viscosity because it contains large, linked organic molecules. These molecules of mucus (mucopolysaccharides), proteins, and even nucleic acids are tenuous, diaphanous, hyaline, and the same density as water, but can still affect the physical structure of water (LaFee, 2000; Long & Azam, 2001). Microscale strands of these molecules form a three-dimensional matrix that acts as a platform for bacteria, which are consequently clumped rather than distributed randomly. A single milliliter of seawater contains a cross-connected mass of thin filaments that, if lined up end to end, would extend as much as 5900 kilometers. This matrix is imperceptible to direct human observation, but it nevertheless is important to microbial ecology and ecosystem functioning.

The physical phenomenon of viscosity affects the body shape and behavior of aquatic organisms. This interaction is discussed further in chapters 3 and 4.

Surface Tension

The tendency of water to form hydrogen bonds with itself and its attraction to or avoidance of surfaces produce surface tension, a force that causes water to creep up small tubes and allows water striders to walk on water (figure 2.3). This surface tension produces a distinct boundary—the surface film between air and water. Hutchinson (1967) described whole communities of invertebrates, algae, and protozoans that are specialized for living on, in, or under the surface film.

The odd thing about surface tension is that its force is proportional to length. Consider the foot of a water strider (figure 2.3). The outer covering of the foot repels water, all along the edge that is touching the water's surface. The water molecules tend to stick together, so the repellent force is expressed as an upward force—the "surface tension"—supporting the strider on the surface. Longer feet (with more total edge) exert more upward force. If the force due to surface tension is greater than gravity, the animal will stay on the surface. If gravity is greater than surface tension, the animal falls into the water.

The practical break point between gravity and surface tension is reached for animals that weigh less than a gram. This is because, if the length of an animal doubles, the weight increases by eight. To support the weight increase, the larger animal must either have feet eight times as long, or live with less of a safety factor. After a point, long feet interfere with motion and mobility. Water striders compromise—the larger species have disproportionately long feet and a weight that is only slightly less than the surface tension can bear.

A few minutes' observation of a group of large water striders reveals that mating is a time when animals are most likely to fall through the surface. When the male jumps on the back of the female, both often fall through the surface. During mating, the female supports the weight of both partners. Her legs make deep dimples into the surface, indicating the surface is stretched to its maximum.

Of course, any weight can be supported by surface tension, if the feet are long enough. A person who weighs 70 kg could at least stand on water, with feet about 25 km long. Walking would be inconvenient. In other words, surface tension is scale dependent and is only biologically important at the centimeter scale and below.

Solvent

Water molecules stick to other polar molecules and ions. The association between the water and another chemical allows the chemical to move freely in—that is, dissolve in—liquid water. Water is a **solvent** (dissolves things) that dissolves a wide variety of polar solutes (substances that dissolve). Examples of chemicals that are **soluble** in water include acids, bases, salts, sugars, and alcohols. Common chemicals that are not particularly soluble in water include oils, waxes, pesticides, and petroleum products.

The amount of a chemical dissolved in water can be expressed in at least three ways. The first is **concentration,** expressed as mass per unit volume. For example, lake water might have a concentration of calcium (Ca^{++}) of 14.2 mg per liter. Second, the amount could be expressed as **molarity,** which is the concentration divided by the weight mass of the element or molecule. The molarity of a solution of 14.2 mg Ca^{++} per liter is 0.35 millimoles, because the atomic weight of Ca^{++} is 40.08 grams per mole. A *mole* is the weight in grams that equals a specific number of atoms, ions, or molecules—this number is 6.023×10^{23}. (You can look up the atomic weight of most chemicals on the Internet.) Third, the chemical molar **equivalents** of a solution is the molarity value multiplied by the valence (ionic charge) of the chemical (this only works for ionized chemicals). A 0.35-millimolar solution of Ca^{++} can also be expressed as a 0.70 milli-equivalent solution. For practice with these concepts, see the Chemical Equivalents Exercise at the end of this chapter.

Most aquatic animals are wet—water molecules readily adhere to the outer surface of the animal—for example, fish and aquatic plants have wet surfaces. A few aquatic organisms live in water but are not wet. *Daphnia* are covered with a nonpolar, **hydrophobic** (water-hating) exoskeleton. Because the covering is nonpolar, it has no attraction to water and it does not get wet. By repelling water, this covering keeps *Daphnia* dry even though it lives in water. The functional advantage of having a hydrophobic surface is that it discourages external parasites from attaching to the *Daphnia.* However, if *Daphnia* comes in contact with the surface, surface tension pops it out of the water—a distinct disadvantage!

Solubility of Gases

Atmospheric gases, such as oxygen and nitrogen, dissolve in water. The higher the temperature, the less gas is dissolved. Within the range of biologically relevant temperatures (zero to 35°C), oxygen differs in concentration by a factor of 2.1 (figure 2.4).

Air or water pressure also affects the saturation concentration. Lakes at different elevations have significant differences in saturation concentrations. High-elevation lakes contain as little as half the oxygen found in lakes at sea level, due to the lower air pressure at high elevations.

A layer of water 10 meters deep weighs as much as the entire atmosphere above it. Thus, if temperature is

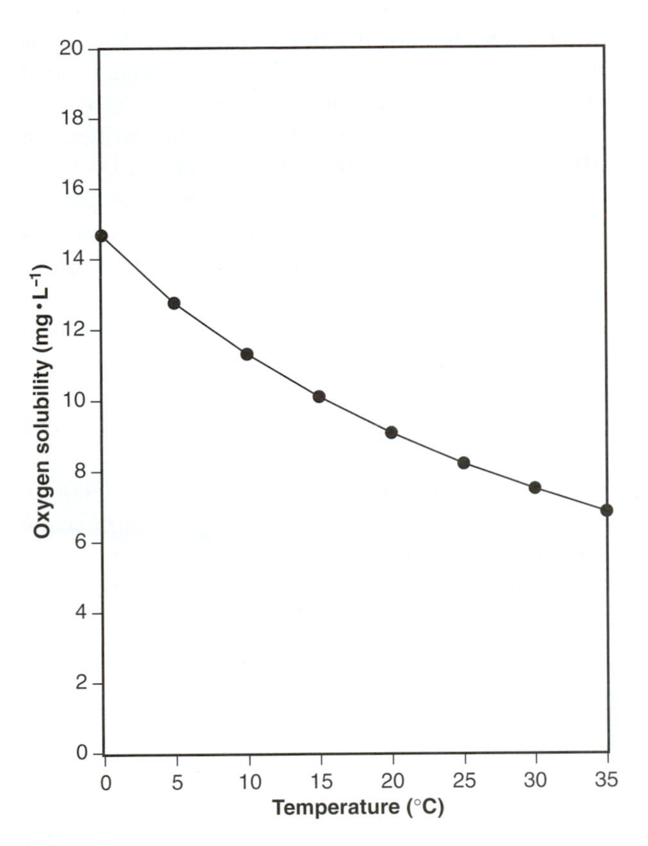

FIGURE 2.4 The relationship between oxygen solubility (at saturation) and water temperature over the ecologically relevant range of temperatures for limnology. These levels are for pure water at sea level. Decreased pressure or solutes dissolved in the water would reduce the amount of oxygen that can be dissolved in water. **Source:** Data from APHA, 1988.

constant, a liter of water at a depth of 10 meters can hold twice the amount of gas (such as oxygen) as a liter at the surface of a lake. A liter at 100 meters will hold 10 times as much gas. Except in a few unusual cases, we do not find huge amounts of gases dissolved in deep water because most of the gas input comes through the surface layer, and thus is in equilibrium with surface conditions. Occasionally, there can be a source of gas at depth. When this gas comes out of solution, it can, depending on the scale, produce **bubbles, bends,** and **eruptions.**

Bubbles rising from **anoxic** (no oxygen) bottom sediments are composed mostly of carbon dioxide and methane, produced in the sediments by bacteria breaking down organic material. Small bubbles of oxygen are sometimes seen rising from aquatic vegetation in bright sunlight.

Water can actually hold several times the saturation concentration. Supersaturation, caused by photosynthesis or intense mixing below dams, can produce bubbles. Supersaturated water will tend to lose oxygen by diffusion into the atmosphere from the surface, or by forming bubbles.

Bends are caused by gas coming out of solution in blood. SCUBA divers who spend time at depths greater than about 10 meters run the risk of putting a great deal of gas into their blood. The source of the gas is the air tank. When the diver returns to the surface, pressure decreases and the supersaturated gas forms bubbles in the blood and tissues, causing painful and even lethal health problems.

Some lakes have significant inputs of carbon dioxide from volcanic action at great depths. These lakes can erupt if any mixing brings saturated water up to a shallow depth—the gas forms bubbles, which rise, forming a water current that brings more deep water toward the surface, resulting in a geyser of water and gas. (See the section on Carbon Dioxide in chapter 10 for a more complete discussion of the eruption of killer Lake Nyos.)

Besides temperature and pressure, the solubility of gases also depends on the concentration of solutes (such as sodium chloride) dissolved in water (APHA, 1998). Saline water holds less oxygen than does pure water. There is enough salt in the oceans to decrease the saturation amount of oxygen at 4°C from 13.1 mg per liter (in pure water) to 8.6 mg per liter. Even at cold temperatures, salt lakes (such as the Great Salt Lake, Utah) have only a few milligrams of oxygen per liter. Wurtsbaugh and Berry (1990) reported that Great Salt Lake water at about 10°C contained about 60 grams of salinity in the relatively dilute, mixed upper layer in contact with the atmosphere. This mixed layer contained only about 6 milligrams of oxygen per liter (whereas about 11.3 mg would be expected in saturated pure water).

Dissociation and pH

Limnologists often use a simple model for the chemistry of pure water. The model states that the molecule water can break up (**dissociate**) into two ions: the **hydrogen** ion (H^+) and the **hydroxyl** ion (OH^-):

$$H_20 \rightleftharpoons H^+ + OH^-$$

The two ions are in chemical equilibrium with the nonionized water molecule. The relative concentrations of the molecule and the ions depend on random processes of dissociation and recombination, determined by the strength of chemical bonds in the molecule, the tendency for the ions to attach to each other, and the concentrations of the ions and water molecules. By "equilibrium," water chemists mean that a change in the concentration of one of the members of this system will affect the other members in a predictable manner. For example, if H^+ is added to pure water, increasing the H^+ concentration, then some of that H^+ will combine with OH^- to make water molecules. In this system, random processes dictate that the concentration of H^+ and OH^- will be in equilibrium in the sense that the product of their concentrations is 10^{-14} equivalents per liter. The same random processes governing the dissociation of water are, in effect, in more complex chemical systems, such as the inorganic carbon reactions discussed in chapter 10.

These two ions, H^+ and OH^-, can be thought of as being dissolved in water. Although H^+ ions are typically present at rather low concentrations, they are very active chemically. Therefore, knowing the H^+ concentration (concentration is written with square brackets as $[H^+]$ by convention) gives the limnologist an indication of the general state of water chemistry.

The hydrogen ion concentration is often written as a **pH** value. To calculate pH, we first express the hydrogen ion concentration as 10 to an exponent. For example, the hydrogen ion concentration 0.00001 equivalents can also be written as 10^{-5} equivalents. The units are "equivalents," or moles of hydrogen ions per liter, and this concentration is less than 1 mole per liter, so the exponent is a negative number. The pH is then the exponent times (-1):

$$\text{pH} = -\log[H^+].$$

Example: The pH of a solution with 0.00001 equivalents of hydrogen ion is 5. The pH of an aqueous solution can best be measured electronically.

Humans experience a wide range of pH values in their food and environment each day (table 2.1).

Table 2.1 Approximate pH of Some Common Substances

Substances	pH
Gastric juice	0.9–2.0 (weaned human)
Battery acid	1.0 (hydrochloric acid solution in water)
Lemon juice	2.3 (mostly citric acid solution)
Vinegar	2.8–3.0 (mostly a weak acetic acid solution)
Cola drinks	3.0 (mostly a strong carbonic acid solution)
Apple juice	3.1 (includes various organic acids)
Gastric juice	4.5–5.5 (nursing human)
Banana	4.6
Urine	4.8–7.5
Bread	5.5
Unpolluted rainwater	5.6 (pure water in equilibrium with carbon dioxide)
Meats	6–7
Saliva	6.3–6.8
Cow's milk	6.5–6.9
Human milk	6.6–7.6
Distilled water	7.0 (free of carbon dioxide)
Human blood	7.35–7.45 (has an inorganic carbon buffer system)
Seawater	8.0–8.5 (has an inorganic carbon buffer system)
Milk of magnesia	10.5 (saturated magnesium carbonate in water)
Ammonia solution	12 (cleaning solution)

Source: Data for foods is from Lund et. al., 2000.

Table 2.2 Neutral pH Is Not Always at pH 7.0

The pH of neutrality depends on temperature.

Temperature (°C)	K_w	Neutral pH
0	0.113×10^{-14}	7.47
25	1.008×10^{-14}	7.00
35	2.089×10^{-14}	6.84

Source: Data from APHA, 1998.

Depending on the concentration of other chemicals in water, the hydrogen ion concentration changes, from more concentrated in **acid** solutions (pH $<$ 7) to less concentrated in **alkaline** solutions (pH $>$ 7). At pH = 7 (at 25°C), H^+ and OH^- ions are equally abundant and both have a concentration of 1×10^{-7} equivalents per liter.

The pH of neutrality depends on temperature, but is near pH 7 at biologically relevant temperatures (Hutchinson, 1957). In table 2.2, the K_w is a constant that is the product of the concentration of H^+ and OH^- ions, expressed as molecular equivalents.

The pH of lake and river water covers a huge range of values (pictures of some of the extreme habitats can be found at the website address given at the end of this chapter). The most acidic lakes (pH between 0 and 2) are often associated with active volcanoes. Bog lakes, with high concentrations of organic acids, range from pH 2 to 6. The majority of lakes in the temperate and tropical zones have a pH in the range of 7 to 9. Desert lakes, saturated with carbonates, can have a pH of 10 or higher.

Extreme pH water is off limits for many kinds of organisms. The most acidic water is home to specialized bacteria and fungi. Some specialized plants and animals can live in bog lakes. Mollusks are restricted to water with a pH higher than 8 because their carbonate shells dissolve in lower-pH water. At the other end of the scale, water with the highest pH values is again dominated by bacteria and fungi.

Water's ability to absorb hydrogen ions when acid is added is called its *acid neutralizing capacity* **(ANC),** also called **alkalinity.** If the water is able to absorb a lot of acid, without changing pH, then the water is well **buffered** (see the discussion of the carbonate buffering system in chapter 10 for more information on how buffers work). The most common buffer is the inorganic carbon system. Acid rain or acid mine drainage entering lakes and streams can be neutralized (buffered) by carbonates dissolved from limestone bedrock. Lakes and streams on granite bedrock have low carbonate concentrations and low buffering capacity.

Redox Potential

Environmental chemists and microbial ecologists use information about electron activity in an environment, such as in a layer of bottom sediment, to predict whether the environment is suitable for specific chemical reactions. For example, the environment must be strongly reducing if sulfate reduction is to occur. Aquatic biologists who study higher organisms seldom use data about oxidation-reduction, or "**redox,**" reactions because these organisms require oxygenated water that produces a strongly oxidizing environment. However, when there is no oxygen present, a series of chemical reactions becomes possible—both spontaneous reactions and reactions that are catalyzed by bacteria.

To understand redox reactions, it is useful to remember that:

$$\text{Oxidation} = \text{loss of electrons}$$

For example, if soluble ferrous iron (Fe^{++}) is oxidized, it loses one electron per atom (one unit of negative charge) and becomes relatively insoluble ferric iron (Fe^{+++}).

Electrons are not actually in solution, but they can be thought of as having a concentration related to their activity or availability (just as we think of hydrogen ions having a concentration). An oxidizing chemical environment is one that has few available electrons—electrons tend to flow into the environment. A reducing environment has high electron activity.

Electron activity is typically measured as a voltage (electric potential) difference between an environment (which might be a solution in a beaker or in a lake) and a standard electrical instrument that is like a battery. Electron activity is abbreviated Eh, and it has volts as units. An Eh value can be converted into a concentration of active electrons, and the negative log concentration of electrons is $p\epsilon^-$, which is analogous to pH for hydrogen ion activity (Snoeyink & Jenkins, 1980). Just as a lower pH indicates a higher hydrogen ion concentration, a lower $p\epsilon^-$ indicates a greater electron activity (concentration) and a higher Eh value.

Limnologists measure the overall oxidation-reduction (redox) condition of an aquatic environment

Table 2.3 Examples of Oxidation-Reduction Reactions Commonly Seen in Aquatic Habitats

Reaction	**Electron Activity**	**pE**
Aerobic respiration (oxygen oxidizing organic material)	Lowest	Highest
Reduction of soluble ferric iron to less-soluble ferrous iron		
Anaerobic nitrogen fixation (nitrogen gas converted to ammonia)		
Sulfate reduction to sulfide		
Anaerobic respiration (production of methane)		
Photosynthesis (reduction of carbon dioxide)	Highest	Lowest

Source: Based on information in Snoeyink & D. Jenkins, 1980.

using a standard meter, which measures electron activity by the rate of flow of electrons into or out of a platinum electrode that is part of a battery-like circuit. If the environment has a higher electron activity than the meter, electrons will flow into the meter—the environment is oxidized and the meter is reduced. The environment is "reducing" the meter. Positive Eh values (of the meter relative to the environment) indicate an oxidizing environment; negative values indicate a reducing environment (Hutchinson, 1957). Lake water typically has Eh values of around 500 millivolts in well-oxygenated habitats, such as moving water in streams or an epilimnion. Anoxic water, as in hypolimnia, will have an Eh value near zero or even in the negative range.

Oxygen dissolved in water combines readily with electrons. Even if there is a very low electron activity, oxygen will combine with electrons and hydrogen ion to produce water. Many reactions that require high electron activity, such as nitrogen fixation or methane production, cannot take place as long as oxygen is present because the oxygen strongly depletes electron abundance.

Respiration removes oxygen from water. When oxygen combines with organic material, the organic material is oxidized, using up free oxygen. As oxygen concentration in water decreases due to decomposition of organic material, the electron activity increases. If oxygen is at low concentration or absent, the electron activity can rise above the level that would be seen if oxygen was abundant. This allows reactions to occur that would not occur in the presence of oxygen (table 2.3).

Ferric (Fe^{+++}) iron, also written as Fe(III), is able to attract electrons at an activity level just a little higher than the minimum produced by oxygen. If oxygen is present, iron will be in the insoluble Fe^{+++} ferric form, because electron activity is kept at a low level by the oxygen. In the absence of oxygen, ferric iron is converted (reduced) to the soluble ferrous (Fe^{++}) form by gaining an electron.

After all the ferric iron is reduced, electron activity can again rise until there is enough activity to reduce (ferment) organic matter to alcohols. At still higher electron activity, we see nitrogen fixation, sulfate reduction, methane production and nitrogen fixation, and at the highest level of activity, photosynthesis. Methane production, nitrogen fixation, and photosynthesis require higher levels of electron activity than occurs in fresh water, which means these reactions will not occur spontaneously. They can only occur in the chemically specialized interiors of bacteria and plants.

Limnologists often use a special voltage meter to measure electron activity in a specific freshwater habitat. The electron activity level gives an indication of what chemical reactions will occur in the water of that habitat. The reactions listed in table 2.3 will occur approximately in the order they are listed, depending on temperature and concentrations of the various reactants. For example, if a chemical limnologist measures the redox potential of water at the bottom of a swamp, and if she knows the concentrations of the reactants, then she can predict which chemical reactions will take place.

Chemicals Dissolved in Water

Besides hydrogen ions and gases, water typically also contains dissolved ions (salts). **Salinity** is measured as the total concentration (mg per liter) of ions dissolved in

Table 2.4 Examples of Conductivity Values for Different Kinds of Water

Water Source	Specific Conductance (μS)
Rainwater with dissolved CO_2 and small amounts of other chemicals	ca 10–15
Low-productivity mountain lakes in granite	ca 10–30
Lake Mendota, a nutrient-rich, high-productivity lake in a limestone watershed	250–390 (highest in the summer hypolimnion)
Lake Mead, Colorado River reservoir	850
Atlantic Ocean	43,000
Great Salt Lake	158,000

Source: Data from Hutchinson, 1957.

water. The global average for fresh water is about 120 mg per liter. Ions common in fresh water include **calcium** (Ca^+), **magnesium** (Mg^+) and **bicarbonate** (HCO_3^-), with lesser amounts of other common ions such as sodium (Na^+), potassium (K^+), sulfate (SO_4^{-2}), chloride (Cl^-), nitrate (NO_2^-), phosphate (PO_4^{-3}), and silicate (SiO_4^{-2}). The ions dissolved in a specific lake depend on patterns of water flow, evaporation, atmospheric input, contributions from soil and bedrock, and biological activity. In fresh water, calcium bicarbonate is often the most abundant salt, especially if there is any limestone in the watershed. Sodium chloride is the major salt in the sea but a minor component of most fresh waters, except saline lakes such as the Great Salt Lake, Utah.

Salinity (which depends on the concentration of ions in water) is measured as **specific conductance** (also called **conductivity**). The more ions in water, the better it conducts electricity. Conductivity is measured by the rate of flow of current between two platinum electrodes, and the preferred units of electrical conductance are micro-Siemans (μS). Table 2.4 gives examples of conductivity readings in different kinds of water.

The conductivity of water is sometimes used as a surrogate measure that indicates the level of **primary productivity** (the average rate of photosynthesis) in water (Prepas, 1983). In fresh water, salinity depends on a number of ions, including nitrate and phosphate, two important plant nutrients. There is a general correlation between total conductivity and the concentration of plant nutrients.

Salinity can be distinguished from **total dissolved solids, hardness,** and *alkalinity,* all terms used to describe solutes commonly found in fresh water. Total dissolved solids (**TDS**) includes all organic and inorganic salts as well as the nonionized components of fresh water. Organic compounds (such as sugars and organic acids) contribute to the total dissolved solids. This organic material (or **DOC**—dissolved organic carbon) sometimes stains water a dark tea color (as discussed in chapter 9, under the Carbon Cycle).

Water with high concentrations of dissolved calcium and magnesium carbonates is **hard.** Water from wells in limestone or sandstone bedrock is often hard. Calcium and magnesium carbonates have an unusual solubility pattern. Most chemicals have higher solubilities at higher temperatures. However, calcium and magnesium carbonates are less soluble in hot water than in cold water. Thus, when hard water is heated, the carbonates precipitate—as a crusty scale in hot water pipes and tea kettles. As desert lakes dry, a white band of carbonate salts precipitates along the shore (plate 7). In addition to the scale problem, the calcium and magnesium in hard water combine with soap and prevent it from being an effective dirt remover. For these reasons, hard water is often converted into **soft water** (softened) by passing it through an ion exchange column. Resins in the column are saturated with sodium chloride by exposing them to a concentrated salt solution. In the presence of concentrated sodium chloride, the sodium ions replace magnesium and calcium on the resin. When the resin is exposed to hard water, the sodium is replaced by calcium and magnesium. The sodium is released and calcium and magnesium attach to the resin. The purpose of this resin exchange is to replace calcium and magnesium bicarbonate with sodium carbonate, which is very soluble and does not precipitate on plumbing or cooking utensils. Also, sodium does not cause soap to precipitate. One problem with softening is that it makes water taste saline and somewhat unpalatable.

Density

In physical limnology, *density* means the weight of a substance per unit volume. For example, 1 milliliter of pure water at 3.98°C weighs 1 gram; the density is 1 gram per milliliter at 3.98°C. The density of a substance depends on molecular composition and the distance between molecules. The average distance between molecules is affected by their shape and by the strength of attractions between molecules. For example, in ice, each molecule is attached by hydrogen bonds to three other molecules. In liquid water, each molecule can be momentarily bonded (hydrogen bond) with one or two other molecules. In the liquid, short chains rapidly form and break. Chain length depends on the energy (heat) content of the liquid. Molecules are not attached in the gases.

Water's temperature-density relationship is unusual (figure 2.5). The liquid is more dense than the solid. Thus, ice floats on liquid water. Liquid water is most dense, not at the freezing point, but at 3.98°C.

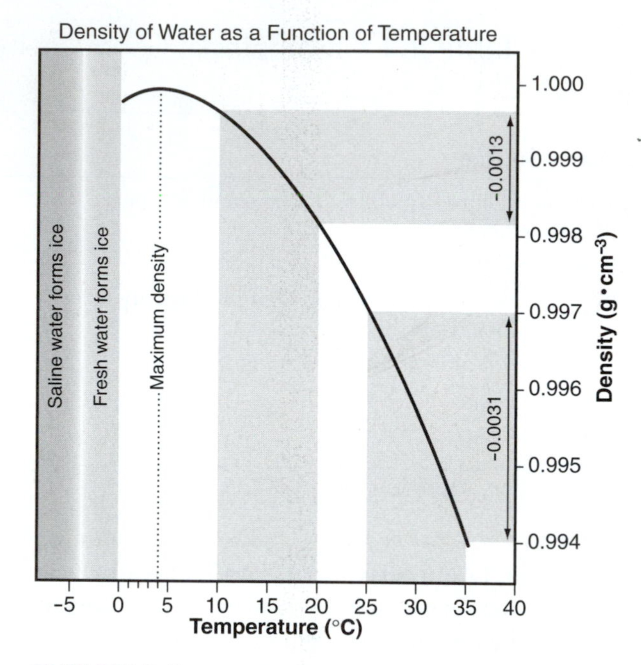

FIGURE 2.5 Water's temperature-density relationship. The maximum density is at about 4°C. A small change in temperature in warm water results in a larger decrease in density than the same amount of change in cool water. **Source:** Data from Hutchinson, 1957.

Moving water is much more powerful than moving air because of the difference in density. Air has a mass of about 1.2 grams per liter, while water is 1000 grams per liter. A breeze of 10 miles per hour (430 cm per second) is barely detectable, but a water current of the same velocity is a force to be reckoned with because of water's higher density.

The power output an organism needs to produce to stay in one place is proportional to drag (which is a force) and the current velocity. Drag depends on the organism's cross-sectional area perpendicular to flow and to the square of current velocity (Vogel, 1983). Organisms that must live in moving currents, such as stream insects and fish, typically have a streamlined body form that lessens drag by minimizing the area presented to the current (Peckarsky et al., 1990). Stream organisms also tend to live in the minimum flow that will just meet their needs for oxygen and optimize their net caloric intake. There is a tradeoff here: A faster current brings more food but may require more of the organism's energy to avoid being swept away.

The combined effects of drag and current velocity suggest that the power required to resist a current is proportional to velocity (V) to the third power. If the velocity of a current doubles, an organism needs to produce eight times (or 2^3) as much power to remain static (because power is proportional to the V^3). This is why floods can be catastrophic and why meteorologists warn motorists to avoid driving a car into even shallow moving water.

Heat

Heat and temperature are sometimes used as synonyms, but there is an important distinction between them. Heat is a form of energy, measured in units appropriate to energy, such as calories or joules. One **calorie** is defined as the amount of heat required to raise the temperature of water from 14.5°C to 15.5°C. Temperature is a sensation of hot or cold. Heat and energy are related in that adding heat to a substance usually increases the temperature, except that heat can also be added to cause phase changes (e.g., solid to liquid) with no change in temperature. *Phase* refers to the physical form of a substance. In the temperate zone, we experience the three phases of water annually: solid (ice), liquid (water), and gas (vapor, as humidity, or steam).

The coldest phase-form of water is solid ice (plate 5). When enough energy is added to ice (about 80 calories per gram of water for ice at 0°C), the frozen water melts as most of the hydrogen bonds are broken. The breaking bonds allow the molecules to move closer together and to move about more freely relative to one another. Because the liquid-phase molecules are closer together than the solid-phase molecules, liquid water is about 8% denser (more weight per ml) than ice. Water is densest at about 4°C (closer to 3.98°C; see figure 2.5). Above 4°C, water density decreases steadily (but minutely) until 100°C is reached. At this temperature, additional heat (540 calories per gram) is needed to break the remaining hydrogen bonds, and suddenly the unconnected, furiously vibrating and spinning water molecules find themselves drifting in space as independent components of steam (plate 4).

Cooling and Evaporation

Water cools when it evaporates. **Evaporation** occurs when molecules of liquid water escape through the surface and go into the vapor phase. Water can evaporate because in even the coolest water, there are a few molecules at the surface that are vibrating fast enough to break all hydrogen bonds. For each gram of water that evaporates, 540 calories are removed from the liquid water. This is a tremendous amount of heat, compared to the amount of heat that is required to warm water from 0°C to 100°C!

Cooling by evaporation is a significant factor for lakes and streams—cooling the surfaces of lakes and streams, just as it cools the sweating limnologist. Lakes lose a large part of their water through evaporation—about 50–70% as much water evaporates from Lake Mendota as flows out through the surface outlet (Brock, 1985). Surface evaporation contributes to turbulent mixing in the upper water layers of a lake. Water evaporating from the surface takes heat with it, leaving cooler water behind. This cool water tends to sink down to cooler layers deeper in the lake (see figure 2.10). The sinking, cool water mixes several meters of the upper water layers.

Summary of Critical Limnological Numbers

It is useful to remember a few critical numbers from chemical and physical limnology in order to easily understand limnological stories. The numbers for temperature and density and temperature and O_2 saturation in table 2.5 are so important to limnology that it is worth memorizing them.

VERTICAL STRATIFICATION IN LAKES

"In spring the water of the pond is like blue wool, endlessly tossing. The heavy, cold water has sunk to the black bottom of the pond and, struck by this weight, the bottom water stirs and rises, filling the pond's basins with wild nutrition. . ."

Mary Oliver, 1991, from "The Ponds" in "Blue Pastures"

The study of **stratification** is the study of layers. Knowledge of vertical stratification in a lake provides a powerful tool for understanding how lakes work in terms of chemistry, physics, and biology. One thing that limnologists tend to have in common the world over is a first-hand experience of measuring a lake's temperature and oxygen concentration at different

Table 2.5 Summary of Physical Characteristics of Water at 1 atmosphere of pressure

Temperature (°C)	Density (gm/cm^3)	Viscosity (centipoise)	O_2 Saturation (mg/liter [ppm])	Surface Tension (dynes/cm)
0	0.9998679	1.79	14.6	75.6
3.98	1.0000000	1.57	13.1	74.9
20	0.9982323	1.00	9.1	72.7
30	0.9956756	0.80	7.5	71.2

Data is from Hutchinson 1957.

depths (see figure 2.10). If you haven't been on a lake in a boat and measured temperature and oxygen, then do so as soon as possible! Students who have this experience find the study of limnology more comprehensible and exciting than students who have not been on a boat (plate 1).

Vertical Profiles of Temperature

A **profile** is a vertical pattern. Temperature, because of its relationship with water density, is the major cause of vertical stratification. Thermal stratification is the cause of other kinds of vertical stratification, such as chemical, light, and biological.

Measurement of Temperature in Lakes

The first measurements of temperature deep in lakes were made by researchers who lowered a thermometer in a bottle to different depths. Limnologists now typically use an electrical device, a **thermistor** probe. Thermistor is a contraction of the words "thermal resistor." This device is made of metals whose electrical resistance depends on temperature. A change in resistance to an electrical current indicates a change in temperature. The thermistor probe is lowered into the lake by its electrical cord; the cord is usually calibrated with tape to indicate meters or decimeters of depth. The thermistors are accurate to within one-tenth to one-hundredth of a degree Celsius.

Oceanographers have used a probe that resembles a brass rocket. Ingenious and simple mechanical devices inside the metal casing allow the simultaneous recording of temperature and depth as the probe is dropped through great depths. Researchers then haul the probe back to the surface and record the temperature and depth curves into a computer.

Patterns of Vertical Temperature Profiles

A graph of the temperature-versus-lake depth is called a **vertical temperature profile.** As mentioned earlier, the temperature profile (figure 1.8) is a graphical logo of limnology.

Water's temperature-density relationship (figure 2.5) produces thermal stratification in lakes (figure 2.6). Fresh water is densest at 4°C. Less-dense water floats on denser water and resists mixing. Thus, 4°C-water will always tend to sink and remain on the bottom of a lake. Above 4°C, warmer water floats on cooler water, and below 4°C, colder water floats on relatively warmer water.

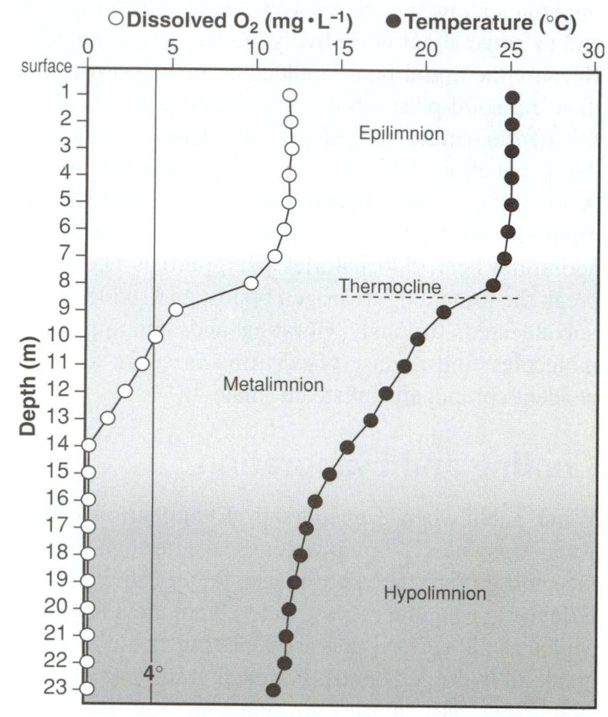

FIGURE 2.6 Lake stratification for Lake Mendota, Wisconsin, a productive dimictic temperate lake in the middle of summer, on 17 July, 2000. Stratification is indicated by the vertical profiles of temperature, dissolved oxygen, which define major layers of the lake and the thermocline. **Source:** Data from the NSF Long Term Ecological Study North Temperate Lake Data Base.

If the lake water is the same temperature from top to bottom, it is said to be **isothermal.** Under isothermal conditions, wind can mix the entire lake because water from all parts of the lake has the same density. However, if there is a vertical temperature difference along the depth gradient, the lake is considered to be thermally stratified. The upper layer, which is less dense, is called the **epilimnion;** the lower layer is called the **hypolimnion.** The transition zone between the warm and cold water is sometimes called the **metalimnion.** Within the metalimnion is the **thermocline,** the depth at which temperature is changing the fastest. Figure 2.6 is an example of a temperate lake that is thermally stratified under summer conditions.

Limnologists sometimes graph stratification data for multiple dates onto a single graph. Equivalent temperatures (or concentrations) are then connected by a line, and the collection of lines gives an impression of how stratification changes over time (figure 2.7). These

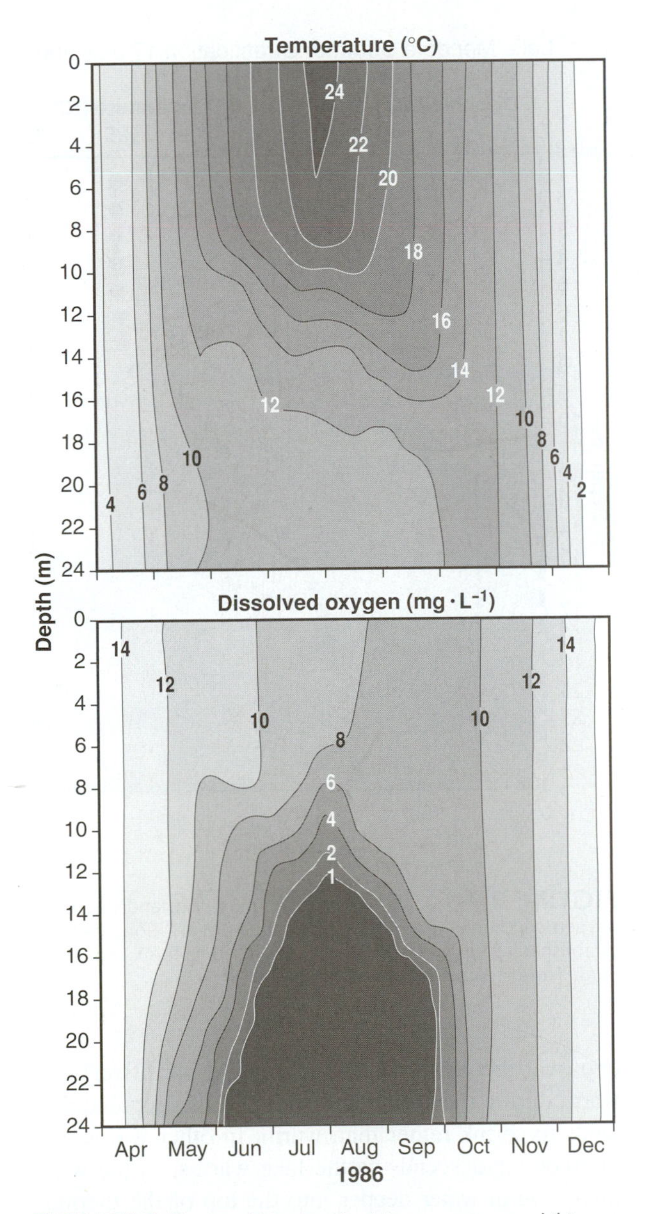

FIGURE 2.7 Isopleth plots for (a) temperature and (b) oxygen concentration for Lake Mendota (Wisconsin) in April through December of 1986. Isothermal periods of temperature are indicated by the vertical lines and summer stratification by the curved lines. The numbers on the upper graph indicate temperature and those on the lower graph indicate oxygen concentration. **Source:** Redrawn using data from Lathrop, 1992.

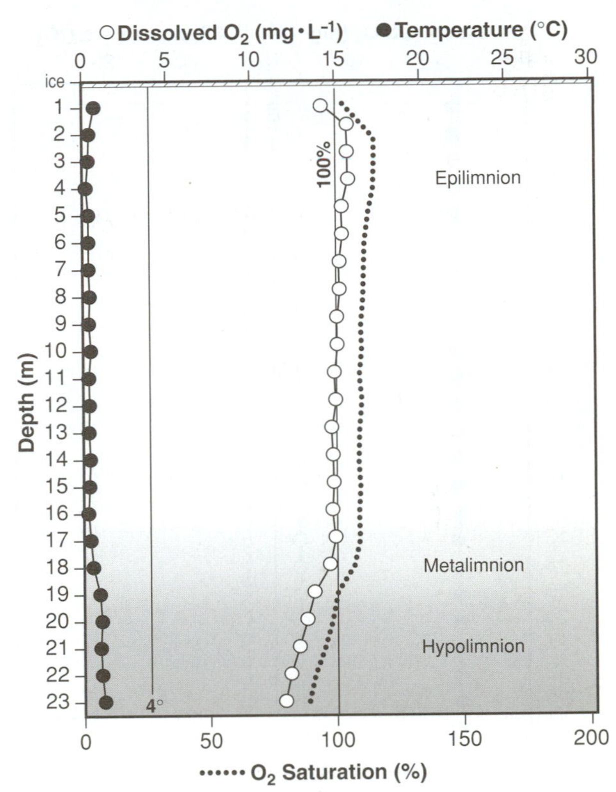

FIGURE 2.8 Lake Mendota temperature and dissolved oxygen conditions under 17 cm of ice, on 17 January, 2000. The lake is nearly isothermal, with slight winter reverse stratification—the epilimnetic 0.4°C water is floating on slightly warmer hypolimnetic water (up to 1.6°C). During the previous fall and early winter, the lake was isothermal and mixed by wind, so that the entire water body cooled to less than 1°C before ice formed on the surface. The deepest water is warmed slightly by the bottom sediments. The % of oxygen saturation values are not adjusted for the depth-pressure effect in this or the next three graphs. **Source:** Data from the NSF Long Term Ecological Study North Temperate Lake Data Base.

lines of equivalent temperature (or concentration) are called **isopleths.** The isopleths in figure 2.7a give a strong impression of how warm water floats on top of cooler water during the summer in a stratified lake.

Typical annual pattern in deep temperate zone lakes Lake Mendota, Wisconsin (United States) is a typical deep, midtemperate zone lake. Lake Mendota is covered with ice for two months or so each winter. During this time, the lake is stratified, with a cold epilimnion floating on a slightly warmer hypolimnion (figure 2.8). Stratification is subtle under the ice because the lake was mixed by wind as isothermal water cooled to less than 1°C the previous fall and early winter.

Ice melts from Lake Mendota in mid-April (although see the paper by Magnuson et al., 2000 for information on the effects of global warming). As the lake warms in the spring, it becomes isothermal and circulates as wind mixes lake water from top to bottom. The

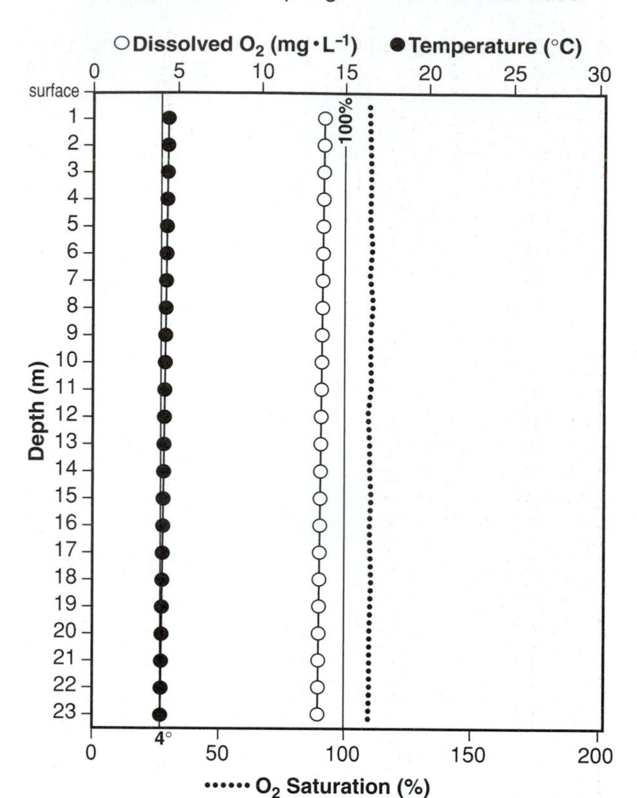

FIGURE 2.9 Lake Mendota temperature and dissolved oxygen conditions during the spring overturn, 29 March, 2000. Note that the temperature is isothermal at about 4°C, with a range of variation of only 0.3°C. **Source:** Data from the NSF Long Term Ecological Study North Temperate Lake Data Base.

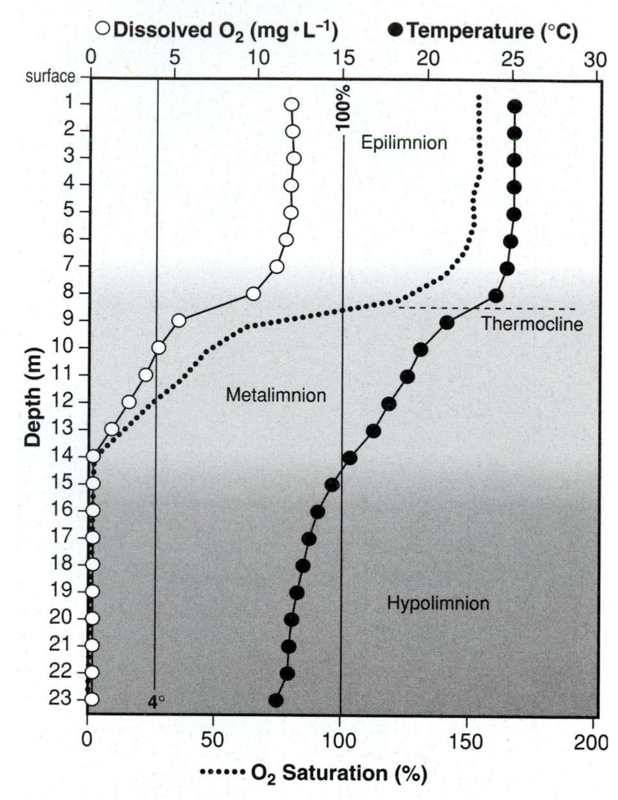

FIGURE 2.10 Lake Mendota temperature and dissolved oxygen conditions on 17 July, 2000 after stratification has been established. **Source:** Data from the NSF Long Term Ecological Study North Temperate Lake Data Base.

lake is isothermal as it warms (figure 2.9). Note that the lake does not need to be at 4°C to circulate; it just needs to be isothermal. An isothermal lake that experiences warm, windy weather during the spring can warm isothermally, so that it is, for example, 11°C from top to bottom.

The isothermal lake, on a still and warm day in the spring, will collect heat near the surface. The floating, warm water will mix with itself rather than the deeper, colder water, resulting in a warm, relatively isothermal epilimnion floating on a colder layer of more dense and relatively isothermal hypolimnion water (figure 2.10). The lake warms as solar light and heat enter during the day and as convection with the warm air transfers heat to the water. Summer storms tend to keep the warm epilimnion isothermal, and the thermocline moves downward during the summer. During the summer, the hypolimnion warms only slightly, typically a degree or two over the course of the season. As stratified lakes warm during the spring and summer, the hypolimnion tends to shrink rather than warm. In other words, the thermocline descends as the lake warms, as the wind mixes warm water deeper into the top of the thermocline. The depth to which a lake mixes during the summer increases with lake size, water clarity, and average temperature (Fee et al., 1996). The summer thermal structure of the metalimnion can be complex (with a step-like thermal profile) due to a sequence of storms and quiet, warm periods.

Toward autumn, cooler weather and shorter days begin to cool the surface water. The epilimnion cools as cool surface water falls downward, until the epilimnion is the same temperature as the hypolimnion (usually at a temperature greater than 4°C). When the lake is isothermal, as shown in figure 2.11, it mixes. The lake will then mix isothermally and cool until the entire lake

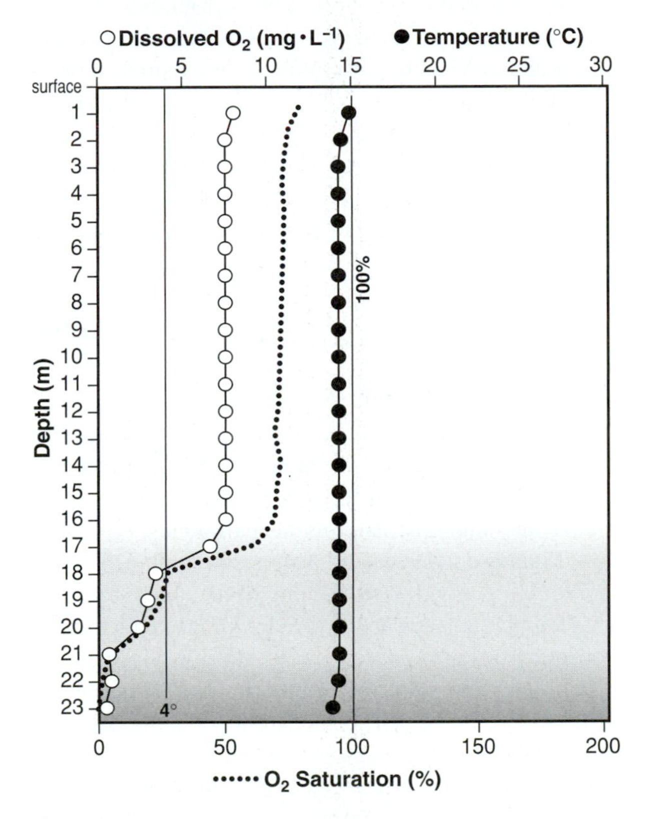

FIGURE 2.11 Lake Mendota temperature and dissolved oxygen conditions at the beginning of the autumnal overturn, on 24 October, 2000. Thermal stratification is no longer evident, but the deeper layers of water are still anoxic due to the relatively warm water and the rain of dead, organic material produced during the summer. **Source:** Data from the NSF Long Term Ecological Study North Temperate Lake Data Base.

is at 4°C or colder. If the lake continues to cool past 4°C, it can develop a cold epilimnion (**inverse stratification,** as shown in figure 2.8). The coldest water is in contact with the air and, as the weather cools, the lake can eventually freeze.

The phenomenon of thermal stratification depends on several factors, including:

- **Time of year:** The annual cycle is as discussed previously for a deep, temperate zone lake. Different patterns of annual and daily variation in temperature, as are found in more tropical or more arctic areas, have a great effect on stratification.
- **Lake depth:** Shallow lakes tend to be easily stirred by the wind, so stratification is more easily destroyed by a wind storm. Shallow lakes typically spend more time in the isothermal condition than do deep lakes. If a shallow lake is not as deep as the thermocline of deep lakes in the same region, the shallow lake will resemble the epilimnion of the deep lake.
- **Wind fetch: Fetch** is the distance wind travels uninterrupted over water. The longer the fetch, the more likely the wind can mix a lake. However, because fetch is often correlated with depth (large lakes tend to be deep), a long fetch more often has the effect of mixing the epilimnion deeper than mixing the whole lake.
- **Topography:** Adjacent geographic features (such as a bluff or cliff) that shield the lake from wind will contribute to early stratification and to a shallow thermocline.
- **Solutes:** Salts (or organic solutes) in lake water make the water denser; enough solutes can make water dense enough to resist mixing. Shallow, saline lakes like the Great Salt Lake, Utah or Mono Lake, California can be mixed by strong wind, but deep, saline lakes tend to have a layer of fresh water floating on a deep, salty layer.

Annual Mixing Patterns

Hutchinson (1957) classified thermal stratification and annual lake-mixing patterns:

Dimictic: Most lakes in the cooler parts of the temperate zones are dimictic; they stratify twice a year, once in the summer with a warm epilimnion and once in the winter with a cold epilimnion relative to the hypolimnion. Lake Mendota, Wisconsin is a good example of a temperate zone, dimictic lake (Brock, 1985; see figures 2.6 to 2.11).

Monomictic: Lakes in mild climates may only stratify once a year, in the summer. The lakes cool in the fall, become isothermal and mix, but do not cool enough to re-stratify in the winter. Lake Washington, in Seattle, Washington, is a good example of a cool, temperate zone, monomictic lake (Edmondson, 1991). The lake is stratified in the summer, with a warm epilimnion (up to 24°C) and cools to about 5°C in the winter. The cold, isothermal water cools throughout the winter months, before the lake begins to warm in the spring. Lake Oglethorp, near Athens, Georgia, is a warm (up to 28°C), shallow (8.5 m maximum depth), monomictic lake that stratifies from April through October and mixes during the cooler part of the year when the water cools to about 4°C (Pace & Orcutt, 1981).

Shallow lakes in the Arctic and at high elevations stratify in the winter and freeze but stay isothermal in the summer. An example of a cold, monomictic lake is Char Lake, a well-studied lake on Cornwallis Island in the high (almost 75° N latitude) Canadian arctic (Kalff & Welch, 1974).

Amictic: Ice cover protects lakes from mixing by the wind, although springs, river currents, and convection also mix water. Antarctic, Greenland, and mountain lakes with permanent ice show little evidence of mixing. Large Lake Vanda near McMurdo Sound in Antarctica is covered with 2–3 meters of ice. It is thermally stratified but does not mix. In summer, the surface water is near freezing, temperature increases slowly with depth to about 10°C at 55 meters, and the dense and salty, lower layer down to 66 meters is much warmer, up to 20°C. The salty water comes from hot springs. In this case, nonmixing does not imply isothermal.

Polymictic: Most shallow lakes experience frequent bouts of mixing. Any lake less than 3 meters deep is likely to stratify between storms. Many tropical lakes (such as Lake Amatitlán in Guatemala and Mexico's Lake Chapala) are shallow and large polymictic lakes (plate 33). These lakes may mix each afternoon as the wind rises and then stratify in the morning as the sun warms the upper water. Lake Thonotosassa, Florida, is a shallow, subtropical lake (average depth 3.5 m) that stratifies occasionally during the warmest weather for a few days but mixes with a little wind (Cowell et al., 1975).

Meromictic: Salinity affects water density much more than does temperature. If enough ions or organic chemicals are dissolved in the hypolimnion, it will be dense enough to resist mixing with the epilimnion. If the deeper layer of water fails to mix, it continues to collect salts or organics and becomes more and more resistant to mixing. A meromictic lake is one that contains a deep layer of nonmixing water. The deepest layer is called the **monimolimnion** and the depth of greatest change in water density between the monimolimnion and the upper water is called a **pycnocline.** The pycnocline is often associated with a **chemocline,** a depth of greatest change in solute concentration. Water above the monimolimnion (above the pycnocline) can be stratified into a hypolimnion and an epilimnion, as in a normal dimictic lake. Meromictic lakes are generally unusually deep or protected by surrounding geography such as bluffs or trees. An example is Mary Lake in Vilas County, Wisconsin, a small (200-m diameter) and deep (20 m) lake. Enough organic chemicals (from decay of plant materials) are dissolved in the deepest waters to make Mary Lake meromictic. The much larger Mono Lake (about 12 km across and 38 m deep) in California, has had, since 1995, a deep, saline monimolimnion that is dense enough, relative to the upper freshwater layer, to inhibit mixing (MacIntyre, 1999).

Ice Cover and Lake Size

Larger lakes have less surface per unit volume than small lakes. Heat is lost or gained over the surface and stored in the volume of the lake. Therefore, large lakes freeze later and thaw later than small ponds.

Magnuson (1990) and Magnuson et al. (2000) report a steady trend of shortening of the amount of time that temperate lakes are covered with ice. This trend is most obvious when decades or centuries of ice data are examined. The El Niño-Southern Oscillation in the tropical Pacific Ocean surface temperature has a cycle of warming and cooling every 3 to 7 years. This oscillation is reflected in records of ice duration on continental lakes and rivers. Also, compared to 1846, lakes and rivers in the Northern Hemisphere now freeze about 6 days later and thaw about 7 days sooner, reflecting an increase in average air temperature of about 1.2°C per 100 years—evidence of long-term global warming. Some longer data sets suggest the warming trend began as early as the 16th century, with an increase in the rate of warming after about 1850.

Vertical Profiles of Oxygen

Oxygen concentrations depend on thermal stratification and biological activity. Because oxygen is a reactive chemical required by so many organisms, oxygen concentrations reveal a great deal of information about water chemistry and the distribution and abundance of organisms.

Measurement of Oxygen in Lakes

Oxygen concentration was initially measured using a chemical assay. Researchers hauled water up from the desired depth (using a line calibrated in meters) and treated it with a series of chemicals to determine the oxygen concentration. Currently, most limnologists use an electrical probe that contains an element whose resistance is affected by the oxygen concentration. Both the chemical and electronic methods are typically accurate to about 0.1 milligram of oxygen per liter.

Patterns of Vertical Oxygen Profiles

Thermal stratification sets the stage for patterns of vertical chemical stratification in lakes. Oxygen is a biologically critical chemical that dissolves in water and shows vertical stratification. There are three major patterns of concentration variation with depth: orthograde, clinograde, and heterograde.

An **orthograde** pattern results from oxygen being saturated throughout the lake. The warmer the water, the less oxygen the water holds (see figure 2.4; also see the line for 100% oxygen saturation in figure 2.9).

If a lake is saturated with oxygen at all depths and has a warm epilimnion, it will have a higher oxygen concentration in the hypolimnion than in the epilimnion. This condition is usually seen in the spring in low-productivity lakes that have recently stratified. Biological activity tends to modify the orthograde pattern.

A **clinograde** pattern results from bacterial decomposition (and the associated oxidation) of organic material in the hypolimnion. In the summer, a stratified, productive lake will typically have oxygen concentrations near or exceeding saturation (at least during the day) in the (relatively isothermal) epilimnion, and little or no oxygen in the hypolimnion (see figures 2.6, 2.7, 2.8, and 2.10). This is because **phytoplankton** (small, photosynthetic organisms suspended in the water) supply oxygen to the epilimnion, which is mixed, and decomposition removes oxygen from the unmixed hypolimnion. As productivity increases due to algal growth, lake water becomes more **turbid** (cloudy, opaque), light penetrates less deeply, and more organic material falls into the hypolimnion. Bacterial metabolism of this organic material during the summer can reduce hypolimnetic oxygen concentrations to zero.

A **heterograde** pattern is a profile with a peak (maximum oxygen concentration) at an intermediate depth (for example, the observed oxygen curve shown in figure 2.12). The peak in oxygen concentration at a middle depth is called an oxygen anomaly. This pattern is characteristic of low-productivity lakes, in which the algae community is so rarefied that light penetrates into the metalimnion or hypolimnion (Fee, 1976). Algae in the unmixed hypolimnion flourish, receiving light from above, and nutrients such as phosphate from the sediments. The algae often form a

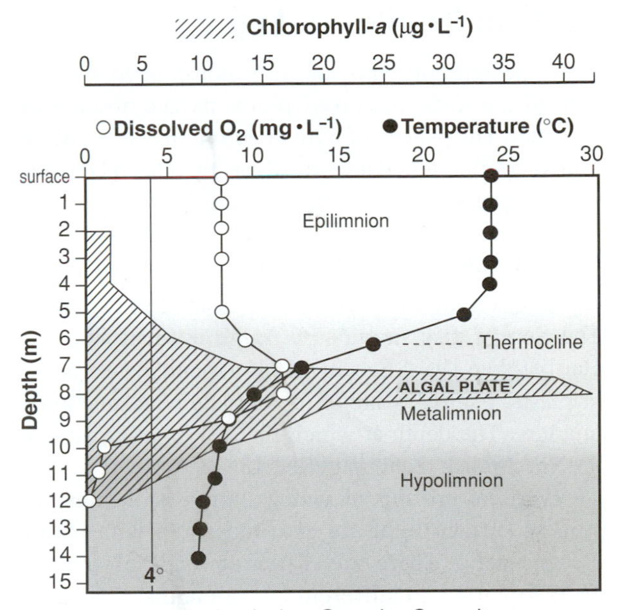

FIGURE 2.12 A vertical profile of chlorophyll *a* (a measure of algal abundance), oxygen concentration, and temperature for oligotrophic Jack's Lake, Ontario, Canada. Note that the algae are most abundant below the thermocline, and that oxygen is also most concentrated below the thermocline. The arrow on the Y-axis indicates the depth of 1% light penetration (at a depth just below the peak in algal density). Note the different scales for the X-axis. **Source:** Redrawn from Pick et al., 1984.

visible but thin and insubstantial layer, an **algal plate** (plate 18) that is about 1 meter thick, deep in the lake. A SCUBA diver descending into the depths may mistake the opaque plate for the lake bottom. During the day, algae in the plate produce oxygen, which tends to be present at or above saturation levels (see figure 2.12). Above the algal plate, oxygen will show a more-or-less clinograde pattern. Oxygen concentration is typically less below the algal plate. The depth of the algal plate depends on the depth of the thermocline, which depends on factors such as lake size, depth, and amount of wind. In the summer, Jack's Lake, Ontario had its algal plate at about 9 meters (see figure 2.12), while Lake Michigan typically had a layer of algae at about 20 meters—in each lake the algal plate was just below the thermocline.

The isopleth plot of oxygen concentration for a stratified lake gives an impression of how stratification pattern changes with time (see figure 2.7b). This plot shows the development of the anoxic hypolimnion during the summer in productive Lake Mendota, Wisconsin.

Vertical Profiles of Light

Light penetration into lakes is constrained by suspended particles, dissolved pigments and absorbance by water. The pattern of light penetration is often determined by the abundance of algae or other photosynthetic organisms.

Measurement of Light in Lakes

Light penetration into water is measured in two ways. The **Secchi disk** provides a rough estimate of water clarity (plate 19). A Secchi disk is a round plate of metal or plastic painted white or in alternate white and black quarters. The depth at which the disk is just visible is the Secchi disk depth (figure 2.13).

Vertical profiles of water clarity or of transmission of surface light are also measured with photoelectric cells. There are different kinds of sensors, each sensitive to different wavelengths. The most common sensor measures light in the range of wavelengths used for human vision and photosynthesis. Light in the range of wavelengths plants can use for photosynthesis is called *photosynthetically active radiation,* or **PAR.**

Light penetration (intensity and color) depends on depth (figure 2.14) as well as on dissolved and suspended materials found in water. As one descends into a lake of clear water, the light becomes less intense and more green and blue.

Light has been measured in a number of different ways (Kirk, 1994). A common unit used today is the micro-einstein, μE, per m^2 per second of PAR light. An einstein is a specific number of photons (light quanta). The actual number of photons that make up an einstein of radiant energy is a "mole" of photons, which is 6.023×10^{23} photons. A micro-einstein is a millionth of a mole, or 6.023×10^{17} photons.

On a sunny, summer day, a temperate zone lake receives at least 2000 μE m^{-2} sec^{-1} at the surface. Most algae grow best at about 200 μE m^{-2} sec^{-1}, and most algae need at least 20 μE m^{-2} sec^{-1} (that is, about 1% of the maximum surface light level) to support enough photosynthesis to balance their use of energy in metabolic activity (respiration). It is just possible for a person to read at about 0.05 μE m^{-2} sec^{-1}. The upper portion of the lake that contains enough light for net accumulation of energy by photosynthesis is called the **euphotic zone.** The bottom of the euphotic zone occurs at the depth to which 1% of surface light penetrates. This depth corresponds to about *three times the Secchi disk depth.*

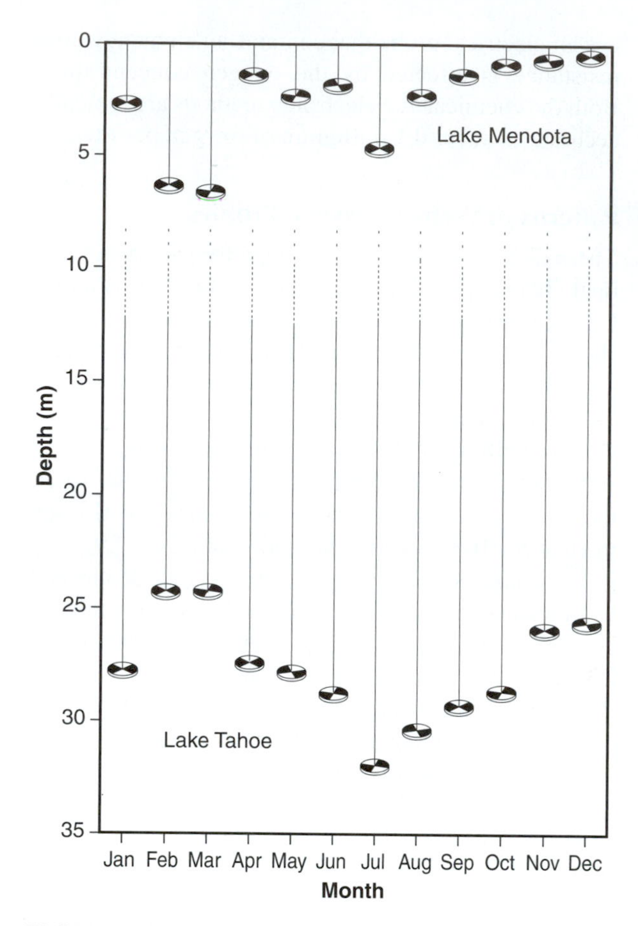

FIGURE 2.13 The disks represent Secchi disk depth over a typical year in a productive algae-rich lake (Lake Mendota, Wisconsin) and an unproductive, algae-poor lake (Lake Tahoe, California). **Source:** Lake Mendota data from Brock, 1985; Lake Tahoe data from Horne and Goldman, 1994.

At the bottom of the euphotic zone, there is just enough light for photosynthesis to balance the energy needs of algae. This balance occurs at the **compensation depth.** Water below the euphotic zone is in the **aphotic zone.**

Pigments

Water may be stained with pigments, such as the yellow and brown stains from decaying vegetation. This dissolved organic carbon is especially evident in the "tea-stained" water of cold bogs. The pigments readily absorb light (Williamson et al., 1996; Klug and Cottingham, 2001), significantly reducing the PAR intensity at all depths.

Scattering

Scattering and *reflection* both refer to a change in direction of light as it interacts with matter. Light is

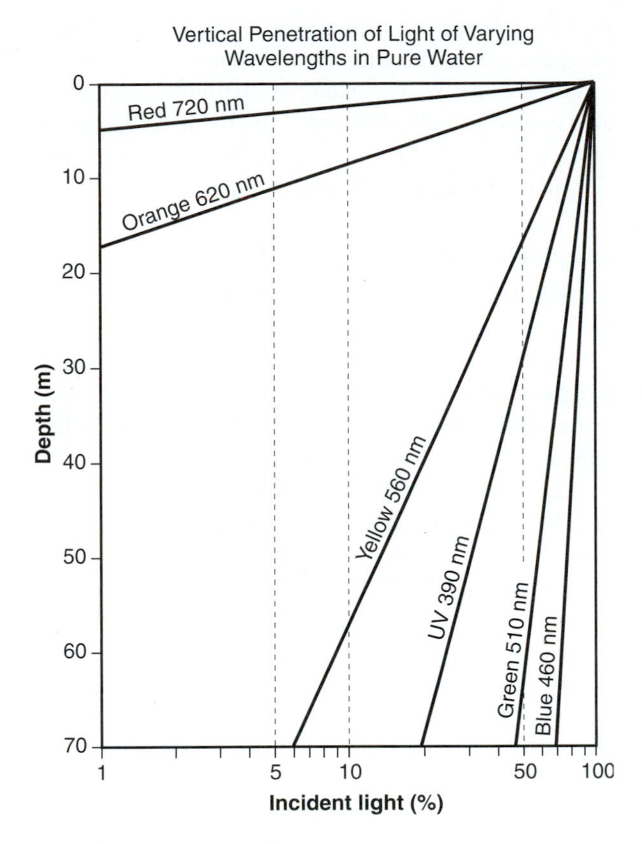

FIGURE 2.14 Penetration of different colors of light. 720-nm light is red, while 460 is at the blue end of the spectrum. Note that the X-axis is logarithmic.

scattered from molecules or small particles (dust, sediment particles, algae) and reflected from surfaces (the liquid or frozen lake surface, crystal surfaces). When the sun is overhead, little light is reflected by a lake's surface, but almost all the light is reflected when the sun is near the horizon. Light from the sky or sun that enters the lake is either reflected back out again or absorbed by the water or matter in the water. A small portion of the light that is reflected or scattered will be again scattered from particles in the air and re-enter the lake. Blue light is scattered much more than red light. This phenomenon accounts for the blue color of the sky and of lakes.

Absorbance

Infrared, red, and ultraviolet (UV) light are **absorbed** most strongly by water; blue and green light are absorbed the least (figure 2.14). Thus, as a diver goes deeper into the lake, colors fade until only drab blues and greens can be distinguished. Light quality at any specific depth in a lake depends on what is absorbed out of sunlight by the atmosphere and the water. Flathead Lake, Montana is an unusually clear lake that, on a sunny day, allows penetration of light deep into the water. Jack Stanford, Director of the Yellow Bay biological station on Flathead Lake, often SCUBA dives into the lake. He says that at noon on a sunny day, the light at 30 meters is dim, brownish, and eerie.

Water's absorption of ultraviolet light helps protect aquatic organisms from this destructive, high-energy radiation (Williamson et al., 1996). In the first few meters of surface water, especially in very clear lakes, UV light affects both longevity and fecundity of aquatic organisms, and therefore significantly affects the direction of natural selection.

Water's absorption of infrared light and its high concentration in Earth's atmosphere make water the most important greenhouse gas (Kiehl & Trenberth, 1997).

Light absorbed by lake water and suspended particles is converted to heat. This absorbed energy is the largest source of heat for most lakes, generally more important than radiative heating from the air or the earth or from inputs of streams or springs.

Vertical Profiles of Living Organisms

The distribution and abundance of organisms in lakes depend, in part, on the vertical profiles of temperature, oxygen, and light.

Algae Patterns

Algae are single-celled or colonial photosynthetic organisms. The algae species that float freely in the open water of lakes or rivers are called *phytoplankton.* Phytoplankton in the summer epilimnion of stratified lakes are often mixed through a wide range of light intensities. Especially in productive lakes, the euphotic zone is restricted to the epilimnion, and is not as deep as the thermocline. To live in these conditions, epilimnetic algae require a physiology that can work at high light intensities near the surface and low light intensities deeper in the epilimnion, and have a capability for using stored energy when there is not enough light for photosynthesis when the algae are carried below the euphotic zone by water currents or simply by sinking. Low-productivity lakes may have so few algae suspended in the water that the euphotic zone can extend deeper than the thermocline. When enough light reaches the hypolimnion, algae often are densest in a thin algal plate in the still water of the hypolimnion (see figure 2.12). These algae are specialized on low light levels and will die if brought up into full sunlight.

Macrophyte Patterns

Macro-algae and flowering plants (water plants or **macrophytes**) attached to the bottom can have a vertical distribution in a lake, based on the influence of light penetration and wave action. Macrophytes are limited to the depth of the euphotic zone in the season in which they start growing from spores, seeds, roots, rhizomes, or bulbs. This is the late spring in temperate zone lakes, often during the **clear water phase** (a time when the water is very clear, the population of spring algae have declined, and the summer algae have not yet become abundant).

The greatest variety of species and growth forms of macrophytes is found where wave action is minimal, as in a protected bay.

A common pattern is to find **emergent** macrophytes (sedges, cattails, rushes) in the shallowest water, *floating* and *nonemergent* macrophytes (lily pads, milfoil, macro-algae, water weeds) in deeper water, and short, slow-growing mosses or attached algae in the deepest water near the bottom of the euphotic zone.

Bacteria Patterns

Vertical patterns of bacterial distribution are especially dramatic in two conditions: (1) in lakes with a euphotic zone extending to the top of an anoxic hypolimnion or monimolimnion, and (2) in the sediments.

Lake water with little or no oxygen, some light, and a source of organic sulfur compounds provides conditions optimal for photosynthetic sulfur bacteria. These bacteria can form a thin but dense layer, similar to the algal plate, at the intersection of the light, oxygen, and sulfur gradients. In such cases, photosynthetic sulfur bacteria can account for the majority of the primary production in a lake (Culver & Brunskill, 1969).

Bacteria in the sediments sometimes have sharp, vertical zonation over a distance of a few centimeters. In lakes with some oxygen dissolved in water next to the sediments, there will be sharp gradients of decreasing oxygen and decreasing redox potential, and increasing concentrations of produces of decomposition a short distance into the bottom sediments. Different species of bacteria can specialize on these chemical differences.

Zoobenthos Patterns

Invertebrates such as worms, crustaceans, and insects that live and feed in or on sediments at the bottom of lakes and streams (the **benthic zone**) are called **zoobenthos.** Like bacteria in the sediments, these invertebrate organisms experience sharp gradients of oxygen concentration over a few millimeters or centimeters. Different species of zoobenthos show different sensitivities to anoxic conditions. For example, the tubificid worm is able to feed in sediments, while many insect larvae are restricted to the sediment surface where there is marginally more oxygen (Vandebund et al., 1995).

Zooplankton Patterns

Zooplankton are small animals (in inland waters, mostly nonphotosynthetic protozoans and invertebrate animals) that live in open water (the **planktonic** or **pelagic zone**). Zooplankton can live in either standing water (lakes, reservoirs, ponds) or in slowly-moving, large rivers (Thorp et al., 1994).

Zooplankton vertical distribution is the result of the distribution of physical factors such as temperature, light, and oxygen, and the distribution of other species, such as algal food and predators. In dimictic lakes in the summer, zooplankton can show **diel** (daily) vertical migration (**DVM**). The zooplankton move down into dimly lit deeper water during the day (and at night when the moon is full) to avoid fish predation. Fish hunt zooplankton visually. The vertical oxygen profile can limit the depth to which zooplankton migrate. Most zooplankton need at least 2–3 milligrams of O_2 per liter. At night, the zooplankton migrate upward to warmer water where growth and reproduction is faster (Lampert, 1993). If DVM is not observed in a lake with fish, the larger zooplankton will, in any case, tend to live deep in the lake day and night.

Fish Patterns

Fish also have vertical distribution patterns. Like zooplankton, small (especially larval) fish may show DVM. Larger fish that eat other fish or zooplankton will be limited in their depth distribution by the amount of light they need to see prey. Some fish, such as the whitefish (called the "cisco"), require cold water and high oxygen. In productive Lake Mendota (Wisconsin), cisco are restricted to the hypolimnion during the summer, and if the hypolimnion becomes anoxic, the cisco die from a combination of heat stress and asphyxiation (Kitchell, 1992).

Vertical Profiles of Primary Productivity

Primary productivity depends on photosynthesis and respiration of algae and aquatic plants. Photosynthesis uses light, carbon dioxide, and water to produce oxygen and energy-rich carbohydrates, while respiration uses oxygen to burn carbohydrate fuel, producing useful chemical energy, heat, carbon dioxide, and water. Primary productivity can be expressed as a short-term rate, such as the milligrams of carbon fixed beneath a square meter of water surface per hour, or as a long-term rate of grams of carbon fixed beneath a square meter per year.

Primary productivity also depends on the concentration of inorganic nutrients, such as nitrate and phosphate. The highest productivity occurs in water that is well lit, warm, and rich in inorganic nutrients. These conditions typically occur together in the summer epilimnion of temperate zone lakes in agricultural or urban watersheds.

The word **trophic** indicates energy or food. Thus, productivity is a trophic process. Lakes with a high rate of primary productivity are called **eutrophic;** those with a low rate are called **oligotrophic** (table 2.6). **Mesotrophic** lakes have intermediate rates of primary productivity. A trophic (energy-related) classification sorts lakes by their level of primary productivity—a fundamental classification scheme used by limnologists.

Small lakes in closed basins and cool climates tend to collect darkly pigmented organic chemicals (also called *dissolved organic carbon,* or **DOC**) that stain the water a dark brown (plate 15). These **dystrophic** lakes typically have very low productivity because the small watershed provides few nutrients, cool temperatures slow photosynthesis, and the DOC absorbs light. The DOC absorbs light before it can reach the hypolimnion, so the hypolimnion tends to be anoxic and without a plate of

Table 2.6 Characteristics of Eutrophic and Oligotrophic Lakes

Characteristic	Oligotrophic	Eutrophic
Lake location and land use	High in the watershed, away from agriculture	Lower in the watershed, with some agricultural land use
Primary productivity	Low, occurs throughout the epilimnion and into the hypolimnion	High, confined to upper regions of the epilimnion
Oxygen in the summer hypolimnion	Present, > 2 mg/l^{-1}	Anoxic
Secchi disk depth	Several meters, often into the metalimnion	Few meters, in the epilimnion
Photic zone	Extends to the hypolimnetic algal plate	Includes upper part of the epilimnion, due to dense algae
Inorganic nutrient inputs	Low	High, especially nitrate and phosphate input
Water quality	High	Low
Lake size	All sizes, including the largest and deepest lakes	Small- to medium-size lakes, often shallow
Water appearance	Clear	Murky, green or brown
Fish production	Low	High
Odor	Fresh water	Fetid algae
Water texture	Fresh water	Slimy
Human development of the watershed	Low	Intense agriculture or urban
Examples	Lake Tahoe, CA Trout Lake, WI Crater Lake, OR (plate 3) Dove Lake, Tasmania (plate 2)	Lake Mendota, WI (plate 1) Clear Lake, CA (plate 20)

photosynthetic algae. Dystrophic lakes have extremely low levels of primary productivity, low fish production (Hasler et al., 1951), and an abundance of predaceous plants (that get their nutrients by digesting insects).

Interaction of Heat and Light with Organisms

Data from a large survey of Ontario lakes (Mazumder et al., 1990) suggest that algal density affects the heating of lakes. Algae affect lake temperature by reducing water clarity. Light is reflected back out of the lake by algae, and the light is therefore not available for heating the water. The data show that lakes with the least amount of algae had higher transparency (measured using a Secchi disk), a deeper thermocline, and a warmer epilimnion. Algal density is controlled (at least in part) by zooplankton grazing, and zooplankton are the food supply for many fish species. The more zooplankton-eating fish in a lake, the fewer zooplankton. Fewer zooplankton often mean more algae. Thus, there is an ecological mechanism by which fish predation can cool lakes.

A warming climate (global warming) will have a number of effects on lakes and streams. Schindler et al. (1996) studied lakes in the Experimental Lake Area of southwestern Ontario. They found that during a period of prolonged warm weather over several years, there was an increase in forest fires, a decrease in stream flow, and less nutrient transport. Probably as a result of less erosion and fewer nutrients for algae, lakes became clearer. They were also warmer and had higher concentrations of most common inorganic chemicals, the result of concentration by increased evaporation due to the higher temperature. Summer thermoclines deepened and hypolimnia had less oxygen because of higher temperatures.

WATER MOVEMENTS

Processes of water movement are scale dependent. On a microscopic scale (millimeter or less), water molecules diffuse or move short distances because of their kinetic energy, which depends on temperature. At larger scales, water moves in two ways: either from one place to another as a current or package (mass transport) or to another place and back in a regular manner (waves or periodic movement). The energy for currents and waves comes from either the wind or gravity. Wind blowing across water creates both currents and waves. Gravity causes currents in streams or in ground water to flow downhill. Gravity also contributes to periodic movement of water disturbed by wind.

Diffusion

At the smallest molecular scale, individual molecules in a liquid or a gas move at high speed at temperatures in the biological range (between freezing and boiling temperatures for water). They frequently bump into and rebound off of one another. The fast movement, collisions, and ricochets result in a random-movement pattern of molecules. This random movement means that as a substance dissolves in water, the substance molecules will spread out through the water, or **diffuse.** The rate of spreading is faster for small molecules at high temperatures.

Diffusion is a significant physical phenomenon in water at small scales of a millimeter or smaller. For example, gases dissolve in the surface film of water and diffuse downward. Diffusion of oxygen into water can keep the concentration in the surface film at saturation. However, it would take years for diffusion from the atmosphere into still water to increase oxygen concentration by 1 mg $liter^{-1}$, for any depth more than a few millimeters.

Mass Transport

At scales above the molecular level, water demonstrates bulk properties that are very different from diffusion.

Laminar Flow and Zero Flow

Whenever a mass of water flows from one place to another across a surface, there is a velocity gradient perpendicular to the direction of flow. There is **zero flow** at the solid surface, such as the bottom of a lake or the stream bed (for more information on hydrodynamics, see Vogel, 1983). Just above the surface, the water flows slowly and evenly, as if it were a sheet. This laminar flow moves fluid without mixing. Laminar flow requires low velocities to be stable. Water flowing far from a surface tends to flow at the highest speed that conditions permit, and it is often turbulent.

Turbulent Flow

Turbulence occurs at higher velocities than does laminar flow. Turbulent water does not all move in the same direction. A velocity gradient might still exist in a turbulent lake or stream current, but only as an average velocity at each depth. Turbulent flow is **chaotic** (with no apparent

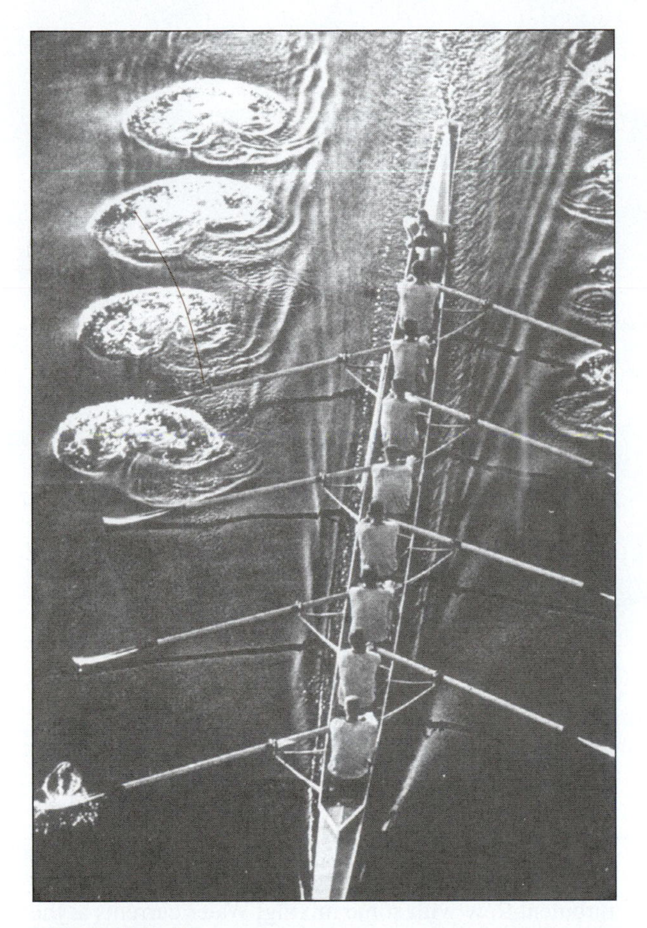

FIGURE 2.15 Small patches of turbulence are created in still lake water by the oars of a rowing shell. **Source:** Photo from "Datelines" 7:27, a 1974 newsletter of the University of Wisconsin Communications Department.

order) as small parts of the water move in different directions. For example, as oars push against water to propel a boat, the oars create patches of turbulence (figure 2.15). Turbulence can be seen in almost any body of open water. Turbulent mixing of water makes white-water kayaking exciting and hot tubs soothing.

Reynolds Number

There is a simple mathematical ratio, called the *Reynolds number (Re),* that simplifies thinking about laminar and turbulent flow. The Reynolds number takes into account the effects of scale and velocity to predict whether a current will be laminar or turbulent. The Reynolds number is a ratio of two forces: viscosity and inertia.

The internal friction, or stickiness, of water is called *viscosity.* Viscosity, or viscous force, tends to keep a mass of water together and moving in the same direction. Therefore, viscous force maintains laminar flow.

Inertial force is related to the momentum (mass times velocity) of water. **Inertia** resists viscosity and keeps water moving. When water with momentum encounters a surface, or when two currents meet, the water changes direction. Inertial force produces turbulence and breaks currents up into patterns of turbulent mixing.

To calculate the value of Reynolds number for a specific situation, it is necessary to know a velocity, a length, and a coefficient representing the physical nature of the water: the kinematic viscosity. The kinematic viscosity depends on both the density and the viscosity of the water, and can be taken as 0.01 cm^2 sec^{-1} for water at 20°C.

Reynolds predictions can be made for either: (1) two currents of water moving past one another (will they mix?), or (2) an object moving through water (will the movement cause turbulence?). The equation for the calculation is:

$$R_e = (\text{inertial forces/viscous forces}) = Ud/\nu$$

where:

U = difference in water velocities (cm/sec) between two currents or the speed of an object in water

d = distance between the two points where U was measured, or the characteristic size of an object moving through water—the length or diameter of an object is sufficient, and there is no need to be too exact

ν = kinematic viscosity of water, use 0.01 cm^2/sec

For an organism swimming in water, the appropriate value of "d" is the length of the organism. For example, for a human swimming in water, the velocity might be about 200 cm/sec and "d" (the length) is 200 cm.

If $R_e < 1$, water (or any fluid) will be more distorted than flowing. For low Reynolds numbers (less than 1.0), viscosity is more important than inertial force.

If $R_e < 500$, flow will be laminar, moving water smoothly without mixing or turbulence.

If $R_e > 2000$, flow will be turbulent, often with circular-appearing mixing patterns called *eddies* or *vortices.* Plates 16 and 21 and figures 2.15, 2.16, and 2.17 give examples of turbulence patterns expressed at different scales.

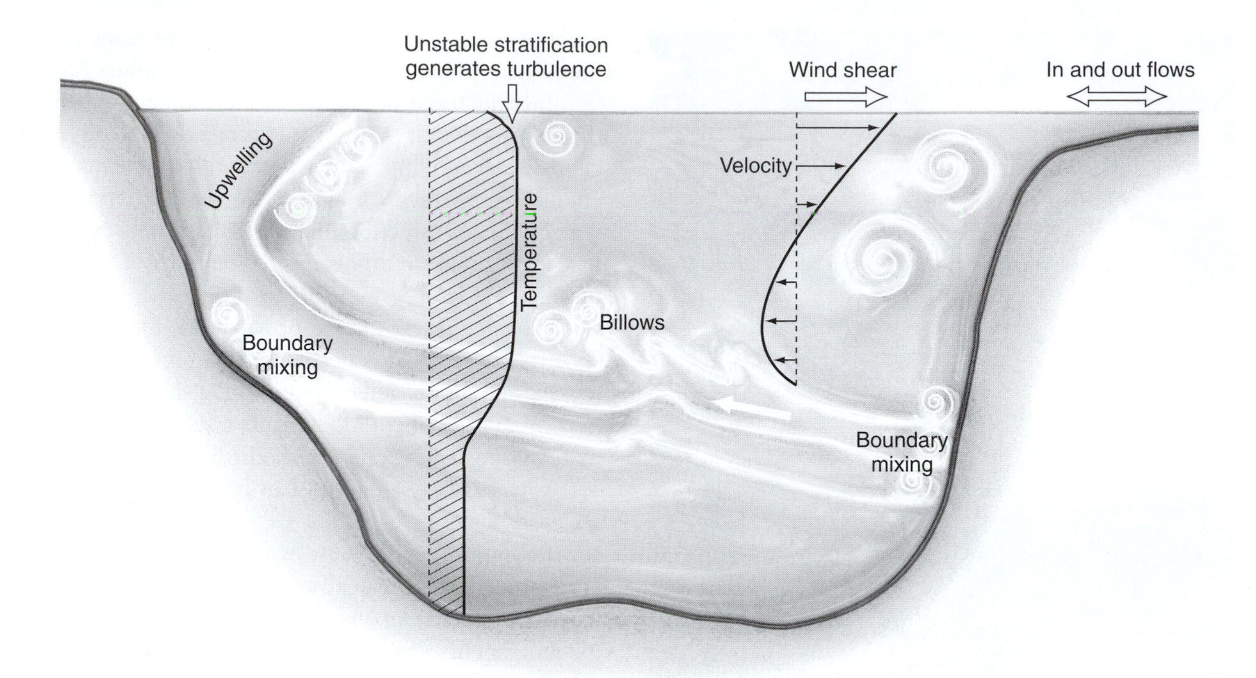

FIGURE 2.16 Turbulent mixing events (indicated by spirals) occur throughout the lake at boundaries between waters of different densities (or different temperatures) and especially where these boundaries intersect the lake bottom. **Source:** Redrawn based on a figure in MacIntyre and Jellison, 2001.

FIGURE 2.17 A NASA LandSat-7 (satellite photo) near-infrared image of Lake Mendota, Wisconsin, 31 October, 1999. The white streaks indicate floating masses of cyanobacteria (probably *Aphanizomenon*). Lake Mendota is about 8.5 km in diameter, so the eddy suggested by the swirl of floating algae is about 4 km in diameter. The inlet (Yahara River) is at the north end of the lake, and the (canalized) outlet to Lake Monona is to the southeast. **Source:** Landsat-7 satellite photo.

Horizontal Currents

Mass transport of water occurs as either laminar flow or turbulent flow with some mixing. Water currents at the surface of a lake have a velocity of about 2% of wind speed, and are often only a few cm sec^{-1} (Ragotzkie & Bryson, 1953). Winds are seldom greater than 50 km hr^{-1}, or about 1400 cm sec^{-1}. This wind speed would be associated with a lake surface water speed of about 28 cm sec^{-1}.

Streams or springs entering the lake, and heat from sediments are potential causes of horizontal currents in lakes, even if the lake is ice covered and not subject to wind.

We can use the Reynolds number equation to get an idea of the kind of flow shown by horizontal currents. We will choose a slow surface current of 2.8 cm sec^{-1}. Compare this surface speed to that of water at a depth of 1 meter (100 cm), which has not yet begun to move. The difference in depth is 100 centimeters and the difference in velocities between the surface current and the water at 100 centimeters is 2.8 cm sec^{-1}, so the Reynolds number for this system is 28,000. This large value of Re suggests the upper layers of the lake experience significant tur-

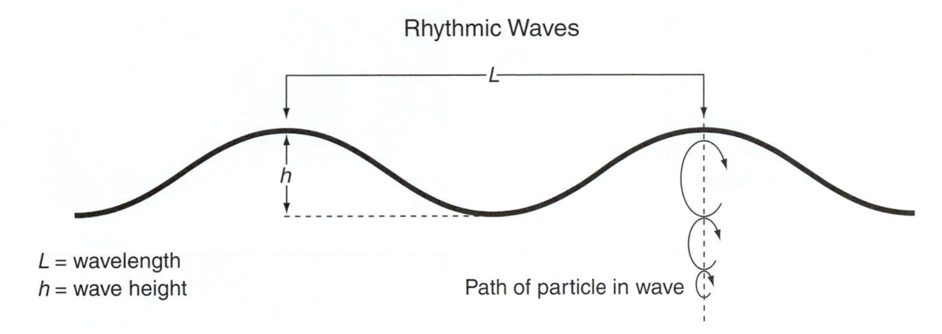

FIGURE 2.18 Model of surface periodic (rhythmic) waves showing periodic motion of water and the dampening of the oscillation at greater depths.

bulent mixing, even for slow, horizontal currents (figure 2.16).

Circular turbulence patterns, or **eddies,** occur over a wide range of scales, from centimeters to thousands of kilometers. Small eddies (**billows**) occur between layers of water with different densities (at a pycnocline, centimeter to meter scale) (MacIntyre, 1998). Strong eddies (on the 10-meter scale) can be observed in rivers.

Floating algae give evidence of large-scale (kilometer) spiral currents in Lake Mendota (figure 2.17). These large eddies are called *gyrals,* or **gyres,** and are hundreds of meters to hundreds of kilometers in diameter. Large gyrals (100-km scale) occur in the North Atlantic, where the warm Gulf Stream rubs against colder coastal water (plate 21).

Periodic Movements

Surface waves appear to be moving across the lake or river, but this is largely an illusion. The water is mostly moving up and down in a periodic motion. The mass moves only a short distance and returns to the starting place. For example, a fishing bobber may drift slowly with the wind, but most of its motion is up and down, even when large waves are present.

Periodic movements are described by wave height, amplitude, period, and wavelength (figure 2.18). Wave height is the vertical distance an individual water molecule travels. At the surface, wave height is the difference in height between the wave's crest and trough. Waves also occur at pycnoclines, and wave height is inversely proportional to the difference in densities. The **amplitude** of the wave is half the height. **Period** is the time it takes for a water molecule to complete one vertical cycle (frequency, the number of crests per unit time, is the inverse of period). **Wavelength** is the distance between the crests of two waves.

Surface (Progressive) Waves

Wave action is a common phenomenon of lake and stream surfaces. In deep water, the interaction of wind energy and gravity results in periodic movements in which water molecules move up and down (**oscillate**) a few centimeters or even meters, but with little or no net horizontal movement. The water associated with these waves is not going anywhere, unless there is also a current (directional movement) present.

Vertical movement (amplitude) attenuates with depth: Amplitude decreases by half for each increase in depth of one-ninth wave length. This means that deeper waters of lakes are relatively unaffected by surface wave action.

Maximum wave height is proportional to the square root of the wind fetch. An empirical formula predicts maximum wave height as a function of wind fetch (Hutchinson, 1957):

$$h = 0.105\ (x)^{1/2}$$

where:

h (wave height) and x (fetch) are in centimeters.

For example, Lake Michigan (one of the North American Great Lakes) is about 141 kilometers or 1.4×10^7 cm across from west to east. The predicted maximum wave height, given a strong wind from the west, is therefore about 390 centimeters, or about 4 meters. The smaller Lake Mendota, about 8.5 kilometers across, has a predicted maximum wave height of about 1 meter. Both wave heights have been observed in their respective lakes, but a 4-meter wave has never been seen in Lake Mendota (plate 22).

The distance (wave length) between crests depends on factors such as wind force, how long the wind has been blowing, and water depth. In general, wavelength is about 20 times amplitude. Thus, the 4-meter waves of Lake Michigan were about 80 meters apart; the 1-meter waves of Lake Mendota were about 20 meters apart.

The oscillation period (time between crests) of a wave is proportional to the square root of the wavelength (distance between crests) (figure 2.18). This relationship suggests that the smaller waves of Lake Mendota will crest (or crash against a boat or the shore) twice as often as the large waves of Lake Michigan.

Viscosity and wave action My students who sail report that **whitecaps** (surface waves that are breaking in open water) form at a lower wind speed in the summer than in the fall. During the summer, whitecaps form at wind speeds of 23 kph. In the fall, winds up to 28 kph mph will not form whitecaps. The tendency for the whitecap threshold to be higher in the winter is probably due to the higher viscosity of water at lower temperatures. The same viscosity-temperature relationship probably is responsible for the difference in the sound of breaking waves in the summer and winter.

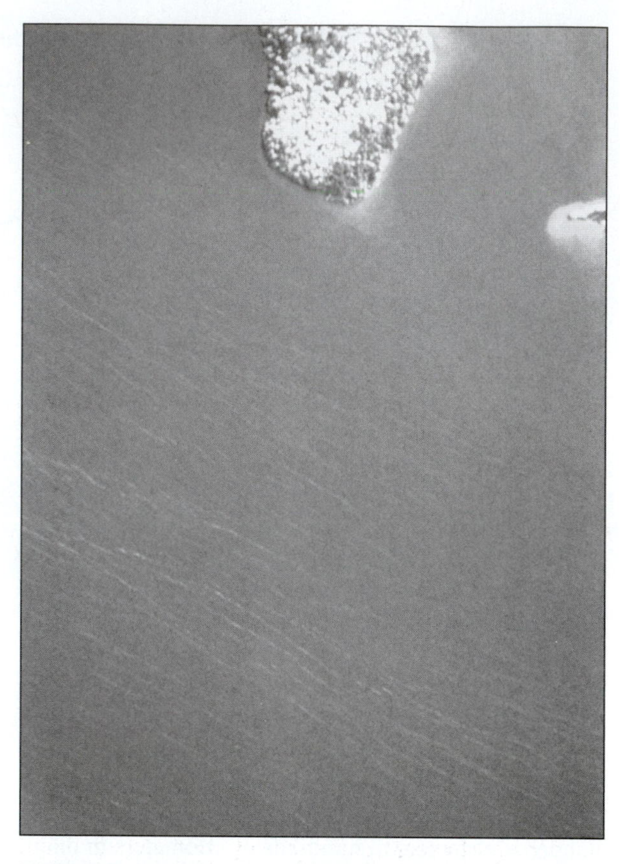

FIGURE 2.19 Aerial photograph of foam streaks on the surface of Big Muskellunge Lake, Vilas County, Wisconsin. The wind is blowing from the upper left. **Source:** Photo compliments of the University of Wisconsin Trout Lake Biological Station, Wisconsin, USA.

> I will arise and go now, for always night and day
> I hear lake water lapping in low sounds by the shore.
> As I stand on the roadway or on the pavement grey,
> I hear it in the deep heart's core.
>
> *W. B. Yeats,* The Lake Isle of Inisfree

Foam streaks **Foam** is a frothy mixture of air, water, and organic compounds that remain stable for hours or days. On a windy day, it is common to observe light-colored foam streaks on the water surface (figure 2.19). These streaks were first described by Langmuir (1938) during an ocean voyage. As he observed the streaks from high above the ocean on the deck of a cruise ship, he developed the theory that the streaks are the result of a complex, meter-scale, helical circulation of surface water. He suggested that the foam streaks are the result of convergent, parallel surface currents of downwelling water. All limnologists know this theory, and the graph of the helices is one of the most recognizable in limnological literature (see Wetzel, 2001 or Brönmark & Hansson, 1998). Although there have been hundreds of papers published that assume the existence of Langmuir's circulation, the circulation itself has not been observed directly in lakes. Also, modern hydrodynamical theory makes long-term, helical, laminar circulation seem unlikely in the extreme. Of course, surface turbulence does exist at the meter scale (figure 2.16), but the regular "Langmuir spirals" are probably nonexistent

Close observation reveals that foam streaks are the result of foam bubbles produced when breaking waves—whitecaps—mix water and air. Individual bubbles appear to be attracted to one another on the water's surface and to adhere to form patches. Foam patches are then stretched out into long, roughly parallel streaks, in the direction of the wind. The streaks are on the order of a centimeter wide and several to many meters long. Longer streaks are seen to branch and are not evenly distributed—the distances between streaks vary from centimeters to meters. In the opinion of this author, there is no evidence that these surface streaks indicate anything about subsurface water currents in lakes, but are merely a surface phenomenon (for examples of other surface phenomena, see Camazine, 2003).

Waves breaking on shore Surface wave oscillations near shore are deformed by interaction with the shore, so that the vertical movement is translated into lateral movement. Friction with the bottom or shore can cause the wave to break (the top of the wave surges forward in the air) if the water is shallow enough and the horizontal velocity is great enough (plate 22).

Seiches (Standing Waves)

The periodic drying out and reflooding of shallow basins attached to large lakes, as the water in the lake oscillates back and forth, is called **seiches.** This term, like "limnology," was coined by Forel, who observed a regular oscillation of lake level in Lake Geneva. The oscillation had a period of about 73 minutes, during which time lake level changed by as much as 1.5 meters! You can use a mathematical model on a website to estimate seiche height and period for a lake (see the end of this chapter for directions to the web site).

Seiches are caused by sharp differences in air pressure from one side of a lake to the other. Differences in air pressure are often associated with violent storms. Water is blown toward the lee shore. When the storm is over, the water falls back down, pressing on adjacent water, and forcing the water at the other end to rise. The momentum of the water causes a pendulum-like movement of water sloshing back and forth across the lake. The wavelength of the standing wave is twice the length of the lake, and the amplitude depends on the differences in air pressure. The fluctuations, over a few minutes or hours, are often large enough (up to a meter or more) to cause problems near lake shores—beaches can quickly appear or disappear.

Seiches are most obvious in large lakes, such as Lake Geneva, that are long, deep, and have steep sides with relatively straight shores, like a bathtub. In such a lake, there may be a **uninodal** oscillation—the lake surface resembles a plane that is slowly rising at one end, then falling, with a node in the center of the lake (where water movement is only horizontal, not vertical). Harmonics of the uninodal oscillation can also occur, leading to more complicated wave patterns. Complex standing waves are caused by complex basin shapes. A **binodal** oscillation has two nodal points, each a third of the way across the lake, and so on.

The seiche oscillation continues, with a period of minutes to hours, depending on the lake size (see the website at the end of this chapter), until all the energy is dissipated by friction within the water or against the bottom. The average water molecule in the lake only moves a short distance, perhaps tens of meters, but water motion is greatly amplified at the end of the sloshing water, such that spectacular changes in water level can be observed in shallow bays.

Seiche wavelength is related to lake diameter, the period is independent of the wave length, and the amplitude varies according to the forcing function and the density difference (table 2.7).

Movement of the surface water can induce seiches in the thermocline or at a pycnocline: An internal seiche (see figure 2.16) descending into the depths is affected by the difference in density of the two layers—internal seiches (water-water contact) have greater amplitude and longer periods than do surface seiches (water-air contact). These internal seiches are detected using vertical strings of sensitive thermisters: Rapid changes in water temperature indicate internal water movements. Internal seiches can be important to lake productivity if they bring deep, nutrient-rich water up into the euphotic zone (MacIntyre and Jellison, 2001).

Modern technology allows simultaneous measurement of water height at many points on a lake. These studies show that there is a lot of large-scale turbulence

Table 2.7 Data for Lakes of Different Sizes

The relationship among lake size, the amplitude of the surface seiche, and its period.

Lake	Lake Length (km)	Seiche Amplitude (m)	Seiche Period (min)
Lake Erie, North America	400	1–2	840
Lake Geneva, Switzerland	38.6	1.87	73
Lake Mendota, Wisconsin	9.1	A few cm	26

Source: Hutchinson, 1957.

in lakes and surface seiches are probably more complex than originally thought. The lake surface is not observed to move as a single plane oscillating back and forth, especially if the lake has nonbathtub shape. Attempts to interpret a lake's response to energy inputs (wind) in terms of stable basin, standing waves may be inappropriate (Bohle-Carbonell, 1986). A study of the large Swiss lake, Lake Geneva, suggested that even this lake's response to wind storms was episodic, short-lived, and complex. For lakes that are long and narrow along the direction of the prevailing wind (e.g., Lake Erie), it is possible that periodic seiches do exist (Korgen, 1995).

Lunar Tides

Ocean tides are a familiar phenomenon along a marine coastline. These **tides** are a rising and falling of ocean water twice a day of (usually) a meter or more, and the alternately submerged and dry zone provides the setting for a diverse and ecologically important **intertidal** community of organisms. The same tidal phenomenon occurs in inland waters, but on a much smaller scale. The amplitude of the tide depends on the size of the lake. Large Lake Superior (a North American Great Lake) has a lunar tide of about 2 centimeters (Hutchinson, 1957), and smaller lakes have even smaller daily tidal fluctuations. Probably because freshwater tides are so small, we do not see an intertidal community living along the margins of lakes.

SUMMARY

This chapter contains a great deal of basic information about the nature of water on micro and macro scales. Some of the concepts, such as pH, chemical cycles, eutrophication, and biological stratification, will be further explored in later chapters.

Now that the stage of the freshwater ecological theater is set, we will consider the organisms that, to use Hutchinson's metaphor, act out an evolutionary play in the limnological context. Perhaps it would seem natural to consider water chemistry next after physics. However, because water chemistry is so dependent on living organisms, the aquatic organisms are described next in chapters 3, 4, and 5 before we move to their role in the chemistry and ecology of lakes and streams.

Study Guide

Chapter 2 Setting the Limnological Stage: Water as an Environment

Questions

1. Which physical characteristic of water has the greatest effect on vertical patterns of temperature and chemical concentrations in stratified lakes?
2. A current of 8 kph is capable of doing a certain amount of damage through the power it exerts on its surroundings. What current speed would create twice as much damage?
3. You see a picture of a car mostly submerged in water. It is night, and the car's headlights are burning brightly. You know the location of the water is either the ultra-oligotrophic Toolik Lake, Alaska or the saline Great Salt Lake. Which is the most probable location, based on this information?
4. Do waves breaking on the shore of a lake or river sound the same in winter and in summer? (This is a long-term project.) Assuming there is a difference in the sound, which of the temperature-dependent physical aspects of water causes this difference?
5. Which are more important, viscous or inertial forces, for a human swimmer? Do people produce a laminar or a turbulent wake?
6. What is the difference between mass movement and periodic movement?
7. Can you draw an isopleth diagram for annual temperature data? (An example of a completed thermal isopleth diagram is given in figure 2.7a.) To find out if you understand how this graph is drawn and interpreted, construct your own. Find a set of vertical temperature profiles taken at several different times of year for a specific lake. Data from aquatic sites can be found on the Internet (see addresses at the end of this chapter) at Water on the Web, the LTER-NTL site, Lake Access, or the website of the Ecological Society of America (*http://wilkes.edu/~kklemow/eco-research.html*). For example, the site for the Long-Term Ecological Research—North Temperate Lakes has physical data freely available, starting at the home page: *http://limnosun.limnology.wisc.edu/* Then, use the data to create an isopleth diagram.
8. Can isopleth lines cross?
9. How do you categorize your favorite lake, in terms of annual thermal stratification pattern?
10. How would you test Langmuir's model about the relationship between foam streaks and circulation patterns in surface waters?

Words Related to the Aquatic Environment

absorbance
acid
algae
algal plate
alkaline
alkalinity
amictic
amplitude
ANC
anoxic
aphotic zone
bends
benthic zone
bicarbonate
billow
binodal
bubble
buffer
calcium
calorie
chaotic
chemocline
clear water phase
clinograde
compensation depth
concentration
conductivity
density

diel
diffuse
dimictic
dissociate
DOC
DVM
dystrophic
eddy
emergent
epilimnion
equivalents
eruption
euphotic zone
eutrophic
evaporation
fetch
fluid
foam
gyre
hardness
hard water
heterograde
hydrogen
hydrogen bonds
hydroxyl
hydrophobic
hypolimnion
inertia
intertidal
inverse stratification
isopleth
isothermal
macrophyte
magnesium
meromictic
mesotrophic
metalimnion
milieu
molarity
monimolimnion
monomictic
oligotrophic
orthograde
oscillation
PAR
pelagic zone
period
pH
phase
phytoplankton
planktonic zone
polar (charge)
polymictic
primary productivity
profile
pycnocline
redox
salinity
scale dependent
scattering
Secchi disk
seiche
soft water
soluble
solvent
specific conductance
stratification
TDS
thermistor
thermocline
tide
total dissolved solids
trophic
turbid
turbulence
uninodal
vertical temperature profile
viscosity
wavelength
whitecap
zero flow
zoobenthos
zooplankton

Major Concepts to Understand

The relationship between the power of a current and the density of the fluid and its velocity.

The effect of drag and current velocity in the vertical profile of a stream.

The concept that power is proportional to the third power of the velocity.

The calculation of the Reynolds number, and what the number tells us about viscosity and inertia:

Re = diameter times velocity divided by 0.01 cm^2 per sec (the units of diameter and velocity are centimeters and seconds).

The calculation of pH: $pH = -\log[H^+]$

The annual cycle of thermal stratification in a temperate lake.

The relationship between the vertical oxygen profile and factors such as water temperature and biological activity.

The major patterns of vertical distribution of aquatic organisms in lakes, as related to stratification patterns of temperature and chemicals (including oxygen and nutrients).

The relationships among Secchi disk depth, algae abundance, and the euphotic zone of a lake.

The hypolimnetic algal plate at the intersection of light and nutrient gradients in the hypolimnion of an oligotrophic lake.

The distinction between mass transport and periodic movement.

The relationship between wavelength and wave oscillation period.

The predicted period for a uninodal seiche in Lake Mendota (hint—check out the website under seiche, given in the Exploring the Internet section).

Major Examples and Species Names to Know

Lake Mendota, Wisconsin, annual patterns of temperature profiles—a dimictic, eutrophic lake

Lake Washington, Washington, an oligotrophic, monomictic lake

Lake Vanda, Antarctica, an amictic, hyper-oligotrophic lake

Mary Lake, Wisconsin and Mono Lake, California, meromictic lakes

Lake Amatitlán, Guatemala, a large, eutrophic, polymictic lake

Caspian Sea, the largest inland water body; it is brackish, but many limnologists still consider it to be a lake

Cladophora, a green alga that attaches to rocks, also called "green rock hair"

Exercises

Reynolds Number Calculations Exercise

For a city bus moving down University Avenue,

a. What is the speed in miles per hour? ______

b. What is the speed in cm per sec? ________

c. What is the length of the bus in m? ________

d. What is the length of the bus in cm? _________

e. What is the Reynolds number for the bus? _________

f. Which is a more important force for the bus, viscous or inertial force? Explain your answer.

pH Calculation Exercises

1. Fill in the blanks in the following table, using what you know about the dissociation of water into hydrogen and hydroxyl ions.

$[H^+]$ concentration as molar equivalents	$[OH^-]$ concentration as molar equivalents
10^{-7} molar equivalents per liter	
10^{-3}	
	10^{-4}

2. pH calculations:

a. Write the concentration 0.0001 grams per liter as 10 raised to a power. __________

b. What is the log (base 10) of 0.0001? __________

c. If the hydrogen concentration in water is 0.0001 grams per liter, the pH is ___________.

d. If the hydrogen concentration in water is 0.000001 equivalents per liter, the pH is ______.

e. If you mix 1 liter of pH 3 lake water and 1 liter of pH 7 lake water, the pH of the 2-liter mixture will be closest to (circle one choice):
   1. 2
   2. 4
   3. 6
   4. 8

   Explain your answer (don't worry about the possible effect of buffers for this answer).

Chemical Molarity and Equivalents Exercise

The molarity of a solution of an ion is the concentration of the ion in grams (or milligrams, etc.) per liter, divided by the molecular weight of the ion (or atomic weight, if the ion is a single ionized element).

The equivalents of the ion is the molar concentration (molarity) multiplied by the valence (number of positive or negative charges) per ion.

The ion concentrations are what might be found in the water of Trout Lake, Wisconsin.

	Symbol	Concentration (**mg**/l)	Molecular Weight	Molarity (**millimoles**)	Valence	Equivalents (**meq**/liter)
Cations						
	Ca^{++}	14.2				
	Mg^{++}	3.45				
	Na^{+}	1.91				
	K^{+}	0.89				
	H^{+}	0.0001				
Anions	HCO_3^{-}	11.6				
	SO_4^{--}	3.28				
	Cl^{-}	1.65				
	NO_3^{-}	40				
	PO_4^{---}	3				
	OH^{-}					

The equivalents for all cations added together should be very similar to the total for all anions. Is this the case?

Cation equivalents: __________

Anion equivalents: __________

What accounts for the difference between the cation and anion totals?

Crossword Puzzle

Use the following crossword puzzle to review terminology used in the first two chapters of this text. Words are drawn from the lists of important words given at the end of each chapter in the Study Guide.

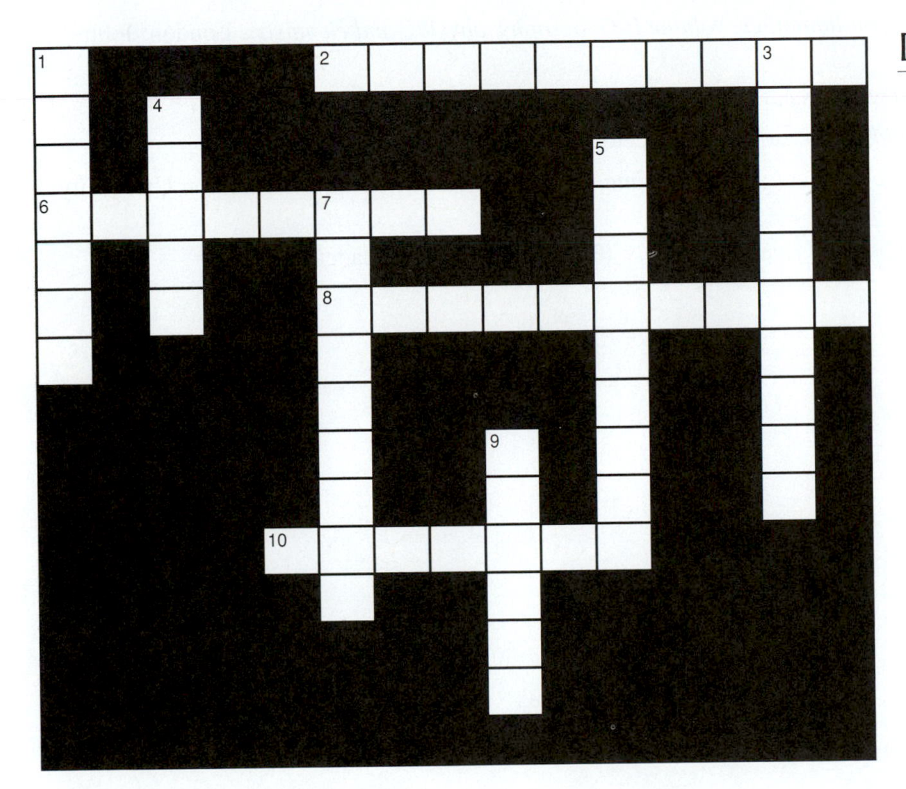

Down

1. The genus of a small crustacean (invertebrate, arthropod) called the "water flea"
3. The same temperature from top to bottom
4. The "Father of Limnology" who wrote a monograph on Lake Geneva
5. The study of inland waters, including lakes, streams, and wetlands
7. An ecological system, often represented by a box and arrow graphical model
9. Surface or internal large-scale waves in lakes

Across

2. In this kind of lake, the top layers (but not the lowest layer) mixes during the year.
6. Depends on the concentration of calcium and magnesium ions in water
8. Water boils (at sea level) at this number of degrees in the Celsius system (two words)
10. Weight per unit volume (it is about 1 gram per liter for water)

Additional Resources

Further Reading:

Horn, A. J., and C. R. 1994. Goldman. *Limnology.* (2nd ed.) New York: McGraw-Hill.

Hutchinson, G. E. 1957. *A treatise on limnology. Volume I: Geography, physics, and chemistry.* London: John Wiley & Sons. 1015 pp.

Kalff, J. 2002. *Limnology: Inland water ecosystems.* Englewood Cliffs, NJ: Prentice-Hall.

Vogel, S. 1983. *Life in moving fluids.* Princeton, NJ: Princeton University Press.

Vogel, S. 1988. *Life's devices: The physical world of animals and plants.* Princeton, NJ: Princeton University Press.

Wetzel, R. G. 2001. *Limnology: Lake and river ecosystems.* (3rd ed.) New York: Academic Press.

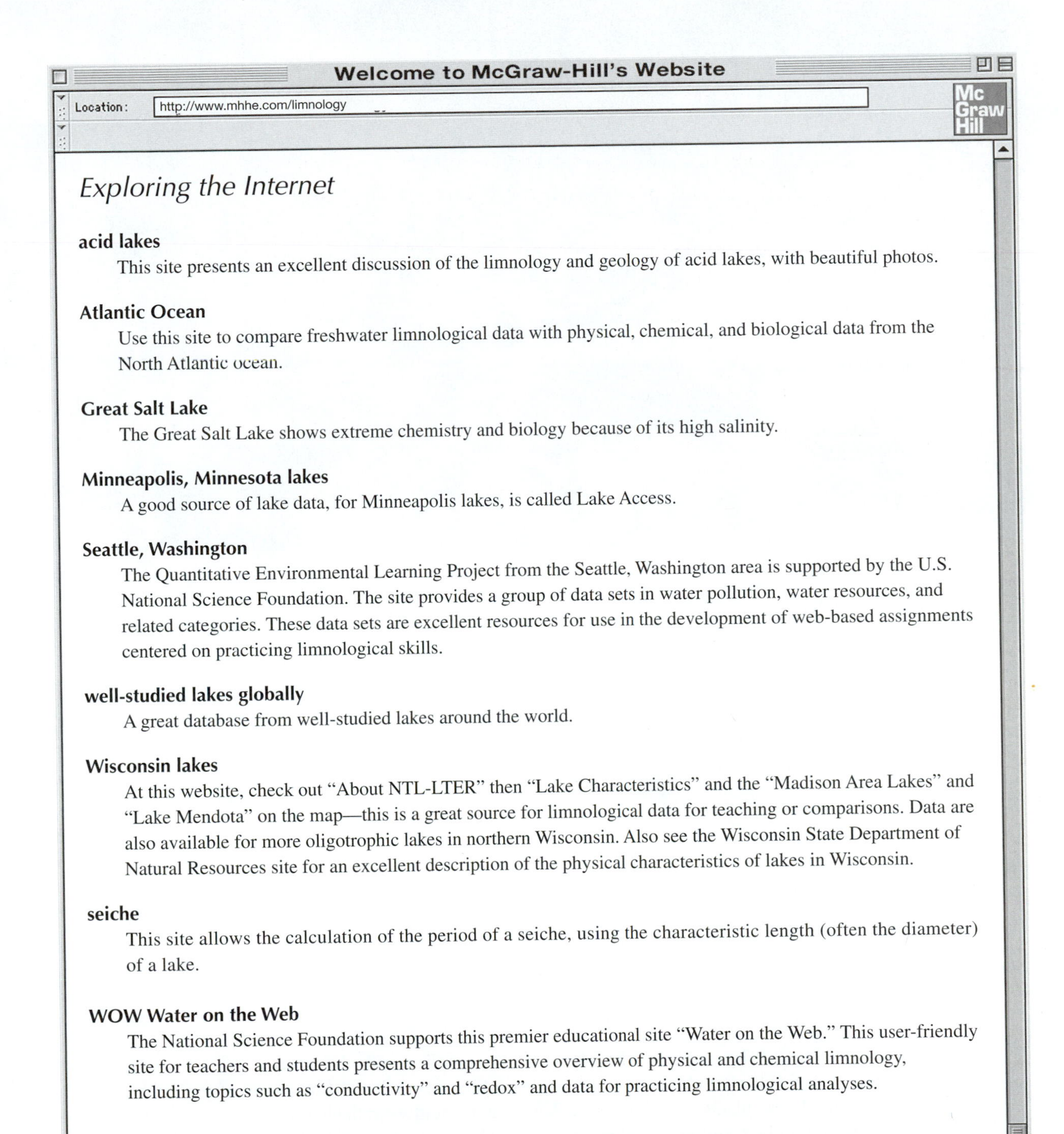

Welcome to McGraw-Hill's Website

Location: http://www.mhhe.com/limnology

Exploring the Internet

acid lakes

This site presents an excellent discussion of the limnology and geology of acid lakes, with beautiful photos.

Atlantic Ocean

Use this site to compare freshwater limnological data with physical, chemical, and biological data from the North Atlantic ocean.

Great Salt Lake

The Great Salt Lake shows extreme chemistry and biology because of its high salinity.

Minneapolis, Minnesota lakes

A good source of lake data, for Minneapolis lakes, is called Lake Access.

Seattle, Washington

The Quantitative Environmental Learning Project from the Seattle, Washington area is supported by the U.S. National Science Foundation. The site provides a group of data sets in water pollution, water resources, and related categories. These data sets are excellent resources for use in the development of web-based assignments centered on practicing limnological skills.

well-studied lakes globally

A great database from well-studied lakes around the world.

Wisconsin lakes

At this website, check out "About NTL-LTER" then "Lake Characteristics" and the "Madison Area Lakes" and "Lake Mendota" on the map—this is a great source for limnological data for teaching or comparisons. Data are also available for more oligotrophic lakes in northern Wisconsin. Also see the Wisconsin State Department of Natural Resources site for an excellent description of the physical characteristics of lakes in Wisconsin.

seiche

This site allows the calculation of the period of a seiche, using the characteristic length (often the diameter) of a lake.

WOW Water on the Web

The National Science Foundation supports this premier educational site "Water on the Web." This user-friendly site for teachers and students presents a comprehensive overview of physical and chemical limnology, including topics such as "conductivity" and "redox" and data for practicing limnological analyses.

3

"I would also emphasize how fantastically complicated the lacustrine microcosm is likely to be. There is probably no almost complete list of species of animals and plants available for any lake, but it would seem likely from the several hundred species of diatoms and insects known from certain lakes that a species list of the order of a thousand entries may be not unusual."

G. Evelyn Hutchinson, 1965

Diversity of Aquatic Organisms: The Single-Celled and Colonial Organisms

outline

THE DIVERSITY OF LIMNOLOGICAL ORGANISMS

The major players on the freshwater ecological stage include a number of groups from all the kingdoms of living organisms. The purpose of this chapter is to introduce the cast of characters and, as a starting place, to introduce the reader to the smaller aquatic organisms. The introduction (naming) of the organisms is accompanied by a description of their relationships and diversity, what they look like, where they are most often found in the limnological theater, and what they do in terms of movement, feeding, and life history. A final section for each group discusses the intersection between the ecology of the organisms and human economic or conservation issues.

In chapters 3, 4, and 5, we will meet the organisms that have the most important roles in the limnological theater; organisms that will be mentioned again in the later chapters. These are the organisms limnologists choose when they give examples about water chemistry, population dynamics, ecological interactions, or distribution and management. The major groups include the:

Archaea

Bacteria

Protists (algae and protozoa)

Fungi

Rotifers

Annelids (Oligochaeta)

Branchiopod crustaceans

Copepod crustaceans

Insects

Mollusks

Fish

Other vertebrates

Plants (macrophytes)

A consequence of this strategy of focusing on major ecologically important groups is that some fascinating groups of freshwater organisms will not be mentioned, including mosses, liverworts, ferns, flatworms, nematodes, sponges, gastrotrichs, tardigrades, coelenterates, and bryozoans. Information about these organisms can be found in specialized texts or on the Internet. The chapters in Thorp and Covich (2001) provide excellent information on many invertebrate groups. Additional sources are cited at the end of the next three chapters.

To give structure to the descriptions of aquatic organisms, the same outline is presented for each of the major groups discussed in the three chapters on organisms (chapters 3, 4, and 5):

Taxonomy and Diversity

External Structure, Appearance, and Anatomy

Habitat

Swimming and Escape Behavior

Feeding Preferences and Behavior

Life Cycle and Diapause

Economic Importance and Management

Taxonomy and Diversity

In biology, **taxonomy** refers to the systematic classification of living organisms. Because of the general nature of this text, the taxonomy will focus on higher levels of classification, such as phyla, divisions, classes, and orders.

It is important to realize that classification is a dynamic process, in constant flux and with little agreement among taxonomists. Students are advised that different authors tend to have slight differences in their classification, so be prepared for inconsistencies when you use the Internet or other resources.

In ecology, **diversity** refers to the number of different kinds of organisms (see chapter 7). As an index of the diversity within each major group, this text indicates the number of known species for that group.

External Structure, Appearance, and Anatomy

It is important to know an organism's approximate size and what it looks like. The outer form of an organism is its **morphology. Anatomy** is the internal structure of an organism.

One of the most important morphological characteristics of an organism is its size. Many ecological aspects are scale-dependent, including diet, locomotion, and life history. Plankton (organisms living in open water) are categorized according to their body size (table 3.1).

Habitat

A habitat is where an organism lives. Among all aquatic organisms, each kind of organism has a preferred location and an optimal environment in a lake or stream. The concept "ecological **niche**" refers to an organism's relation to its environment, including food and ene-

Table 3.1 Categories of Plankton (organisms suspended in water) According to Their Size

There are 1 million micrometers (**μm**) in 1 meter and 1000 μm in 1 millimeter (see table 1.1).

Category	Approximate Size (diameter)	Representative Organisms
Zooplankton	50–5000 μm	Protists, microcrustaceans
Phytoplankton	Less than 0.5 μm to 500 μm	
Net plankton	50–500 μm	Protists: protozoans and algae
Nannoplankton	10–50 μm	Protists: protozoans and algae
Picoplankton	0.5–10 μm	Smallest protists and large bacteria
Femtoplankton	Less than 0.5 μm	Bacteria and viruses

Size categories are generalizations based on data in Hutchinson (1967).

mies. The term is adapted from art history, in which a *niche* is a shallow recess in the wall of a building, usually for the purpose of sheltering a statue.

Swimming and Escape Behavior

Many organisms are capable of movement and make a habit of moving about in specific ways in or among aquatic habitats. There is always a strong link between size and form, on the one hand, and size and locomotion on the other. As we saw in chapter 2, size strongly constrains the way an organism perceives its watery environment. For example, the scale dependent ratio between viscous and inertial forces determines optimal swimming behavior and, therefore, aspects of form such as whether there is a benefit to being streamlined.

Feeding Preferences and Behavior

All organisms collect energy in one form or another, so in the most general sense, all organisms (including algae and plants) have feeding preferences and behavior. The general term trophic refers to food or energy. The terms relating to trophic condition of a whole lake (eutrophic, etc.) were introduced in chapter 2. There is also a complex classification of energy (or trophic) strategies for groups of organisms (Hutchinson, 1967; Wetzel, 2001). For example, algae and green plants are **autotrophic** organisms that collect their energy as sunlight using photosynthesis. Protozoans, animals, and fungi are **heterotrophic** organisms that extract energy from the chemical energy stored in organic substances, by processes such as consumption and decomposition.

Life Cycle and Diapause

Life cycle describes the various developmental stages, the timing of development, and the pattern of time and energy allocation over the life span of an organism. Within a species, the rate of growth and amount of time spent in each developmental stage is the result of an evolutionary process in which development is adapted to the opportunities and hazards of the environment. Adults produce as many offspring as soon as possible, given their specific ecological context. In juveniles, energy that is taken into the body is allocated toward maintenance and growth, and for adults, energy is also allocated toward reproduction. Different organisms display very different patterns of time and energy allocation.

The **diapause** condition is a special physiological state in which metabolism is almost completely suspended. A specific environmental signal (such as a change in temperature or day length) is needed to break diapause—to start up the metabolic process again. Diapause often occurs in spores, eggs, and early embryos in higher organisms. The diapausing stage can persist for several or even many annual cycles, waiting for the correct environmental signal. The diapause stage can be an important dispersal stage because the diapausing form is often small and resistant to damage. Equally important, diapausing stages are critical in maintaining persistence through unfavorable periods in aquatic habitats. Aquatic habitats can freeze, dry up, become anoxic, or suddenly be filled with predators. Resistant stages stored in sediments (a seed bank or egg bank) provide a way to bridge these unfavorable times.

Economic Importance and Management

Aquatic organisms often have direct or indirect links to issues of human importance. Aquatic autotrophs directly affect water quality and provide the energy (food) to support economically important populations of vertebrate organisms such as fish and waterfowl. Bacteria, algae, and protozoans include species that present health concerns for wildlife and humans. The larger aquatic organisms and their habitats are the focus of major conservation efforts and management programs.

SIZE

This text's description of organisms is organized according to size. The smallest organisms are described first, here in chapter 3. These bacteria, protists (algae and protozoans), and fungi are commonly referred to as **microbes,** or microscopic organisms. Although they are in the micrometer size range, their abundance makes them of great ecological importance. Zooplankton (rotifers and microcrustaceans) and the major insect groups are featured in chapter 4. Chapter 5 presents information on the larger organisms—mollusks, fish, and aquatic plants. The smaller organisms, described first, provide the ecological basis for the existence of the larger charismatic organisms.

PROKARYOTES (BACTERIA AND ARCHAEA)

"These are the "bugs that run the world"—They're more powerful than Alan Greenspan, George W. Bush, or even Madonna" (quoted in *Science* 293:399—see the website address at the end of this chapter). Microbes hold sway over such vital planetary matters as the composition of the atmosphere and the productivity of ecosystems (Leslie, 2001). (For general information and beautiful images of bacteria at all scales, see the Microbial World website at the address given at the end of this chapter.)

Taxonomy and Diversity

Prokaryotes are commonly called "bacteria." Actually, the term **bacteria** more precisely refers to one major group of prokaryotes. Individuals from the other major group of prokaryotes, the **Archaea,** are bacteria-like in structure but very different in the details of genetics.

Prokaryotes are distinguished from the eukaryotes by their lack of intracellular organelles (figure 3.1). Prokaryotic genetic material is not enclosed in a nuclear membrane. Prokaryotes also have distinct macromolecules (such as genetic material, enzymes, and structural proteins) that are different from those of eukaryotes.

Early microscopists originally classified prokaryotes by their appearance (figure 3.2). Early bacteriologists also classified prokaryotes according to the kinds of media on which the organisms grew. Today, prokaryotes are classified principally on the bases of their biochemistry and molecular biology. Different prokaryotes have different ways of making a living by degrading energy-rich (usually organic) compounds. There are two entire domains of prokaryotes. A *domain* is the largest possible classification category of life—there are currently three: two for prokaryotes (bacteria and Archaea) and the Eukaryota. The prokaryotic domains are distinguished by different metabolic biochemistry, composition of the cell wall, and by the pattern of similarity in structure of RNA polymerase and a ribosomal protein.

There are at least 10,000 named species of prokaryotes. Most of these species are in the domain bacteria. However, the taxonomy of prokaryotes is still in its infancy. Most of what we know about prokaryotic diversity is based on those forms that can be cultured in the laboratory. Recent molecular studies suggest that only a small percentage of the total bacterial diversity can be cultured (Pace, 1997).

The bacteria group includes most known prokaryotes and includes cyanobacteria, the green and purple **phototrophic** sulfur bacteria, and bacteria in the guts of planktonic organisms. These prokaryotes have a cell wall made of **peptidoglycan,** a complex polymer made of sugars cross-linked with short polypeptides. The peptidoglycan wall is, because of the cross-links, a single, giant molecule that surrounds and protects the cell and can withstand great internal pressure.

The Archaea group is distinguished from bacteria in that Archaea cell walls lack peptidoglycan.

External Structure, Appearance, and Anatomy

Prokaryotes are among the smallest living organisms. Individual cells are typically 0.1–2.0 μm long. Individuals are typically microscopic—too small to see unless they are in groups. The largest bacteria cells, about

FIGURE 3.1 An idealized dissection of a prokaryotic cell.

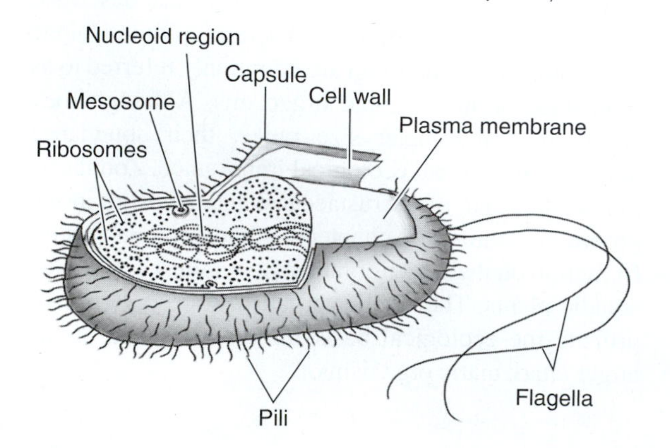

FIGURE 3.2 Shapes of different bacteria. The individual coccus cells are about 1 μm in diameter.

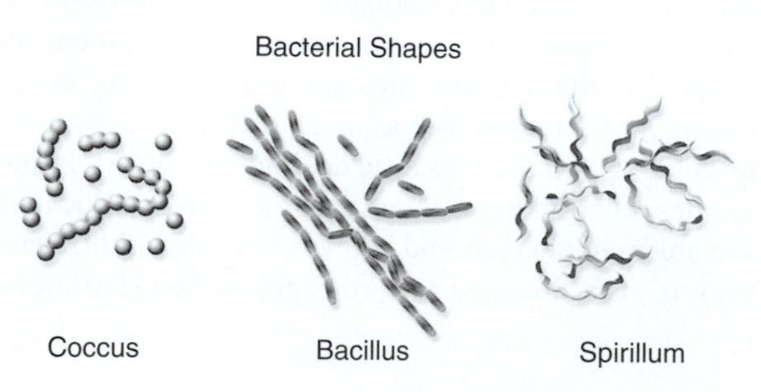

1 mm long, are actually visible to the unaided eye (Angert et al., 1993).

Filaments, or masses, of attached cells can be millimeters or even centimeters long. For example, some of the cyanobacteria (photosynthetic blue-green bacteria, also called blue-green algae) and colorless sulfur bacteria form filaments, or masses, that are easily visible. The large cyanobacteria **colonies** that form **stromatolites** can be a meter or so across (stromatolites are layered, calcareous fossil colonies that are some of the earliest evidence of life on Earth).

Prokaryotes often have smooth, rounded shapes. D'Arcy Thompson (1917) suggested that bacterial shape is dictated by small size. Very small objects have a surface that is tightly curved. The tight curve produces a high surface tension, which makes projections, or infoldings, impossible to maintain.

The small size of prokaryotes gives them the largest *surface-to-volume ratio* of living organisms. Consider a spherical bacterium, with a radius, r. Its surface area is proportional to r^2 and its volume is proportional to r^3. This means that the ratio of the surface to the volume of the sphere is proportional to $1/r$. As a bacterium grows, both its surface area and volume increase. However, the ratio of surface to volume actually decreases directly proportional to the increase in size. For example, if a bacterium grows so that it increases its radius by a factor of two, we say it has doubled its size. Along with the doubling of the radius, the surface area will be four times as large, and the volume of the cell will be eight times as large. However, the surface-to-volume ratio will be half what it was when the cell started growing.

The surface-to-volume ratio has important implications for the ecology of organisms. For example, assume that a small subsection of a bacterial cell requires a certain minimum energy input, and that the requirement for this small subsection is independent of the size of the cell. The total energy requirement of the cell depends on the number of subsections and the requirement of each subsection. The energy requirement is met by movement of nutrients across the cell surface. Thus, supply of nutrients depends on the surface area, but the energy requirement depends on the cell volume. Because of the nature of the surface-to-volume ratio, larger cells will have less surface for each unit of volume, and each subsection of the cell will receive less energy to meet its requirements. The ecological implication is that smaller cells (with the larger surface-to-volume ratio) will be able to live in habitats with lower nutrient concentrations. If nutrients are limited, the smallest cells will persist the longest (all other factors, such as predation, being equal).

As stated in the introduction to this section, prokaryotes are often found as groups of attached individuals in colonies (masses of cells) or filaments (strings of cells attached end-to-end). These attached cells can be either poorly organized, slimy masses or more organized colonies (e.g., some of the cyanobacteria form long chains of cells, plates, or even three-dimensional masses of bacteria embedded in a hard, gelatinous matrix; see figure 3.3). A colony is distinguished from a tissue in that each of the individuals of a colony is potentially able to live independently, while the cells of a tissue cannot live independently.

Prokaryote colonies are often colored, but individual bacteria are too small to give an impression of color. Individual bacteria are about the same length as the wavelengths of visible light. For example, blue light has a wavelength of about 0.47 μm and red light has a wavelength of about 0.65 μm. The similarity in lengths means that visible images of prokaryotes are often indistinct and fuzzy. Colony color is the result of the metabolic specialization of the bacteria. Colonies of prokaryotes can be brightly colored: red, orange, pink, blue-green, yellow, and so on. These bright colors are due to pigments involved in photosynthesis or electron transport.

Because bacteria are often about the size of a single wavelength of light, optical microscopy is inefficient. Because of their small size, it is very difficult to see individual prokaryotes, even with the highest magnification. In the 1970s, microbial ecologists developed filtration and fluorescent-staining techniques to help in the visualization of bacteria (Hobbie et al., 1977; Porter & Feig, 1980). These new techniques made it possible to observe and count live prokaryotes more easily: The material is stained with dyes that attach specifically to live bacteria and then fluoresce when illuminated with ultraviolet (UV) light. Thus, the individual organisms are seen as bright spots on a dark background (the UV light is invisible to human observers). To the surprise of microbiologists and ecologists, these new techniques resulted in an approximately 10-fold increase in estimates of the number of prokaryotes found in natural habitats.

Habitat

Prokaryotes occur everywhere—in all habitats and on and in all parts of the biosphere. Bacteria are more abundant than Archaea in most easy-to-sample places

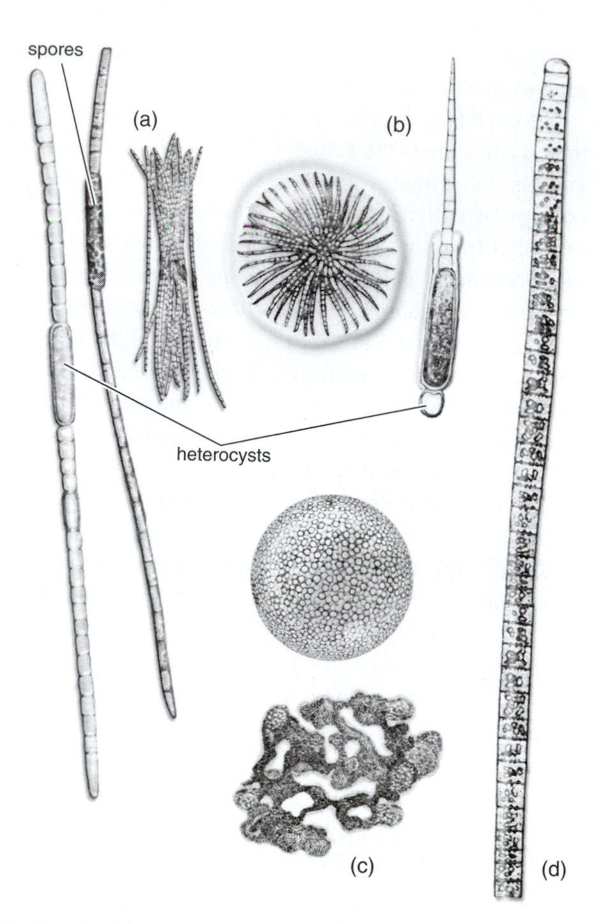

FIGURE 3.3 Examples of cells and colonies of cyanobacteria. All individual cells are about 1 μm across. (a) *Aphanazomenon.* The top figure is a colony of filaments at low magnification, with examples of individual filaments below (with spores and heterocysts); (b) *Gleotrichia.* The spherical colony (low magnification) is as much as 1 mm in diameter, while the high-magnification image of an individual filament shows the heterocyst at the end of the filament, next to a large, resting spore. (c) *Microcystis.* Two colonies—the left colony is at higher magnification. Note there are no heterocysts and the amorphous colony at the left is at low magnification. (d) *Oscillatoria.* A single filament. Note that there are no heterocysts. *Aphanizomenon* and *Gleotrichia* (both have heterocysts) can fix nitrogen. *Aphanizomenon* and *Microcystis* produce gas vacuoles, allowing them to float near the surface (plates 19 & 20). *Gleotrichia* and *Microcystis* produce large amounts of gelatinous matrix.

where other organisms live, such as fields and streams, cities and wilderness areas, and in and on other organisms. Different freshwater habitats do have different kinds of bacteria (Fisher & Triplett, 1999).

Archaea are usually more abundant in more extreme or less accessable habitats where no other life occurs. The domain includes methane bacteria, **halophiles** (salt-loving species), and colorless sulfur bacteria (which are also **thermophiles** requiring hot water). Many Archaea species are capable of **anaerobic** life in anoxic water such as that found in bottom sediments. Favorite harsh habitats of **extremophile** (lovers of the extreme) are boiling or near-boiling water in hot springs (plate 4) and submarine vents, and the salty brine evaporation ponds used to produce salt (plate 8).

Many bacteria species live on or in other organisms. When the relationship is not beneficial to the host organism (the living habitat), these bacteria are considered disease organisms. However, there are many examples of **symbiotic** relationships, which are beneficial to both species.

> "I would like to point out that we [humans] depend on more than the activity of some 30,000 genes encoded in the human genome. Our existence is critically dependent on the presence of upwards of 1000 bacterial species (the exact number is unknown because many are uncultivable) living in and on us; the oral cavity and gastrointestinal tracts contain particularly rich and active populations. Thus, if truth be known, human life depends on an additional 2 to 4 million genes, mostly uncharacterized. Until the synergistic activities between humans (and other animals) with their obligatory commensals has been elucidated, an understanding of human biology will remain incomplete" (Davies and Govedich, 2001).

Despite their small size, bacteria account for a significant portion of the organic material in a lake (table 3.2).

Bacteria are often present in huge numbers. In the open water of lakes, there are between 0.2 million and 0.7 million bacteria per milliliter in clear-looking, oligotrophic lakes, and between 2 million and 400 million bacteria per milliliter in eutrophic lakes. These huge numbers go unnoticed because of the small size of each individual. The mass of 400 million bacteria, each 1μm in diameter, occupy less than 1/1000th of 1 milliliter. Colonies of bacteria, containing billions of individual cells, are often visible as floating scum (plate 19), flecks in the water, or mats on stones or plants.

Swimming and Escape Behavior

Swimming is accomplished via *flagella,* which are composed of a double strand of microtubules composed of various proteins, depending on the group. The fla-

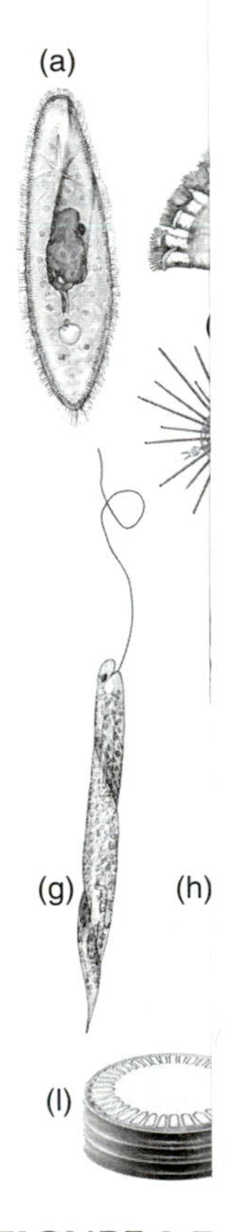

FIGURE 3.7
(a) *Paramecium*
(c) *Vorticella,* a
zopoda); (e) a h
flagellate (Zoon
chloroplasts (Eug
phyta); (i) *Cosma*
noflagellate (Di
(Chrysophyta); (
atom (Chrysophy
(n) the diatom *A*

Followin
of protists (ex

Ciliophora

These organis
with thousan
The cilia mov
feeding curre
tion organism
(figure 3.7b)
large, gelatin

chapter for addresses). Techniques
cyanobacteria are discussed in chapter

PROTISTA

The Eukaryota are distinguished from
that they have cells containing **organell**
cleus and mitochondria and/or chlorop
membrane that contains cholesterol (f
karyotic organisms include organisms
four kingdoms: plants, animals (metazo
the protists (protozoa and algae) (figure

Taxonomy and Diversity

Protista is a kingdom of one-celled o
either as single cells or as simple coloni
tozoa are animal-like protists and **alga**
thetic protists.

Like prokaryotes, protist species s
colonies. Each cell of the multicellula
its individuality and its ability to live
However, the protist colony is often ma
timeters to meters in diameter or length
logically important unit.

FIGURE 3.5 An idealized cross section
photosynthetic protist cell. (b) A plant or phot

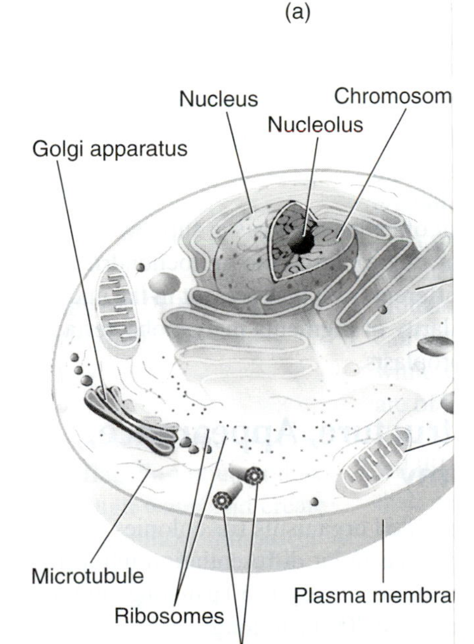

Table 3.2 Representative Distribution of Organic Carbon into Various Categories in a Mesotrophic Lake

Form of Organic Carbon	Density as g/m^{-2} Below the Lake Surface = Biomass (= weight per unit area)
Dissolved organic carbon (such as amino acids and organic acids, but because of limits of filter technology, also includes the small bacteria < .5 μm that pass through filters)	About 150
Dead particulate matter	About 15
Bacteria (except cyanobacteria)	0.1–10
Phytoplankton (including cyanobacteria)	1–100
Zooplankton	1–100
Fish	About 1–10

Source: Estimates from Brock, 1985.

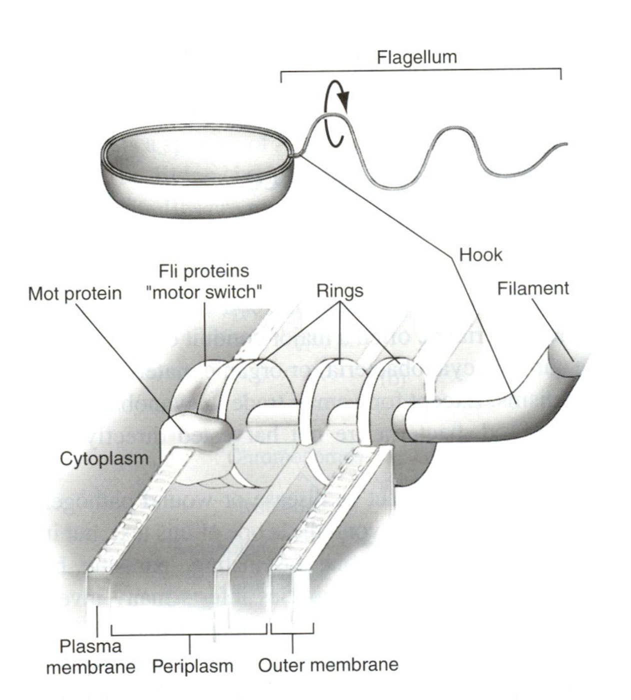

FIGURE 3.4 An illustration of the rotary mechanism used by bacteria for swimming.

gellum works as a wheel—it is anchored in the plasma membrane, goes through the cell wall, and spins (figure 3.4). The bacterial flagellum moves through water like a corkscrew through cork, and the flagellum drags the bacterial cell along through the sticky water.

Bacteria sometimes swim toward or away from a stimulus. This directed movement is accomplished by modifications of turning behavior. If swimming includes few turns, the organism goes in a relatively straight line. Turning frequency increases when the organism gets to a place where it no longer detects a stimulus. Directed movement might be toward food or light or away from a toxic chemical. Some bacteria with magnetite crystals may be able to detect and use the Earth's magnetic field as a navigational aid. In bacteria, turning is accomplished by reversing the rotation of the flagellum; the animal then moves in an unpredictable direction, often backwards.

Bacteria have only two speeds: go and stop. They do not have brakes; the viscosity of the water makes brakes superfluous. Thus, the only escape response for bacteria is to change direction. For a more detailed discussion of bacterial functional morphology and location, see Purcell (1977).

Bacteria can move at about one body length per second (which is a general measure for all organisms). This is just fast enough to move out of the zone of surrounding water that is depleted by diffusion. This speed of 1 μm per second amounts to 8.64 centimeters per day—although bacteria typically do not move in straight lines, but rather in a seemingly random pattern called a **random walk.**

Feeding Preferences and Behavior

While this heading works well for many organisms, for bacteria, it is more realistic to title the section "Energy Capture."

Bacteria that require an energy sour
of organic molecules, which they take
cell wall, are called *heterotrophs.* Bact
transport to take up organic food source
sugar glucose), while algae (protists) and
imal cells use diffusion for the same purp
Hobbie, 1966). Bacteria have an uptake
low concentrations of substrate because
port is more efficient than diffusion at l
tions. However, the active transport cost
and the system is saturated at relatively l
tions. Diffusion is free (there is no energ
efficient (relative to active transport) at
concentrations. Diffusion does not get sat
concentrations of substrate. Thus, bacteri
take strategy that is relatively most efficie
centrations of substrate; algae have an u
that is most efficient at high concentratio
that use light energy to facilitate uptake an
of organic molecules dissolved in water a
toheterotrophs, in contrast to **chemol**
which receive energy and carbon only fro
of organic molecules taken up from the w

Microbes that convert light energy i
energy using photosynthesis are called *au*
ergy (ATP) is generated by oxidizing in
stances such as hydrogen sulfide, ammon
(Fe^{++}) iron, or some other reduced inorga
Species of bacteria are often quite spec
which inorganic molecules they use in ox

Photoautotrophs use light energy to
thesis of organic compounds from car
These are mainly cyanobacteria (blue-gree
as *Aphanizomenon* and *Microcystis*, and t
purple phototrophic sulfur bacteria. Actu
the photosynthetic bacteria are *photo*
which use light to generate ATP but can al
in organic form as an energy source to ge

Cyanobacteria are characterized by a
of advantageous specializations (adaptatio
izations of these ecologically important bac
their ability to both take up organic che
erotrophy) and perform photosynthesis,
(*Aphanizomenon* and *Gleotrichia*), produ
uoles to float (and therefore shade competit
become toxic, and assume a colonial life
hibits algae-eaters, such as *Daphnia.* Thes
tions are most effective at conditio
temperature, intense light, heavy grazing,
trient concentrations.

Stamer
Chrys
Diator
Brown
Oomy
Opalir

E

FIGU

the organelles
include the foll

Folded prokary
otic nucle
Golgi appa
membranes
preted as
plasma me
sequence o
volume rati

Eukaryotic **mit**
brane)—Th
house of the
represents
prokaryote

Chloroplasts (
ganelles m
endosymbi
invaded to
tions seen t

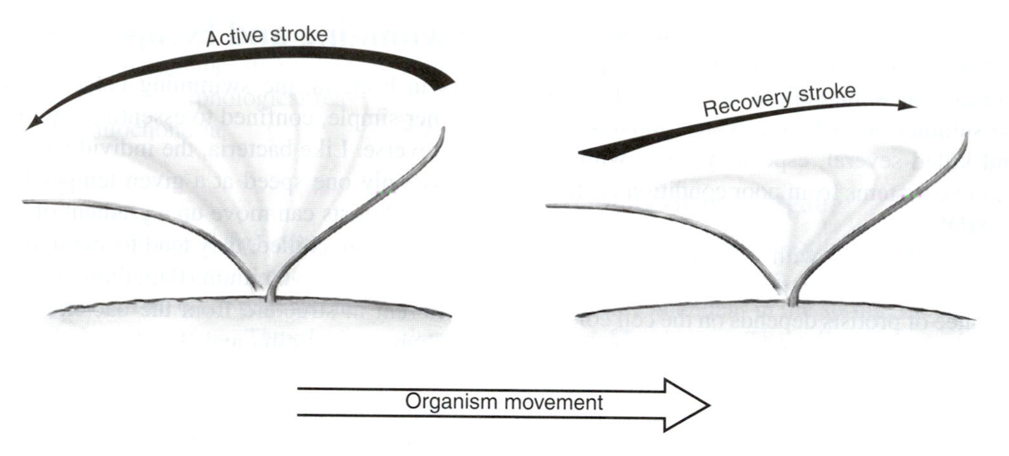

FIGURE 3.8 The movement pattern of a eukaryotic cell flagellum.

their peak density, were consuming more bacteria than nonpigmented flagellates, ciliates, rotifers, and crustaceans combined.

Ciliates have complex patches of cilia for capturing food and complex mouth regions for ingesting particles—usually viruses, bacteria, and small algae. In some genera, several cilia fuse to form stiff, spine-like structures, which help the organism to crawl over surfaces.

A few protists, such as the predaceous ciliate *Didinium,* attack and swallow large prey such as other smaller protists.

Life Cycle and Diapause

Many protists can reproduce asexually by simple cell division—the process of **mitosis.** The descendants of one mother cell produce a **clone,** or group of genetically identical cells. Colonies are generally clones.

To gain the advantages of sexual recombination, protists also use **meiosis** (chromosome number reduction) to produce haploid gametes. With a few exceptions (such as the Euglenophyta), protists can reproduce sexually. Gametes are almost always motile: using flagella or pseudopodia to creep through the water. Gametes fuse together to produce resistant **diploid** spores or the active diploid life-history form.

One of the most complex mating behaviors is seen in *Paramecium* species. In each species, there are many (up to 20 or so) mating compatibility strains. A strain can mate with any other strain, but not with its own genotype. Thus, it is as if there are many sexes in each *Paramecium* species. To mate, two individuals of compatible mating strains align side by side and fuse. They exchange chromosomes and then separate.

Most protists produce diapausing spores. These spores are often the result of sexual reproduction. A common method for culturing a diverse group of protists is the "hay infusion." The culture is started by pouring water over dried hay and a pinch of soil. In a few days, the resulting infusion will be a rich mixture of bacteria and protists, all of which came from diapausing stages associated with the dried hay and soil. This infusion can be used to feed laboratory cultures of some of the phagotrophic and filter-feeding organisms discussed in the present and following chapters.

Economic Importance and Management

Protists are major components at the bottom of the food chain. Protozoans (the animal-like, heterotrophic protists) are relatively unimportant in terms of energy transfer. Protozoans, as part of the "microbial loop," can interfere with efficient energy transfer from the bottom to the top of the food chain (as discussed in chapter 8). Also, protozoans can be significant parasites of other aquatic organisms and even of terrestrial animals and humans that drink surface water. Pigmented (photosynthetic) protists—the algae—are a major source of palatable food for zooplankton and other aquatic herbivores. Although of great ecological importance, it is common for these organisms to be considered a nuisance when they become too common. Management strategies of algae are discussed in chapters 7 and 9, and pictures and discussions of the ecology and management of algae can be found on Internet sites (examples are given at the end of this chapter). When algae are not abundant enough, they can be encouraged by adding inorganic fertilizers (nitrogen and phosphate) to lakes. For example, Stockner and Macisaac (1996) showed that salmon production could be increased in oligotrophic lakes in British Columbia by whole-lake fertilization.

FUNGI

Although now seriously outdated, Sparrow's (1959) text provides an excellent general overview of the classification, morphology, and ecology of aquatic fungi. Kendrick (2000) explores all aspects of the study of fungi (**mycology**), from ecology to classification to medical and economic issues, and touches on aquatic fungi.

Taxonomy and Diversity

Aquatic fungi include several groups of true fungi as well as morphologically similar but distantly related, protist-like water molds (Oomycota) and slime molds (related to several groups of protists) (see figure 3.6). There are about 60,000 named species of true fungi known in terrestrial and aquatic habitats combined, 700 species of water molds, and 800 species of slime molds (Hawksworth et al., 1995). As with bacteria and protists, modern molecular techniques allow the identification of many more fungal species than does identification of gross, morphological characters. The named species of terrestrial and aquatic fungi probably represent no more than 10% of the true number of species.

Individual fungal filaments, or cells, which are only a few tens of micrometers in diameter, are nearly impossible to identify taxonomically. Fruiting bodies, gametes, and spores (resting stages) have traditionally been used to classify fungi, but molecular techniques are showing that this approach gives only a faint indication of fungal diversity. Fungal spores are common in fresh water—some are produced by aquatic forms and some are produced on land and washed into streams or lakes by surface flow.

External Structure, Appearance, and Anatomy

At the morphological level, fungi are morphologically depauperate (their morphology is simple). Their growth form is either single, spherical cells (yeasts, a morphological term) or, more often, long thin filaments of single cells joined end to end (**hyphae**). The mass of threads that makes up most fungi is called the **mycellium.** Yeasts can occur as either spherical cells or filaments, depending on their genetics and environmental conditions. The cells in filaments are open at each end in some groups, so that cytoplasm and even nuclei move (stream) back and forth through the mycellium. In some cases, such as when food is abundant or when the fungi form fruiting bodies, the filaments grow together to produce a solid tissue, or **thallus.**

An individual fungus, composed of a diffuse (fuzzy) network of microscopic threads, can nevertheless, in huge aggregates weigh tons. Thus, because of their distinctive geometry, they are both microscopic and among the largest of organisms on Earth.

> A fungus is "a multinucleate cytoplasmic mass, motile (by cytoplasmic streaming) in a system of tubes" (Langeron, 1945).

Habitat

In aquatic systems (as in terrestrial systems), fungi occur wherever there is sufficient organic substrate for growth. Fungi compete with bacteria and tend to be most successful when the food supply is difficult-to-decompose plant or animal polymers, such as wood or chitin. Fungi are a component of the aufwuchs community and are often parasitic, growing in or on living plants and animals.

Swimming and Escape Behavior

Mycellia are nonmotile. Motile spores occur in the chytrid water molds. Individual slime mold cells can crawl like amoebae. Spores of aquatic fungi are typically motile, powered by one or more flagellae. This motility is at a microscopic scale; wind and water currents are also important dispersal agents of spores at the macro scale.

Feeding Preferences and Behavior

All fungi are heterotrophic, living by absorbing organic compounds from the surrounding environment. Aquatic fungi are **saprobic** (decomposing dead organic material) or symbiotic, living on or in other living organisms (as mutualists or parasites). Freshwater fungi are key players in the degradation and conversion of resistant biopolymers such as the cellulose of wood and dead leaves in aquatic systems (Fryar et al., 2001).

Life Cycle and Diapause

Fungi grow by asexual division and fragmentation of the mycelium. Many forms also produce motile spores that may act as gametes. Hyphae of genetically different individuals of the same species can also fuse, forming a genetically diverse individual.

Economic Importance and Management

Fungi are able to decompose tough, organic molecules such as chitin, lignin, and cellulose (Fryar et al., 2001) and dead, organic material in general. Fungi decomposing dead leaves in streams are an important food source for invertebrate stream insects that shred leaves, such as the immature stages of stone flies, caddis flies, and midges (Paul & Meyer, 1996). Fungi play an important role in the decomposition of plant material in aquatic systems and are sometimes added to sewage assist with decomposition in sewage treatment plants (Jones, 1976). On the other hand, fungi can become a nuisance by growing on wooden timbers in cooling towers or other wood in contact with water.

Fungi are also capable of growing on living organisms, becoming parasites or wound pathogens of living plants or animals. For example, *Saprolegnia* (a "water mold") grows in fish wounds, producing a mass of white threads (Post, 1983). Infection of living organisms tends to occur at high densities of the host population and can actually modify population dynamics of significant aquatic organisms of all kinds. Examples of well-studied effects of fungi on aquatic populations include studies of diatoms (*Asterionella:* Canter & Lund, 1951) and midge larvae (Martin, 1991). It is possible that fungus-infected fish (trout) grown in hatcheries can infect natural amphibian populations when the fish are used for stocking streams (Kiesecker et al., 2001).

SUMMARY

These small organisms—bacteria, protists, and fungi—might be insignificant individually, but in their numbers, they are key players in the limnological theater. These microscopic organisms provide food for many of the larger organisms to be considered in chapters 4 and 5. As such, microbes are essential members of aquatic communities and ecosystems. Microbes are also important living components of most chemical cycles, to be discussed in chapter 9. Morel and Antoine (2002) make a strong case for the importance of unicellular, photosynthetic organisms at the global scale.

Study Guide

Chapter 3 Diversity of Aquatic Organisms: The Single-Celled and Colonial Organisms

Questions

1. What are the ecological relationships of the major groups discussed in this chapter? To answer the question, draw a box-and-arrow diagram for the major groups (including the protistan phyla), showing which groups are photosynthetic (one box can be labeled "sun") and who eats whom. Compare diagrams in class.
2. Limnologists often ask, "What is the dry weight of bacteria in a specific water sample?" Answer this question, given the following information: From counting with the aid of a microscope, you know that there are 100,000 bacteria per milliliter (= cm^3) in a specific sample from Lake Erie. You estimate that the average bacterium in your sample is spherical, with a radius of 0.5 μm. You also estimate that each bacterium is 90% water. Using this information, what is the dry weight of bacteria in your water sample?
3. D'Arcy Thompson remarked that protozoans appear to zoom across the field of vision of a microscope at "freight-train speed." You have probably had the same experience—of just catching a glimpse of a protozoan as it streaks across the field of view. How fast is this protozoan really moving? To answer the question, assume you are viewing a protozoan with a length of 10 μm. The field of view of the microscope is 40 μm across, and the protozoan takes 4 seconds to cross the field. What is the protozoan's velocity in kilometers per hour? These units will allow you to compare the velocity of the protozoan to that of a freight train, which moves at speeds of up to roughly 100 kilometers per hour.
4. What are the essential morphological differences between prokaryotes and eukaryotes?
5. How does endosymbiosis relate to the evolution of eukaryotes?
6. Does the surface-to-volume ratio explain why bacteria use active transport to take up nutrients while larger algae depend on diffusion?

Words Related to Bacteria, Protists, and Fungi

(Note: Important taxonomic names are listed in the next section.)

algae
anaerobic
anatomy
asexual
aufwuchs
autotrophic
axopodia
biomass
carotenoid
chemoheterotroph
chlorophyll
chloroplast
cilium (cilia)
clone
colony
cosmopolitan
diapause
diploid
diversity
endosymbiosis
extremophile
filament
gamete
halophile
haploid
heterotrophic
hyphae
interstices
meiofauna
meiosis
microbe
μm
mitochondria
mitosis
morphology
mycellium
mycology
niche
organelle
parasite
peptidoglycan
periphyton
phagotrophic
photoautotroph
photoheterotroph
phototrophic
plasma membrane
protozoa

pseudopodium	spore	thallus	xanthin
random walk	stromatolite	thermophile	
saprobic	symbiosis	tubulin	
sessile	taxonomy	undulapodium	

Major Examples and Species Names to Know

To speak the language of limnologists, it is necessary to know some common or scientific names of a few frequently encountered or important organisms:

Archaea

Bacteria, especially the cyanobacteria genera *Aphanizomenon* and *Microcystis* and the parasitic *Wolbachia*

Protists:

Amoeba, Centropyxis	Rhizopoda
Ciliates (*Paramecium, Ophridium, Vorticella*)	Ciliophora
Cryptosporidium, Plasmodium	Apicomplexa
Diatoms (*Asterionella, Cyclotella*), *Dinobryon*	Chrysophyta
Dinoflagellates (*Ceratium*)	Dinoflagellata
Flagellates (*Euglena*)	Euglenophyta
Green algae: desmids, *Chlorella*	Chlorophyta
Heliozoans and radiolarians	Actinopoda
Giardia	Zoomastigophora
Rhodomonas	Cryptophyta

What Was a Limnological Contribution of These People?

Lynn Margulis

D'Arcy Wentworth Thompson

Additional Resources

Further Reading

Graham, L. E., and L. W. Wilcox. 2000. *Algae.* Upper Saddle River, NJ: Prentice-Hall.

Koneman. E. W. 2002. *The other end of the microscope: The bacteria tell their own story.* Washington, DC: ASM Press.

Madigan, M. T., J. M. Martinko, and J. Parker. 2000. *Brock biology of microorganisms.* 9th ed. Upper Saddle River. NJ: Prentice-Hall.

McNeill Alexander, R. 1979. *The invertebrates.* Cambridge, UK: Cambridge University Press.

Overbeck, J., and R. J. Chróst. 1994. *Microbial ecology of Lake Plusssee.* Volume 105 of Ecological Studies. New York: Springer-Verlag.

Patterson, D. J. 1998. *Free-living freshwater protozoa: A colour guide.* New York: John Wiley & Sons.

Pennak, R. W. 1989. *Freshwater invertebrates of the United States: Protozoa to Mollusca.* 3rd ed. New York: John Wiley & Sons.

Reid, G. K. 1987. *Pond life: A guide to common plants and animals of North American ponds and lakes.* New York: Golden Press.

Thorp, J. H., and A. P. Covich. 2001. *Ecology & classification of North American freshwater invertebrates.* 2nd ed. San Diego, CA: Academic Press.

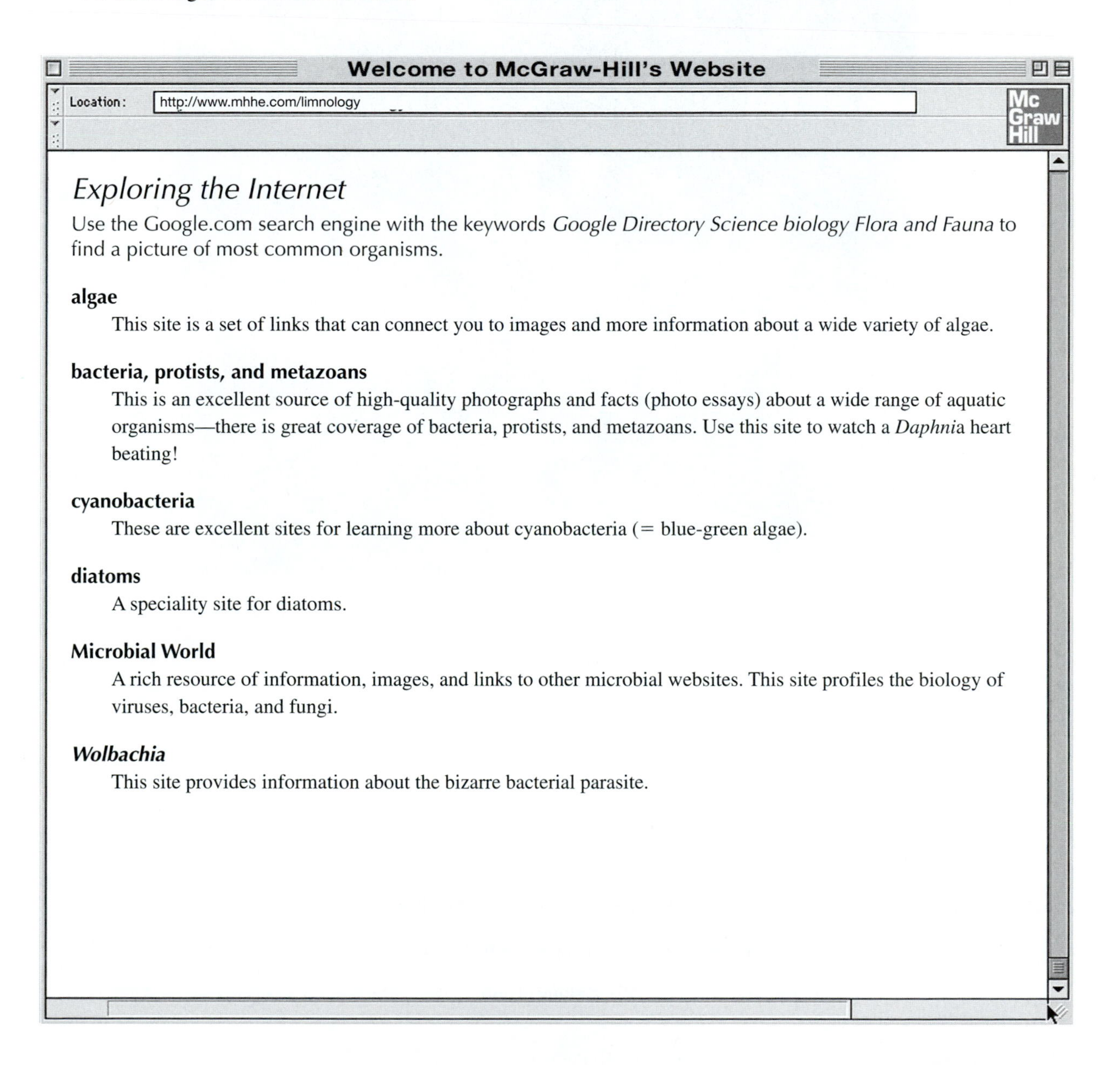

Exploring the Internet

Use the Google.com search engine with the keywords *Google Directory Science biology Flora and Fauna* to find a picture of most common organisms.

algae

This site is a set of links that can connect you to images and more information about a wide variety of algae.

bacteria, protists, and metazoans

This is an excellent source of high-quality photographs and facts (photo essays) about a wide range of aquatic organisms—there is great coverage of bacteria, protists, and metazoans. Use this site to watch a *Daphni*a heart beating!

cyanobacteria

These are excellent sites for learning more about cyanobacteria (= blue-green algae).

diatoms

A speciality site for diatoms.

Microbial World

A rich resource of information, images, and links to other microbial websites. This site profiles the biology of viruses, bacteria, and fungi.

Wolbachia

This site provides information about the bizarre bacterial parasite.

4

"The Nauplius is a wobbly thing, a head without a body;
He flops about with foolish jerks, a regular tom-noddy."

W. Garstang, 1966

Diversity of Aquatic Organisms: Rotifers, Annelids, and Arthropods

THE SMALL INVERTEBRATES

The organisms described in this chapter are small, metazoan animals that are common members of the zooplankton as well as benthic and interstitial, meiofauna communities, or members of the aufwuchs, living on aquatic plants or on mud or stone surfaces. These small animals are easily overlooked—some are visible to the unaided eye only as minute specks. However, because of their numbers, they are of major importance to aquatic ecology. These organisms are the primary consumers of the microbes that were discussed in chapter 3, and they are the favorite food of larger organisms, such as fish (see chapter 5).

ROTIFERS

Taxonomy and Diversity

Rotifers (figure 4.1) are multicelled animals made up of eukaryotic cells (Wallace & Snell, 2001). Cilia patches on the head give the appearance that the head is rotating—hence, the name of the group. Rotifers include the smallest multicellular animals, some no more than 0.1 mm long.

There are two major groups of freshwater rotifers in the Phylum Rotifera:

Class Bdelloidea: Approximately 360 species—all females that practice only **obligate parthenogenetic** (asexual) reproduction and have paired ovaries. The body is often worm-like. Most bdelloids live in association with a surface, creeping along surfaces of plants or stones, or on or in the bottom sediments.

Class Monogononta: Approximately 1600 species—probably all **facultative** parthenogenetic species that produce asexually or sexually. Each female has only a single ovary. Species in this group show a range of morphologies and behaviors, from free-swimming to sessile, and including some creeping forms.

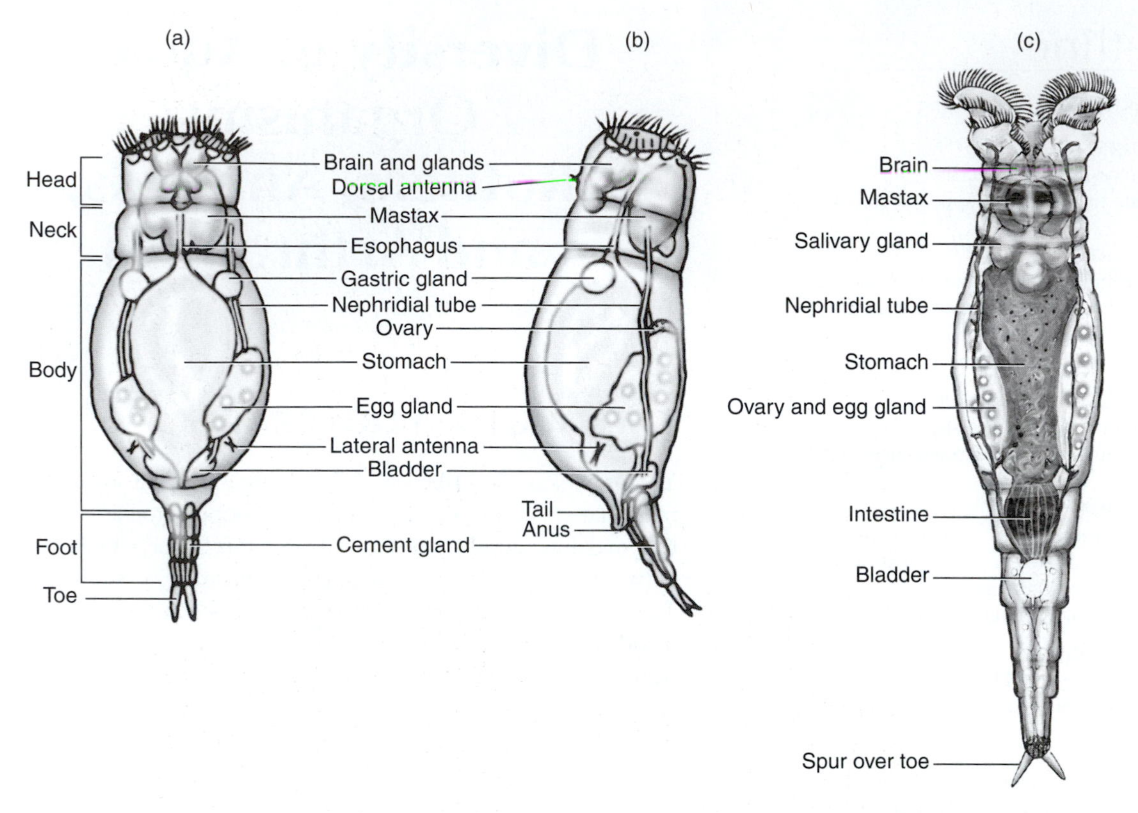

FIGURE 4.1 General rotifer anatomy. (a) and (b) Dorsal and lateral views of a monogonant. (c) Dorsal view of *Philodina,* a bdelloid rotifer. **Sources:** (a, b) From Ward & Whipple, 1966; (c) from Pennak, 1989.

Monogonanta is more diverse than Bdelloidea. The three Monogonata orders are defined mainly by the structure of jaw-like structures. Members of each order share common morphology and life forms. Members of the order Collothecacea are large (several millimeters long) and mostly sessile. The Flosculariacea are smaller sexual animals that make up the majority of planktonic rotifers (individuals or colonies and either sessile or pelagic); members of the Ploimida are sexual species that are mostly benthic and pelagic.

Rotifer species are probably not as cosmopolitan as are the microbe species. While rotifer species might appear (morphologically) to be the same species on different continents, molecular evidence suggests that individual species are restricted to local regions (Gomez et al., 2002).

External Structure, Appearance, and Anatomy

Rotifers are small and transparent. The rotifer *Ascomorpha minima* (figure 4.2a) is the smallest **metazoan** (multicellular organism), with an adult size of 80 μm. The largest pelagic rotifers, in the genus *Asplanchna* (figure 4.2b), are up to 1.5 mm long. Some sessile rotifers may be longer but are more slender than *Asplanchna.*

Rotifers have a fixed number of cells (species-specific, cell constancy), usually about 900 to 1000, which is far fewer than most metazoans. The number of cells is constant. The lineage of each cell in the adult can be traced back to the original diploid egg (Waneczek, 1930). There is no regeneration of tissue or healing of wounds. If a cell is killed during development, a portion of the adult rotifer will simply be missing; other cells have no ability to compensate for accidents during development.

Juvenile rotifers are nearly as large as the adults (the juveniles come from relatively large eggs). Juveniles grow by expanding cells and there is no molting. Males are smaller than females and morphologically simplified.

The rotifer body is covered with an epidermis. In some species, hard plates of protein lie under the epidermis to form a shell called the **lorica.** Usually somewhat flexible, the lorica can be quite stiff in species such as *Brachionus* (figure 4.2c), a rotifer often used as food for

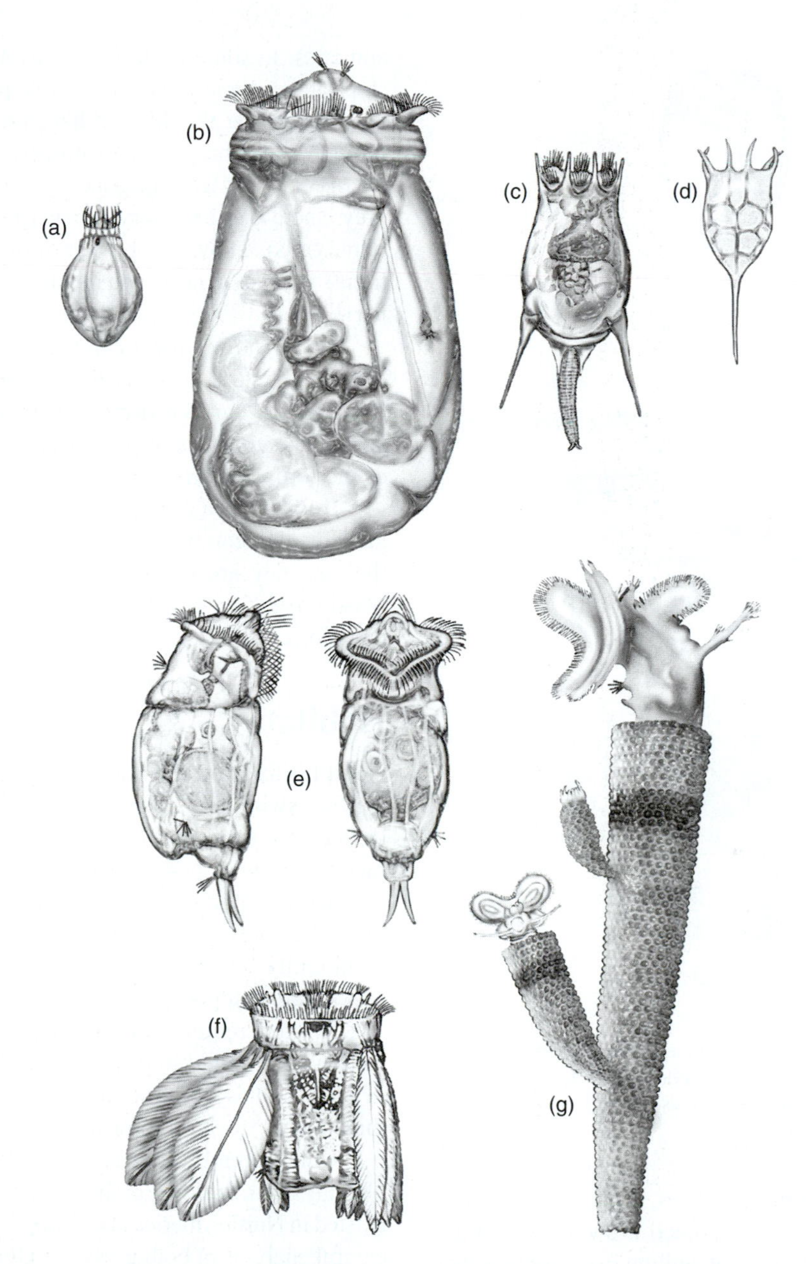

FIGURE 4.2 Examples of monogonant rotifers. (a) *Ascomorpha,* the smallest metazoan, as small as 0.1 mm long. (b) *Asplanchna,* one of the largest rotifers. (c) *Brachionus* with a prominent lorica and foot. (d) *Keratella,* with a lorica, but no foot. (e) *Cephalobdella* in side and ventral views. (f) *Polyarthra,* with its wings. (g) *Floscularia,* a sessile rotifer—adult with two juveniles.

young fish in aquaculture) or *Keratella* (figure 4.2d), a loricate rotifer that lacks a foot.

Some rotifers exhibit phenotypic plasticity—their body form depends on which environmental signals they receive during development. Gilbert (1966) demonstrated that annual changes in morphology seen in *Brachionus* are examples of phenotypic plasticity (also termed **cyclomorphosis**). *Brachionus* collected in winter or spring, or grown in the laboratory, have short spines on the lorica, while their offspring can have long spines if exposed to water from a culture of the predaceous rotifer *Asplanchna* (figure 4.3).

Rotifers show no sign of body segmentation (see figure 4.1). Appendages are not jointed. The appendages are more or less movable spines, modified parts of the lorica, or paddle-like outgrowths of the

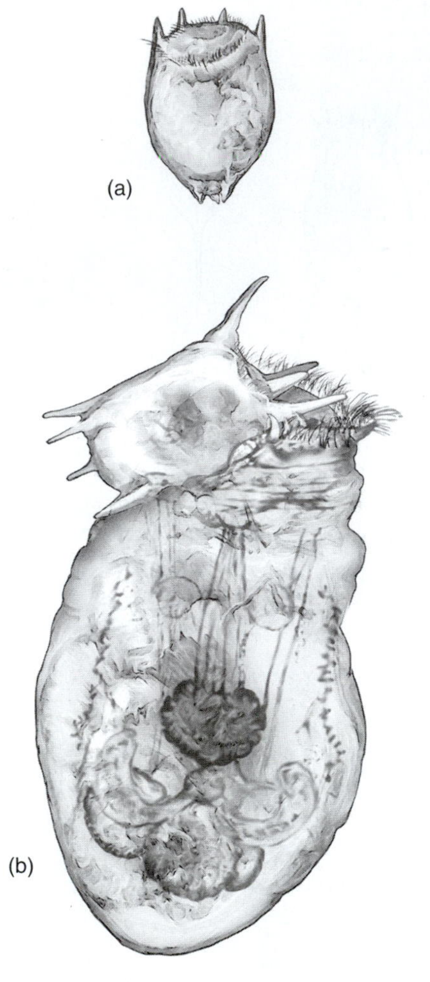

FIGURE 4.3 (a) A juvenile *Brachionus* with short spines. (b) A photograph of the large, predaceous rotifer *Asplanchna* having difficulty eating an adult *Brachionus* with long spines. **Source:** Drawn from Gilbert, 1966.

body wall (Stemberger & Gilbert, 1987). The body is divided into a head, a body (sometimes with swimming appendages), and often a foot, ending in one or two toes and a cement gland. The mouth is on the head; the anus (if present) is at the base of the foot. Cement allows attachment of the rotifer to a surface or attachment of eggs to the foot.

Sensory structures include three or fewer antennae, a simple eye (associated with a nerve ganglion), and a possible chemotactic organelle at the mouth opening.

Internal organs can usually be seen through the cuticle and lorica (see figure 4.1). The organs are typically made of a few number of cells, and always the same number. Organs include a gut (with esophagus, muscular pharynx or **mastax**), stomach and digestive gland, cloaca, and anus. In some rotifers, such as *Asplanchna,* the gut stops at the stomach, so indigestible particles must be regurgitated. The **trophi** (rod-like, intracellular plates of protein in the pharynx) are variously modified (in different species) for tearing, grinding, pumping, or grasping prey. There is an excretory system (protonephridium with flame cells), an ovary (or a pair of ovaries in bdelloid rotifers) with associated glands, and a set of muscle and nerve fibers.

Some rotifers form colonies, which secrete a mucous matrix to hold individuals together. The colonial habit makes small rotifers into large masses, and the larger size gives some protection from predators, such as the rotifer *Asplanchna.*

Rotifers, as a group, are the smallest freshwater organisms that can be seen with the unaided eye. Nevertheless, they are essentially transparent, like a clear, plastic bag. Except for the red eye spot, rotifers tend to be colorless.

Habitat

Adult rotifers live in a range of habitats. Pelagic forms are free-swimming. Sessile species are permanently attached to a substrate. Creeping forms move over surfaces or among sand grains or sediment particles, and are therefore part of the meiofauna. These three groups of rotifers are all important members of the aufwuchs community (described in chapter 3), and a few rotifer species are ectoparasites.

Rotifers require several milligrams of O_2 per liter, so they are limited to oxygenated habitats. They occur in most bodies of static and flowing fresh water, and there are a few marine forms.

Until recently, it was assumed that rotifers were globally cosmopolitan, in the sense that the same species existed in North America and Europe. However, results of careful analyses of both gross morphology and of macromolecules suggest that different species (that are morphologically similar) occur on different continents.

Swimming and Escape Behavior

Rotifers were named in reference to the band of cilia on the head, which appears to rotate. Rotifers swim and capture food particles using the cilia on their head. Swimming speed is generally a slow cruise (ca. 1 body length/sec^{-1}) (Santos-Medrano et al., 2001).

There are many sessile species of rotifer (figure 4.2g). These animals can swim as juveniles, which then glue themselves down and remain attached as adults.

Cephalobdella forficula (figure 4.2e), the subway rotifer, is a free-swimming rotifer that builds a tube out of fecal pellets and mucus and farms bacteria on the inside of the tube. It spends its life swimming back and forth within its tube.

Most of the pelagic forms show some sort of avoidance behavior when they bump into an object. Some species, such as *Polyarthra* (figure 4.2f) and *Hexarthra,* have paddles or arms that allow short jumps, with bursts of up to 50 mm/sec^{-1}.

Keratella is able to increase its swimming speed from .5 mm/sec to 1.8 mm/sec^{-1}, for about 1–2 seconds, using its coronal setae. Escape response is induced by either contact or close approach to *Asplanchna* or *Daphnia.* The signal *Keratella* uses is perhaps a chemical "smell" produced by the predaceous rotifer *Asplanchna.* Alternatively, the signal that induces escape might be a disturbance in the water's flow field caused when a larger organism approaches.

Sessile forms, such as *Floscularia,* build tubes that provide defense against small predators. The walls of the tube are made of fecal pellets (figure 4.2g). Some sessile rotifers can withdraw their body into the tube. Similarly, the loricate rotifers can withdraw their body into the lorica. These withdrawals are a short-range escape response.

Feeding Preferences and Behavior

Rotifers show a wide variety of feeding specialization, as reflected by the structure of the trophi. Trophi can be plates for grinding or forcep-like rods for grasping prey. Species that eat microbes (*Ascomorpha, Floscularia, Brachionus, Keratella, Polyarthra, Synchaeta*) filter very small particles from the water (mostly 1 μm or smaller), using the setae on the head. They can be selective, taking only the most nutritious and least toxic algae. These algae are then ground up in the pharynx, using plate-like trophi.

Some predaceous forms use elongated trophi that can be extended outside the mouth to catch prey. Prey are engulfed whole. For example, *Asplanchna* eat smaller rotifers, such as *Brachionus* (a rotifer also widely used for aquaculture because it grows in both saltwater and fresh water).

Within a lake, rotifers are at highest abundance when bacteria and or picoplankton are most plentiful. Among lakes, species diversity is highest in oligotrophic lakes, but total numbers are greatest in eutrophic waters.

Life Cycle and Diapause

Rotifers develop from a single egg cell into an adult organism with a few thousand cells, which make up distinct tissues and organs. Adult rotifers have a constant number of cells per individual and are not able to regenerate parts that are damaged.

One or a few eggs are produced at a time. These are typically all female, diploid eggs that are genetically identical to the mother. Thus, the offspring of a single adult constitute a genetic clone. The eggs are usually attached to the outside of the female, near her "foot." In some species, eggs develop into young inside the mother's body.

Much development takes place in the egg, which is a significant fraction of the adult mass. The young, when released, grow for a few days only before producing more eggs parthenogenetically (asexually; see figure 4.4).

Males are produced in suboptimal, often crowded conditions. Under such environmental conditions, the female produces haploid, unfertilized eggs. If haploid eggs are not fertilized, they develop into haploid males that produce haploid sperm. If the haploid eggs are fertilized by males, a diploid, resting (diapausing) egg results. Males are smaller than females and usually do not feed—they live on stored energy only to reproduce.

FIGURE 4.4 The life history strategy of rotifers. (a) Parthenogenetic development of eggs. (b) Given the correct environmental signal, the adult female produces haploid eggs instead of diploid eggs. (c) If the haploid eggs are not fertilized, they develop into haploid males. (d) If a male finds a female carrying haploid eggs, and fertilizes them, the result is a diploid egg in diapause. (e) Diapause is broken by the appropriate environmental signal.

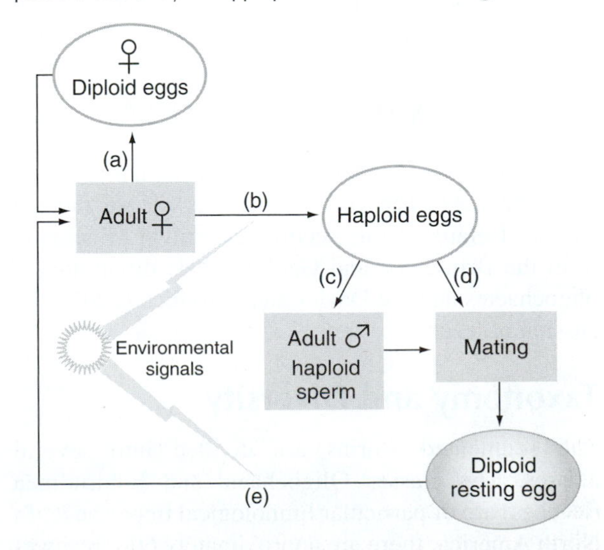

Both sexual reproduction and asexual reproduction have advantages. Sexual reproduction provides increased genetic variability in offspring, as a result of sexual recombination. Asexual reproduction produces only females, which means the rate of reproduction is twice that of the sexual strategy (if there is an even sex ratio).

Rotifer resting eggs (the result of sexual reproduction) enter embryonic diapause. The diapausing embryos are resistant to drying, freezing, low oxygen concentrations, and high temperatures. These resting embryos (often referred to as *resting eggs*) are important in dispersal and in dealing with suboptimal environmental conditions. Diapause is broken by favorable environmental conditions. Bdelloid rotifers display **cryptobiosis**—they dry out and can be revived with water. Cells and organelles shrivel, but the dried-out animal can rehydrate and revive in a few minutes.

Economic Importance and Management

Rotifers are not directly exploited as a natural resource, either for human food or for the aquarium trade. Rotifers (*Brachionus*) are grown as fish food by aquaculturalists. There is no conservation issue related to rotifers.

In productive lakes with abundant populations of size-selective, plankton-eating fish, it is possible that rotifers provide the major energy conduit between bacteria and algae on the one hand, and larger zooplankton or fish on the other hand. The predator-prey interactions of zooplankton are discussed in chapter 7. In chapter 9, we will see that rotifers participate in the "microbial loop," which diverts energy to protists and invertebrates and actually reduces fish productivity in oligotrophic lakes.

ANNELIDA (OLIGOCHAETA)

Several different kinds of worms are abundant in freshwater habitats. The first of these is the segmented worms. Details in this section are drawn principally from the Brinkhurst and Gelder (2001) discussion of oligochaetes and the Davies and Govedich (2001) discussion of leeches.

Taxonomy and Diversity

The segmented worms are divided into several classes. The classes Oligochaeta and Euhirudinea (leeches) are of particular limnological importance. In North America, there are approximately 600 freshwater oligochaete species (Madill et al., 1992) and about 73 leech species (Davies & Govedich, 2001). The three major groups of freshwater oligochaetes are sludge worms (such as *Tubifex*), delicate naidid worms (such as *Aelosoma*), and earthworm-like forms (such as *Lumbriculus*). Species of oligochaetes are typically distinguished by general morphology, anatomical details, and the fine structure of spines (chaetae) occurring in bundles on either side of each segment. Leeches, which have rudimentary chaetae, are identified using general appearance and details of internal anatomy.

External Structure, Appearance, and Anatomy

Worms are elongated animals, much longer than wide. Many invertebrates are worm-shaped, including most of the bdelloid rotifers. The three major kinds of freshwater worms that can be seen with the unaided eye are oligochaetes, nematodes, and dipteran (fly, often midge) larvae (figure 4.5). Freshwater worms are typically small—a few millimeters in length, although some oligochaetes and leeches can be several centimeters long. The three groups of worms (figure 4.5) can easily be distinguished by their gross appearance, especially when they are alive (table 4.1).

FIGURE 4.5 Examples of the different kinds of worms found in fresh water. (a) Segmented worm. (b) nematode. (c) chironomid insect larva, segmented, and with a head capsule and feet. Source: From Merrit and Cummins, 1986.

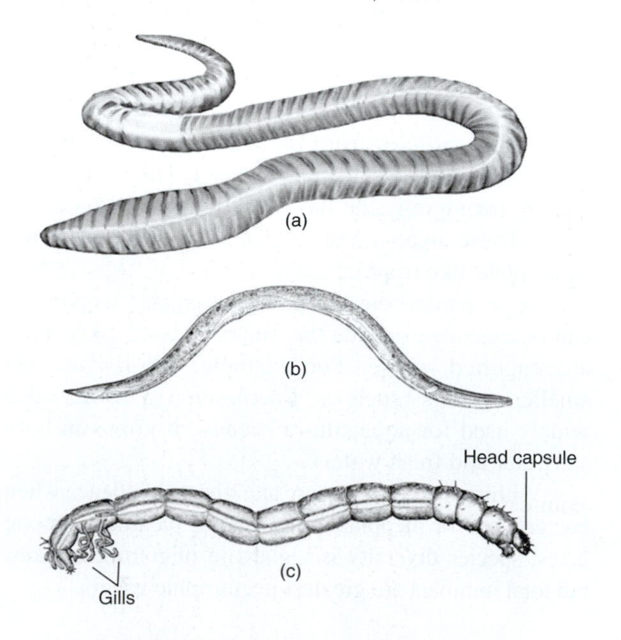

Table 4.1 Distinguishing Characteristics of Freshwater Worms

Use this guide to observe worms with the unaided eye.

Taxonomic Group	Body Appearance During Movement	Head Appearance	Color
Annelida (Oligochaeta; figure 4.5a)	Body lengthens and shortens during crawling; segmentation may be visible	Pointed or blunt, made of same material as the body; not segmented	Opaque to translucent, clear, brown, or black, sometimes red; segmentation may be visible
Nematoda (figure 4.5b)	Body keeps same length; whips back and forth in crescent-shaped arcs; no appendages	Sharply pointed, pin-like; made of same material as the body	White and translucent
Insecta (Diptera; figure 4.5c)	Body keeps same length as it crawls; tends to curl up and extend; stubby legs may be visible near the head and near the back end	Blunt; the hard, shiny head capsule differs from the rest of the body	Many possibilities: clear, translucent, green, brown, black, or bright red

(Summarized from Pennak, 1989).

The body of segmented worms is a long string of connected muscular compartments. Many organ systems (muscular, nervous, excretory) are replicated in most segments. Most segments also possess one or more pair of bundles of stiff spines (**chaetae**). At a gross level, the groups of oligochaetes can be distinguished by size and shape. Tubificid and naidid worms are relatively cylindrical and are less than 1 cm long, while leeches tend to be flattened and adults are often more than 1 cm long (figure 4.6).

Worms living in or near anoxic sediments are often bright red in color due to high concentrations of **hemoglobin,** the red respiratory pigment in the blood.

Habitat

Many oligochaete species move through mud, silt, and organic material, feeding as they burrow (figure 4.6b and c). Other species, such as *Tubifex* (sludge worms), live partly in mud in a vertical tube built of mucus (figure 4.6a). The tail is extended out of the tube above the mud and waved about as the body undulates. The movement circulates water into the tube, assisting in respiration in what is often an anoxic sediment. Meanwhile, the head probes about in the sediment. Worms are restricted to the top few centimeters of bottom sediments because they require access to at least low levels of oxygen.

Leeches attach to surfaces (figure 4.6d) or host organisms, or swim through water or soft sediments (figure 4.6e).

Swimming and Escape Behavior

Except for leeches, oligochaete worms swim rarely and poorly. More often, they burrow through soft sediments or they are effectively sessile. Movement is affected using a **hydrostatic skeleton,** longitudinal muscles, and circular muscles. Contraction of the circular muscles squeezes the internal fluid, causing the body to lengthen. The worm then anchors its anterior end with chaetae and relaxes the circular muscles while contracting the longitudinal muscles, causing the tail to be drawn toward the head. The worm then anchors the tail end and starts the cycle over again by contracting circular muscles.

Most leeches are active swimmers, undulating their body through the water. Leeches also attach to substrate (or host) using suckers at either end of the body (figure 4.6d) and can creep over a surface by attaching the suckers alternately.

The escape response for all oligochaetes is a sudden contraction of the body and a change in direction.

Feeding Preferences and Behavior

Oligochaetes (except leeches) feed by ingesting a mixture of bacteria and dead, decaying, or excreted organic material (**detritus**). These worms constantly swallow mud and digest out the organic material.

Many leeches live at least part of their life attached to some host organism, such as a fish or turtle, from which they suck blood. Some predaceous leech species

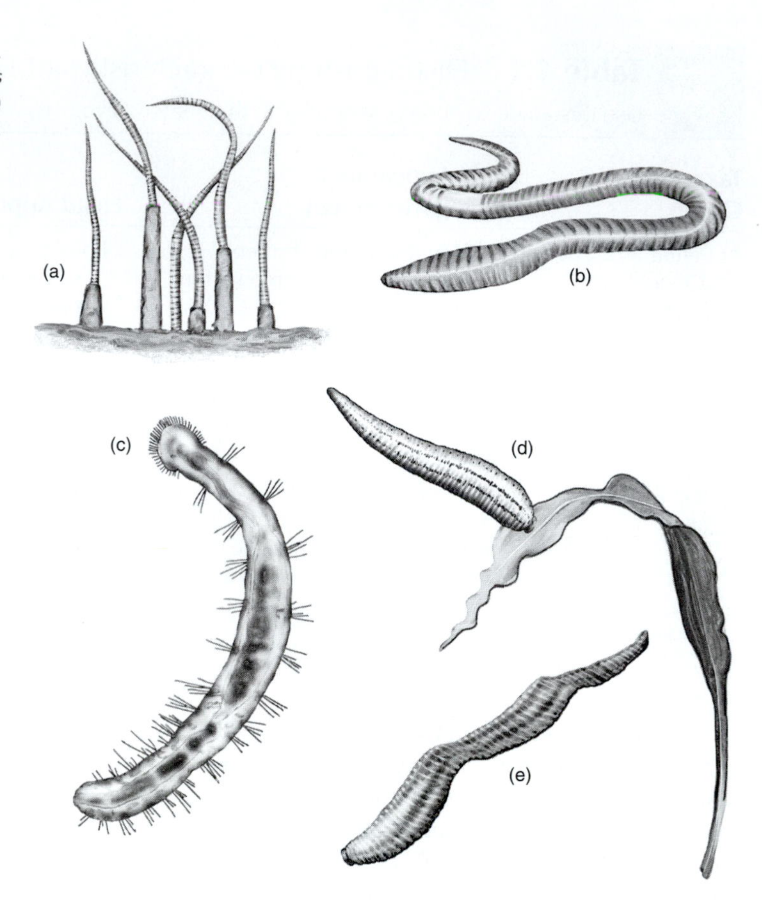

FIGURE 4.6 Diversity of oligochaetes. (a) *Tubifex* (5 mm) in their tubes. (b, c) *Lumbriculus* (2 cm). (c) Aelosoma (5 mm). (d) *Erpobdella* (3 cm) attached to a water weed with its posterior sucker. (e) A swimming *Erpobdella*.

spend part or all of their life swimming through water or in superficial soft sediments, scavenging small, dead animals or ingesting small-animal prey.

Life Cycle and Diapause

Segmented worms are **hermaphroditic**—each individual possesses both male and female organ systems. Self-mating is rare. Typically, an individual will seek out another individual in order to mate. These worms produce one or a few generations per year. Some naidids can reproduce asexually by growing a head in the middle of the body, which then separates into two animals (Learner et al., 1978).

Eggs are typically enclosed in a mucous cocoon (Pennak, 1989), which may allow some protection and facilitate dispersal from place to place and survival for extended periods in moist soil or leaf litter.

Economic Importance and Management

Aquatic oligochaetes occupy a central position in aquatic systems. They are abundant consumers of bacteria and organic material, and constitute a significant component of the diet of fish and waterfowl (Pennak, 1989). Aquatic oligochaetes have long been used in environmental monitoring (Chapman, 2001). These are widely distributed, abundant, and ecologically important animals that have a long history of use in pollution monitoring, environmental assessment, and toxicity testing. Different species sensitivities range from those restricted to the purest water to species tolerant of extreme pollution.

BRANCHIOPODS

These animals are the first of several groups of **arthropods** that are important in freshwater systems. Arthropods are metazoan invertebrates that have a segmented body plan, an external skeleton, and jointed appendages. Crustaceans and insects are arthropods.

Crustaceans, a subphylum of Arthropoda, are characterized by having five pairs of appendages (not counting the compound eyes) on the head: two pairs of **antennae** and three pairs of mouthparts (**mandibles** and the first and second **maxillae**). The thorax has several pairs of legs and, in some groups, there are also pairs of legs on the abdomen. Crustaceans typically have several times as many appendages as do insects, the other major group of aquatic arthropods.

Taxonomy and Diversity

Branchiopods are small crustaceans—most species are in the size range 0.2 to 5 mm. Branchiopods are a major component of the plankton of open water and of the invertebrate fauna that live on aquatic vegetation and in and on sediments in lakes and streams (Dodson & Frey, 2001). They are also called **microcrustacea** or **water fleas.**

Branchiopods (Class Branchiopoda) are a diverse group of about 400 species worldwide (figure 4.7) that are common in freshwater habitats. The species are classified into several orders (table 4.2). The Branchiopod species (and species of all other metazoans discussed in this book) show little or no cosmopolitanism—species are restricted to local regions, not spread over different continents (Frey, 1982). Even though morphologically similar populations can occur on two continents, breeding trials and molecular analyses typically suggest that these are actually different biological species (see chapter 7 for a discussion of species concepts).

External Structure, Appearance, and Anatomy

The arthropods are complex animals with a segmented body plan, a chitinous exoskeleton, and jointed appendages. **Chitin,** a nontoxic, biodegradable polymer of high molecular weight used as a protective body cover, is an animal analogue of the plant structural polymer cellulose. Chitin is made up of a linear chain

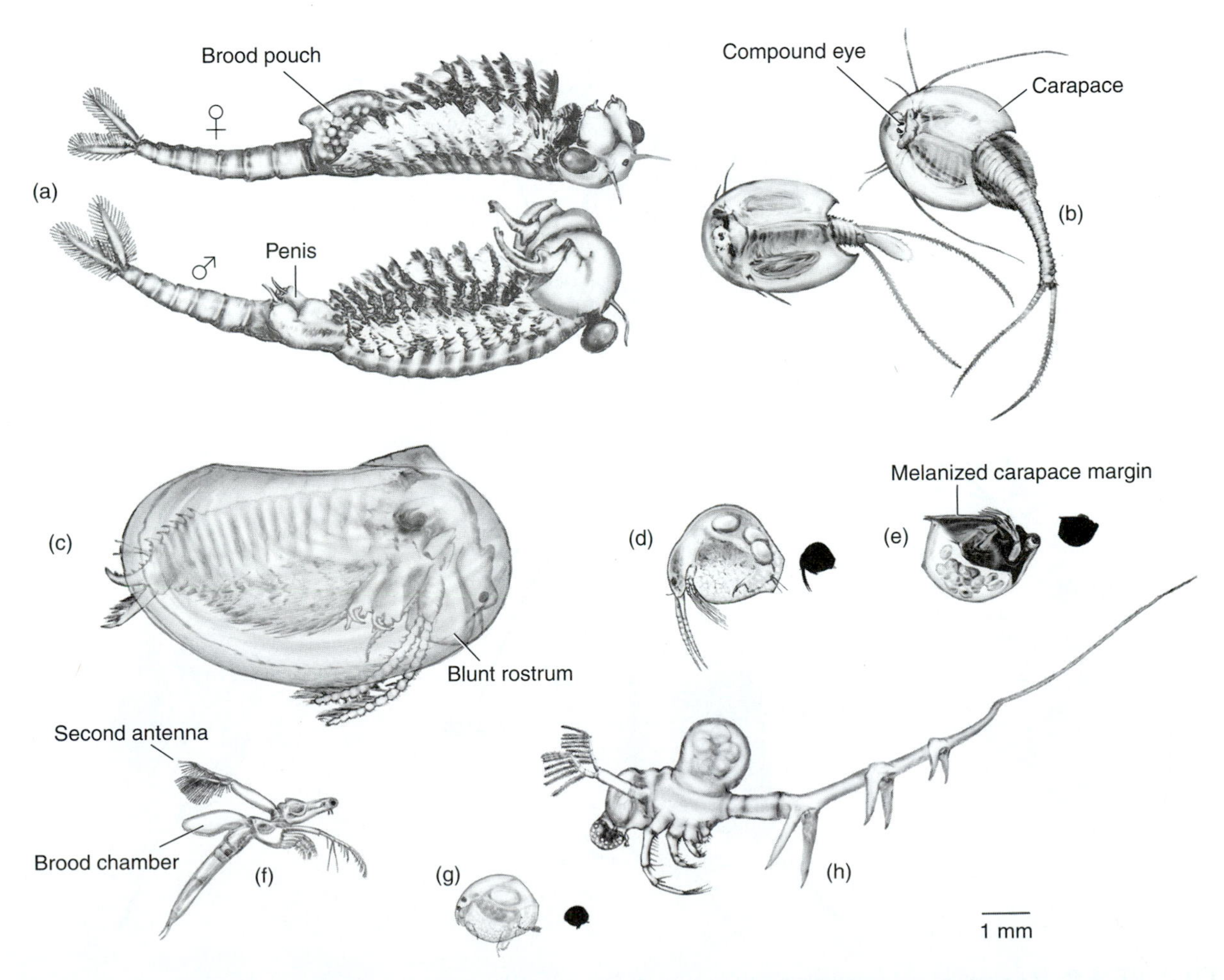

FIGURE 4.7 Examples of branchiopods. (a) A male and female fairy shrimp. (b) Two species of tadpole shrimp. (c) A clam shrimp with the shell (carapace). (d) *Bosmina,* a very small water flea. (e) *Scapholeberis,* a small water flea that swims upside down. (f) *Leptodora,* a large predaceous water flea. (g) *Chydorus,* a small chydorid water flea. (h) *Bythotrephes,* the spiny water flea. All animals are drawn to the same relative size (note size bar) except (d), (e), and (g), whose silhouettes represent their size relative to the others.

of modified, sugar-like molecules called *acetylglucosamine groups.* Arthropods grow by shedding their exoskeleton, a process called **molting,** or **ecdysis.** All cuticular structures shed when the animal molts, including linings of the foregut and hindgut. After the animal molts, it absorbs water to increase its size rapidly and then forms a new exoskeleton.

Crustacean appendages (antennae, mouthparts, and legs) can be branched and often end in a claw or pincher. Theoretically, there is one pair of appendages per segment, although in many branchiopods, external evidence of segmentation is obscured or absent. As in insects, the body is divided into three major regions: **head, thorax,** and **abdomen.** Even the smallest arthropods have, as adults (except in the more extreme parasitic species), a complete set of internal organs, including the neuromuscular, circulatory, reproductive, and digestive systems. These small crustaceans, such as *Daphnia* (figure 4.8), have similar external and internal parts found in larger crustaceans such as lobsters and crabs.

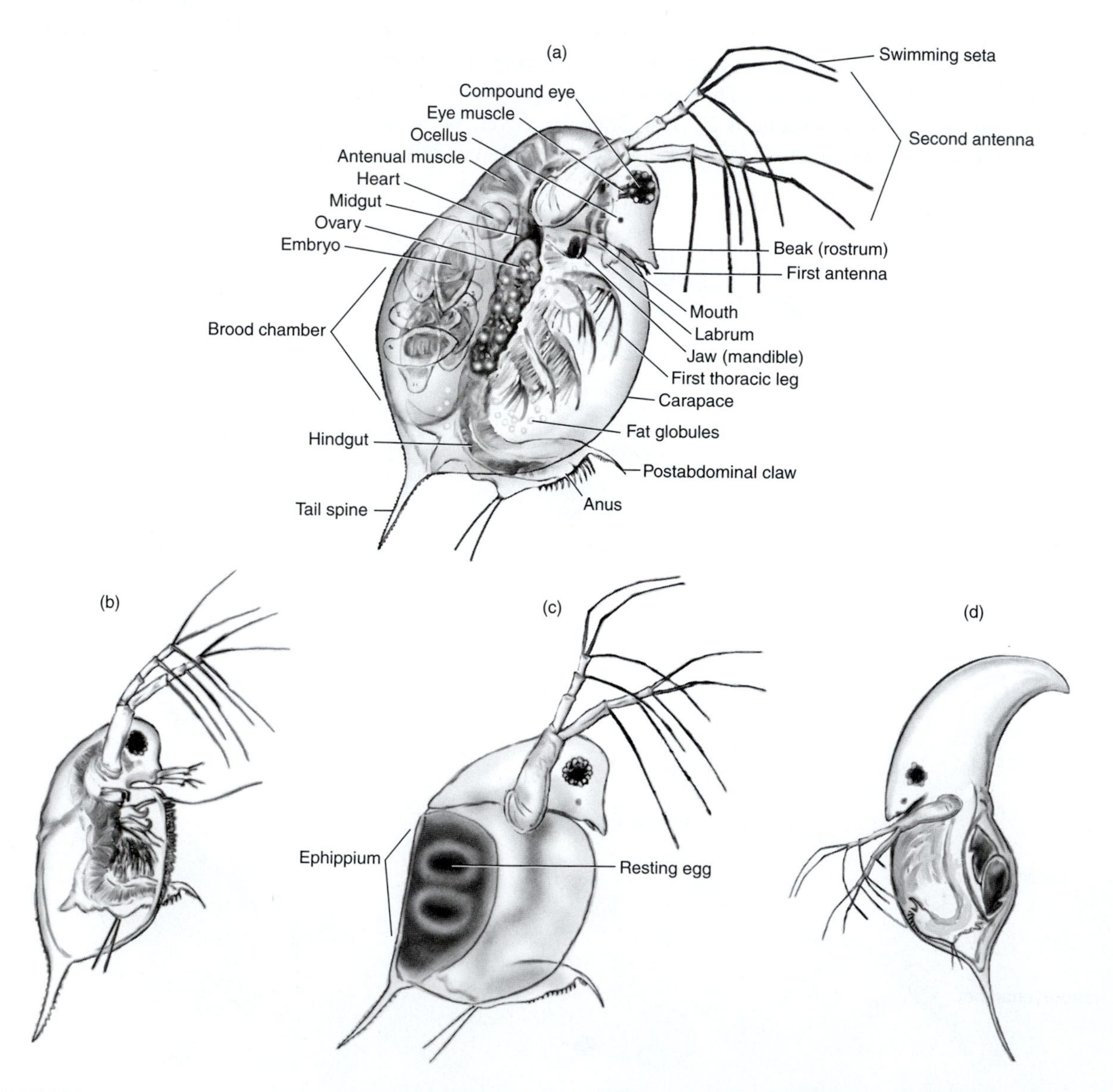

FIGURE 4.8 Anatomy of and morphology of *Daphnia.* (a) General adult female anatomy. (b) A male with long first antennae. (c) An adult female with an ephippium containing two resting eggs. (d) An adult female with an elongated helmet and tail spine.

Branchiopods are small crustaceans (about 0.5 mm to 2 cm long) that typically have compound eyes; flattened, leaf-like legs; and a **carapace,** a shell-like outgrowth of the back of the thorax. The carapace folds over the back to cover part or all of the body. The form of the carapace and the length of the spines are variable from one generation to the next, often becoming longer in the summer generations than in the winter generations (see the discussion of cyclomorphosis in chapter 7).

Branchiopod classification is controversial. At this writing, morphological characters suggest the branchiopods are divided into eight orders, as discussed in Dodson and Frey (2001). This classification is used in table 4.2.

The eight orders of branchiopods have some morphological similarities, but comparisons of taxa using molecular techniques suggest the eight orders are not closely related in an evolutionary sense. For example, tadpole shrimp, fairy shrimp, and water fleas are quite distinct and have probably been so throughout the evolutionary history of arthropods (for at least the last 550 million years—for the big picture, see Gould, 1990). Some of the water flea groups may have separated from conchostracans about 500 million years ago.

The best-known branchiopods are the water fleas (including the genus *Daphnia*). The water fleas, also called *cladocerans,* are classified into four orders (see table 4.2), which have some superficial morphological

Table 4.2 Members of Class Branchiopoda

Order	Common Name	Carapace	Pairs of Thoracic Legs	Typical Diet and Feeding Mode	Approximate Number of Living Species Worldwide
Anostraca (figure 4.7a)	Fairy shrimp	Lacking	11–19	Algae and detritus by scraping or filtration; small prey by grasping	100
Notostraca (figure 4.7b)	Tadpole shrimp	Over head and thorax	34–70	Organic matter by scraping and grubbing through sediments	10
Spinicaudata & Laevicaudata (figure 4.7c)	Clam shrimp ("Chonchostraca")	Over most of body	10–32	Algae by filtration	25
Anomopoda (figures 4.7d, e, g and 4.8a)	Water fleas (including *Daphnia, Moina, Bosmina, Scapholeberis, Macrothrix,* and the **chydorids,** a group of small species such as *Chydorus*	Covers thorax and abdomen	4–6	Algae by scraping or filtration	250
Ctenopoda	Water fleas (including *Sida, Penilia,* and *Holopedium*)	Covers thorax and abdomen	6	Algae by filtration	10
Haplopoda (figure 4.7f)	Water fleas (*Leptodora*)	Reduced to a dorsal brood sack	6	Small zooplankton by predation	1
Onychopoda (figure 4.7h)	Water fleas (including *Polyphemus, Evadne, Bythothrephes,* and *Podon*)	Reduced to a brood chamber	4	Small particles, mostly zooplankton, by predation	15

Based on the classification scheme in Dodson & Frey, 2001.

similarities, such as a reduced number of appendages (no obvious appendages on the abdomen), very small second antennae, and little or no indication of body segmentation. It is controversial whether these orders are (1) only distantly related groups of crustaceans in an evolutionary sense (Fryer, 1987), based on an analysis of morphological characters, or (2) monophyletic, as suggested by an analysis of molecular similarity (Crease & Taylor, 1998).

The carapace (see table 4.2 and figure 4.8) varies in its development in the various groups of water fleas. The maxillary gland exits on the carapace, in the anteriodorsal region. It secretes a glue in some of the water fleas. The chemistry of the carapace is important because cladocerans tend to be **hydrophobic** (they do not get wet).

Branchiopods are usually transparent or a nondescript color. Tadpole shrimp are often olive-green and fairy shrimp can be faintly iridescent. The smaller pelagic species tend to be transparent. If a red color is present, it is due to hemoglobin. Synthesis of this respiratory pigment is induced in many species by low oxygen in the environment (about 2 parts per million [ppm] or less). Hemoglobin is dissolved in the blood (**hemolymph**) of crustaceans. Black pigmentation is due to **melanin,** which is probably induced by exposure to blue or **ultraviolet** light during synthesis of the inner part of the exoskeleton. For example, the water flea *Schapholeberis* lives in shallow water, often feeds on the surface, and is pigmented black. Only the parts of *Schapholeberis* that face the sun are pigmented (figure 4.7e) (Siebeck, 1978).

The black pigmentation probably screens out dangerous solar radiation (Luecke & O'Brien, 1983; Siebeck, 1978). Experiments in shallow enclosures (ca. 3 cm) showed that the dark forms of *Daphnia* from ponds survived significantly better than did the light forms from lakes. In sunlight, the dark form had a significantly higher survivorship, and if the enclosures were covered with a pane of glass, the mortality of the clear forms was reduced, suggesting that the UV component of the light may be a factor (because glass is not particularly transparent to UV light).

Blue, green, and orange colors are due to **carotenoid** pigments (such as carotene). Carotenoids dissolved in fat droplets are orange. Blue and green colors are produced by carotenoids bonded chemically to proteins. These carotenoids are extracted from algae.

In well-oxygenated water, cladocerans tend to be clear, except for carotenoids in fat bodies and eggs (often green or blue), and the black compound eye and sometimes a simple eye, or **ocellus.**

Juvenile branchiopods resemble adults, with a few exceptions. For example, fairy shrimp hatch as a **nauplius,** a stage that has only a few pairs of legs, only one simple eye, and little outward evidence of segmentation. When it is ready to grow, the nauplius sheds its exoskeleton (**molts**). The developmental stages between molts are called **instars.** The first instars are the most simple and, at each subsequent molt, additional segments, spines, and setae are added. In most cladocerans, the juveniles are just simplified versions of the adult. Males are also similar to females, the main differences being that males are smaller and have elongated first antennae (sensory?) and elongated first thoracic legs, used for holding onto females during mating (figure 4.8b).

Habitat

Branchiopods are typically found in fresh water or saline ponds and lakes. Only a few species are marine (*Evadne, Penilia, Podon*), occurring mostly near shore or in **estuaries** (near-shore marine environments characterized by high algal food concentrations). These animals are characteristic of still water, but they do occur in rivers and streams as wash-outs from upstream lakes. Branchiopods occur on all continents except Antarctica. Habitats range from wetlands, to large rivers, to all still water (temporary and permanent), ground water, and water captured in leaves of tropical, epiphytic plants.

Most branchiopods live in either pelagic (open-water) habitats or they move along surfaces of plants or sediments in the **littoral** zone (in shallow water) and in the benthic zone (bottom sediments). Some of the smallest species are part of the meiofauna—small animals (such as rotifers, protists, and branchiopods) 0.05 to 1.0 mm long, living in spaces in sediments.

Most branchiopods require at least 2 ppm oxygen and temperatures between 0°C and about 35°C. Different species have different tolerances to salinity. For example, some of the fairy shrimp (Anostraca) can live in water saturated with salt; other fairy shrimp are found in the nearly distilled water of alpine ponds. The brine shrimp, *Artemia,* a kind of fairy shrimp is typically found in salterns (ponds used to evaporate seawater to produce salt) or desert ponds that become salty as they dry (plates 7 and 8). Branchiopods found in soft-water, oligotrophic lakes (conductivity 50 μS) typically cannot be cultured in hard water (conductivity 450 μS).

Swimming and Escape Behavior

Tadpole shrimp grub smoothly along the muddy bottom of ponds. They move their many thoracic legs back and forth, giving the impression of waves of movement from back to front. Fairy shrimp swim smoothly through the water, using the same leg motions as tadpole shrimp. Pelagic water fleas species have a rowing or hopping motion, the result of the use of the large, second antennae. Species associated with surfaces (the bottom or leaves of aquatic plants) swim in short jerks or creep over the substrate. Some water fleas (e.g., *Sida, Simocephalus*) tend not to swim at all, but glue their back to a substrate.

Swimming speed in water fleas (especially *Daphnia*) depends on the scale at which speed is measured—the measure of swimming speed is scale dependent. This is because the animals hop as they swim, with several hops per second (figure 4.9). If swimming speed is measured at a small time scale (such as over intervals of 0.03 sec), then a greater total distance is measured in 1 second because of the hops. However, if speed is measured with time intervals of 1 or more seconds, the hops are missed and the distance measured will be shorter. Ordinary swimming speed measured at 0.03-sec time intervals is about one to five body lengths per second (Dodson & Ramcharan, 1991).

Many species show **tropisms** (attraction or avoidance) to light, temperature, and perhaps food concentration. In most species, there is a tendency to swim toward dim light and away from bright light.

Some branchiopods, especially fairy shrimp and water fleas, show swarming behavior. It is common to see masses of *Daphnia* (or other zooplankton) along the edge of ponds, in patches of lights, or over contrasting dark or light patches of bottom sediments. The significance of swarming is controversial because of lack of evidence. Swarming may be the result of individual responses to environmental cues, such as a response to a combination of food patches, water currents, water depth, and light intensity. That is, the environment may concentrate individuals that have no social interaction. Alternatively, some evidence suggests that the swarms may be adaptations related to predator avoidance or mating. While branchiopods probably do not use pheromones (sexual chemical signals), males do have swimming behaviors that tend to concentrate them in the vicinity of females (Brewer, 1998).

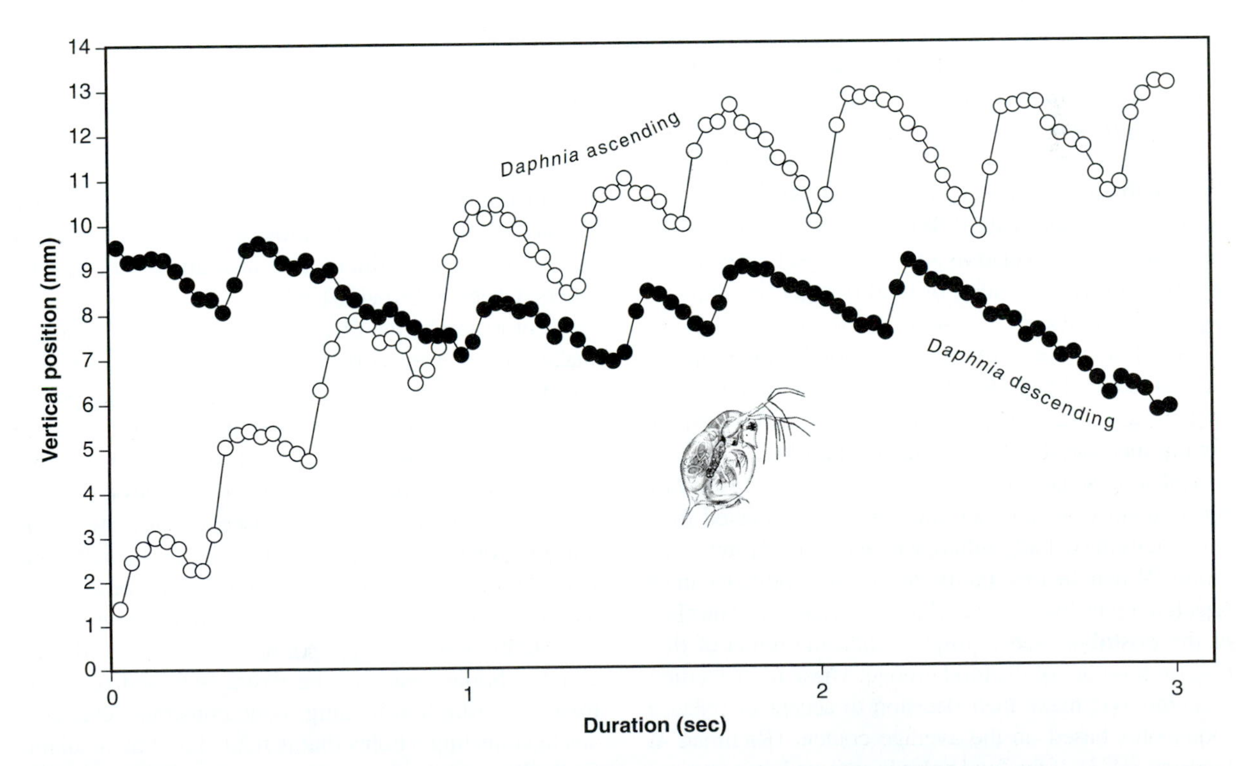

FIGURE 4.9 The movement of a *Daphnia* measured in time intervals of 0.03 sec. The animal is hopping. Source: Data from Dodson & Ramcharan, 1991.

Brachiopods have well-developed escape responses. The free-swimming branchiopods show a rapid change in motion and/or looping motion when disturbed by touch, light, or temperature. Cladocerans can achieve maximum speeds of about 2 cm/sec^{-1} (Kerfoot et al., 1980).

The small water flea *Bosmina* shows a *dead-man's response* (Kerfoot, 1978). When attacked by a copepod predator, the *Bosmina* stops feeding and swimming and begins passive sinking. This means it stops making any noise (vibration) and disappears from the predator's sensory field. The *Bosmina* sinks for a few seconds, often avoiding capture, and then resumes feeding and swimming.

Feeding Preferences and Behavior

Most pelagic branchiopods are filter feeders (see table 4.2), using their thoracic legs to filter particles out of the water (figure 4.10). Their diet includes algae, the occasional small rotifer, bacteria, and detritus filtered out of the water (see Dodson & Frey, 1991; Fryer, 1987; Lampert, 1994). Preferred particles are from less than 1 μm to about 25 μm, although larger species filter larger particles as well. The filtering mechanism is relatively insensitive to food concentration; it runs at about the same speed over a wide range of food concentrations.

In *Daphnia* (see figure 4.8a), thoracic legs flip back and forth several times per second, moving water through the space between the third and fourth legs (food groove) (Gerritsen et al., 1988). The first two legs and setae on the carapace margin are used to eject large or distasteful particles. Particles of food (mostly bacteria and algae) stick to the long setae of legs three and four (figure 4.10), due to electrostatic attraction between the setae and the food particles. These setae resemble feathers by having short, spine-like extensions (**setules**), which greatly increase the surface area of the filter comb (the row of long setae; figure 4.10). As the legs move, they brush against one another and comb the collected food and mucus into a ball (**bolus**), which is moved toward the mouth. When the food bolus reaches the mouth, it can be rejected if it is distasteful or if the gut is too full. The claw of the **postabdomen** is used to flick the bolus of distasteful food out of the food groove. These filter-feeding branchiopods make their decision to accept or reject a food bolus based on the average content (Richman & Dodson, 1983). If the food bolus is accepted, it is chewed by the mandibles and swallowed. Because of the small

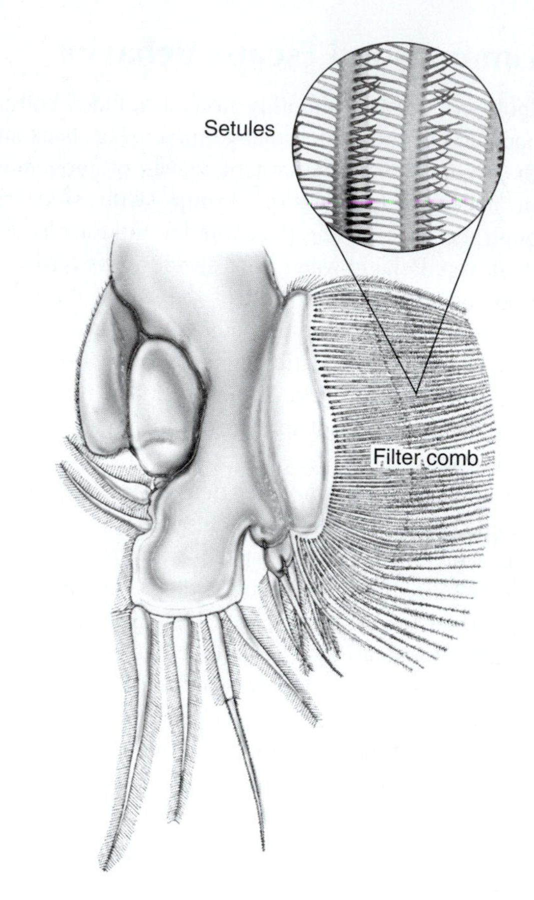

FIGURE 4.10 Filtering leg of a *Daphnia*. Source: Based on Lilljeborg, 1900

size of the filters and the rapid movement of the legs, Reynolds number considerations suggest that these legs cannot act as sieve filters, but must act as electrostatic filters (Cheer and Koehl, 1987).

For example, the *Daphnia* third thoracic leg might make one sweep every 0.1 sec, through a distance of 0.01 cm. Thus, its velocity is about 0.1 cm/sec^{-1}. The "pores" (spaces between the feathery setules on each filter seta) are about 0.001 cm in diameter. The Reynolds number for these pores is the width of the pores times the velocity, divided by the kinematic viscosity of water (0.01 cm^2/sec), or Re = 0.01 (see chapter 2). This small value indicates that viscous forces are dominating the flow and that little water is likely to move through the pores.

Herbivorous zooplankton face many dietary choices. Some algae are the wrong size and some are toxic. The filtration feeding mode constrains cladocerans to collecting a bolus that is made up of all available food. When toxic algae are abundant, cladocerans will reject each bolus as they collect it. They can therefore face food limitation, even in productive lakes. In their study of Lake Constance, Hairston et al. (1999) showed that *Daphnia* adapt rapidly to changes in algal composition. Toxic cyanobacteria appeared in Lake Constance about 1960. The researchers collected resting eggs from the sediments, dated the sediments using radioactive techniques, and then hatched the eggs. *Daphnia* clones from eggs produced about 1960 thrived on green algae, but grew poorly when fed a mixture of green algae and cyanobacteria. *Daphnia* clones from the late 1970s and mid-1990s grew significantly better than the older clones when fed the mixture of algae and cyanobacteria.

Several kinds of branchiopods, including some fairy shrimp and some water fleas, are predaceous, capturing small zooplankton (protozoans or immature crustaceans) or animals living in sediment. These predators lack the long filter combs typical of the herbivores and instead have long, spine-like setae on their thoracic legs. These legs are used for grasping prey. Together, the spiny, thoracic legs resemble a basket that can be opened and closed to hold prey (see figure 4.7f and h). Even the typical herbivores, such as *Daphnia,* will occasionally eat protozoans and small rotifers (Jack & Gilbert, 1994).

Many of the branchiopods, including species that are primarily filter feeders, can be seen scraping away at surfaces or digging in superficial sediments. Tadpole shrimp are specialists at scavenging in mud, but many fairy shrimp and water fleas also scrape or grub.

Life Cycle and Diapause

Asexual reproduction is very common in branchiopods. Tadpole shrimp can be hermaphroditic or parthenogenic (asexual). Fairy shrimp are mostly obligate sexual, but there is at least one asexual species.

Conchostracans and water fleas tend to be either asexual or show facultative sexual reproduction (figure 4.11). Females are present for most of the year, and reproduction is asexual as long as environmental conditions are close to optimal. In asexual reproduction, mothers produce offspring from eggs that develop immediately into daughters (eggs that develop immediately, without a diapause, are called **subitaneous**). In *Moina* (a water flea found in temporary ponds), the developing subitaneous eggs are nurtured by a placenta-like function of the inner wall of the carapace. Because *Moina* embryos depend on this energy from the mother, they will not develop outside the brood chamber. Most water fleas have embryos that can develop even if removed from the brood chamber

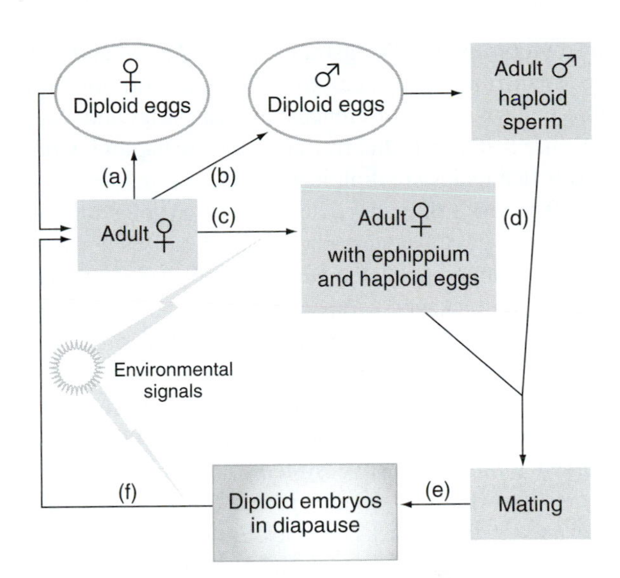

FIGURE 4.11 The life history strategy typical of many conchostracans and cladocerans, including *Daphnia.* (a) Parthenogenetic reproduction. (b) Parthenogenetic reproduction. (c) A developmental change induced by low food? (d) Males mature and locate females. (e) The result of mating is a diploid egg. (f) The diapausing egg begins developing when the appropriate environmental signals are received.

because all the energy necessary for development is already stored in the egg.

For many water flea species, such as *Daphnia,* reproductive strategy depends on the environmental signals the adult receives. These signals (such as day length and chemical signals related to crowding) are correlated with suboptimal environmental conditions. The signals often precede changes in the environment, which spell the end of the growth season (figure 4.11; see Kleiven et al., 1992). The response to the environmental signals is to begin sexual reproduction, which will result in resistant eggs that are able to withstand extreme environmental conditions. In sexual reproduction, diploid males are produced first (asexually). Adult females molt to produce a modified carapace morphology. The signals initiating these morphological changes are not completely understood, but include a sudden change in food concentration.

The carapace over the brood chamber is darkened with melanin and thickens, with one or two egg-holding bulges. The female produces haploid eggs, which lie in the bulges. The diploid males produce haploid sperm when they mature. The males find the females and mate. The males hold onto their mate's carapace, insert the postabdomen into the brood chamber, and release haploid sperm. (The spermatozoa of cladocerans are bizarre in shape, somewhat like heliozoans, with rods in the cytoplasm; see Wingstrand, 1978.) *Moina* and *Sida* spermatozoa display the most extreme morphologies. (Note that a major difference between rotifer and water flea life histories is that rotifer males are haploid, while water flea males are diploid.)

The result of mating is one or two **zygotes** (diploid fertilized eggs) that start development and then shortly go into diapause (called *resting eggs*). When the female molts, the thickened carapace (holding the two resting eggs) closes over the diapausing embryos, acting as a protective cover. The thickened carapace is called an ***ephippium*** (figure 4.8c), a Greek word meaning "saddle." The ephippium can survive being dried, frozen, or passed through digestive systems. Laboratory experiments show resting eggs are viable after passing through the guts of trout, ducks, a black-crowned night heron, and laboratory rats. The resistance and hatching rate of ephippia depend on several factors, including the mother's genotype, the age of the egg, and environmental conditions such as temperature, photoperiod, and oxygen concentration (Cáceres & Schwalbach, 2001).

The black coloration (melanin) of the ephippium attracts visual predators such as fish, and thereby increases the probability of dispersal (Mellors, 1975)—fish eat ephippial adults more readily than adult *Daphnia* with light-colored ephippia or subitaneous eggs.

Diapausing resting eggs are produced by most branchiopod groups, although they are not necessarily protected in an ephippium. Fairy and tadpole shrimp simply broadcast the resting eggs.

Diapause can last for days or decades, until the resting eggs receive the proper signal to resume active metabolism and continue development. Diapause is broken by an appropriate environmental signal—for example, exposure to freezing and total darkness for a couple of days, followed by warm temperatures and long day length, or by exposure to constant light. In water fleas, resting eggs usually produce females; a notable exception is that resting eggs of *Moina* (adapted to temporary ponds with short growing periods) produce both males and females.

Tadpole shrimp and fairy shrimp hatch in the form of a **nauplius**—developmental stages that have only a few segments and poorly developed appendages. The tadpole shrimp nauplius is quite similar to the copepod nauplius. The nauplii molt several times and go through a **metamorphosis** (a radical morphological change) to become juveniles, which resemble the adults. Several additional molts are necessary as the juvenile grows into an adult.

Except for *Leptodora,* the naupliar stage in water fleas is hidden during development of the embryo. When the water flea female molts, she releases the **neonates** (newborn), which resemble adults. There are usually four immature instars, although other numbers can occur. The instar stage before adulthood exhibits developing eggs in the ovaries. Adults usually molt several times, with a clutch produced each instar if food and temperature permit.

Economic Importance and Management

Branchiopods often provide a major energy conduit between the base of the food chain (bacteria and algae) and fish. As such, they are of great, indirect economic importance. Branchiopods are also of direct economic value in the aquaculture business. Resting eggs of brine shrimp are collected in dump-truck quantities at salt lakes, so that the nauplii can be used to nurture larval fish and macrocrustaceans in aquaculture. Branchiopods are sometimes collected from nature to feed fish in aquaculture or the aquarium trade. Dried *Artemia* and *Daphnia* can be bought in most pet stores.

Branchiopods are not significant from the perspectives of human health and conservation. Branchiopods are not known to transmit disease or parasites to humans with the possible exception of a study that suggests *Daphnia* may be a source of cholera in drinking water (Chiavelli et al., 2001). Some of the larger branchiopods (fairy shrimp and tadpole shrimp) are officially recognized as endangered species. These species have restricted distributions (as small as a single pond) and are typically found in ephemeral ponds in dry climates. The species and their habitats are protected by the U.S. Federal Endangered Species Act (administered by the U.S. Fish and Wildlife Service; see the end of the chapter for a web site address). There are no officially recognized endangered water flea species.

COPEPODS

Taxonomy and Diversity

Like the branchiopods, copepods are microcrustaceans that are important members of freshwater communities (Williamson & Reid, 2001). Copepods are shrimp-like crustaceans with spiny legs, no carapace, simple eyes, and several pairs of thoracic legs (figure 4.12).

Damkaer (2002) provides an excellent overview of the study of copepod taxonomy and morphology, in a historical context, in addition to a treasure trove of bibliographic information about early copepod scientists.

There are more than 10,000 species of copepods. Most live in marine habitats. Copepods are one of the most abundant kinds of zooplankton in freshwater lakes. There is a total of 10 copepod orders, but only three orders of mostly free-living copepods are common in fresh water: Calanoida (600 species), Cyclopoida (490 species), and Harpacticoida (300 species).

External Structure, Appearance, and Anatomy

Copepods have a segmented body covered with a jointed, chitinous exoskeleton. Copepods start life as a zygote, formed as the egg (oocyte) is pushed out of the body. The egg meets stored sperm, one of which fertilizes the egg. Fertilized eggs are either broadcast into the water or carried in sacs attached to the adult female. The zygotes begin cleaving immediately. The structures commonly called *egg sacs* are actually embryo sacs, and what are commonly called *resting eggs* are actually diapausing embryos.

All copepods hatch from the egg membranes as a larval form called a nauplius (figure 4.13). The nauplius that emerges from the eggshell typically starts with three pairs of appendages: the first and second antennae and the mandibles. The nauplius has a short, rounded, segmented body with a mouth, anus, and simple gut, and muscles to make the appendages move. Copepods sense light with a single simple eye (ocellus) and lack the complex compound eyes found in branchiopods and insects.

Adult free-living copepods are characterized by an elongated, shrimp-like body (see figure 4.12; plate 23). Calanoids are typically elongated and torpedo-shaped; cyclopoids are typically teardrop-shaped. Harpacticoids are relatively long and slender.

The copepod head has long first antennae that are covered with receptors for chemicals and motion (Strickler & Bal, 1973). Behind the first antennae are

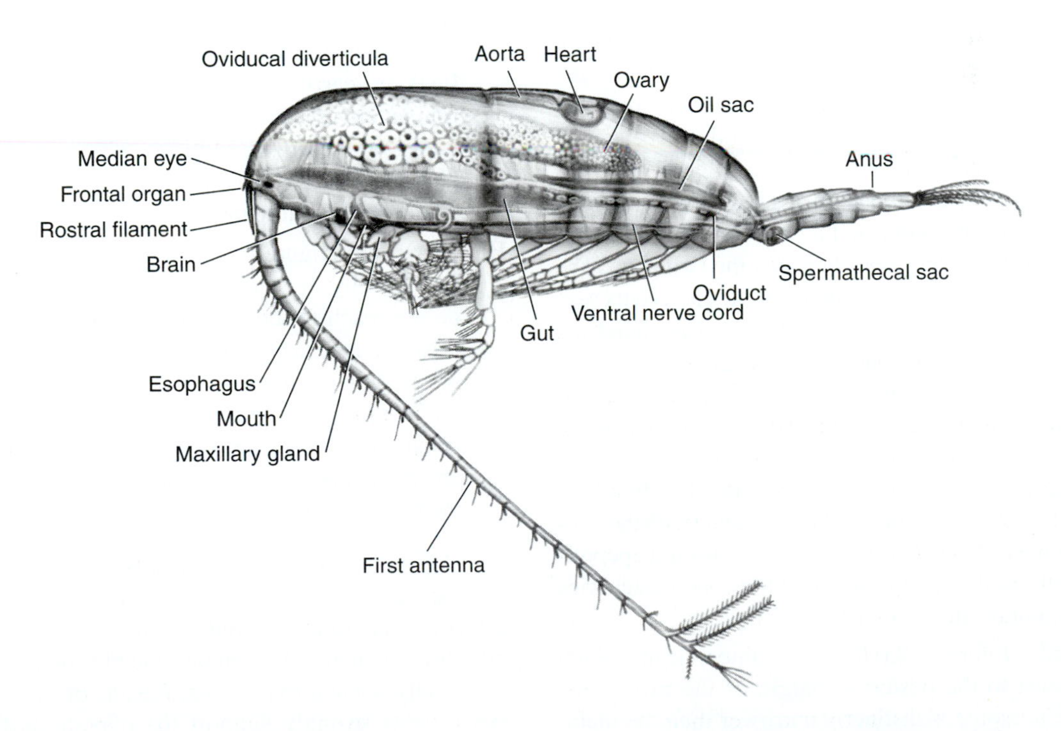

FIGURE 4.12 The anatomy of a copepod. Notice the long first antenna. Source: From Marshall & Orr, 1972.

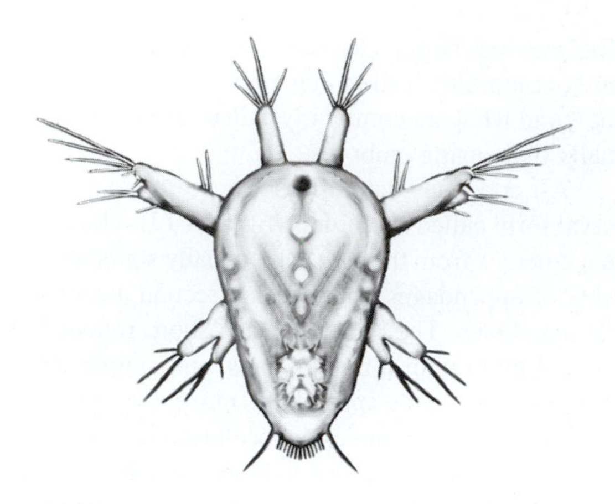

FIGURE 4.13 The copepod nauplius. **Source:** From Alexander, 1979.

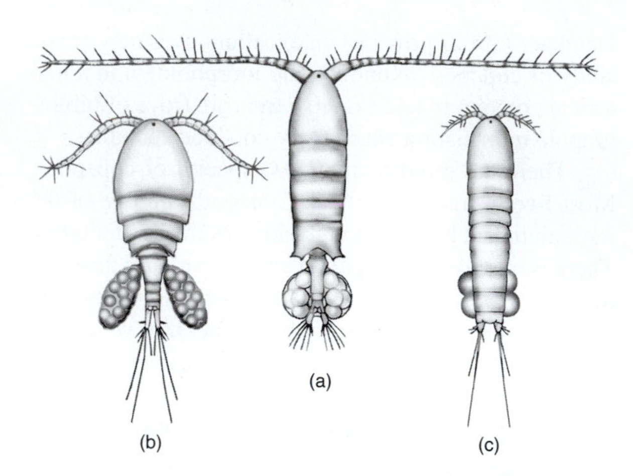

FIGURE 4.14 External forms of three copepod orders. (a) Calanoida. (b) Cyclopoida. (c) Harpacticoida. **Source:** From Ward & Whipple, 1966.

several pairs of distinct mouthparts (second antennae, mandible, first and second maxillae, and a well-developed **maxilliped**). The thorax has several (about five) pairs of legs. The first four thoracic legs are distinctly jointed, usually branched, and ornamented with setae, but lack the filter combs characteristic of branchiopods. The fifth and even sixth pair of legs are much reduced to one to three segments and a few setae or spines. The abdomen lacks legs, but there is a pair of small, posterior, terminal segments (the **furca** = branched part) with long setae. The long setae are probably important for swimming.

The three groups of common freshwater copepods can be distinguished by their general body shape and by the way the females carry eggs (actually embryos) (figure 4.14). The copepod body in all three orders is divided into an anterior **prosome** and a posterior **urosome** with a distinct articulation (joint) between the two parts. The most anterior "segment" of the prosome represents several (eight) segments fused together, and is called the **cephalothorax** (or cephalosome or cephalic segment); the segments between the cephalothorax and the articulation are called the **metasome.** The metasome is often flared out into sharp points, or "wings" (metasomal wings), especially in calanoid copepods. The first segment of the urosome is the genital segment (with the genital openings). Note that the body divisions in copepods do not fall easily into the standard arthropod categories of head, thorax, and abdomen.

Calanoida (figure 4.14a) have first antennae that reach at least to the posterior margin of the metasome. The urosome is distinctly narrower than the metasome. Eggs are carried in a single mass. The right, male first antenna is modified for holding onto the female during mating—it is flattened, with enlarged segments and spines and a joint near the tip. Mouthparts are complex (highly segmented and covered with spines and setae) and used for capturing particles one at a time.

Cyclopoida (figure 4.14b) have first antennae that reach to about the end of the cephalothorax. The urosome is distinctly narrower than the metasome. Eggs are carried in two lateral masses. The male first antennae are modified (on both sides), with a joint near the tip. Mouthparts, which are less complex than those of calanoids, are used for scraping and capturing large particles.

Harpacticoida (figure 4.14c) have first antennae that are typically shorter that those of cyclopoids—they do not reach to the end of the cephalothorax. The urosome is nearly as wide as the metasome. Eggs are carried in a single mass. Male first antennae are not obviously modified, but sexual dimorphism occurs in several pairs of appendages. Mouthparts are relatively simple (though still distinctly segmented) and are used for scraping surfaces.

Copepods, especially calanoids, are often brightly colored: red, orange, blue, or green (plate 23). These colors are due to carotenoids (from algae) that are red to yellow when dissolved in fats and blue or green when chemically bonded to proteins. Results of lab and field experiments strongly support the hypothesis that the pigments have two important functions: The pigments confer **photoprotection** against damaging blue and ultraviolet light but they also attract the attention of predaceous fish (Luecke & O'Brien, 1983). This is an excellent example of a **trade-off** between two characteristics.

Cyclopoids and harpacticoids tend to show less pigmentation than do calanoids, being often rather colorless. The cyclopoid species *Homocyclops ater* is conspicuously colored dark blue-green. It lives in littoral areas and, perhaps, mimics aquatic mites, which live in the same habitat. *H. ater* swims unlike other copepods (smoothly, with movement not punctuated with jumps) and rather more like mites. These mites are distasteful and fish avoid them (Kerfoot et al., 1980).

The copepod eye is often pigmented. The visual pigment is invisible (or a faint purple), but the masking (photoprotection) pigments associated with the eye are often bright red.

Habitat

Cyclopoid and calanoid copepods are diverse members of the freshwater and marine zooplankton. In freshwater habitats, they are often as ecologically important, diverse, and abundant as branchiopods, the other major group of arthropod zooplankton. In marine plankton, they are much more important than branchiopods and often are the dominant planktonic animals. Marshall and Orr (1972) suggest that a common marine calanoid copepod, *Calanus finmarchicus,* is perhaps the most common, nonparasitic metazoan organism on Earth.

Many species of calanoid copepods prefer the pelagic zone, although some species rest on aquatic vegetation. Cyclopoid species occur in all possible aquatic and moist, terrestrial habitats. This includes in open water, among aquatic vegetation, within interstices in benthic sediments (including subsurface groundwater), in moist soil and vegetation, and in small pools of water in terrestrial plants such as bromeliads. Species have been found in permanent and ephemeral lakes and streams, wetlands, subterranean water, water trapped in plant leaves (**phytotelmata**), humid soils, leaf litter, and city water supplies. Diapausing juveniles of several species of copepods can persist for months in dry soil of dried, temporary pools. If the soil is rehydrated, the copepods emerge within a week or so. Harpacticoid copepods tend to live on surfaces or in substrate: Favorite habitats are the benthic, littoral, and interstitial zones.

Swimming and Escape Behavior

Copepods exhibit a wide range of swimming behaviors. They are typically more agile and faster than branchiopods of the same size.

Harpacticoids have perhaps the least-interesting swimming behavior. They typically creep or slither along surfaces and show little in the way of an escape response.

Calanoid and cyclopoid copepods perform truly impressive escape responses. Jumps are accomplished by slapping the first antennae against the body as the thoracic legs are whipped backwards. Calanoid and cyclopoid copepods can reach speeds of 35 cm/sec^{-1} for short bursts (Kerfoot et al., 1980) and have reaction times of a few milliseconds (Lenz & Hartline, 1999). Fish often fail to capture calanoid and cyclopoid copepods, but the slower-moving cladocerans are easy prey (Drenner et al., 1978). The copepod escape response is costly, using large amounts of stored energy.

When not escaping, calanoids move slowly through the water for minutes at a time. They swim using the second antennae, which are also used to create a feeding and an anti-sinking current. Calanoids swim on their back, holding the first antennae at right angles to their body (figure 4.14) and pointing the thoracic legs toward the head. Calanoids can also hang suspended in open water, balancing their feeding current against the pull of gravity.

Cyclopoid species swim or hop with their back upward. They jump more often than do calanoids—every few seconds. Because of the water's viscosity, copepods swim with the same movements they would use to move through mud. Cyclopoid copepods do not swim in the same way larger animals do, by forcing water backwards. Instead, they "walk through water" (Strickler, 1984). Many cyclopoid species are loosely associated with the bottom or substrates (aquatic vegetation).

Like other zooplankton, copepods do not swim at random. They show strong reactions (tropisms) to light and currents. Predaceous copepods occasionally display a small, looping pattern of swimming (described in the Branchiopod section in reference to *Bosmina* escape behavior). There is some evidence that male planktonic, calanoid copepods can follow pheromone trails left by females of their species.

Feeding Preferences and Behavior

All copepod nauplii eat small particles such as algae, protozoans, and bacteria. As the animals grow and develop, their feeding behavior often becomes more specialized. Adult copepods, depending on the species, feed by filtration, grasping, or scraping. Most copepods are more or less omnivorous, specializing more on particle size and taste than on the source (plant, animal, protist, bacteria) of the particle.

Filter-feeding mode is developed to the greatest extent in calanoid copepods. This is not filter feeding in the same sense as the behavior shown by branchiopods—it is instead a kind of micrograsping behavior. The functional morphologist Professor Mimi Koehl has shown mathematically that combs of hairs and setae act more like paddles than like sieves (Cheer and Koehl, 1987). Professor Rudi Strickler has provided beautiful movies of the calanoid feeding mechanism (the movies are not easily available, but see his web site listed at the end of the chapter; see also Doall et al., 2002). Calanoids use their second antennae to create a current of water that flows past the mouthparts. The current first passes over the first antennae sensory pits and setae. The copepod uses chemical signals to time the opening of its mouthparts and is able to capture (catch and grasp) individual particles of desired food particles. The mouthparts look and act something like a miniature baseball catcher's mitt. Once an individual food particle is captured, the particle is held and the surrounding (viscous) water is pressed through the "fingers" and setae of the mouthparts. The smaller calanoids, such as *Diaptomus* species up to 1.5 mm long, eat small particles 0.5 to 25 μm in diameter, such as algae and other protists, rotifers, and nauplii (Williamson & Butler, 1986). The great advantage of this mode of feeding is that it allows the copepod to select each particle it ingests (Richman & Dodson, 1983). The disadvantage of this mode of feeding is that it is much less efficient than the branchiopod method of filter feeding, which does not select individual particles.

Some of the larger species of calanoid copepods use this grasping (predaceous) feeding mode to eat smaller zoo plankton (figure 4.15). The larger cyclopoid and calanoid copepods catch and shred particles up to about 1.5 mm. Particles in this size range include rotifers, branchiopods, copepods, and small insect larvae. For example, the 2.5-mm-long *Diaptomus shoshone* catches and eats smaller (1.2-mm) adult calanoid copepods, such as *Diaptomus coloradensis.* Cyclopoid copepods sometimes grasp long filaments of algae and start ingesting them at one end. When copepods are feeding in this manner, some can be seen with a large, coiled filament in the gut. Some of the larger cyclopoid species, especially in the genus *Mesocyclops,* readily catch and eat mosquito larvae. These species show promise for biological control of mosquitoes that carry malaria or dengue fever (Schaper, 1999). Cyclopoid copepods, using the grasping feeding mode to its extreme, can become nuisance predators (ectoparasites) of fish. The copepod grasps a fish fin and begins chewing.

Scraping feeding mode is especially characteristic of harpacticoid copepods. These animals eat aufwuchs or organic detritus in sediments. This food source includes mats of bacteria, fungus, or algae and other protists. In this mode, it is not necessary to catch the food, merely to locate it. Food is scraped off a surface with spiny mouthparts.

FIGURE 4.15 A predaceous calanoid copepod, *Epischura lacustris,* grasping a smaller copepod, which it will shred and consume.

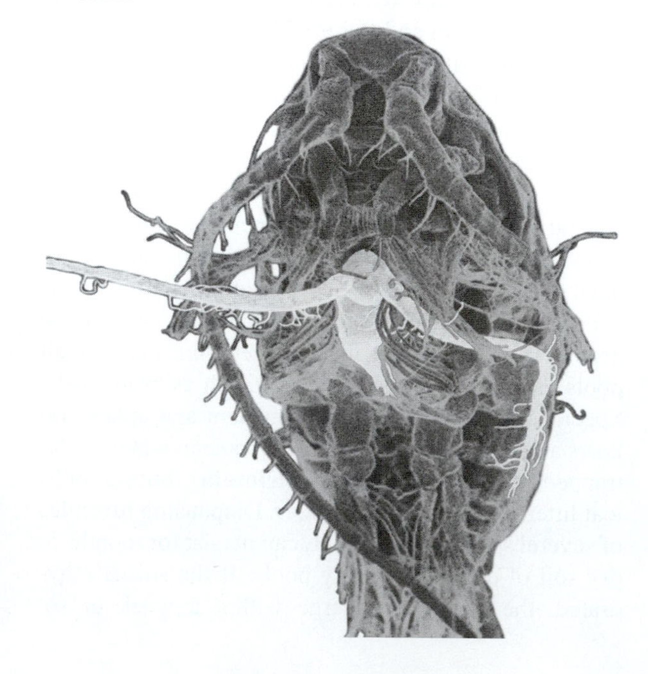

Copepods produce fecal pellets: Feces are enclosed in a sack made from the tissue lining the gut. This is perhaps an adaptation to eating diatoms, which have glass cases that produce sharp fragments when chewed. Protection of the gut is particularly important in copepods because the adults do not molt. Cladocerans, which also eat diatoms, renew the gut lining each time they molt, which is every few days. Copepods can live for weeks or months and reproduce several times, all on a diet that includes ground glass.

Life Cycle and Diapause

With a few exceptions in the harpacticoids, freshwater copepods are strictly sexual (unlike many cladocerans, the other major group of crustacean zooplankton). There is some sexual dimorphism of adults, especially among the cyclopoid species. Copepods typically develop from egg to adult through a number of developmental stages (instars) (figure 4.16): There is typically a total of 13 stages, including the egg (embryonic) stage, 6 naupliar stages, 5 juvenile copepod stages, and the sixth copepodid stage, which is the adult. Adults do not molt again.

Nauplii typically start life with a large amount of stored energy (from the egg) in the form of oil droplets. As the nauplius molts from instar to instar, it adds body segments at the posterior end and increases the segmentation and complexity of its appendages and musculature. The sixth naupliar stage undergoes a profound metamorphosis, changing into a shrimp-like **copepodid** form that resembles the adult, only less

FIGURE 4.16 A representative life cycle of a calanoid copepod, *Eudiaptomus vulgaris.* The diagram shows the egg, six naupliar stages, and six copepodid stages, with the sixth stage being the adult.

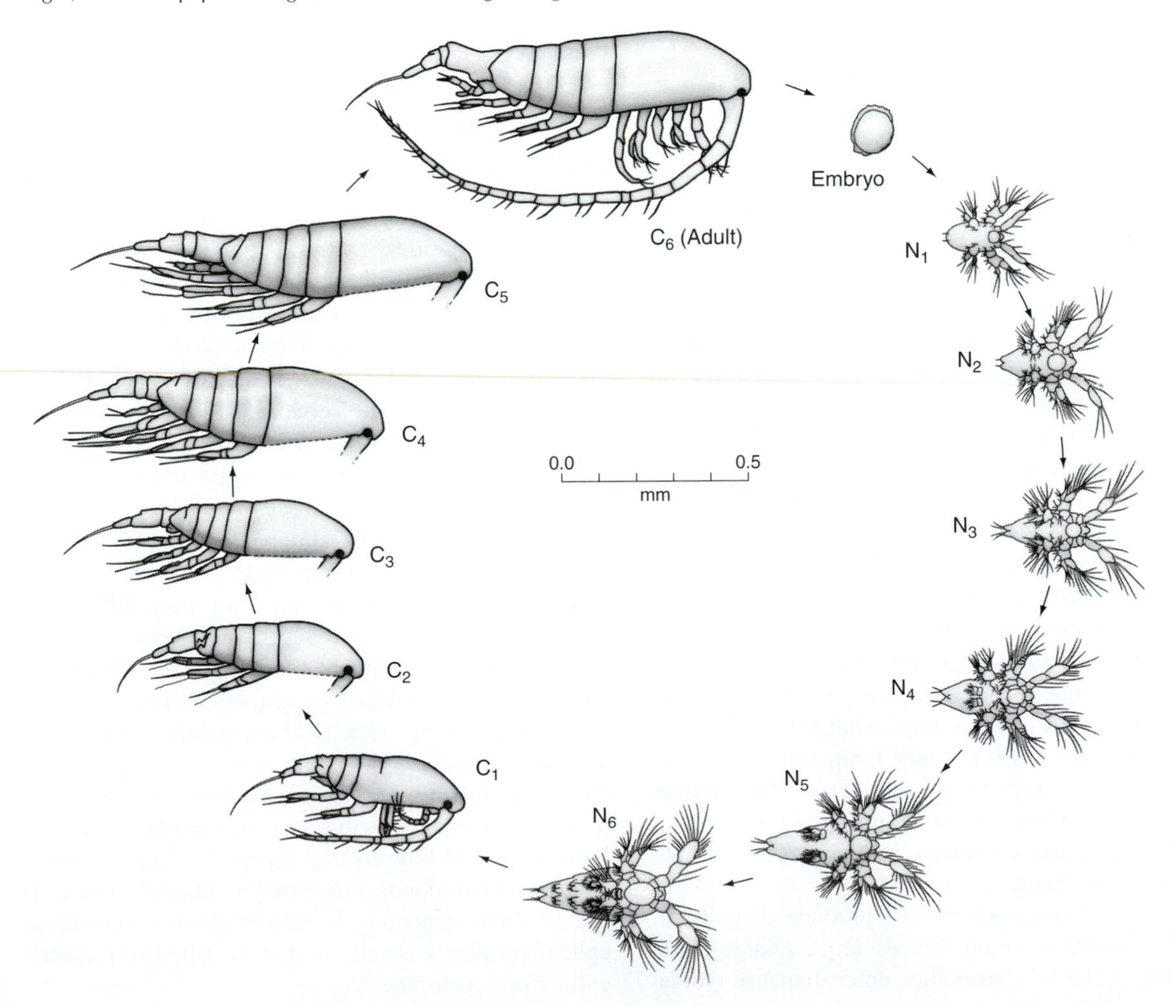

complex. The first stage copepodid molts, adding segments, spines and setae. This process continues until the sixth copepodid stage, the adult, is reached.

The cyclopoid life cycle is similar to that of calanoids, but often faster, allowing cyclopoids to reproduce as rapidly as cladocerans, and in some cases faster (Maier, 1992)—cyclopoids can grow from egg to adult in as little as one week. On the other hand, if the animal diapauses, it may live for 2 years as a copepodid or decades as an egg.

Calanoid and harpacticoid copepods can produce both subitaneous and diapausing eggs. The resting eggs are actually embryos, arrested in an early stage of development, but commonly called eggs. The shift in egg physiology is related to the annual environmental cycle (for example, the timing of pond thawing in the spring or the onset of intense fish predation when young-of-the-year fish appear). Diapausing eggs are very similar in structure to subitaneous eggs, except that the diapausing eggs have an outer membrane that is a little thicker.

Wyngaard et al. (1991) showed that different species of cyclopoid copepods living in a North Carolina temporary pond break diapause under different conditions. Some species appeared immediately after the pond refilled and other species did not appear until months later. Because of large interannual variation in filling time and duration, the species composition varied from year to year.

The calanoid copepod *Diaptomus sanguineus* lives in permanent and temporary freshwater ponds in Rhode Island (Hairston & Olds, 1984). All of these ponds become uninhabitable at some time during the year, when they dry up, but the nature and timing of the harsh period varies both spatially and temporally. Females produce discrete clutches of either subitaneous eggs that hatch immediately or diapausing eggs that hatch the following season. The two egg types show distinct chorion morphologies under the light microscope or under transmission electron microscopy (Lohner et al., 1990). In permanent ponds, the copepods start making diapausing eggs in March, one month before rising water temperatures induce planktivorous sunfish to become active. Depending on environmental conditions within a pond or lake, diapausing eggs are produced in the spring or fall (Nilssen, 1980; Santer et al., 2000).

Cyclopoid copepods do not produce diapausing eggs. However, in some species (e.g., *Diacyclops*), copepodid stage IV copepodids enter diapause (some species of harpacticoid copepods also can diapause in late copepodid stages). Juvenile diapause may be an adaptation to drying of shallow ponds or to surviving periods of warm water in lakes. *Cyclops* can potentially persist in alpine lakes stocked with trout, whereas *Daphnia* and *Holopedium* cannot because the eggs of the *Cyclops* pass undisturbed through the trout's gut, while the subitaneous cladoceran eggs are easily digested (Gliwicz & Rowan, 1984).

Fishless lakes often contain one or more species of large, brightly colored calanoid copepods. These large copepods are extirpated from such lakes within a year or two after fish are stocked—a common practice in oligotrophic, alpine lakes (McNaught et al., 1999; Tyler et al., 1998).

Cyclopoid and harpacticoid copepods live in an astonishing variety of aquatic and humid, continental environments and microhabitats (Reid, 2001). Associated with the different habitats are distinct life history responses to environmental signals such as filling of a pond (Wyngaard et al., 1991).

Economic Importance and Management

Like branchiopods, copepods are critical members of aquatic communities. They are often important grazers of algae and an important item in the diet of many fish.

Copepods often carry immature forms of parasites such as tapeworms, flukes, and nematodes, and can be an important vector for parasites that affect populations of fish and other aquatic vertebrates. Several important human parasites (in tropical climates), including the guinea worm and various tapeworms, are transmitted by copepods.

Copepods are of less direct economic importance than cladocerans. While copepods are often a critical link in aquatic food chains, they are more difficult to maintain in captivity for aquaculture purposes. The noncannibalistic species of Cyclopoids are the easiest copepods to rear and may be adaptable to aquaculture.

There are no officially recognized endangered species of copepods in North America. Nevertheless, there are probably species of the large, brightly pigmented calanoid species that are in danger of extinction—for example, the blue and red copepods of alpine ponds, ephemeral grasslands, and woodland ponds. The danger to these copepods is habitat destruction—these ephemeral ponds are the first to be filled in for agriculture or development.

A reserve has been established for the endangered calanoid species *Limnocalanus grimaldii* variety *macrurus Sars,* in Lake Richi, northwestern Belarus (Suschenya et al., 1981). This hypolimnetic species lives in oligo- and mesotrophic glacial lakes. The copepod is endangered via cultural eutrophication and pollution with industrial and agricultural chemicals and wastes.

MALACOSTRACANS

Taxonomy and Diversity

Malacostracans are a class of mostly marine crustaceans, so they have many morphological and anatomical features in common with their smaller-sized relatives, the branchiopods and copepods. Several chapters in Thorp and Covich (2001) provide an excellent overview of their morphology and ecology. Freshwater malacostracans include four large and ecologically important Orders: Mysidacea, Amphipoda, Isopoda, and Decapoda (figure 4.17). The first three groups are rather homogeneous, but the last group, the decapods, is much more variable. The major freshwater decapod groups are crayfish and shrimps. In North America there are 300–400 species of mysids, amphipods, and isopods, and about the same number of crayfish and freshwater shrimps. Mysids, although represented by only a few freshwater species, are ecologically important. (Rare species, such as the occasional freshwater crab, will be passed over in this discussion.)

External Structure, Appearance, and Anatomy

Freshwater malacostracans are unique in having more thoracic legs associated with the head than is the case for branchiopods or copepods. The four major groups are distinguished by differences in body shape and the arrangement of their appendages (details can be found in Thorp & Covich, 2001), but members of all four groups are more or less shrimp-like in general appearance (figure 4.17).

Malacostracans are larger than the branchiopods and copepods. Adult mysids are about 1 cm long, and the largest malacostracans, such as the largest freshwater shrimp and giant crayfish of Australia, are about 10–30 cm long. Amphipods and isopods are typically 1 cm or so long, although there are larger species (several centimeters long) in the deep-water habitat of Lake Baikal, Siberia.

FIGURE 4.17 (a) Isopoda. (b) Amphipoda. (c) Mysidacea. (d) Decapoda, freshwater shrimp. (e) Decapoda, crayfish.

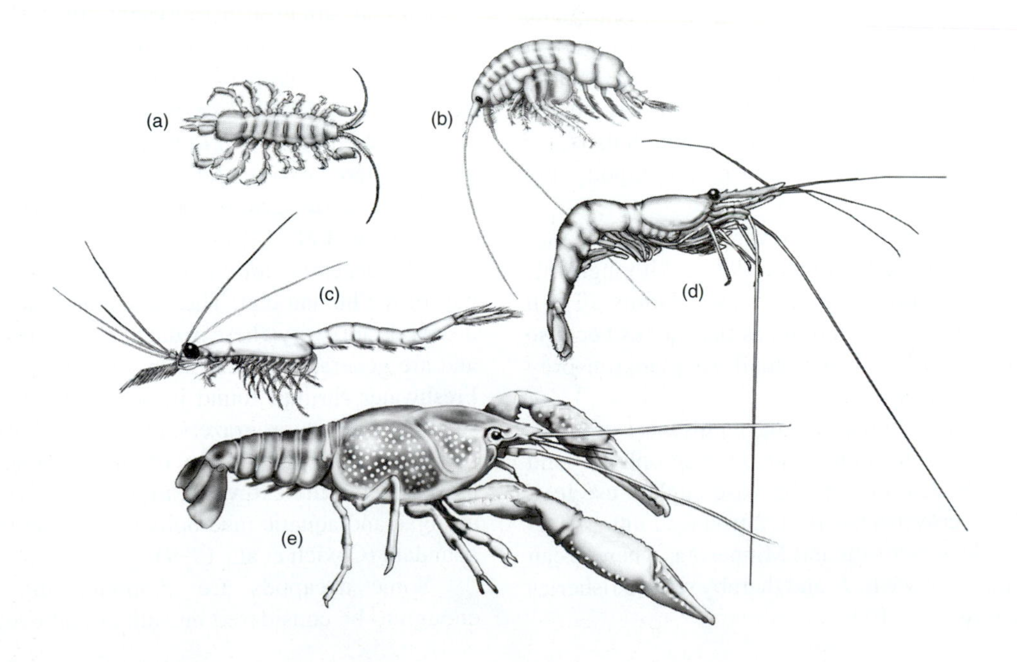

Habitat

Perhaps as a result of their relatively large body size (attractive to fish), most freshwater malacostracans live on the bottom of lakes and streams, often in coarse cobble or in aquatic vegetation.

Mysids show the greatest tendency to leave surfaces for open water—several species are planktonic, living in deep water of oligotrophic lakes such as the Laurentian Great Lakes where they are indigenous, or in lakes such as Flathead Lake (Montana) and Lake Tahoe (California), where they have been introduced (Spencer et al., 1999).

These larger crustaceans often display a diel behavior pattern—hiding in refuges during the day and moving about during twilight and darkness. In clear lakes, such as the North American Great Lakes, mysids migrate vertically over tens of meters (Grossnickle, 1982).

Swimming and Escape Behavior

These relatively large organisms have the ability to walk, swim, and escape rapidly. They are most at home in water, but some crayfish spend a significant amount of time walking on the ground. Because they are relatively large animals, they have correspondingly fast rates of movement.

Feeding Preferences and Behavior

Malacostracans tend to be omnivores, eating organic material of all sorts. They all eat organic detritus (if it isn't too nutrient-poor), carrion, aquatic vegetation, and aufwuchs. The larger crayfish also eat snails. Freshwater forms are not parasitic (although there are some spectacularly parasitic marine amphipods and isopods).

Freshwater mysids eat a variety of small zooplankton (Cooper & Goldman, 1980). Although initially introduced into large lakes as a promising fish food, their presence often reduces fish stocks because they compete with fish for small zooplankton prey (Spencer et al., 1999).

Omnivorous crayfish can make nuisances of themselves by over-consuming aquatic macrophytes and macroinvertebrates. Perhaps because anglers use it as bait, the rusty crayfish has been introduced into a myriad of lakes in Wisconsin and Minnesota, where it can decimate macrophyte beds and thereby reduce fisheries (Byron & Wilson, 2001).

Life Cycle and Diapause

Malacostracans have a life cycle similar to that of the copepods, with separate sexes and a long series of juvenile stages. The adult female often carries the eggs, and the early stages of development (naupliar stages) are passed in the eggshell. Individuals typically live for about a year, depending on body size.

Malacostracans do not have diapausing stages. Some species of freshwater crayfish are able to be dormant in moist burrows during periods of low water and high temperature. Dormancy is characterized by decreased activity, but not the extreme metabolic minimization typical of diapause.

Economic Importance and Management

Several of these species are economically and ecologically important. The smaller species provide an important ecological service by scavenging in freshwater environments. Malacostracans are also major diet items for many aquatic vertebrates, from fish to mammals. These bottom-dwelling crustaceans are as important to their food chain as the plankton crustaceans are to the open-water food chain.

Humans have little direct contact with the smaller malacostracans. An exception is the use of several of the smaller species, especially the estuarine mysids and shrimp, in ecotoxicology research. The animals are used as representative test organisms to evaluate the significance of both freshwater and marine contaminants to invertebrates in general (Ingersoll et al., 1999). The predators *Mysis relicta* (a mysid) and *Pallasea quadrispinosa* (an amphipod) are considered to be endangered species in Belarus, at risk because of agricultural and industrial pollution (Suschenya et al., 1981).

The decapods are large enough to be a desirable part of the human diet. The largest and most desirable decapods are everywhere in danger of overexploitation, and are generally protected, to some degree, by laws. Freshwater shrimp, found in tropical and subtropical streams, are major grazers in the food web. When shrimp are removed (as they often are, either legally or by poachers), the entire stream ecosystem can change as algae and aquatic macrophytes become much more abundant (Covich et al., 1999).

Some decapods are abundant and voracious enough to be considered nuisances and even ecological disasters. For example, one species of moderately large crayfish, the "rusty crayfish," is expanding its range into lakes of northern North America (Byron & Wilson, 2001).

INSECTS

Insects, a subphylum of Arthropoda, have three pairs of appendages (not counting the compound eyes) on the head: one pair of antennae, one pair of mandibles, and one pair of maxillae. Insects have fewer appendages than do the crustaceans. There are three pairs of walking legs and up to two pairs of wings on the thorax. There are no legs on the abdomen, although there can be pairs of gills or other extensions.

Taxonomy and Diversity

Copepods and branchiopods are the dominant and diverse small invertebrates in the open water of lakes. Insects are more diverse and abundant than microcrustaceans in streams. Both groups (small crustaceans and insects) are common and diverse in wetlands and littoral beds of aquatic plants.

Like the microcrustaceans, insects are arthropods, with segmented bodies and appendages and with complex internal anatomy. There are thousands of species of aquatic insects, and at least 5000 species that spend most of their life in water in North America. Aquatic insects are grouped into at least 14 orders. Table 4.3 summarizes the most common groups.

External Structure, Appearance, and Anatomy

The appearance of an insect depends on its developmental stage. Immature aquatic insects (table 4.4; figure 4.18) can be very different in appearance compared to the adults (table 4.5; figure 4.19). Many insects hatch from eggs as **larvae** (for example, Coleoptera and Diptera). These worm-like forms often lack true legs, and resemble an adult only by having a hard head capsule with one pair each of antennae, mandibles, and maxillae. The larva grows by molting several times, then it molts into a more-or-less inactive stage, the **pupa.** Inside the pupa, the tissues reorganize, and an adult emerges from the pupa.

Other insects hatch from the egg as juveniles that superficially resemble the adults, except that they are smaller, lack (functional) wings and reproductive systems, and have a few special structures, such as **gills,** necessary for life in water.

Habitat

Aquatic insects occur anywhere there is water. Some insect larvae breathe air (such as some fly larvae, beetles, and bugs) and can therefore live in habitats with no oxygen, such as sewage treatment ponds.

Aquatic insect species (or genera) are widely used as indicators of water quality. Some species are specialists that live in only the purest water; some are adapted to acidic, anoxic, or muddy environments. By knowing the species that live in a stream or lake,

Table 4.3 Some Major Orders of Aquatic Insects

Order	Common Name	Approximate Number of Living Aquatic Species, Worldwide
Coleoptera	Beetles	More than 1100
Diptera	Flies	Thousands
Ephemeroptera	Mayflies	575
Hemiptera or "Heteroptera"	Bugs	217 (plus 107 living on water)
Odonata	Dragonflies and damselflies	415
Plecoptera	Stoneflies	515
Trichoptera	Caddisflies	1340

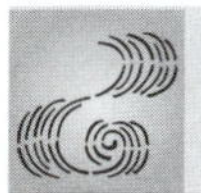

Table 4.4 Morphological Characteristics of Common Immature Aquatic Insects (see figure 4.18)

Order	Developing Wings	General Appearance (all have three pairs of jointed legs, unless noted)
Coleoptera	Internal (not visible)	Very diverse forms. Usually with a distinct, hard head, pincher-like jaws, and three distinct pairs of legs. If there are gills on the abdomen, they are filamentous.
Diptera	Internal	Worm-like (maggot), without jointed legs
Ephemeroptera	External (visible as pads on the back)	Leaf-like gills on abdomen, two or three long tail filaments
Hemiptera	Internal	Juveniles resemble adults (oval body with three pairs of legs), with the mouth at the end of a sharp needle. Gills and wings are absent.
Odonata	External	Somewhat like the adult, but with wings as pads on the back of the thorax. The mouth is covered by a spoon-like lower lip on the underside of the head. Damselflies: abdomen ends in three long, leaf-like gills Dragonflies: abdomen ends in three short, triangular scales.
Plecoptera	External	Juveniles (naiads) resemble the adults, but the wings are absent or represented by small pads on the back of the thorax. Abdomen ends in two, long tail filaments.
Trichoptera	Internal	A worm-like larva with a hard head capsule, jointed legs near the head, and an elongated abdomen ending in a pair of hooks. Many species of Trichoptera live in mobile tubes made of stones or vegetation as larvae. They also pupate inside the tube.

we can know the average water quality in that body of water (plate 16, also see chapter 12). Species in the different orders favor different parts of the aquatic environment (table 4.6).

Swimming and Escape Behavior

Larvae show a wide variety of creeping, walking, and swimming behaviors (table 4.6). Escape behaviors are primarily faster swimming in the opposite direction from an unpleasant stimulus. Peckarsky and Penton (1988) report that when some mayfly nymphs encounter predaceous stonefly nymphs, they freeze and assume a defensive posture, with their tail bent over the back and the tail setae pointed forward—this is called the defensive "scorpion" posture.

To resist the power of water currents, stream insects exposed to currents tend to be streamlined. They have smooth curves, no projections, and short, powerful legs that can grasp the substrate. Insects that prefer still water are much more angular, often with longer and weaker legs.

Feeding Preferences and Behavior

Aquatic insects exhibit a variety of feeding habits (table 4.7). While aquatic insects tend to be omnivorous, they show large differences in where and how they collect food (Merritt & Cummins, 1984). The wide variety in feeding preferences and behaviors may be the ecological factor that allows so many different species to live together in the same patch of stream or lake. Important ways of obtaining food include shredding, scraping, collecting, predaceous capture, and piercing. Shredders chew, tear up, and eat large pieces of vegetable material, such as decaying leaves. Many of the case-building caddisflys (Trichoptera) are shredders. Scrapers harvest the layer of

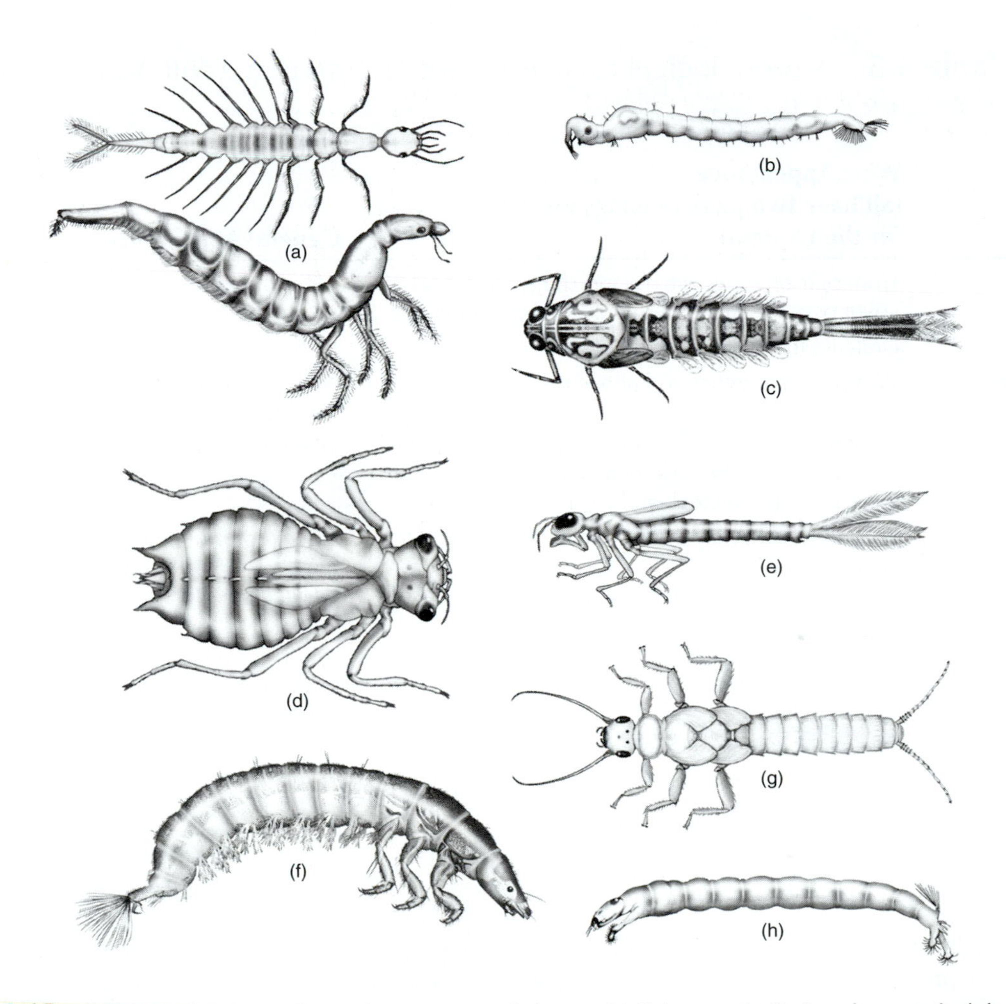

FIGURE 4.18 Examples of immature forms of common aquatic insects. (a) Coleoptera (a *Dytiscus* larva on the left and *Copto-tomus* larva on the right. (b) Diptera (phantom midge or *Chaoborus* larva). (c) Ephemeroptera (*Baetis* nymph). (d) Odonata (dragonfly naiad, *Epicordulia*). (e) Odonata (generalized damselfly naiad). (f) Tricoptera (*Hydropsyche,* a caddis fly that doesn't build a larval case). (g) Plecoptera (stonefly nymph, *Alloperla*), (h) Diptera (*Chaoborus,* a phantom midge larva).

aufwuchs from the surface of stones and plants. Mayfly nymphs (Ephemerella) are a good example of shredders. Collectors use their legs or web-like nets to filter fine particles out of a current or to sieve food from fine sediment. Examples of collectors are *Hydropsyche* caddis fly larvae (Trichoptera) and black fly larvae (Diptera).

Predators catch and grasp prey, which they chew. *Dytiscus* larvae (Coleoptera) and *Chaoborus* larvae (Diptera) are ecologically important, invertebrate predators in freshwater habitats. Piercers grasp prey or plants and suck out nutrient juices. Common piercers include the water boatmen (*Corixa*) and backswimmers (*Notonecta*) (both are in Hemiptera).

Life Cycle and Diapause

In temperate freshwater habitats, aquatic insects typically produce one or two generations a year, depending on temperature and whether their habitat is a temporary pond or stream or is permanently wet. Some aquatic insects have a nonmotile stage, such as the egg (as in mosquitoes) or pupa (as in some beetles), that is resistant to freezing and drying.

As an insect develops from an egg to an adult, it goes through several developmental stages. The stages can resemble each other closely or can be wildly different. For example, larvae are more or less worm-like, pupae are usually nonmotile and look like small, carved

Table 4.5 Morphological Characteristics of Common Adult Aquatic Insects (see figure 4.19)

Order	Wing Appearance (all have two pairs of wings except for the Diptera)	General Appearance
Coleoptera	Front pair of wings is hard and shiny, used as a cover over the membranous second pair of wings	More or less rounded and shiny
Diptera	Delicate flies with one pair of wings	Includes midges, mosquitoes, and gnats
Ephemeroptera	Wings held together and above the body	Abdomen ends in two or three filaments as long as the body; delicate animals
Hemiptera	Front pair of wings cross over each other, forming a "v" pattern at the base of the wings; thick at the base and membranous at the tip.	Like beetles, but not shiny
Odonata	Dragonfly wings are held horizontally, at right angles to the body. Damselfly wings are held together and vertical, pointing back along the abdomen.	Large, colorful flying nails; wings are membranous, but often with striking designs or colors
Plecoptera	Wings are held flat (horizontal) and pointed back along the abdomen.	Adults have two, long filaments projecting from the back of the abdomen; more chunky than mayflies
Trichoptera	Wings are held at an angle, on either side of the abdomen.	Moth-like, with slender antennae as long as the body

pills, and adults may have wings and jointed legs. The type of development can give clues as to the identification of immature forms (table 4.8).

Economic Importance and Management

Although insects are small in size, their huge numbers contribute to their major economic and ecological importance. On the positive side, insects are a major food source of fish and waterfowl. Microcrustaceans are a major link in the planktonic food chain, but in shallow water, insects take over the role of transferring energy through the middle links of the food chain. Humans tend to avoid eating insects (although there are a few exceptions), but we love to eat the fish that depend on a diet of insects.

Some aquatic insects, such as dragonflies and tiger beetles, have attracted the attention of conservationists. Rare and endangered insects have been instrumental in the preservation of habitats. For example, the Nature Conservancy "Mexican Cut" nature preserve and research area in western Colorado was protected from mining exploitation, partly because it is the habitat of a rare dragonfly. Trout anglers (who use artificial insect baits) have been instrumental in restoration and conservation of many lake and stream habitats and their associated species because trout depend on a healthy community of aquatic insects, including caddis flies, mayflies, and midges (see the web site address at the end of this chapter for the Trout Unlimited organization).

On the less-positive side, aquatic insects are often the focus of concern about environmental quality and even human health. For example, midges of Clear Lake, California are nonbiting, but can, by their numbers, make camping and picnicking a miserable experience. Biting insects (biting flies of various kinds, including horse flies, snipe flies, black flies, and mosquitoes) can be an intense annoyance. Mosquitoes carry

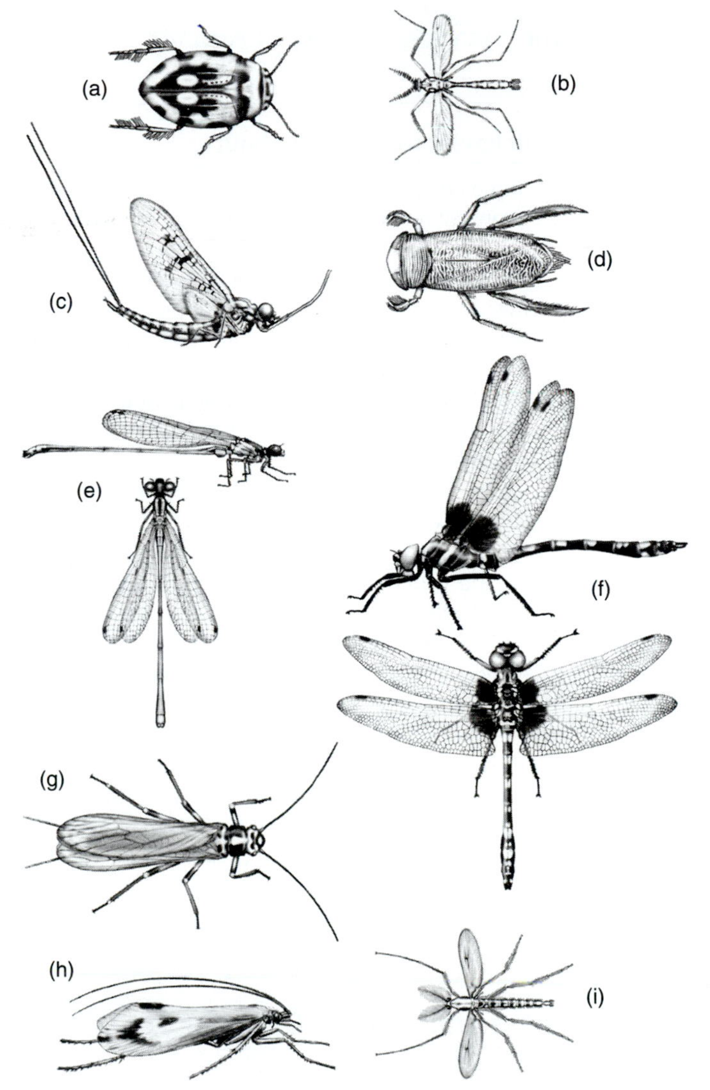

FIGURE 4.19 Examples of adult forms of common aquatic insects. (a) Coleoptera (an aquatic beetle). (b) Diptera (a phantom midge *Chaoborus*). (c) Ephemeroptera (mayfly). (d) Hemiptera (water boatman). (e) Odonata (damselfly). (f) Odonata (dragonfly). (g) Plecoptera (stonefly). (h) Tricoptera (caddisfly). (i) Diptera (*Chaoborus,* phantom midge).

diseases of major significance to humans and wildlife, such as malaria and dengue fever.

Modern chemical technology is very effective in killing insects that are vectors of disease, but, unfortunately, pesticides used in aquatic ecosystems are seldom specific—pesticides tend to affect all aquatic and even terrestrial organisms (Colborn et al., 1996). Pesticide effects tend to fall heavily on predaceous insects (such as bugs and dragonflies), resulting in a reduction of natural control on pestiferous insects such as mosquitoes. Integrated pest management strategies, which combine judicious use of pesticides with biological control, are sometimes effective and less damaging to aquatic systems than control by the use of pesticides alone. Chapter 10 provides an example of the effectiveness of various insect control strategies—The Clear Lake story. As management strategy shifts from wholesale use of pesticides, and as water quality improves from massive pollution (in response to clean water legislation), insects extirpated from aquatic habitats are making a reappearance. For example, the large *Hexigenia* mayflies of Lake Erie once appeared in huge clouds each summer. In the last few decades, the mayflies were absent, but as Lake Erie

Table 4.6 Typical Habitats of Species in the Different Orders of Aquatic Insects

Order	Typical Habitat	Behavior for Moving About
Coleoptera	Larvae and some adults: bottom sediments, stream beds, vegetation, and open water of small ponds	Some creep along surfaces; some are moderately strong swimmers that blunder into prey in open water or on surfaces.
Diptera	Larvae are ubiquitous; adults are terrestrial	Mostly poor swimmers. Even those larvae that live in open water tend to be stationary most of the time, with a quick movement of the body for escape or for capturing prey.
Ephemeroptera	Larvae only: bottom sediments, stream beds, vegetation, and open water of small ponds	Mostly crawling species, although a few are fast swimmers over surfaces. They scrape food (aufwuchs) from surfaces.
Hemiptera	Adults and larvae: diving and swimming in shallow water or lying in wait for prey in vegetation	Strong swimmers that pursue prey or swim over surfaces to gather aufwuchs
Odonata	Larvae sit on surfaces (vegetation, rocks, or the bottom), waiting for living prey (small insects, crustaceans) to appear.	Sit-and-wait predators; they sit on a surface quietly, then when a prey item approaches, they flip out the long lower lip to make a capture.
Plecoptera	Larvae usually associated with rocks and gravel, especially in streams	General body movements resemble those of mayflies, but they also make quick movements when they capture prey, such as small mayflies.
Trichoptera	Larvae in most species build elaborate cases of plant material or sand; crawl over substrates in both lakes and streams.	These animals creep along surfaces, usually carrying a case that is used as an escape refuge. They either scrape surfaces or grasp smaller insects.

Table 4.7 Characteristic Feeding Preferences and Feeding Behavior of Common Aquatic Insects

Order	Typical Diet and Feeding Mode
Coleoptera	Both adults and larvae can be aquatic; different species collect food in all possible ways. Some species are voracious predators, some scrape rocks for aufwuchs, some filter particles out of water, and some shred plant material.
Diptera	Larvae mostly eat algae and detritus by scraping surfaces to gather aufwuchs. Some forms collect particles and some are predaceous on smaller animals.
Ephemeroptera	Immature forms mostly scrape aufwuchs (algae and detritus) off surfaces.
Hemiptera	Predaceous or herbivorous, using needle-like mouthparts to pierce prey or suck up algae and detritus.
Odonata	Larvae are predaceous.
Plecoptera	Larvae are mostly predaceous.
Trichoptera	Mostly herbivorous, some predaceous. Larvae scrape surfaces for aufwuchs, shred plant material, or spin silken nets to collect food from the water. Some species are predators.

Table 4.8 Typical Developmental Patterns of the Orders of Aquatic Insects

Order	Common Name	Type of Development
Coleoptera	Beetle	Perfect (development with larva, pupa, and adult)
Diptera	Fly	Perfect
Ephemeroptera	Mayfly	Imperfect (development with immatures are called "naiads" or nymphs" that resemble adults)
Hemiptera	Bug	Imperfect
Odonata	Dragonfly	Imperfect
Plecoptera	Stonefly	Imperfect
Trichoptera	Caddisfly	Perfect

water quality improves, the mayflies are returning (Masteller & Obert, 2000).

> "I came down to the bank [of the Blackfoot River] to catch fish. Cool wind had blown in from Canada without causing any electric storms, so the fish should be off the bottom and feeding again. When a deer comes to water, his head shoots in and out of his shoulders to see what's ahead, and I was looking all around to see what fly to put on. But I didn't have to look further than my neck or my nose. Big clumsy [stone]flies bumped into my face, swarmed on my neck and wiggled in my underwear. Blundering and soft-bellied, they had been born before they had brains. They spent a year underwater on legs, had crawled out on a rock, had become flies and copulated with the ninth and tenth segments of their abdomens, and then had died as the first light wind blew them into the water where the fish circled excitedly. They were a fish's dream come true—stupid, succulent, and exhausted from copulation . . ."
>
> Norman Mclean, 1976 *A River Runs Through It*

Study Guide

Chapter 4 Diversity of Aquatic Organisms: Rotifers, Annelids, and Arthropods

Questions

1. If you were to sample a lake, which groups of these organisms would you expect to see with your unaided eye? What are the fewest number of morphological characters you would use to tell the different kinds of organisms apart? Would you find the same groups and use the same morphological characteristics if you were looking at a sample of organisms from a stream?
2. It is common to find 1000 rotifers per liter in lake water. If each rotifer is spherical (a reasonable approximation), with a diameter of 100 μm, what percent of the liter of water is occupied by rotifer mass?
3. What are escape strategies of the organisms discussed in chapter 3?
4. What is the nature of "worm" and what are possible variations on the theme of "worm"?
5. Compare and contrast the life history strategies, including the reproductive strategies, of rotifers, copepods, and water fleas.
6. What is the relationship between UV light and pigmentation in microcrustaceans?

Words Related to Zooplankton

(Note: Important taxonomic names are listed in the next section.)

abdomen
antenna
arthropod
bolus
carapace
carotenoid
cephalothorax
chaeta
chitin
chydorid
copepodid (adjective)
cryptobiosis
cyclomorphosis
detritus
ecdysis
ephippium
estuary
facultative
furca
gill
head
hemoglobin
hemolymph
hermaphrodite
hydrophobic
hydrostatic skeleton
instar
larva
littoral
lorica
mandible
mastax
maxilla
maxilliped
melanin
metamorphosis
metasome
metazoan
microcrustacea
molt
nauplius
neonate
obligate
ocellus
parthenogenesis
photoprotection
phytotelmata
postabdomen
prosome
pupa
setule
subitaneous
thorax
trade-off
trophi
tropism
ultraviolet
urosome
water flea
zygote

Major Examples and Species Names to Know

To speak the language of limnologists, it is helpful to know the common or scientific names of a few frequently encountered or important organisms:

Rotifers:

Ascomorpha minima

Asplanchna sieboldi

Brachionus

Cephalobdella forficula

Collotheca

Keratella

Polyarthra

Annelida:

Oligochaeta

Tubifex

Aelosoma

Lumbriculus

Erpobdella

(also Nematoda and dipteran larva)

Branchiopoda:

Arthropoda

Crustacea

Branchiopoda

Anostraca (fairy shrimp)

  Artemia

Conchostraca (clam shrimp)

Notostraca (tadpole shrimp)

Cladocera (water fleas)

  Daphnia

  Moina

  Bosmina

Copepoda:

Cyclopoids

Calanoids

Harpacticoids

Insecta:

Coleoptera (beetles)

Diptera (flies, midges, gnats)

Ephemeroptera (mayflies)

Hemiptera or (true bugs)
Odonata (dragon flies and damselflies)
Plecoptera (stone flies)
Trichoptera (caddis flies)

Additional Resources

Further Reading

Damkaer, D. M. 2002. *The copepodologist's cabinet: A biographical and bibliographical history.* Philadelphia, PA: American Philosophical Society. This book is a delight for those with even the slightest interest in copepods or the history of science.

Fenchel, T. 1987. *Ecology of protozoa: The biology of free-living phagotrophic protists.* Madison, WI: Science Tech Publishers (Brock/Springer). 197 pp.

Gould, S. J. 1990. *Wonderful life: The Burgess Shale and the nature of history.* New York: Norton. USA. 347 pp. Madison, WI: Science Tech. Publishers.

McNeill Alexander, R. 1979. *The invertebrates.* Cambridge, UK: Cambridge University Press. This is a good book for finding out about form and function of aquatic invertebrates.

Merritt, R. W., and K. W. Cummings, eds. 1996. *Aquatic insects of North America.* 3rd ed. Dubuque, IA: Kendall/Hunt.

Pennak, R. W. 1989. *Freshwater invertebrates of the United States: Protozoa to mollusca.* New York: Wiley. This text has largely been eclipsed by the Thorp and Covich (2001) text, but it has several sections that have useful identification keys, and the keys are to the species level.

Reid, G. K. 1987. *Pond life: A guide to common plants and animals of North American ponds and lakes.* New York: Golden Press. This little text is handy to carry in the field or to use as a companion at the microscope for rough-and-ready identification of aquatic organisms.

Schmitt, W. L. 1971. *Crustaceans.* Ann Arbor, MI: University of Michigan Press.

Thorp, J. H., and A. P. Covich, eds. 2001. *Ecology & classification of North American freshwater invertebrates.* 2nd ed. San Diego, CA: Academic Press. 1056 pp.

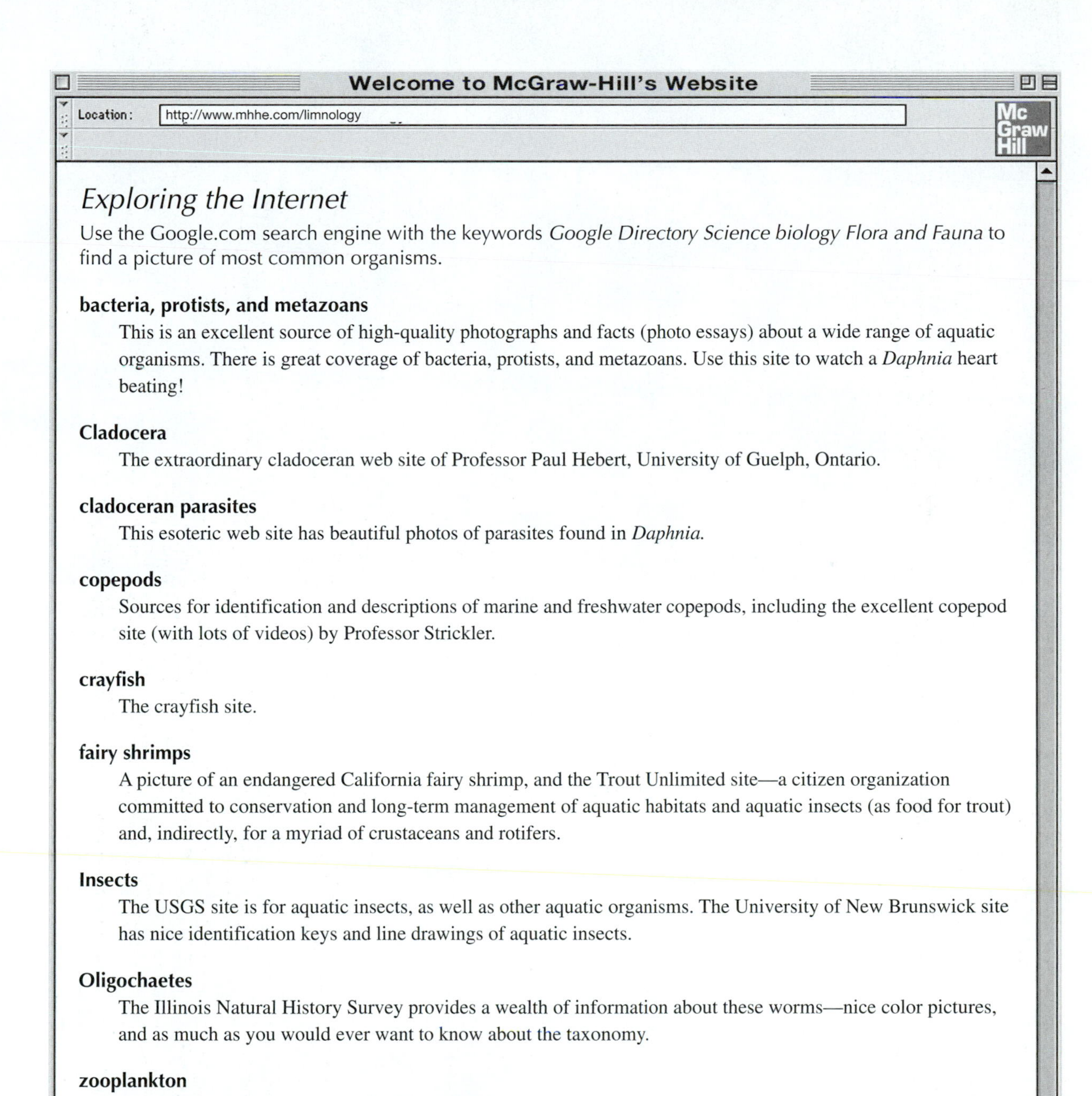

Exploring the Internet

Use the Google.com search engine with the keywords *Google Directory Science biology Flora and Fauna* to find a picture of most common organisms.

bacteria, protists, and metazoans

This is an excellent source of high-quality photographs and facts (photo essays) about a wide range of aquatic organisms. There is great coverage of bacteria, protists, and metazoans. Use this site to watch a *Daphnia* heart beating!

Cladocera

The extraordinary cladoceran web site of Professor Paul Hebert, University of Guelph, Ontario.

cladoceran parasites

This esoteric web site has beautiful photos of parasites found in *Daphnia.*

copepods

Sources for identification and descriptions of marine and freshwater copepods, including the excellent copepod site (with lots of videos) by Professor Strickler.

crayfish

The crayfish site.

fairy shrimps

A picture of an endangered California fairy shrimp, and the Trout Unlimited site—a citizen organization committed to conservation and long-term management of aquatic habitats and aquatic insects (as food for trout) and, indirectly, for a myriad of crustaceans and rotifers.

Insects

The USGS site is for aquatic insects, as well as other aquatic organisms. The University of New Brunswick site has nice identification keys and line drawings of aquatic insects.

Oligochaetes

The Illinois Natural History Survey provides a wealth of information about these worms—nice color pictures, and as much as you would ever want to know about the taxonomy.

zooplankton

Sites that cover several groups of zooplankton.

5

It was the season of waterlilies. . . "I wish I was a fish, said the Wart."
"What sort of fish?" . . .
. . . "I think I should like to be a perch," he said. "They are braver than the silly roach, and not quite so slaughterous as the pike are." . . .
They swam along, Merlyn occasionally advising him to put his back into it when he forgot, and the strange under-water world began to dawn about them. . . The great forests of weed were delicately traced. . . Water snails slowly ambled about on the stems of the lilies or under their leaves, while fresh-water mussels lay on the bottom doing nothing in particular. . . Once the two travelers passed under a swan. The white creature floated above like a Zeppelin. . .

T. H. White, *The Once and Future King,* 1939

Diversity of Aquatic Organisms: Larger Organisms

outline

THE LARGER AQUATIC ANIMALS AND PLANTS

For the general public, the organisms described in this chapter are the most familiar players in the ecological theater. These animals and plants are the largest organisms in lakes, and include the charismatic aquatic megafauna. They are all easily observed and many of the species described here are the focus of avidly interested collectors, anglers, hunters, wildlife observers, and stream and lake managers. These organisms, like the small invertebrates discussed in chapter 4, perform important ecological services, such as cleaning water and providing food for other members of the community, including humans. However, unlike the small invertebrates, the mollusks, vertebrates, and macrophytes are often of direct and significant economic importance.

MOLLUSKS—PHYLUM MOLLUSCA

Taxonomy and Diversity

There are two major groups of mollusks in fresh waters (figure 5.1): Class Bivalvia (270 species of freshwater clams and mussels in North America) and Class Gastropoda (500 species of freshwater snails in North America). Molluscan diversity is relatively low in fresh waters compared to that of the oceans. Marine habitats are home to several additional classes and tens of thousands of species.

In North America, gastropod species fall into two subclasses: the Prosobranchia (about 350 species) and the Pulmonata (about 140 species). Bivalve species fall into four categories: the Superfamily Unionacea (or freshwater mussels, about 230 species), the family Sphaeriidae (or "pea clams," about 40 species), the Asiatic clam *Corbicula fluminea,* and two *Dreissena* species (the zebra mussel *D. polymorpha* and the quagga mussel *D. bugensis*).

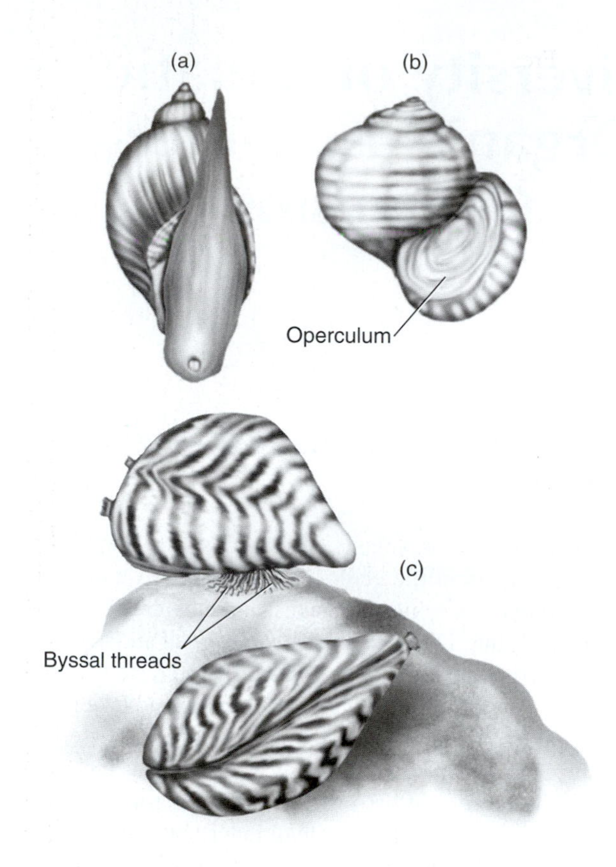

FIGURE 5.1 Representative freshwater mollusks. (a) Freshwater snail; (b) freshwater snail with an operculum shielding the shell opening; (c) two zebra mussels, with byssal threads visible in the side view.

External Structure, Appearance, and Anatomy

Although many mollusks are in the millimeter to centimeter size range, some mollusks are among the largest invertebrates, with species as large as a person's fist. Their abundance and relatively large size make them conspicuous components of aquatic habitats. Freshwater mollusks have either a pair of shells in the bivalves, or a single, coiled shell in the snails. Slugs, which are shell-less snails, require high moisture but they avoid submersion in aquatic habitats.

The shell is a mixture of proteins and crystalline calcium carbonate. The outside can be quite **cryptic** (camouflaged), usually dark gray or brown, and covered with attached algae. The body inside the shell is soft, few hard parts. Such as the microscopic teeth on the tongue-like organ (called the **radula**) of snails. Molluscan bodies are divided into a head, a muscular **foot** (used for locomotion), a visceral mass, and a mantle, which is an outgrowth of the back that secretes shell. The shell (or shells) protect a soft, unsegmented body.

Prosobranch snails obtain oxygen from water using gills, while pulmonates use a lung-like pouch that is an infolding of the mantle. Snails can withdraw their body into the shell for protection. Prosobranchs close off the shell opening with a horny plate, the **operculum,** on the back of their foot. Pulmonates lack an operculum, but can secrete a **mucous** (slimy) mass over the shell opening.

In fresh water, prosobranchs and pulmonates can be distinguished by their shape—snails have a body that is twisted into a spiral either to the right or left. You can determine the orientation by holding a snail so the point of the shell is pointing upward and the shell opening is facing you. If the opening is on the right side of the snail and if there is an operculum (figure 5.1), it is a prosobranch snail. Pulmonate snails have the opening on either the left or right side.

Habitat

Freshwater mollusks occur in nearly all well-oxygenated and permanent aquatic habitats, in both streams and lakes. Because of their need to produce a calcium carbonate ($CaCO_3$) shell, shelled mollusks do not occur in very soft water. However, mollusks are common in fresh water with more than about 5 mg/l^{-1} $CaCO_3$ (Lodge et al., 1987). For harder water, the distribution and abundance of freshwater mollusks is probably the result not of water chemistry preferences, but of biological interactions, especially predation. The presence or absence of predators such as fishes, birds, and crayfishes has a strong influence on the number and abundance of mollusk species in a given habitat (Byron & Wilson, 2001).

Snails and small bivalves ("pea clams") are particularly abundant in beds of aquatic vegetation, and the larger bivalves are specialists at living in the sediments of rivers. Different bivalve species show preferences of substrate type—some prefer cobbles and some prefer soft mud.

Mussels typically live in shallow water, where algae and detritus are accessible and inorganic sedimentation is minimized. Densities of hundreds per m^2 are commonplace, and densities up to 500,000 per m^2 have been reported for the zebra mussel (plate 32).

Swimming and Escape Behavior

Although mollusks are thought of as being phlegmatic or even immobile, most are able to move about in their habitat. Of the mollusks, snails are the most active. Most bivalves show some ability to move through sediment, and only a few species (such as adult zebra mussels) are truly immobile.

Snails glide about on their muscular foot, which is lubricated with mucus. Ripples in the foot move the snail forward. Several species of snail have demonstrated an escape response to specific chemical signals. These snails crawl upward, and even out of an aquarium, when they smell crushed snail (Covich et al., 1994).

Many species of bivalves are also able to move about using their flexible and muscular foot. They can either burrow into sediments or move along the surface of sediments using movements of the foot. The adult zebra mussel (which has recently invaded North America from Europe) glues itself down by attaching strong filaments (**byssal threads**) between its shell and a hard surface (figure 5.1; plate 32). Perhaps the larvae of zebra mussels are free-swimming and planktonic to make up for this stationary habit of the adults.

Feeding Preferences and Behavior

Although there are predaceous marine snails, the freshwater species are herbivorous. Snails use their toothy radula to scrape food off surfaces and even consume delicate, aquatic vegetation. Clams make a living by filtering fine particles of living or dead organic matter out of the water.

An unexpected consequence of the zebra mussel invasion of North American lakes has been an increase in water clarity. In some lakes (such as the western basin of Lake Erie), the mussels are abundant enough to compete with zooplankton for algae (although a minor competitor; Culver, 1992; Yu and Culver, 1999).

Life Cycle and Diapause

Mollusks are typically hermaphrodites, with both male and female reproductive systems active at the same time in adults. Mollusks avoid self-mating, but seek another individual. When two individuals mate, they often enjoy reciprocal mating, with the two individuals exchanging sperm. Clams and mussels broadcast gametes into the water.

Freshwater snails lay many eggs together in clumps of clear, jelly-like mucus. The ciliated **veliger** larvae develop inside the egg, eventually producing a small snail. For a brief time, the juvenile snails swim about inside the egg. However, as soon as they hatch, the larvae settle down and look and act like the adults.

Zebra mussel eggs are fertilized in open water. The zygotes develop quickly into ciliated veliger larvae that go through several morphological changes before growing a shell and attaching to a suitable substrate (Ackerman et al., 1994). Most other freshwater mollusks lack the motile larval stage.

Bivalves have a more complicated life cycle. In unionid clams, the larvae (less than 1 mm long) leave the eggs and swim up into the water to find a fish. The parasitic larvae burrow into the side of the fish and live just below the skin for several weeks. Then they leave the fish, change their shape and begin producing a shell, and settle to the bottom to take up life as a clam or mussel. Pea clams do not have a swimming larva. The juveniles that come out of the eggs resemble the adults.

Clams show a great deal of diversity in life history. Unionid clams typically reproduce several times and have clutches of hundreds of thousands of very small eggs. Pea clams typically produce one clutch of a few, relatively large eggs.

Adult mollusks can withdraw into their shells and outlast periods of dryness. Among the snails, pulmonates are the best at surviving a dry period and are the species most often found in temporary ponds. Pea clams also do well in temporary ponds and are able to withstand weeks or even months of dryness. The ability of zebra mussels to close their shells and live outside water for days has facilitated their recent dispersal to lakes and rivers across the north-central United States.

Economic Importance and Management

Mollusks in North America have supported a modest industry along major rivers. The native unionid mussels are harvested in large numbers as a raw material for buttons, freshwater pearls, and as a source of carbonate. The result of this exploitation, coupled with river pollution, dredging, and impoundment, is the extinction of some species and the reduction of many populations to a few individuals. Currently, more than 70% of our 270 native mussel species are listed as recently extinct, endangered, threatened, or of special concern.

While most native bivalves are conservation issues, a few exotic species (e.g., zebra mussel, quagga mussels) pose serious economic and ecological threats to North American freshwater environments. These mussels grow in and on man-made structures, such as buoys and water intake pipes. Zebra mussels are famous for growing on top of native freshwater bivalves and contributing to the decline of the native species. While individuals of exotic mussel species can be easily killed in small areas, they have proven impossible to exterminate. The United States Geological Service (USGS) has an excellent website for nonindigenous (exotic) species of vertebrates, invertebrates, and plants (the website address is listed at the end of this chapter).

FISHES—PHYLUM CHORDATA, SUBPHYLUM VERTEBRATA

Fishes (Class Pisces) fascinate both limnologists and the general public because they are ecologically and economically important in most freshwater environments. For many people, fishing formed their first impressions of the importance and fascination of water.

Taxonomy and Diversity

Fishes are a kind of vertebrate that live in water. A vertebrate has an internal skeleton of cartilage or bone and a dorsal (spinal) nerve cord protected by cartilage or bone (the **vertebrae**). Vertebrates never have more than two pairs of fins or legs on the sides of the body, and for this reason are called **tetrapods** (animals with four "feet"). Vertebrates living today include the fishes (such as sharks and bony fishes) and the salamanders, reptiles, birds, and mammals. Fishes are distinguished from other aquatic vertebrates by having fins instead of legs, and by being more or less the same temperature as their habitat.

Fishes are the most species-rich group of vertebrates, with more species (over 24,000) than there are of all other vertebrates combined (about 23,500 species). Fishes are grouped into three classes (table 5.1):

- Agnatha—The jawless fishes such as lampreys; not common in fresh water, although of economic importance as parasites.
- Chondrichthyes—Sharks and related fishes with cartilaginous skeletons; not common in fresh water, restricted to large rivers and estuaries near the sea.
- Osteichthyes—Bony fishes; the most diverse in fresh water.

The bony fishes make up the vast majority of freshwater fish groups (table 5.1). Sharks and rays only rarely occur in fresh waters. Most freshwater bony fishes are in the Division Teleostei and have, besides bones, spines in their fins and lip bones that can be extended out in front of the mouth. Teleosts typically have a mouth toward the front, two pairs of lateral fins (pectoral and pelvic), and several fins along the midline of the body: dorsal on top (sometimes divided into a front and a back dorsal fin), sometimes a small fin on

Table 5.1 Examples of Ecologically Important Aquatic Fishes

Group	Representative Common Name	Global Number of Species In or Using Fresh Water
Agnatha	Lamprey, including the invasive sea lamprey	41
Clupeiformes	Herring (also shad, gizzard shad)	80
Cypriniformes	Minnows (also carp, dace, koi, zebrafish, suckers)	2662
Cyprinodontiformes	Mosquitofish (also guppies, mollies, platies, pupfish)	805
Esociformes	Pike (also pickerel, muskellunge, mudminnows)	10
Perciformes	Perch, (also bass, walleye, darters, sunfish, drums)	2185
Salmoniformes	Salmon (also trout, char, whitefish, ciscoes, graylings)	66
Scorpaeniformes (Order)	Sculpins	62
Siluriformes (Order)	Catfish	2,280

top but near the back (the adipose fin), the tail fin, and a fin on the bottom of the body, just behind the anus (the anal fin).

External Structure, Appearance, and Anatomy

Adult size varies from a few centimeters to a few meters. The shape of freshwater fishes varies from the elongated, snake-like eels and lampreys, chunky catfish, and flat-sided sunfish, to streamlined herrings and salmon (figure 5.2). Young, nondescript larvae (still carrying some of the egg yolk from the egg) hatch from eggs. Larvae typically grow at first using energy stored in the yolk of the egg that produced them. As the larvae grow, they are able to begin to feed. Larvae grow and change form and color pattern to eventually become adults.

There is a close relationship between a fish's shape and the kind of food it eats (table 5.2). For example, the lampreys (figure 5.2a) either consume organic bottom sediments and the associated meiofauna, or are parasites with little streamlining of their elongated body. Lampreys also lack paired fins on the sides of the body. The mouth (which lacks a jaw bone) is surrounded with several rows of teeth and is used for sucking up organic material from the sediments or for sucking body fluids out of other fishes (a habit of the sea lamprey). Parasitic lamprey species feed by clamping the mouth onto the body of a host fish, by contracting the muscles around

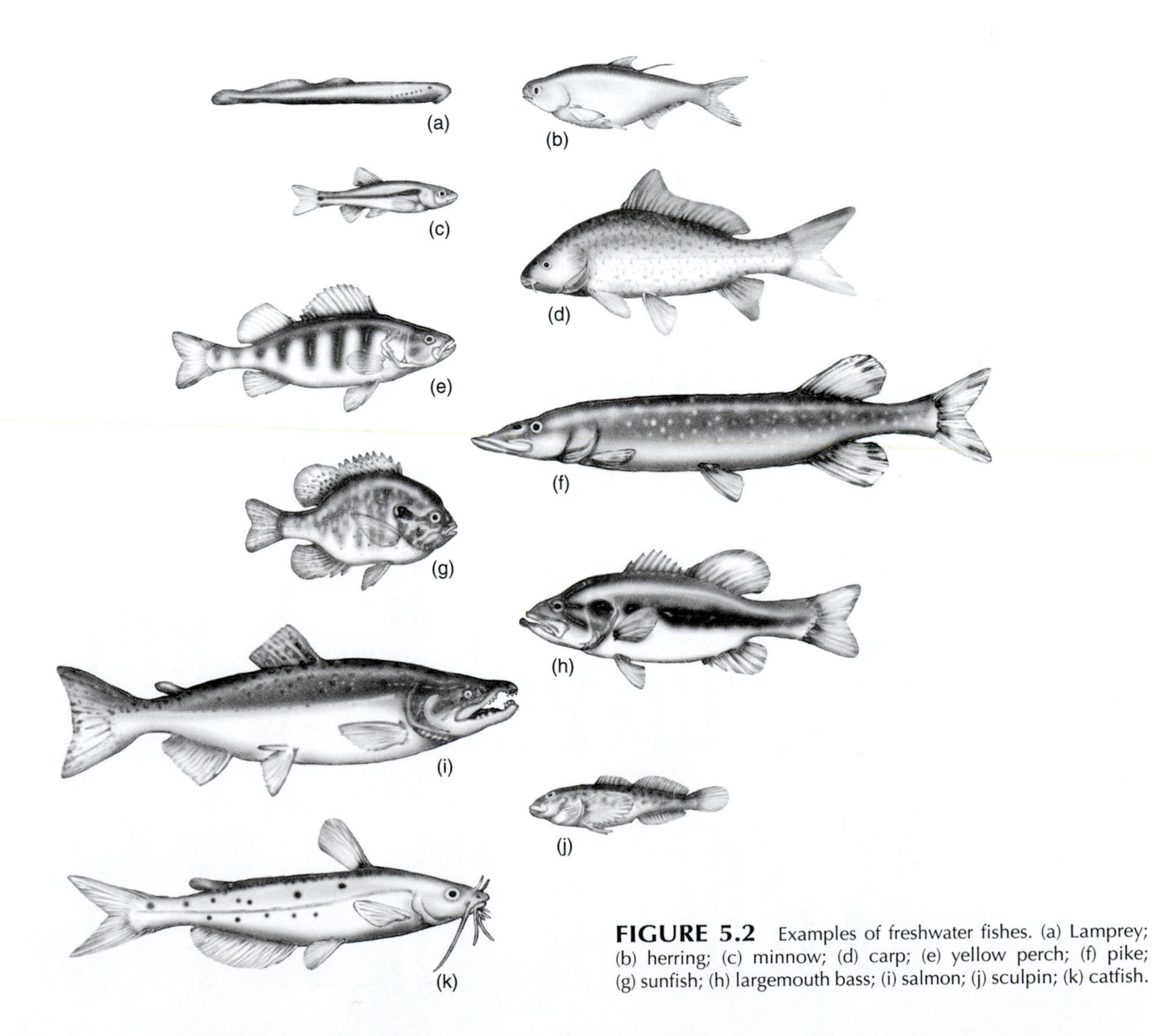

FIGURE 5.2 Examples of freshwater fishes. (a) Lamprey; (b) herring; (c) minnow; (d) carp; (e) yellow perch; (f) pike; (g) sunfish; (h) largemouth bass; (i) salmon; (j) sculpin; (k) catfish.

Table 5.2 Relationships Between the Kind of Food Eaten, Where the Food Occurs, and the Morphological and Behavioral Specializations Fish Use to Capture Food

Feeding Preferences	Fish	Habitat	Morphological Specialization	Swimming Behavior
Parasitic feeding mode—preferred food is blood of prey	As adults, large fish that are external parasites such as the lamprey (figure 5.2a)	Larvae hatch in small streams; juveniles migrate to lakes where they attach to the outside of fish such as lake trout	No jaws; a circular patch of teeth around the mouth for making a suction fit to the host fish and to cut into the host's flesh to drink blood; fins are poorly developed, body is elongated but not streamlined	Can swim between hosts, but prefer to hitch a ride
Zooplankton, usually the larger individuals	Many small (especially larval fish) and larger plankton-eating fish; includes the herrings and salmonids (figure 5.2b, i)	Open lake water or open patches of water in beds of aquatic plants, or flowing water	Elongated, torpedo-shaped; streamlined bodies; mouth near the front of the body; teeth poorly developed; large eyes with some binocular, forward vision; fine gill filaments for sieving plankton out of water sucked into the mouth cavity	Steady swimming especially during the day; fast swimming is possible but slow swimming is more efficient for seeing and catching zooplankton
Zooplankton and algae	A few fish, such as the gizzard shad, are filter-feeding and omnivorous	Open, often turbid water of eutrophic lakes and large rivers	Similar to the fish specialized for feeding on zooplankton	Steady and efficient swimming
Aquatic insects	Most fish, but especially the sunfish (figure 5.2g)	Surfaces of aquatic plants, water surface	Protrusible lips and a large mouth cavity, making a powerful suction; teeth poorly developed; body not streamlined but very maneuverable	Hang in the water and turn on the vertical axis of the body to inspect vegetation
Aquatic plants	No good North American examples: introduced carp (figure 5.2 d) and tilapia	Shallow water	Large, muscular lips; mouth on bottom of head; large pads of teeth for grinding food; body not streamlined	Slow, cruising meandering through vegetation; pausing to eat
Aufwuchs	Especially the mosquitofish group	Surfaces of aquatic plants, stones	Mouths specialized for scraping food off surfaces; small bodies for feeding in small spaces	Slow, erratic movements; hovering over surfaces

Bottom organisms (insects and mollusks) and carrion	Catfish, carp, suckers, sculpins (figure 5.2d, j)	Sediments of lakes and streams	Body dorsal-ventrally flattened; large, muscular lips; teeth and eyes poorly developed, no binocular vision; mouth on bottom of head and often surrounded by chemosensitive tentacles (barbels or whiskers)	Resting on bottom or moving slowly and erratically across the bottom; rooting in the sediments
Other small fish and amphibians	Bass, walleye, pike group, and salmonids (figure 5.2f, h, i)	Open water or among vegetation or rocks, still or flowing water	Large mouth, sharp teeth to hold prey; some binocular vision; body powerful and streamlined for swimming in short, fast bursts	Long periods of immobility punctuated by fast bursts of swimming
Larger fish and other vertebrates (mice, ducklings)	The larger bass, walleye, pike group, and salmonids (figure 5.2f, h, i)	Open water or among vegetation and rocks or the lake or stream surface	Large body and mouth; large, sharp teeth to hold prey; some binocular vision; body powerful and streamlined for swimming in short, fast bursts	Long periods of immobility punctuated by fast bursts of swimming

These generalizations are based on information in Becker, 1983.

the oral cavity. The parasite uses the rasp-like teeth to grind a hole in the host's side and then sucks the host's blood. The relationship between form and function is described for other major fish groups in table 5.2.

Many species of fish (such as sunfish, darters, and trout) are brightly colored. The colors are used for territoriality and mating behaviors.

Gills are an important internal part of the fish. The fish throat has openings that connect the inside of the mouth cavity to the outside water. Gills, associated with these openings, are flat pieces of tissue well supplied with blood vessels. The fish gains oxygen and loses carbon dioxide as water flows into the mouth, through the throat openings, and over the gills. When the water holds no oxygen, some species of fish can gulp air into the mouth to oxygenate the moist gills (Klinger et al., 1982).

Many fish species also filter small particles from water using their mouth cavity and gills. Fishes often capture prey by sucking it into the mouth. The prey item is separated from the water by straining the mouth contents through the gills. Gills catch the particles, which the fish then swallows and, perhaps, grinds them to a fine mush using hard pads in its throat.

Habitat, Swimming, and Feeding Preferences

Fishes are ordinarily difficult to see in their natural habitat. However, they can be collected by nets or, less destructively, with electro-shocking (plate 17). A moderate electrical current stuns the fish, which then floats to the surface. The fish often recovers within a few minutes. Electro-shocking allows the limnologist to easily sample streams and lakes to determine fish population size, reproductive condition, and distribution; to mark fish for studies of growth; and to collect fishes for studies of their diet.

There is a strong relationship between a fish's feeding preferences, the habitat in which it occurs, and its swimming behavior (table 5.2). Different species of fishes eat a wide variety of food items from different locations. Some are specialists:

Planktivore: Eats plankton (this usually means zooplankton)

Piscivore Eats other fishes (many piscivores are cannibals!).

Detritivore: Eats organic material from sediments, including living insects and other animals, and dead, often decaying material

Herbivore: Eats plants, either algae or macrophytes

Benthivore: Eats food on the bottom (benthos)

Omnivore: Eats both plant and animal food

Fishes are selective predators, taking those prey they can detect and catch (O'Brien, 1979). Fish size has an effect on its diet (see table 5.2). Many fish species will eat any animal (alive or dead) that is available and smaller than the mouth. Larval fishes, usually a few millimeters long, are often able to eat only the smallest plankton, such as protists and rotifers. Fishes a centimeter or so long can eat the larger plankton, including copepods and cladocerans. Fishes in the range of 2–20 cm can eat aquatic annelids, mollusks, insects, and the larger plankton. Prey smaller than about 1 mm are relatively safe from fish predators in the 2–20 cm range, perhaps because the prey smaller than 1 mm are difficult to see. Fishes larger than about 20 cm are large enough to take smaller (often larval) fish and larger invertebrates such as crayfish. The largest freshwater fishes (50–100 cm and even larger) take smaller fish, ducklings, amphibians (frogs and salamanders). Larger size means faster swimming speed, which automatically benefits large predators seeking to capture smaller prey.

Fishes often use vision to find prey (O'Brien, 1979; Zaret, 1980). Consequences of visual hunting are that small prey are often overlooked and predation is often restricted to daylight hours. The physiology of vision has a direct influence on feeding times, locations, and appearance of prey.

Fish swimming behavior optimizes feeding (O'Brien et al., 1990). Most fishes exhibit a **saltatory** search behavior, in which they are at rest for short periods of time and search visually for prey, but then move a short distance to a new location. Prey detection is inefficient while the fish is moving, but fishes that do not move have low encounter rates with prey.

Fishes have sensory modes other than vision for locating prey. Bottom-feeding fishes grope for prey in murky water and mud using their well-developed sense of taste. All fishes have a line of pressure-sensitive cells running along each side of the body. These lateral lines of sensory cells can probably detect vibrations made by swimming zooplankton. For example, laboratory experiments showed that the alewife could capture zooplankton in the dark, and even show size-selective choice of prey (Janssen et al., 1995). It is also possible that some fishes can detect prey, in the dark, using the disturbance that the prey creates in the electrical field around a fish. Paddlefish have a long, flat snout (rostrum) covered with sensory cells that detect electrical field strength. Experiments suggest paddlefish use this system to find zooplankton prey (Wilkens et al., 2001).

Herbivory is rare in freshwater fishes (see table 5.2). There are no native North American fishes that specialize in eating vegetation. However, we do have introduced herbivores. The European carp is a good example of a common fish that actually consumes aquatic plants. Algae and aquatic plants are typically safe from direct consumption by fish. Bottom feeders, such as carp, suckers, and catfish, strongly disrupt aquatic plant communities by uprooting plants.

Fishes are an important food web component that directly affect prey species because of intense and selective predation. Fishes also have indirect effects on other organisms. Bottom-feeding fish, such as carp, muddy the water, thereby reducing algal productivity and removing or reducing macrophyte beds. Intense predators not only eat prey, but can stimulate primary productivity by excreting nitrogen and phosphorus fertilizers (Vanni et al., 1997).

Life Cycle and Diapause

A few species (such as lungfish—from Africa or Australia) can survive dry periods, but the vast majority of fishes require the constant presence of water. Thus, freshwater fishes typically lack a diapausing life-history stage.

All kinds of life histories are exhibited among the vast diversity of fishes. In fresh water, the rule is separate sexes and no changing of sex during the lifetime. Adults can reproduce once or many times, but usually only once a year. Females produce anywhere from a few, large eggs to thousands of small eggs. Large eggs develop more slowly than do small eggs and are therefore laid earlier, so that hatching can occur at the beginning of the warm-water season (Gillooly & Dodson, 1999).

Eggs typically hatch immediately after being laid; young of some species are born live. Sometimes adults in elaborate nests care for the larvae with complex, protective behaviors; sometimes the larvae (especially of open-water species) must fend for themselves.

Young are produced during the warmer part of the year and typically take 1 to 3 years to mature. Eggs often hatch at the beginning of the warm-water season. Larval fishes first feed and grow in the protection of littoral zones of lakes or streams. After they have grown large enough to eat zooplankton, larval fishes of many lake species move out of the littoral zone and into the pelagic zone. Females are typically larger than males. Adults live for several years, with the larger species having life spans of decades.

As with most animals, there is strong mortality during early development (Raffetto et al., 1990) and only a small fraction of the eggs live long enough to reproduce (table 5.3). In northern Wisconsin, smallmouth bass males start reproducing when they are 2 or 3 years old and then die. Females start reproducing at age 4 and have good survival for 2 or 3 more years.

Table 5.3 Survival of a **Cohort** of Smallmouth Bass Observed in Nebish Lake, Vilas County, Wisconsin

The numbers represent totals for the entire lake. A cohort is a group of individuals that started life at the same time.

Time	Stage	Total Number	% Survival of Original Cohort
1982—May	Eggs put into nests	200,000	1.0000
1982—June	Larvae coming off the nest	15,000	0.1100
1982—September	Young of the year	2810	0.0140
1983—May	1-year-old juvenile fishes	Approx. 2810	
1984—May	2-year-old juvenile fishes	947	0.0047
1985—May	3-year-old adults	861	0.0043
1986—May	4-year-old adults	387	0.0019

Source: Data courtesy of Nancy Raffetto.

Economic Importance and Management

Fishes are the focus of major economies and management efforts. Because fishes are so valuable, there are myriad restrictions on how they can be harvested, transported, and sold. Even with intensive management, several desirable species, such as lake trout in the North American Great Lakes and several species of western North American salmonids, have been fished to extinction or to endangered levels. On the other hand, avid fish managers and anglers have aided the distribution of desirable fishes (such as largemouth bass and several trout species) to many lakes and streams where the new predator has had a significant ecological impact (see chapter 7).

A few fishes (such as ruffe, round goby, and grass carp) are considered exotic nuisances, in need of either management or extermination. The spread of these species can be slowed by good management practices, such as not releasing bait fish into lakes and streams. Fish species can be poisoned out of lakes, and this management practice has sometimes produced a restored fish community of more desirable species.

The control of the sea lamprey in the North American Great Lakes is a lesson in the expense and difficulty of managing fish populations at a large scale. Sea lampreys are recent invaders to the upper Great Lakes, present in Lake Erie since 1921 and first seen in the upper lakes in the 1930s. These blood-sucking parasites have had a significant, detrimental effect on other large fish species, including native lake trout, burbot, and the larger planktivores in the lake—several species of white fish, including cisco.

The first efforts (in early 1950s) of management agencies to reduce lamprey populations took advantage of the lamprey's migratory behavior. Adult sea lampreys live in the Great Lakes but migrate to fast-flowing, cold-water tributary streams during spring and early summer to lay eggs (**spawn**) in a nest built in gravel. Larvae hatch from the eggs and remain in the nest for several months before migrating back to the lake, to mature in 3 to 4 years. To inhibit these migrations, wire mesh weirs or low dams were built across tributary streams. These measures were somewhat effective, although plagued by ice damage. Weirs also prevented the passage of migrating desirable species (e.g., trout). Effective but expensive lampricides (poisons, especially 3-trifluoromethyl-4-nitrophenol, or TFM) were applied to streams starting in 1958, and the treatment continued each year since, at a current cost of about $10 million per year. A total of 1064 tons of lampricide were used in Lakes Huron, Michigan, and Superior between 1958 and 1978. As a result of this management practice, lamprey numbers were low by the early 1970s and the occurrence of lamprey scars became lower than 5% on trout and salmon caught in the sport fishery. Although TFM is not acutely toxic to most organisms, at the concentrations used in application, there is concern that the chemical will adversely affect growth and development of aquatic organisms (Hewitt & Servos, 2001).

OTHER VERTEBRATES—CLASSES AMPHIBIA, REPTILIA, AVES, AND MAMMALIA

Taxonomy and Diversity

Along with fishes, the other vertebrates are familiar to us terrestrial beings. Because reptiles, amphibians, birds, and mammals are so familiar, many excellent references exist for their identification. For example, the *Peterson Field Guide* series is a good resource for identification. Addresses for useful websites are given at the end of this chapter.

Each vertebrate group plays one or more significant roles in limnology, even though there are relatively few species in any of these groups. Amphibians are most closely related to fishes. There are less than 100 species of amphibians in North America, divided into the frogs, toads, and salamanders and newts. Worldwide, there are 23 species of crocodilians (all aquatic), about 5000 species of frogs and toads, and about 500 species of salamanders (many are aquatic). There are a few hundred aquatic snakes, about 300 kinds of turtles (most are aquatic), and only a few lizards living close to fresh water.

External Structure, Appearance, and Anatomy

Unlike the fishes, the other vertebrates typically have four legs rather than fins, at least in the adult life stages. The notable exception are snakes, whose legs are reduced to internal vestiges. A number of excellent websites are cited at the end of this chapter as a source for overviews of appearance and identification of vertebrate animals.

Habitat

Amphibians are the major group of tetrapods that spend a significant portion of their time in fresh water (ignoring the few species of marine mammals that have reinvaded fresh water). Most juvenile amphibians live a fish-like existence and then leave water as adults. Some freshwater snakes and turtles lead a life under water, but they all breathe air, so they are not entirely aquatic. Birds can feed in water, but much of their life is spent floating on water or away from water entirely. Aquatic mammals swim and dive into water, but also typically spend a significant portion of their time on land. Aquatic tetrapods often use burrows as refuges to hide. Refuge use may be limited to part of the day, or for longer periods to survive hot weather (**aestivation**) or to survive cold weather (**hibernation**), or to take care of young. Some animals, such as the platypus, build elaborate burrows in stream banks. Beavers and muskrats build artificial burrows of coarse, woody debris in streams and along lake shores. The population of beaver has been greatly reduced in North America through hunting and habitat modification. At one time, most streams in North America hosted several beavers per kilometer. The dam-building habit of beavers was once a major force shaping the riparian and even terrestrial habitats (Naiman et al., 1988). The shallow beaver ponds captured sediments and flooded low-lying forest (riparian forest) along the stream.

Swimming and Escape Behavior

Tetrapod vertebrates are relatively large animals, usually between 10 cm and 1 meter or so long. This large body size allows for fast and powerful movement. Familiar images are crocodiles lunging out of shallow water, snakes striking frogs, and ospreys diving into water to catch a fish.

Feeding Preferences and Behavior

Because of their large size, many vertebrates tend to be at the top of their food chain (table 5.4). These are the large, charismatic animals we all know from visiting national parks and watching nature programs on television. Predaceous aquatic species include adult amphibians, all aquatic snakes, and most of the birds.

Life Cycle and Diapause

Aquatic vertebrates have a familiar life cycle. Most species have separate sexes, eggs are fertilized, and the zygote develops into a larval (or embryonic) form, then a juvenile form, and eventually an adult stage.

Eggs of amphibians are typically laid in water, in jelly masses, while aquatic snakes and reptiles lay their leathery-shelled eggs in moist vegetation or sand near water. All aquatic mammals have internal development of the eggs (except the wonderful platypus!), and all birds incubate their eggs out of the water.

These animals do not show true diapause, although many species can enter a low-metabolism condition (hibernation, aestivation, or torpor) to persist through weather that is too hot or too cold for normal activity.

Economic Importance and Management

Aquatic vertebrates are at the forefront of exploitation, management, and conservation. Humans exploit these animals for food and sport. Many of the species (especially frogs, birds, and mammals) are economically important food sources or the target of avid hunting. Many of the species (especially crocodilians and snakes) are considered nuisances and have been hunted to the point of near extinction. On the other hand, were it not for concerned hunters and anglers, there would be far fewer nature preserves. The efforts of goose and duck hunters have probably resulted in more conservation of aquatic habitats than has any other factor. Many of the species are also inadvertently endangered as a result of habitat destruction—anthropogenic changes in land use.

The future is not entirely bright for aquatic vertebrate animals, especially the larger species. Chapter 12 discusses some measures that humans can take to appreciate and conserve these fascinating and ecologically important animals.

PLANTS—AQUATIC MACROPHYTES

Taxonomy and Diversity

Animals absorb food inside a gut and exchange gasses with lungs or gills, which are also often inside the body. Plants put their surface areas on the outside and absorb nutrients using root hairs. They fix energy and exchange gases using their leaves. In this sense,

> "Plants are inside-out animals."
>
> *Professor Tim Allen*

Plants get their energy via photosynthesis, which allows the production of high-energy organic molecules, starting with sunlight, carbon dioxide, and water. A few species of plants do eat animals, but only for the

Table 5.4 Feeding Preferences of Aquatic Vertebrates Other than Fishes

Group	Examples (common names)	Preferred Prey
Crocodilians (figure 5.3g)	Alligators and caimans	All are predaceous, eating smaller fish and animals of all kinds—any animal larger than 1 cm or so in or near water is likely to be eaten if it is not too much bigger than the predator's mouth.
Amphibians (figure 5.3a–d)	Adult frogs and toads; all stages of salamanders and newts	Terrestrial insects, smaller amphibians, aquatic insects, worms, and aquatic microcrustaceans; larval frogs and toads tend to be herbivorous (Dodson & Dodson, 1971).
Snakes (figure 5.3e)	Water snakes and many species, such as cottonmouths, that live near water or lie on branches of trees overhanging water, to drop on aquatic prey	All are predaceous, eating insects, fishes, amphibians, other snakes, birds of all developmental stages, and small mammals.
Turtles (figure 5.3f)	Most turtles are aquatic; are often seen sunning on logs or rocks near water	Turtles are omnivores, eating (depending on the species) vegetation, smaller animals, or scavenging dead animals in or near water; snapping turtles are large and common scavengers.
Birds (figure 5.4)	Fish-eating hawks and eagles (figure 5.4d), kingfishers, dippers (figure 5.4a), swallows (figure 5.4b), and most kinds of ducks and other waterfowl such as flamingos (figure 5.4e), cranes (figure 5.4f), swans, phalaropes (figure 5.4c), and loons	Birds have specialized diets. Raptors such as bald eagles eat only meat, including carrion from fishes and other vertebrates, including other birds. Kingfishers favor small fishes. Cranes and large wading birds tend to eat small fishes and amphibians, although a few wading birds, such as spoonbills and flamingos, specialize on microcrustaceans and algae in estuarine and saline habitats (Arengo & Baldassarre, 1999). Dippers seek aquatic insects under water, and swallows collect flying insects above the water. Waterfowl tend to be omnivorous, although some (such as geese) are mostly herbivorous, while others eat only fishes (loons and some species of ducks). Phalaropes prey on aquatic insects and planktonic microcrustaceans (Dodson & Egger, 1980).
Mammals (figure 5.5)	Many mammals forage near water, but relatively few species specialize on feeding in aquatic systems in fresh water	A few predaceous species of marine mammals that live in fresh water, such as the Ganges River dolphin and the Lake Baikal seal (figure 5.5d), are strict fish eaters. Other predaceous mammals that are particularly aquatic include bears (figure 5.5e), which prefer fishes, and otters, which favor small fishes, mollusks, and aquatic macrocrustaceans. The platypus (figure 5.5b) feeds on stream invertebrates (Serena et al., 2001). On the miniaturized end of the predator scale, water shrews (figure 5.5a) seek small fish, mollusks, and aquatic insects in and under water. Herbivorous mammals include beavers (figure 5.5c), which feed mainly on riparian vegetation and aquatic rodents, from small mice to muskrats and nutria, to the giant capybara (60 kg!), which specializes on aquatic macrophytes. Moose (figure 5.5g), and probably water buffalo, use aquatic vegetation as a salty supplement to terrestrial vegetation (Pastor et al., 1988), while manatees (figure 5.5f), dugongs, and the hippopotamus stick strictly to aquatic macrophytes.

FIGURE 5.3 Representative aquatic reptiles and amphibians. (a) Frog tadpole (larval form) of the amphibian *Rana;* (b) adult *Rana;* (c) salamander larva of the amphibian *Ambystoma;* (d) adult *Ambystoma;* (e) cottonmouth snake (also called water moccasin—*Agkistrodon*); (f) snapping turtle (*Chelydra*); (g) American alligator (*Alligator*).

inorganic nutrients, such as nitrogen, and not to gain energy-rich food.

The large plants in aquatic systems, called **macrophytes,** include both large algae (macroalgae), such as *Chara,* and flowering plants. Macrophytes are the ecologically dominant plants in most lakes and slow-moving streams. These are the plants that most often make up the "weed beds" in shallow water. Other nonflowering plants (mosses, liverworts, and ferns) can be important in some (usually low-light) habitats, but rarely contribute more than a tiny fraction of the total primary production of a lake or marsh.

Macrophytes are a diverse group in aquatic systems, including several forms of macroalgae and at least 2 dozen families of flowering plants contributing several hundred species. Aquatic macrophytes include emergent and floating species that are partly out of water (figure 5.6) and submerged species that live wholly within water (figure 5.7).

This section will use examples from only a few of these major (abundant and diverse) families of aquatic macrophytes (table 5.5). These examples demonstrate the range of diverse growth forms and ecological strategies of aquatic flowering plants.

FIGURE 5.4 Representative species of birds associated with fresh water. (a) Dipper (*Cinclus*); (b) bank swallows (*Riparia*); (c) red phalarope (*Phalaropus*); (d) bald eagle (*Haliaeetus*); (e) American flamingo (*Phoenicopterus*); (f) sandhill crane (*Grus*).

External Structure, Appearance, Anatomy, and Habitat

In macrophytes, there is an especially close connection between morphological form and habitat (Givnish, 1995) (see table 5.5). Macrophytes optimize the tradeoff between energy allocated toward strength of their stems and energy used to build leaf area within the budget of their photosynthetic income. Because of the great power of flowing water, aquatic plants attached to the bottom require either quiet water (as in shallow, protected pools or bays) or

FIGURE 5.5 Representative aquatic and water-loving mammals. (a) Northern water shrew (*Sorex*); (b) platypus (*Platypus*); (c) beaver (*Castor*); (d) Baikal seal (*Phoca*); (e) grizzly bear (*Ursa*); (f) Florida manatee (*Trichechus*); (g) moose (*Alces*). Animals not drawn to the same scale.

great simplicity in morphology so as to present the least resistance to the stream current. Plants found in moving water tend to be short, with strong, often photosynthetic stems.

Plants that rise above the water can shade their competitors, but need expensive, stiff stems. Plants that live rooted in deep water (such as water lilies) need storage containers (rhizomes or bulbs) to store energy for growing leaves toward the surface in the spring, before algae shut off the light supply. Rooted aquatic plants that are submerged in deep, quiet water need only delicate stems and leaves. Floating plants, without

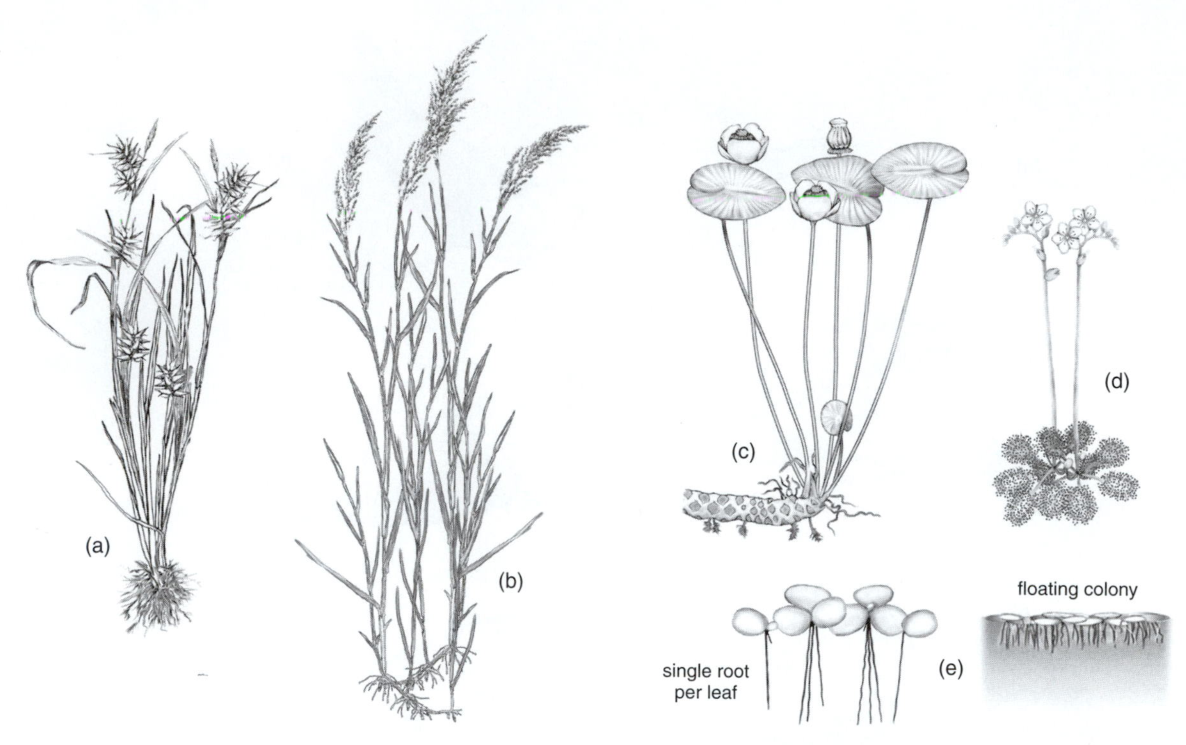

FIGURE 5.6 Representative aquatic plants: wetland, emergent, and floating macrophytes. (a) Sedge (*Carex,* about 1 m tall); (b) grass (reed canary grass, about 1 m tall); (c) water lilies; (d) the sundew (a few cm in diameter); (e) duck weed (only a few mm in diameter).

stems, can avoid the problems of being rooted to the bottom and are free to move with currents.

Swimming and Escape Behavior

None of these macrophytes swims or escapes predators. A few of the carnivorous plants (such as Venus fly trap and sun dew) are able to move their traps fast enough to catch animals. When touched, *Utricularia* bladders inflate suddenly, to suck in microscopic prey (figure 5.7e).

Many plants have developmental stages specialized for dispersal. Seeds are the best example. Aquatic plants typically have seeds dispersed by water (floating seeds) or small, light seeds released from flowers held above the water and dispersed by the wind. Many aquatic plants also reproduce vegetatively (asexually). Pieces broken off the main plant easily produce roots and begin life after being transported by water currents. Entire plants with few or no roots, such as *Utricularia,* can move with water currents.

Feeding Preferences and Behavior

Macrophytes consume nutrients (mostly inorganic) dissolved in the water and from sunlight. Both nutrients and light can be limiting for growth. Nitrogen and phosphorus are the most important inorganic nutrients that limit plant growth. Each year, there is a race between planktonic algae and aquatic macrophytes for light (Scheffer, 1998). As the planktonic algae become abundant, they shade the lake bottom. Macrophytes typically start growing from the bottom, from germinating zygotes (*Chara*) or from seeds or rhizomes (flowering plants). Therefore, macrophytes can only grow as deep as the bottom of the euphotic zone. Thus, eutrophic lakes have shallow beds of aquatic plants growing only 1 or 2 meters deep, and oligotrophic lakes have macrophytes growing at much greater depths, down to tens of meters.

Life Cycle and Diapause

Macrophytes typically reproduce once a year. The macroalgae reproduce by releasing motile male gametes (single cells with one or more flagellae) into the water.

For flowering plants, the flowers (like adult aquatic insects) need to leave the water in order to reproduce. Most aquatic plants, even submerged species, hold their flowers up out of the water on stalks. Larger, aquatic flowers in the air can be pollinated efficiently by wind and animal vectors such as bees, flies, and birds. Seeds mature during the warm season and often enter a diapause stage, which lasts through the cold

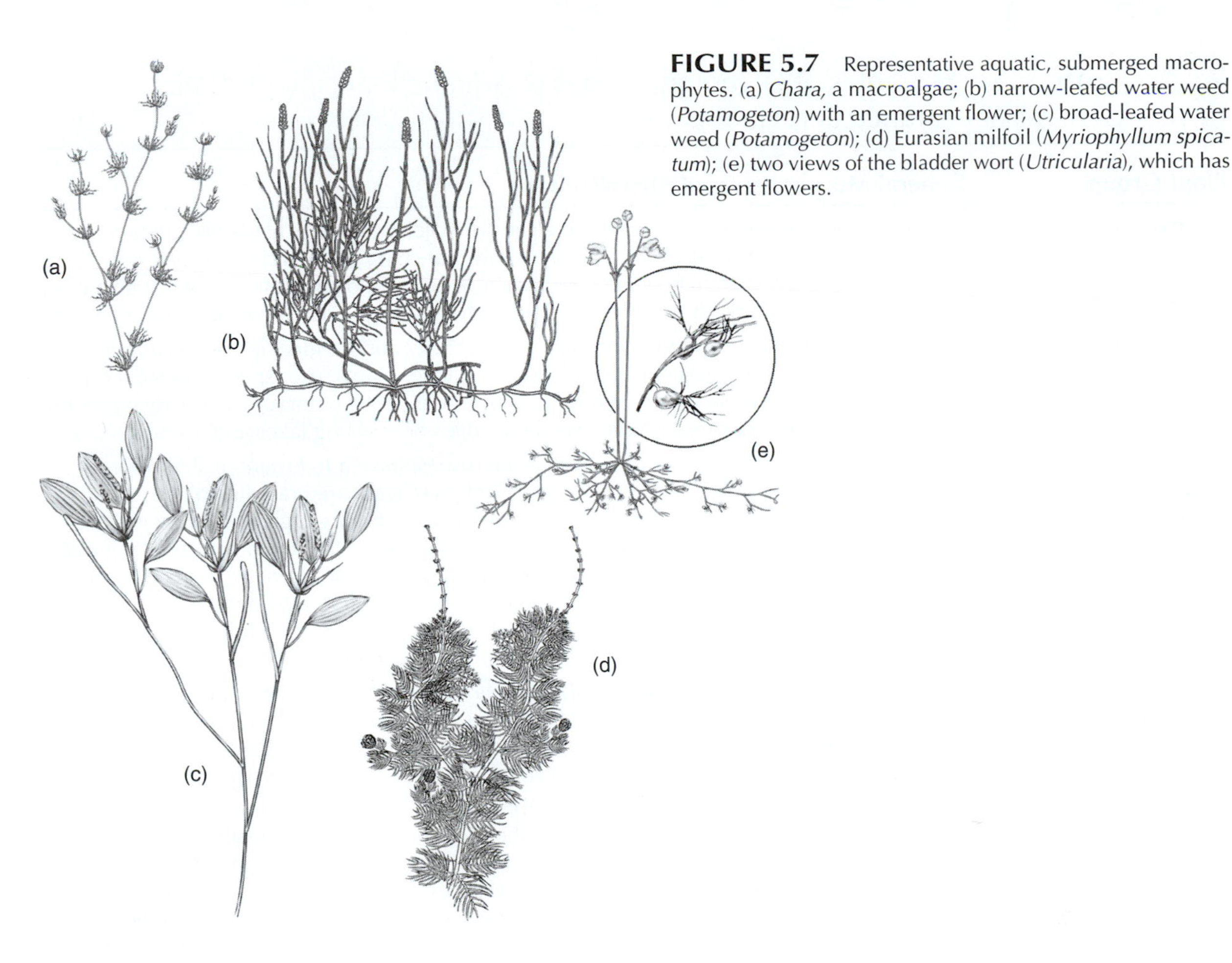

FIGURE 5.7 Representative aquatic, submerged macrophytes. (a) *Chara,* a macroalgae; (b) narrow-leafed water weed (*Potamogeton*) with an emergent flower; (c) broad-leafed water weed (*Potamogeton*); (d) Eurasian milfoil (*Myriophyllum spicatum*); (e) two views of the bladder wort (*Utricularia*), which has emergent flowers.

months. While in diapause, seeds can withstand some drying and freezing, depending on the species, and are therefore the ecological equivalent of the resting eggs of zooplankton.

Seed diapause is broken by a combination of higher temperature and sufficient moisture and dissolved oxygen. Photoperiod can also be a factor. The young plants must be in the photic zone for successful growth.

Economic Importance and Management

Aquatic plants provide important habitat, refuge, and food for fishes and aquatic invertebrates. The plants are also considered nuisances that need to be exterminated in order to improve the appearance and water quality of an aquatic habitat. Aquatic plants are routinely removed by poison and mechanical harvesting, at great economic expense, and less obvious but real ecological cost.

Aquatic plants are of little direct (positive) economic importance. The once-valuable stands of wild rice have been decimated by the feeding activities of mud-grubbing carp. The aquarium trade demands a constant supply of aquatic plants, which most often come more from greenhouses rather than from nature.

Several species of aquatic plants are considered nuisances that cause significant degradation of water quality and even interfere with swimming and navigation—for example, water hyacinth and Eurasian water-milfoil. An address is given at the end of this chapter for a USGS website focused on management of these nuisance (and exotic) species of aquatic macrophytes.

SUMMARY

With this chapter, we conclude the general discussion of common and ecologically important organisms found in freshwater habitats. The next step in becoming familiar with limnology is to consider the dynamic behavior and interactions of these organisms. Chapter 6 addresses population dynamics, and chapters 7 and 8 focus on ecological interactions among organisms, especially predation and competition.

Table 5.5 Examples of Common Groups of Aquatic Plants, with Their General Appearance and Habitat

Plant Group	General Morphology and Microhabitat
Macroalgae *Chara* (figure 5.7a)	This and other freshwater species of macroalgae are typically made of filaments 1 mm or so thick, and up to 1 m or so long. The filaments show branching patterns characteristic of flowering plants, but each joint of the stem is a single cell. The plant is often coated with a crust of carbonate, precipitated during intense photosynthesis. These macroalgae never have flowers.
Flowering Plants Sundew (*Drosera*) (figure 5.6d)	Delicate plants 1 cm or so across, they use special sticky and pad-like leaves to capture insects. Another group of insect-eating plants, the pitcher plants, uses special folded, cone-shaped leaves that hold water and drown insects. Insect-eating plants grow outside water, on moist soil or on the moss (*Sphagnum*) along the edges of acid bog lakes and acidic streams.
Grasses and sedges (figure 5.6a, b and plate 13)	Grasses, sedges, and related groups, such as the rushes, have a luxuriant, if shallow, root system and often rhizomes (horizontal stems). Sedges (*Carex*) are grass-like plants; most species prefer to root in moist soil or under water (at least during high water). Sedges are found in wet meadows or along the edges of lakes and streams. Groups with similar growth forms and habitat requirements include the cattails *Typha,* rushes (*Juncus*), and bulrushes (*Scirpus*). The stems and leaves are slender and stiff, with strong stems holding up the above-water part of the plant. These plants are the rooted and emergent macrophytes. Wild rice is an emergent grass. Many species of *Carex* are emergent sedges.
Water lilies (figure 5.6c and plate 13)	Large tubers along the bottom have shallow roots and send up slender stems to large, floating leaves. These plants occur in water deeper than that preferred by the sedges and rushes, and use flotation instead of stems to support the large leaves. Because these plants are rooted and have floating leaves, they can grow only in relatively quiet water.
Pond weeds (*Potamogeton*) (figure 5.7b, c)	Shallow roots send up slender stems with submerged or floating leaves (or both), depending on the species. All rooted macrophytes face the problem that mud beneath water tends to be anoxic. Thus, the roots are shallow and many species have air ducts for shunting oxygen into roots in anoxic sediments. Floating leaves tend to be broad, while submerged leaves tend to be highly divided or thin filaments. These plants are transitional between the macrophytes with floating leaves and the submerged macrophytes, such as milfoil and bladderwort. These rooted plants can withstand some degree of current or wave action, as long as they are submerged in the water.
Milfoil (including the invasive *Myriophyllum spicatum*) (figure 5.7c and plate 35)	There are dozens of species of rooted and submerged macrophytes, here represented by milfoil. These are delicate plants with thin, highly divided leaves. They can grow in deeper water than can either emergent or floating macrophytes.
Bladderwort (*Utricularia*) (figure 5.7d)	Some submerged macrophytes, such as *Utricularia,* do not bother with roots. These delicate plants have thin, filamentous stems and leaves. Bladderworts are special in that they have bladders about 1 mm in diameter, used to capture small zooplankton and insects.
Duckweeds (including *Lemna*); Water hyacinth (*Eichhornia*) (figure 5.6e and plate 37)	Duckweeds are floating plants that are mostly leaves a few millimeters in diameter, with delicate roots that hang down into the water. Larger, floating plants include the sometimes pesky water hyacinth and water lettuce. Each of these latter plants is 10–20 cm across, with several leaves, and with a few delicate leaves. All floating plants can, in the right circumstances, completely cover a water body.

Study Guide

Chapter 5 Diversity of Aquatic Organisms: Larger Organisms

Questions

1. Do the major groups of aquatic mollusks, fishes, and macrophytes have anything in common besides size?
2. Why are there no whales in large lakes, such as Lake Baikal or the Laurentian Great Lakes?
3. You are a limnologist studying the effect of size-selective predation of fishes on zooplankton. You want to do an experiment in which fish and their prey are put together into an enclosure in a lake. You want there to be some loss of zooplankton to fish predation, but it is important that not all the zooplankton disappear in the first day. How many fishes should you stock in your 1 m^3 enclosures, so that predation intensity will be approximately "natural"? Natural intensity removes about 10% of the zooplankton each day. You plan to stock the enclosures with 30 zooplankton individuals per liter, to mimic the amount of zooplankton commonly found in lakes. The number (N) of zooplankton a fish eats per day depends on the fish's body length (BL, in cm): $N = 100*BL^3$

   How many 4-cm-long fishes can you stock in each 1 m^3 enclosure?

   How many 10-cm-long fishes can you stock?
4. Explain the statement of Professor Tim Allen that "plants are inside-out animals." Hint: Take into account surface-to-volume logic and consider the various organ systems of plants and animals.
5. What are the relationships among form, food preferences, and microhabitat for fishes?
6. What are the relationships between form and microhabitat for the macrophytes?

Words Related to the Larger Freshwater Organisms

aestivation
benthivore
byssal threads
cohort
cryptic
detritivore
foot
herbivore
hibernation
macroalgae
macrophyte
mucus
omnivore
operculum
piscivore
planktivore
radula
saltation
spawn
tetrapod
veliger
vertebra

Major Examples and Species Names to Know

Mollusca

bivalves (unionid clams and mussels; *Dreissena*, the zebra mussel)

gastropods (snails)

- prosobranch
- pulmonate

Fish
lamprey
herring
minnows
pike
perch
salmon
catfish
Other Vertebrates
amphibians (frogs, toads, salamanders, newts)
reptiles (snakes, turtles, lizards)
birds (raptors, wading birds, phalaropes, ducks, geese)
Plants
sundew
sedges and grasses
water lilies
pond weeds
milfoil (*Myriophyllum*)
bladderwort (*Utricularia*)
duckweed

Additional Resources

Further Reading

Borman, S., R. Korth, and J. Temte. 1997. *Through the looking glass: A field guide to aquatic plants.* Merrill, WI: Wisconsin Lakes Partnership. Reindl Press.

McNeill Alexander, R. 1997. *The invertebrates.* Cambridge, UK: Cambridge University Press.

Nelson, J. S. 1994. *Fishes of the world.* 3rd ed. New York: John Wiley and Sons.

Pennak, R. W. 1989. *Freshwater invertebrates of the United States: Protozoa to mollusca.* New York: Wiley.

Reid, G. K. 1987. *Pond Life: A guide to common plants and animals of North American ponds and lakes.* New York: Golden Press.

Thorp, J. H. and A. P. Covich. 2001. Ecology & classification of North American freshwater invertebrates. 2nd ed. San Diego, CA: Academic Press.

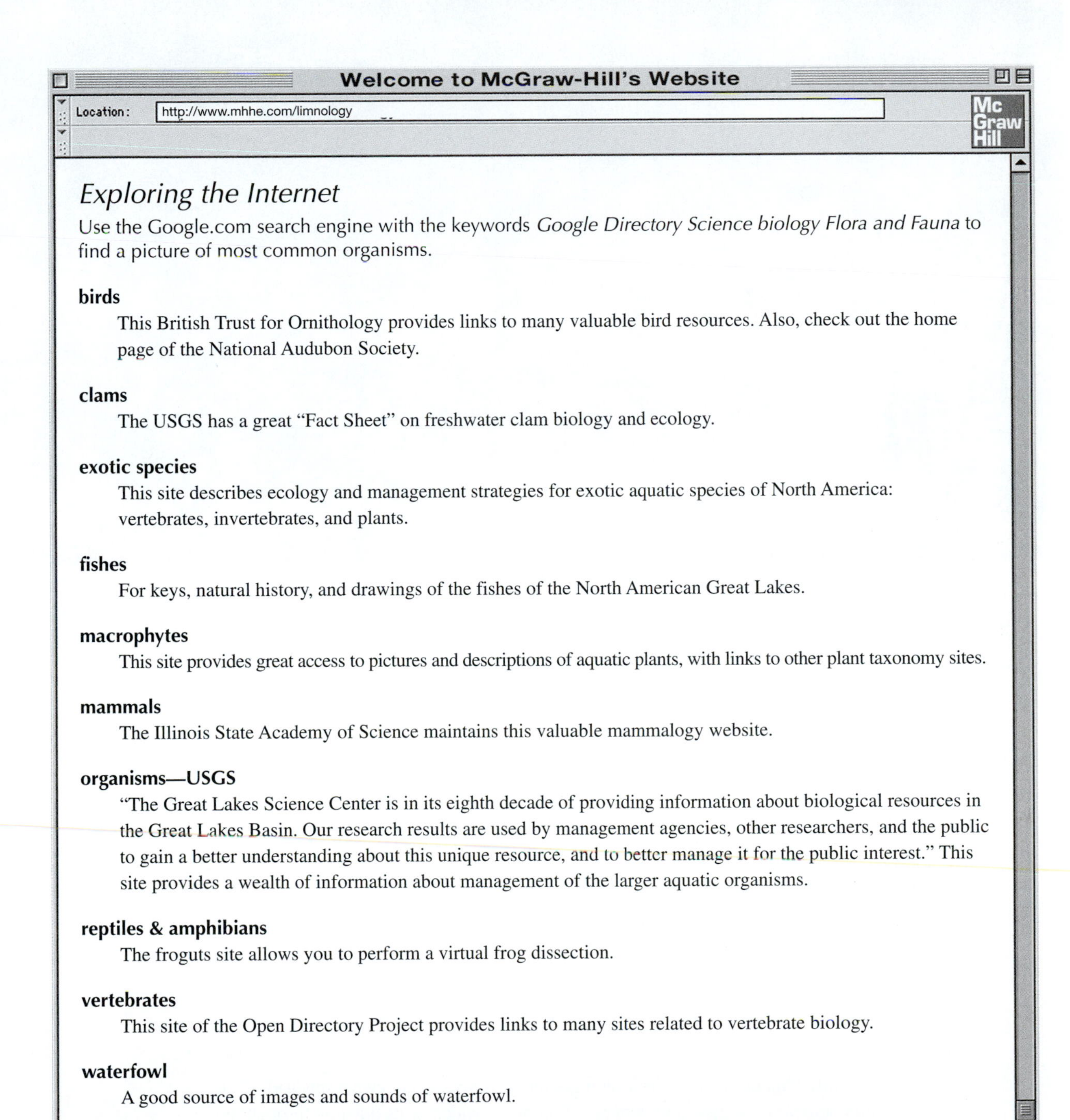

Welcome to McGraw-Hill's Website

Location: http://www.mhhe.com/limnology

Exploring the Internet

Use the Google.com search engine with the keywords *Google Directory Science biology Flora and Fauna* to find a picture of most common organisms.

birds

This British Trust for Ornithology provides links to many valuable bird resources. Also, check out the home page of the National Audubon Society.

clams

The USGS has a great "Fact Sheet" on freshwater clam biology and ecology.

exotic species

This site describes ecology and management strategies for exotic aquatic species of North America: vertebrates, invertebrates, and plants.

fishes

For keys, natural history, and drawings of the fishes of the North American Great Lakes.

macrophytes

This site provides great access to pictures and descriptions of aquatic plants, with links to other plant taxonomy sites.

mammals

The Illinois State Academy of Science maintains this valuable mammalogy website.

organisms—USGS

"The Great Lakes Science Center is in its eighth decade of providing information about biological resources in the Great Lakes Basin. Our research results are used by management agencies, other researchers, and the public to gain a better understanding about this unique resource, and to better manage it for the public interest." This site provides a wealth of information about management of the larger aquatic organisms.

reptiles & amphibians

The froguts site allows you to perform a virtual frog dissection.

vertebrates

This site of the Open Directory Project provides links to many sites related to vertebrate biology.

waterfowl

A good source of images and sounds of waterfowl.

6

". . . breaking of the mere. To a stranger these terms are somewhat misleading as they appear to suggest a violent agitation of the water, or its bursting through its banks, whereas the phenomenon resembles the breaking of wort in the process of brewing, causing a discolouration of the water rendering it unfit for consumption, and spoiling the fisherman's sport. In its normal condition the water is pure and limpid . . . but when it breaks it becomes turbid from the formation of small dark-green bodies in countless thousands, which not only float as a scum on the surface but abound throughout the whole of the water . . ." [a population explosion of cyanobacteria].

William Phillips, *The Breaking of the Shropshire Meres,* 1884.

Population Dynamics in Limnology: Population Size Changing with Time

outline

CHANGE IN LIMNOLOGY

As soon as we meet the players in the freshwater ecological theater, we notice that they are not only moving about, but their numbers are changing constantly. Lakes and streams are inhabited by different kinds of organisms at different times of the year. Dramatic changes occur in the abundance of organisms. For example, in a eutrophic lake, cyanobacteria (such as *Aphanizomenon*) can be undetectable in the winter but are present as a 1-cm-deep scum during the late summer. *Daphnia* may seem to totally disappear during the winter or in late spring but are the most abundant crustacean during mid-spring. This chapter explains various ways to describe populations and gives examples from aquatic habitats. However, description is just the beginning. A major reason for studying populations is to be able to predict changes before they happen.

DEFINITION OF POPULATION DYNAMICS

A **population** is a group of individuals of the same species, living in the same place. For example, the *Daphnia pulicaria* in Lake Mendota, Wisconsin form a population, as do the *Anax junius* (large blue and green dragonflies, similar to those shown in figure 4.18d and 4.19f) whose nymphs live in the ponds, streams, and lake littoral zones of the Yahara River (Dane County, Wisconsin) watershed. Populations have the potential to share the same gene pool.

Population **size** is measured as the number of individuals in a defined area (such as a lake or riffle), as a **density** of individuals per unit area or volume, or as **biomass.** Density means the number of individuals per unit area (as in the number of macrophytes per square meter) or per unit volume (as in the number of copepods per liter), and biomass is the dry or wet weight of the organisms of interest per unit area or volume.

Numerical density is often expressed as the average density, based on a number of samples of an environment. However, Carol Folt (1993; Folt and Burns, 1999) pointed out the importance of perceived density, which can be very different from average density. If individuals are clumped, living in dense clouds or patches, then each individual will perceive a much higher density than predicted by the overall average density. The difference between average density and perceived density can be important for models of population dynamics and interactions among populations (as in predator-prey systems).

Populations change size and structure through time—they are **dynamic.** Many aquatic organisms show marked patterns of seasonal changes. Populations can increase, decrease, or even become locally extinct (extirpation). Predation and competition are two major biological interactions that affect population dynamics. An understanding of population dynamics contributes to our understanding of communities in lakes and streams.

Distribution of a Population

There are two important scales related to the population concept. At the smaller scale, the distribution of an aquatic population is defined by the shores of the lake or stream. For example, the distribution of smallmouth bass nests in Pallette Lake (Vilas County, Wisconsin) (figure 6.1) gives an indication of the size and shape of the population. The number of nests is proportional to the total population size, and the population appears to occupy the whole lake.

Aquatic organisms do not typically occupy an entire lake or stream. (Vertical patterns were discussed in chapter 2.) In the horizontal plane, organisms are often limited to littoral or pelagic, pool or riffle, or other subdivisions of the aquatic habitat. Organisms have specific environmental requirements that produce patchy distributions. For example, aquatic macrophytes are not distributed evenly throughout the lake, but in patches defined by water depth, light conditions, temperature, and wave action factors. Different species of aquatic plants living in the large, shallow Austrian lake Neusiedlersee produce a pattern of patches when seen on a map.

The pattern of distribution of the population within a water body depends on the ecological preferences and life-history stages of the organisms. For example, in the case of zebra mussels (dreissenids) living in Lake Erie, the simplest description of the population distribution is that it occurs in the lake. However, a closer look reveals that, in one sense, the population is restricted to individuals living within 1 meter or so of other zebra mussels. Zebra mussel sperm can probably swim only about 1 meter and still have enough energy to penetrate an egg (David Culver, personal communication). Once the fertilized eggs develop into larvae, however, they can drift great distances throughout the lake, during the 2 weeks

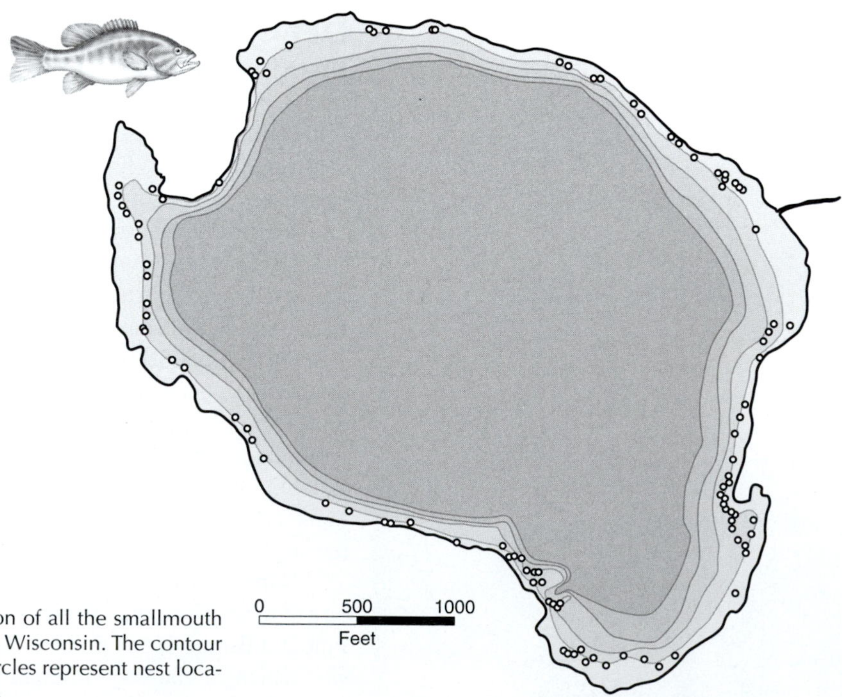

FIGURE 6.1 A map of the location of all the smallmouth bass nests in Pallette Lake, Vilas County, Wisconsin. The contour lines are at 5-foot intervals. The open circles represent nest locations. **Source:** Map data from Jeff Baylis.

or so that they are in the plankton. The larvae then settle and can contribute their specific genotypes to the group of mussels within 1 meter or so of settling.

At a larger scale, the *boundary* of a population might be defined as the collection of all the aquatic sites occupied by individuals of the species. The individual lakes could be considered **metapopulations** that make up the total population. A metapopulation is a *population of populations,* an assemblage of discrete local populations that have their own dynamics, and with migration among the discrete populations (Hanski & Gilpin, 1997). The metapopulation concept makes the most sense for adjacent sites, and is probably less valuable for widely separated sites such as populations of zebra mussels in the Caspian Sea and in North America. A metapopulation can be thought of as the result of a landscape pattern of suitable habitat separated by unsuitable habitat. Dispersal agents such as boats or flowing water could allow enough migration from one site to others, so that if a local population goes extinct, the site can be recolonized from nearby sites. This appears to happen in lake districts because lakes surrounded by many other lakes have a longer species list for zooplankton than do isolated lakes (Dodson, 1992). The term *metapopulation* is most often used by population geneticists and evolutionary biologists—in many cases, ecologists still prefer the simpler term *population.*

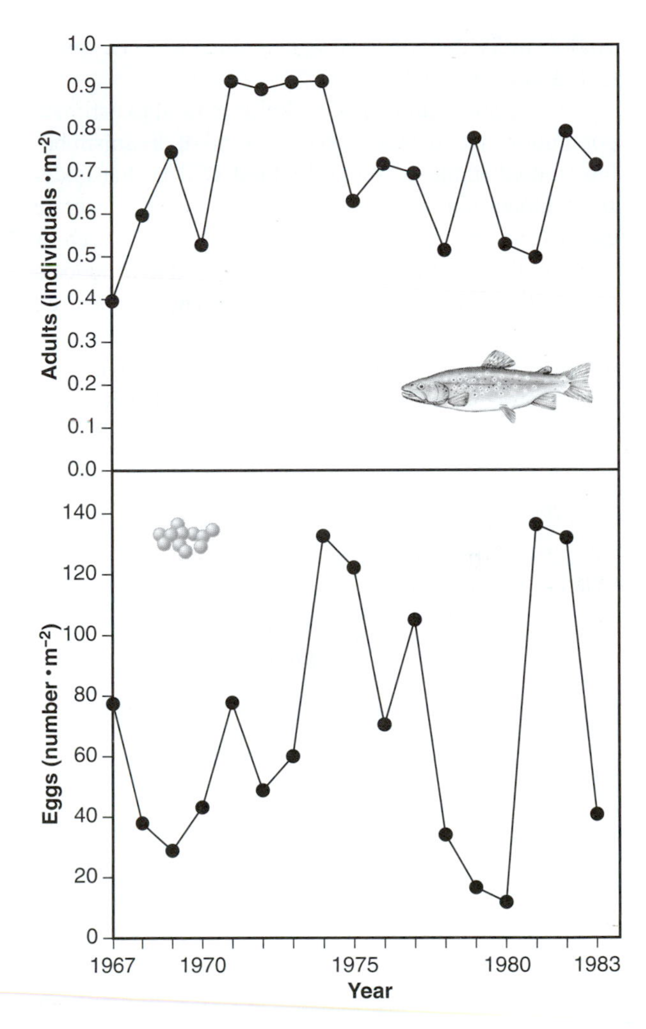

FIGURE 6.2 The density of brown trout eggs and of fish more than 1-year-old, in Black Brows Beck, a tributary of the River Leven in the English Lake District. Note that the scale for the adult trout is much smaller than that for the eggs. **Source:** Data from Elliott, 1984.

Estimating Population Size

In population ecology, there is the question of what to count. If we are interested in a trout population (figure 6.2), it is important to specify if a population size represents all developmental stages (eggs, larvae, juveniles, and adults) or just one stage, such as the adults.

Population sizes can be quite enormous. For example, a lake 1 hectare (ha) in area is a small lake. Imagine this lake has an average depth of 1 meter, so it has a volume of 10,000 cubic meters. Common crustacean zooplankton species occur at abundances of about 10 per liter. Thus, the lake might contain 10 million of each of the common species. The more common phytoplankton species might occur at a level of 10,000 per milliliter, or a total of 10 billion individuals per species in the lake. Because of the huge population sizes of microscopic organisms in lakes and streams, populations are often expressed as numbers per unit volume or area; for example, number of *Daphnia* per liter or number of midge larvae per square meter on the bottom. The actual size of the entire population is often ignored.

Problems may arise with organisms, such as macrophytes, for which the population is not made up of distinct individuals. Macrophytes often have many stems connected by rhizomes lying in the lake or stream sediments. In such cases, it is permissible to use number of stems, or length of stem per square meter, or even weight per square meter, as a surrogate of population size.

Hutchinson (1965) noted that quite large populations of microscopic organisms might be perceived as rare or can even go undetected in a lake. For example, limnologists have difficulty detecting planktonic microorganisms if they are present at a density of less than 1 per cubic meter. This means that, in the previous example, it is quite possible for a population of

10,000 individuals (of, say, *Daphnia*) to exist undetected in a small lake.

There are a number of techniques used to estimate population size, depending on the kind of organism being studied (table 6.1) and the kind of data that is desired. Quantitative samples attempt to give an accurate estimate of the number of organisms per unit area or volume. Qualitative samples are designed to produce a species list. For example, an aquatic dip net can be used for qualitative samples of stream insects, but more sophisticated sampling equipment (such as a grab or sediment corer) is needed to estimated insect densities.

Stability

Stability refers to the tendency of a population to remain at the same size (or density) over time. Calculating the **variance** of population size is one way to estimate temporal stability.

Equation (6.1)

$$\textbf{Variance} = (\text{sum}\ (x_i - \text{average})^2)\ /\ (n - 1)$$

where x_i is the size of the population observed at time i, and n is the total number of observations. The **standard deviation** of the mean is the square root of the variance.

Table 6.1 Examples of the Different Sampling Techniques Used to Study Different Kinds of Organisms

Organism Being Studied	Organism Size Range	Techniques
Bacteria	0.01–1 μm	Collect a few milliliters of water or mud, culture in the laboratory, or analyze for nucleic acid diversity. Count the stained cells under a microscope (smallest cells are invisible). Because most species of bacteria are difficult to culture, this is an imprecise technique, greatly underestimating species diversity.
Phytoplankton	0.1–1000 μm	Collect a few milliliters to a liter or so of water, preserve with gluteraldehyde (health hazard!) or Lugol's solution, allow cells to settle, and count using an inverted microscope.
Protozoa	0.1–1000 μm	Similar to technique for phytoplankton, except protozoa are much more likely to disintegrate upon preservation.
Rotifers	30–3000 μm	Use a fine-mesh plankton net (ca 90 μm) to concentrate animals; preserve with alcohol. Fine mesh is readily clogged with algae, so typically only a few liters are sieved.
Crustacean zooplankton	0.3–5 mm	Use a medium-mesh plankton net (ca. 200 μm to concentrate animals; preserve with alcohol. Typically less than 1 cubic meter is sieved.
Macroinvertebrates	1–20 mm	For a qualitative sample, use a coarse-mesh (ca. 1 mm) dip net to sweep up organisms or to catch them in a water current. For a quantitative sample, sample 1 square meter or so of substrate to a depth of a few centimeters. Take the organisms from the net, put them into a pan of clear water, and individually pick them out of the pan and preserve in alcohol.
Mollusks	1 mm–100 mm	Similar to technique for insect macroinvertebrates, except that more digging may be involved. Wash benthic samples through a 0.5-mesh screen to separate the organisms from the fine sediment.
Fish	5 mm–1000 mm	Use various kinds of large-mesh nets to capture fish that swim into the nets. Fix with formalin (health hazard!), wash, and preserve in alcohol.
Macrophytes	5 mm–2000 mm	Use rakes to gather macrophytes or quadrats (square sampling frames) to sample stems per unit area. Dry representative specimens in a plant press.

The average population size is calculated as:

Equation (6.2)

$$\textbf{Average} \text{ (or } \textbf{mean}) = \text{sum } x_i / n$$

Note that the mean and standard deviation have the same dimensions (they both have the units of "individuals"). When comparing two populations with different means, it is useful to calculate the **coefficient of variation** (CV), which is the standard deviation divided by the mean, for each population. The CV standardizes variability to a per individual basis. Often, the CV is expressed as a percent, so:

Equation (6.3)

$$\text{CV} = (\text{standard deviation}) / (\text{mean})) \times 100\%$$

Populations with low variance have a relatively high stability. For example, consider the population of trout in a stream in the English Lake District (Elliott, 1984). This population of brown trout was studied 17 years, from 1967 to 1983. The number of eggs m^{-2} and the number of fishes aged 1 year and older show different levels of temporal stability (table 6.2).

Table 6.2 Densities m^{-2} of Two Developmental Stages of Brown Trout in a Stream in the English Lake District

Statistic	Eggs	Fish Age 1+ (May–June)
Mean	68.7	0.70
Variance	1712.8	0.03
Standard deviation	41.4	0.17
Coefficient of variation	0.60	0.24

Source: Elliott, 1984

Egg densities were more variable over the 17 years than were the number of adult fish. Notice that the coefficient of variation of the eggs is more than twice as large as that for the adult fish. The key here is that biological processes, such as territoriality and predation, tend to reduce the CV of adult densities, relative to egg densities. Thus, these biological processes are stabilizing.

Stability has a temporal scaling component. For example, copepod populations in Lake Washington (figure 6.3) typically have a high seasonal variance for any single year, with very few copepods present in the winter, a period of exponential population growth in the spring, followed by a population crash in late spring or summer (Edmondson & Litt, 1982). In some years, there is a second peak and crash of the population during the summer or early fall. The within-year variance in copepod abundance is much greater than the variance in the number present at the same time of year, measured over several years, because there is a repeated annual pattern of population growth and crashes.

Populations typically increase and decrease, often with an annual cycle. For organisms like the calanoid copepod Diaptomus ashlandi (Edmondson & Litt, 1982) that have one or more generations per year, the population will typically show short periods of rapid increase and rapid decrease. However, the annual growth rate must average zero if the population is to persist—if the population has long-term stability. Otherwise, populations with a net long-term negative growth rate are headed for extinction, and populations with a net positive growth rate are probably involved in a "population explosion" (or **bloom**) to be followed by a decrease.

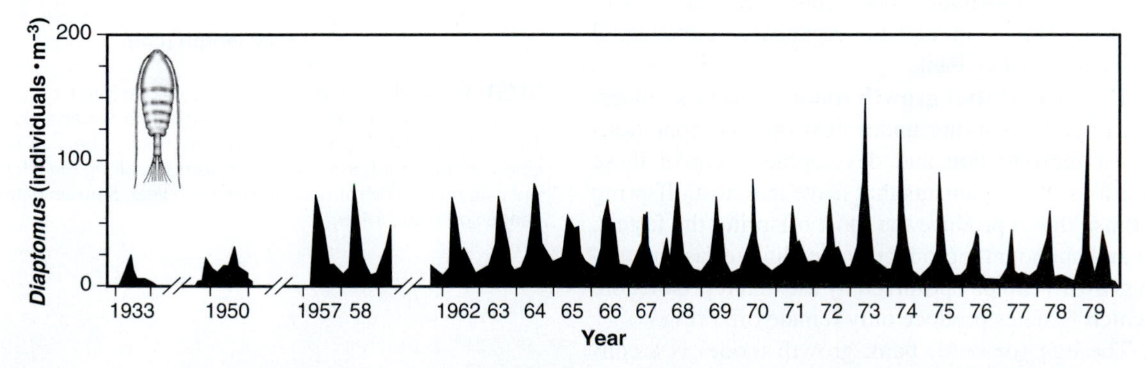

FIGURE 6.3 A graph of number of the calanoid copepod (*Diaptomus ashlandi,* an aquatic population with high variance in numbers over time, with a strong annual component to the variation, Lake Washington. **Source:** Data from Edmondson & Litt, 1982.

Long-lived animals, such as fishes, sometimes appear to have **cycles** in population abundance over several years—cycles are regular oscillations in population size. Cycles in population size are evident in aquatic systems in arctic or temperate climates where extreme annual climate cycles drive biological cycles. Cycles are less obvious or nonexistent in tropical systems with relatively constant climate.

An example of population cycles in a temperate lake is offered by the yellow perch in Crystal Lake, Wisconsin, which exhibited cyclic-like behavior (Sanderson et al., 1999). In this population, a cohort (of one or two **age classes**) dominated the population for about 5 years (figure 6.4). There was a pulse of recruitment about every 5 years, with little or no recruitment in the intervening years. Despite the annual variation in population density, in the long term, the population remained fairly stable, as if it were oscillating around an average value. This stable oscillation is probably the result of predation (cannibalism) of larger perch on perch larvae, so that recruitment is only possible when adult perch are scarce. This is an example of **density-dependent** control of population dynamics.

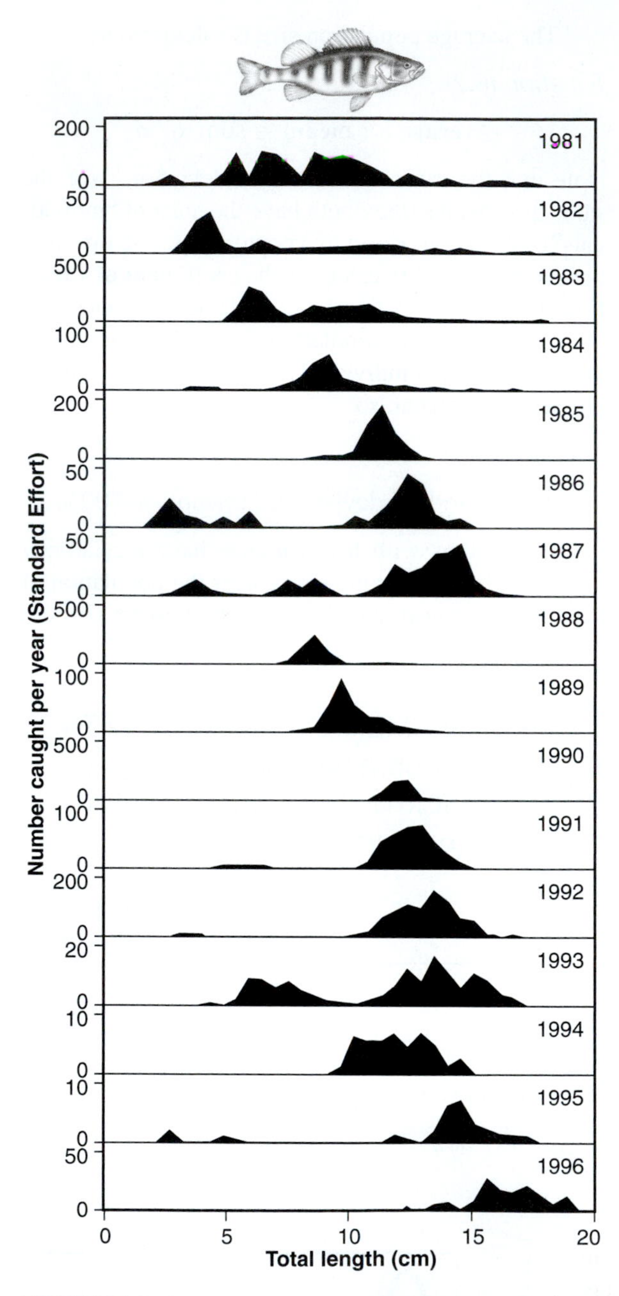

FIGURE 6.4 The number of yellow perch in Crystal Lake, Wisconsin showed cyclic population dynamics over a period of 10 years (annual size-frequency distributions). The year is indicated on the right-hand y-axis. The numbers along the left y-axis indicate relative fish abundance for each year. **Source:** data from Sanderson et al., 1999.

Models of Population Dynamics

Mathematical models allow prediction of future population sizes (the mathematically inclined may wish to look at papers in Tuljapurkar & Caswell, 1997). The most direct application of population models is to predict future populations sizes, in situations like fish management and aquatic macrophyte control, or in predictions of the spread of invading species. Even when a prediction is not realized, the gap between prediction and observation provides important information about biological processes such as reproduction and mortality. When there is such a gap, limnologists have an opportunity to learn something new about how aquatic communities work.

There are two main models for understanding population dynamics in aquatic organisms: exponential growth and the egg bank.

The **exponential growth model** is used to understand organisms living under near-optimal conditions for fast reproduction and development. Under these conditions, the organisms that leave the most offspring are those that reproduce the most offspring the fastest. Fast and abundant reproduction (leading to fast population growth) can be optimized by asexual reproduction, in which females produce only female offspring.

The egg (or seed) bank growth model is a conceptual model used to understand population dynamics of organisms with diapause stages—living under environmental conditions for which fast population growth is not often an option. The egg bank model is appropriate for organisms constrained by resources or time (as in growing season), in which the goal is first to produce **propagules** (dispersal stages, such as diapausing resting eggs, spores, or seeds) that can withstand environments that would kill the adults. In some situations (short growing season because of cold climate, or rapid drying-up of ephemeral water), it may be more important to produce resting stages or propagules rather than having the fastest rate of population growth.

Individual-Based Models

Before the advent of accessible computers, about 1990, population models were simple, using only a few variables and using average rates, such as the population growth rate. Now that powerful computers and software are available, models have become more elaborate. Instead of population averages, **individual-based** models use data for large numbers of individuals. The idea with individual-based models is to simulate the lives of a large number of individuals. Each individual can even be assigned a slightly different value for important life-history parameters, such as age-specific **fecundity** (number of offspring produced per female) and **survivorship** (the fraction of the population surviving from one age to the next). This individual variation is used to calculate what are probably more realistic estimates of population parameters such as growth rate and population age structure.

Individual variation is especially important whenever an unusual event occurs. In real populations, the response in population growth rate will probably not depend on the average response, but on the activity of unusual individuals. For example, in fish populations, annual population recruitment depends not on the average fate of individuals (which is death) but on the success or failure of a few individuals that live to reproduce.

THE EXPONENTIAL POPULATION GROWTH MODEL

The exponential population growth rate can be quantified for growth over a short time interval (usually less than a generation) for populations with overlapping generations using the exponential model equation:

Equation (6.4)

$$dN/dt = rN$$

where N is the population size, t is time, and r is the population growth rate coefficient. The term dN/dt is a derivative, implying a rate of change of the population through time. Equation 6.4 says that the rate of change of the population with time is equal to a constant rate (r) times the size of the population. This equation can also be rewritten:

Equation (6.5)

$$(dN/dt)/N = r$$

which says that the rate of change of the population with time, per individual, is constant. The units of r are per unit time, such as per day or per year.

Equation 6.5 can be integrated to give Equation 6:

Equation (6.6)

$$N_T = N_o * e^{rT}$$

In Equation 6.6, N_o is the initial population size and N_T is the size of the population at the end of time interval T. The e is a constant equal to about 2.7183. A graph of Equation 6.6 shows a pattern of exponential growth. The distinctive characteristic of exponential growth is a graph showing an ever-increasing growth rate (figure 6.5). This

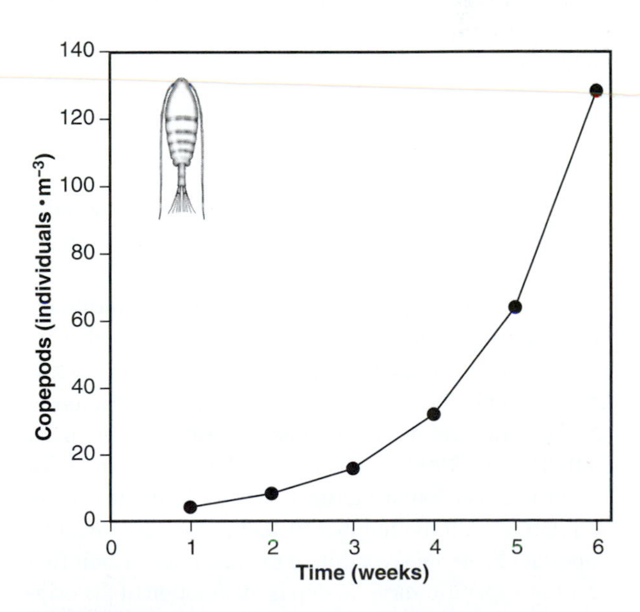

FIGURE 6.5 A model curve for exponential population growth. This simulated population of the calanoid copepod *Diaptomus* is doubling every week. The initial population size is four copepods per m^3.

pattern is often seen in aquatic populations over a time interval of days or weeks. For example, the population of *Diaptomus ashlandi* grows exponentially for several weeks each spring (see figure 6.3).

Equation (6.6) has a number of uses for aquatic ecologists. It can be used to estimate the value of the population growth rate coefficient between two sample times. It can also be used to predict the amount of time it will take for a population to double. When the population doubles, N_T is 2 * N_o, which leads to the equation:

Equation (6.7)

$$T_D = (ln\ 2)/r \text{ or } T_D = 0.69/r$$

where T_D is the time needed for the population to double in size.

Population growth, represented by *r,* is the result of two more basic biological processes:

Recruitment (due to birth or immigration).

Loss (due to mortality and emigration).

In a lake, where migration is minimal, population growth is due to the interaction of birth and death. Each process can be important, and there is often a seasonal component to the relative importance of birth and death.

Understanding population growth means understanding what controls birth and death. The first step is to measure birthrates and death rates. The rates are related to population growth rates:

Equation (6.8)

$$r = (b + d)$$

where *b* is the population birth rate and *d* is the population death rate (a negative number). Both *b* and *d* have the same units as *r.*

Separating the population growth rate into birthrates and death rates allows us to begin to understand what factors control population growth rate and size. For example, the number of eggs produced by a population of smallmouth bass in Nebish Lake (northern Wisconsin; Raffetto et al., 1990) is not correlated with the number of 1-year-old fish found the next spring. Birthrate has no correlation with the number of fish in the lake. What is important is the survivorship of the juveniles in their first year (age-specific survivorship is the fraction of the original cohort—age class—that survives to a specific age). This suggests we need to know more about sources of **mortality,** and that finding out more about egg production could be a lower priority.

Population Data Calculation

W.T. Edmondson (1960) developed an *egg-ratio* technique for estimating rates of population change: *r* (population change), *b* (birth or recruitment), and *d* (mortality) from field samples, plus a little information from the lab. While still an undergraduate student at Michigan State University, Caswell (1972) published an improved mathematical derivation of the technique.

The value of *r* is calculated from field data, which will typically be a sequence of population densities measured at regular time intervals (such as every 2 weeks). The value of *r* can be calculated for each sampling interval. The population is sampled (producing a population size at time zero = N_0) and then after some time interval (T) is sampled again (producing a population size at time $T = N_T$). For example, if a population grows from 10 to 12 individuals in 10 days, then the value of *r* (using Equation 6.6) is 0.018 per day.

The value of *r* typically is different for each time interval because the age-specific survivorship and age-specific fecundity patterns are constantly changing in nature. If the population is increasing, *r* is positive. If the population is decreasing, *r* is negative. A static population has an *r* value of zero.

The Edmondson Egg-Ratio equation for estimating *b* requires information from laboratory cultures: the average time it takes an egg to complete its development (at a specific temperature). The developmental time is estimated by separating individual egg-carrying adults into separate jars and recording the time of egg release or hatching. Adults are chosen at random, so at least some of the individuals will have just produced a new batch of eggs. The longest time taken to release offspring (after excluding sick animals) is taken as the developmental time. The value of *b* is calculated from the equation:

Equation (6.9)

$$b = ln\,(1 + E/♀)/D$$

where $E/♀$ is the number of eggs in a sample divided by the total number of individuals in the population (juveniles and adults), and D is the average time it takes an egg to complete its development (at a specific temperature). The number of eggs per individual is typically known from lake samples. (Note: If you know *r* and *b,* you can calculate mortality rate *d* as the difference between *b* and *r.*)

This Edmondson Egg Ratio Technique has been applied to rotifers and other members of the plankton, such as *Daphnia,* that hold onto their eggs during development and have overlapping generations. For these species, the technique provides useful information about the relative importance of food limitation and predation in controlling population size. The "*E*/♀" variable is affected mainly by food supply. The more food, the more offspring are produced. The *D* variable is affected mainly by temperature. The higher the temperature, the shorter the developmental time and the higher the birth rate.

Don Hall's Surprise

In a ground-breaking study, Don Hall (1964) studied *Daphnia* birthrates and death rates in Base Line Lake, a small lake in central Michigan. His purpose was to use field and laboratory data and a simple (exponential) model to predict the population dynamics of the *Daphnia* population. Base Line Lake has the typical pattern for a mesotrophic, temperate zone lake—algae are scarce during the winter and most dense in the late spring, followed by a dramatic decrease in abundance for the rest of the growing season (figure 6.6). The zooplankton (represented by *Daphnia*) also show a typical pattern of being scarce in the winter and most abundant in the late spring, followed by a dramatic decrease in abundance for the rest of the season (figure 6.6).

Based on previous limnological research, Hall identified two environmental variables—food level and temperature—which were probably important determinants of *Daphnia* population dynamics. He hypothesized that these two factors limited *Daphnia* population growth rate during the summer. He expected to find that the summer *Daphnia* decline was due to low food concentration, which would show up as a decline in reproductive rate, or *b.* During the summer, when the *Daphnia* population is low and not changing (*r* is variable but approximately zero), he expected the birthrates and death rates to be negligible. To test this hypothesis, he compared population predictions based on the laboratory study with actual observations from the lake. Hall's approach was to:

- Measure *Daphnia* and food abundance and temperature over an entire year (1961), sampling approximately twice a month.
- Estimate the population growth rate coefficient, *r,* each time interval, using Equation 6.6 for *Daphnia* abundance data collected during 1961.
- Predict the *Daphnia* birthrate, *b,* in the lake for each time interval, using lake levels of food and temperature and a predictive model developed with laboratory cultures.
- Estimate death rate using Equation 6.8 as the difference between *r* and *b.*

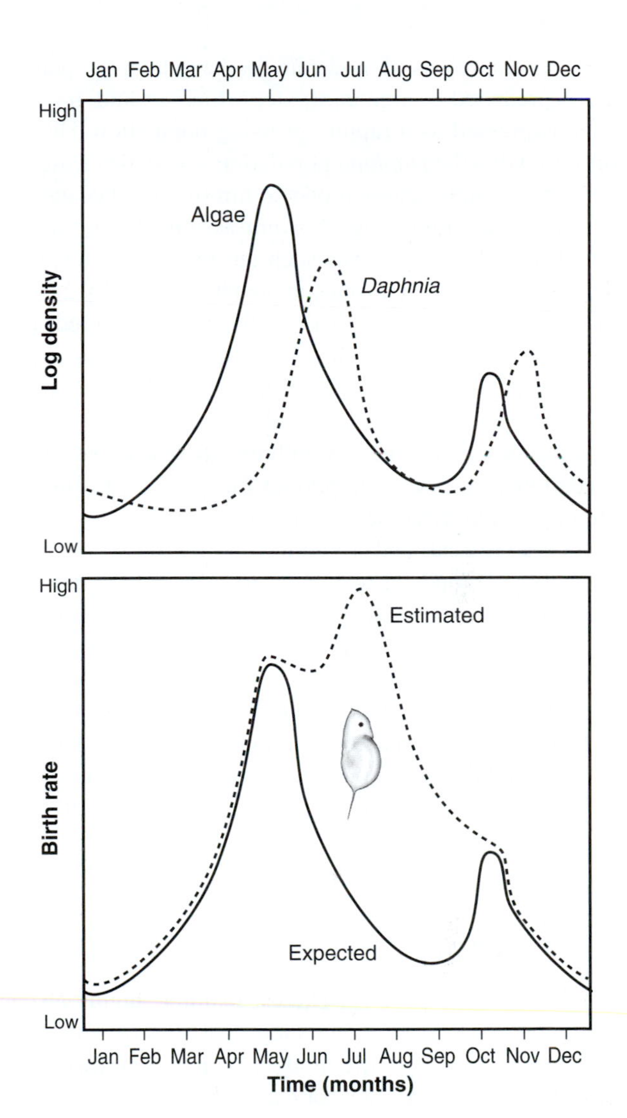

FIGURE 6.6 An idealized interpretation of a *Daphnia* population studied by Hall (1964). The top graph shows the generalized pattern of population dynamics of the algae population and *Daphnia galeata mendotae* in Base Line Lake, Michigan. The lower graph shows the *expected Daphnia* birthrate (the solid line, correlated with algae density in the upper graph) and the actual *observed* rate estimated from the algae and temperature model (dotted line). Hall's surprise was that something besides food appeared to be controlling *Daphnia* birthrate during the summer months.

Earlier research had assumed that food concentration was the main factor that controlled *Daphnia* birthrate. To Hall's surprise, he saw that, based on algal abundance and temperature data from the lake, the estimates of *Daphnia* birthrates remained high during the summer months of June, July, and August (figure 6.6) even when the *Daphnia* population was falling. The summer conditions of moderate food and high temperature predicted a high birthrate, which should have been expressed as a rapidly growing population. Hall proposed that the *Daphnia* population was static during the summer, not because food was limiting, but because predation was removing *Daphnia* as quickly as they could reproduce. This was such an unexpected result that he had collected no data on predation intensity during the study period. However, the result inspired a large number of studies that did estimate the effect of predation by vertebrates (fishes) and invertebrates (various groups) on *Daphnia* population dynamics. This was one of the first studies with results that suggested predation could be an important factor in controlling the population dynamics of an aquatic species.

Chris Luecke and a group of colleagues (1992b) used the Edmondson technique to estimate the total mortality on a *Daphnia* population in Lake Mendota, Wisconsin (figure 6.7). They could estimate how much of this mortality was due to predation by the four most important fish and invertebrate predators, using their knowledge about predator diet and feeding rate. They concluded that (1) fish predation intensity varies by up to eight-fold from year to year, and (2) during all 3 years, most of the annual total mortality was not due to the four most important predators.

Age-Specific Models

Edmondson, Hall, and Luecke used simple, highly abstract models to study population dynamics of natural populations. **Age-specific models,** which use information about birth and death for each age group of a population, are much more flexible and powerful than the simple models.

However, age-specific data are hard to come by—such data often require years of sampling, collecting data on all developmental stages of the organism in question. In aquatic systems, age-specific data are collected most often for crustacean zooplankton, fishes, and mollusks.

Fish populations have been studied in sufficient detail to allow understanding of how the age-specific birth and death schedules result in the observed pattern of population growth. For example, for smallmouth bass living in small lakes in northern Wisconsin, egg production and recruitment of young fishes is not dependent on the size of the population (Wiegmann et al., 1997). Rather, sources of mortality, perhaps predation by odonate nymphs and by fishes,

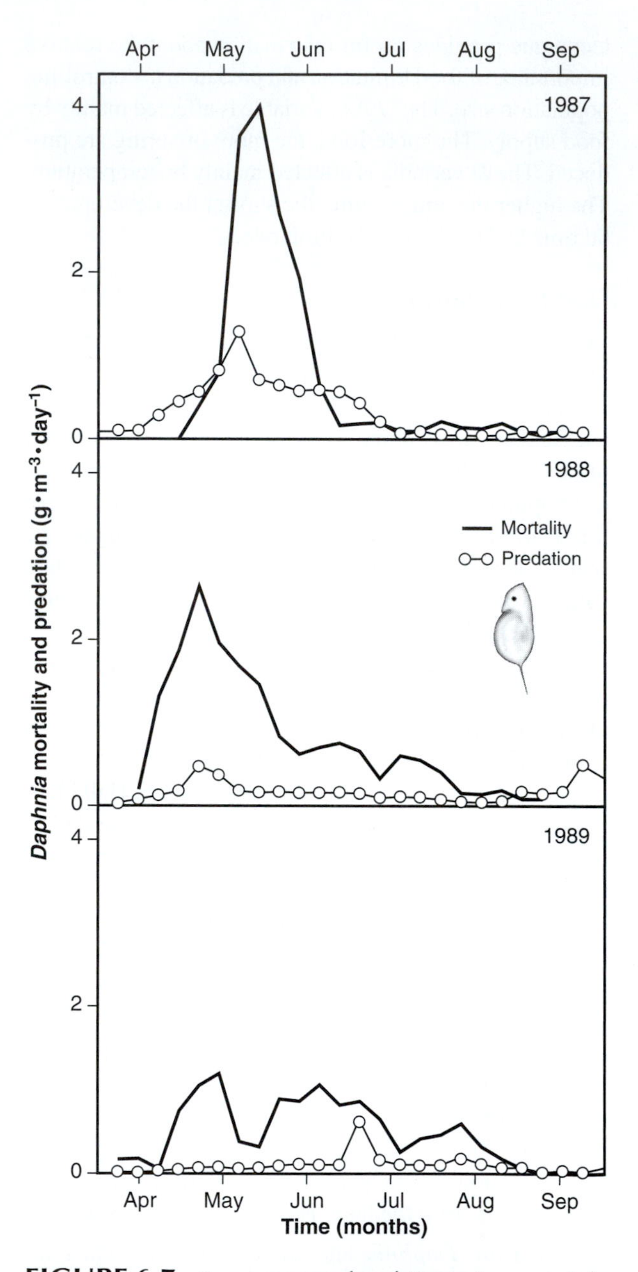

FIGURE 6.7 Two measures of predation in 3 years in Lake Mendota, Wisconsin. The solid line represents total mortality calculated using the Edmondson egg technique, and the line with open circles represents total fish consumption estimated from fish stomach contents. **Source:** Data from Luecke et al., 1992.

are important in determining whether the population grows or declines.

Leptodiaptomus minutus is a small calanoid copepod common in lakes and rivers of central North America. The population dynamics of this copepod are most strongly influenced by the intensity of predation on the nauplius stages (Confer & Cooley, 1977). Mortality was much less on the larger copepodid stages. This is an example of **size-selective** predation. Because size and developmental stage are closely linked, it is often possible to use age-specific models to understand the influence of size-selective predation on population dynamics.

Age-specific survivorship information can be obtained by following a cohort through time. For example, zooplankton eggs might be produced all during the summer, but often hatch synchronously (within a few days of each other), producing a spring cohort. Fishes often produce young only during a short period of time, producing distinct cohorts that can be followed for years. For example, Lake Mendota, Wisconsin, contains a population of a whitefish, the cisco (Magnuson & Lathrop, 1992). Cisco require cold, oxygenated water. This population crashed during the summer of 1987, when the cold hypolimnion became anoxic. A small population remains in the lake, but has had little success reproducing, probably because the decade of the 1990s was the warmest decade in the century. Eventually, environmental conditions will allow reproduction and a strong age class (year class-cohort) will be produced.

Exponential Population Growth and Invasion

People often create new aquatic habitats by building reservoirs or stock ponds. Zooplankton eventually appear in the new habitat, but with a time lag due to the time needed for a single individual to produce a population large enough to be detected. Maximum crustacean population growth rates are about 0.5 d^{-1}, at 20°C. This rate is associated with a doubling time of about 1.4 days. A small pond of 1 hectare, averaging 1 meter deep, has a volume of 10,000 cubic meters, or 10 million liters. Zooplankton are abundant enough to be noticed at about 0.01 per liter using standard collecting techniques. This means the pond population needs to grow to 100,000 organisms. This increase in population size requires about 16.5 doublings. Therefore, it will take an individual invading organism about 23 days, during the summer, to become common enough to be detected by limnologists. Of course, organisms that have an annual life cycle (some copepods and most vertebrates) will take much longer to become abundant after invasion.

THE EGG (OR SEED) BANK POPULATION GROWTH MODEL

The **egg bank** is made up of the resting stages stored in an aquatic habitat (usually in the sediments or along the edge of a stream, lake, or wetland). There are many situations in which understanding zooplankton population dynamics means knowing how many resting eggs are produced at the end of a growing season. Resting eggs (or spores or seeds) can persist for many years and typically break diapause over several or many years (Hairston et al., 1999). If environmental conditions are distinctly suboptimal, or if a population has a severe time constraint, population growth rate during the season may be less important than the number of resting eggs produced at the end of the season. There may even be a tradeoff between the number of resting eggs and the total number of offspring produced.

For example, *Daphnia* in arctic ponds experience about 30 to 60 days of open water (Dodson, 1984). This is enough time to produce one to two generations of *Daphnia*. When the ice melts, female *Daphnia* hatch from resting eggs, grow to adulthood, and then are faced with a decision about reproductive strategy. The goal is to put as many resting eggs as possible back into the pond before the water freezes.

In years with long seasons, there is enough time for the first females to reproduce once asexually, produce several daughters, and for this generation to grow and produce resting eggs. If the seasons were always long, the best reproductive strategy would be to produce all subitaneous eggs the first generation. These offspring could then produce all resting eggs. This strategy results in the most resting eggs possible.

If, however, the seasons were always short, the best reproductive strategy of the initial females would be to produce only resting eggs and no subitaneous eggs. This strategy produces fewer resting eggs than the two-generation strategy can in a long season, but in short seasons, the two-generation strategy produces no resting eggs at all.

The *Daphnia* solution is to produce some resting eggs and some subitaneous eggs in the first generation. If the subitaneous young survive, they produce even more resting eggs. If the subitaneous young do not survive, at least the first-generation resting eggs are in the bank.

During the shortest seasons, the only way for *Daphnia* to have resting eggs is to produce them asexually. These eggs allow the population to persist but do not afford the benefit of genetic recombination. Resting eggs produced by sexual reproduction are only possible in the second generation (during years with long seasons) because the males, produced as young in the first generation, need time to mature. *Moina* species are water fleas that live in temporary ponds. The *Moina* solution is resting eggs that can hatch into either a male or a female.

Not all resting eggs (or spores or seeds) hatch (or germinate) the year after they are produced. Some persist in diapause for several years, decades, or even centuries! Of course, the longer a resting stage is in diapause, the more likely it is to die. However, the benefit of being in diapause for a long time is that diapause can be broken before a particularly good growing season. It is particularly important to produce resting stages with variable times of diapause, if the environment is variable from year to year and if there are often years with unsuitable conditions during the growing season.

BIOLOGICAL FITNESS AND POPULATION GROWTH RATE

Biological fitness is measured by the relative number of fertile and viable offspring that a **genotype** contributes to the next generation. A *genotype* is the inheritable material passed from one generation to the next—the biological instructions for producing a living organism. The physical expression of the genotype is the **phenotype,** and the exact form of the phenotype depends on many factors, such as the genotype, the history of development, and environmental conditions. Fitness has an obvious connection to evolution—genotypes that contribute more than their share of offspring to future generations become more common in the population.

In actual practice, it is difficult or impossible to measure biological fitness in real populations. In many situations, the maximum exponential population growth rate (r_{max}) is used as a **surrogate** (substitute) for an actual measurement of biological fitness (Boersma et al., 1999). This substitution makes some sense, especially for populations, such as *Daphnia* during the summer, that are specialized for fast reproduction. In *Daphnia,* the individuals (genotypes) that can reproduce fastest are most likely to survive predation and to produce the largest population by the time the food supply becomes limiting. However, achieving the fastest population growth rate (by early reproduction) is just one of many possible **life-history strategies** that are common among aquatic organisms.

LIFE-HISTORY STRATEGY

All organisms face a fundamental "general life-history problem" (Stearns, 1992). The problem is how to optimize life-history strategy in order to make the largest contribution to the next generation—or, indeed, any contribution at all to the next generation, in a particular environment.

The success of a particular life-history strategy depends on the environment as well as the genotype. Each set of environmental conditions and each community of interacting organisms **constrains** the range of possible strategies that might be successful. If, as is usually the case, there is a limited resource, life-history strategies are constrained by various **tradeoffs.** The limited resource is often energy (food) or time.

Life-history strategy includes a number of aspects, which will be discussed in general, followed by more specific examples listed in table 6.3.

Seasonal Timing of Life History, Including Reproduction

Timing is everything. Organisms require rather specific environmental conditions for successful completion of each stage of the life cycle (see chapter 8). Natural selection tends to result in organisms that are adapted to a specific seasonal pattern. However, there is typically genetic variation in populations because of climatic variation. A random sequence of warm and cool summers will result in organisms in the population that are adapted to both warm and cool conditions.

Size and Number of Offspring

There is often a tradeoff between number and size of eggs or juveniles—as if there is a set amount of resource (stored energy) that can be used to produce offspring, and this total amount can either be divided up into a few large individuals or many small individuals. Size-selective predation influences the success of this aspect of life-history strategy. For example, fishes eat

Table 6.3 Examples of Different Life-History Strategies and the Environments (Abiotic and Biological) in Which Each Strategy Is Beneficial.

Strategies are for warm (summer or southern) conditions. Fewer generations would be expected under alpine or arctic conditions.

Organism Life History	Life-History Strategy	Optimal Environmental Conditions
Branchiopods such as *Daphnia* in temperate lakes	Reproduce fast and often to keep up with losses to predators; have little ability to escape predators after encounter. Filter large amounts of algae. High mortality balanced by a higher birthrate. Sex ratio is all female.	3–10 generations possible per year, high food abundance
Calanoid copepods in lakes	Reproduce slowly; allocate energy to fast swimming to escape predators; efficiently capture only high-quality food.	1–5 generations per year, low food abundance, cool temperatures
Microcrustaceans in temporary or arctic ponds	Produce resting eggs as soon as possible (population growth rate is low, but persistence from year to year is assured).	Life history constrained by the short time available during the annual climate cycle for growth and development—must reproduce before the water dries or freezes
Macroinvertebrates, macrophytes, and fishes	Reproduce once or twice a year in the temperate zone. In warmer, more constant climates, more generations per year may be possible.	Extreme annual seasonal conditions and timing are very important for organisms that take about 1 year for a life cycle. Summer conditions are typically optimal, when prey are most abundant and temperature is optimal for highest survival of young, and in time for young to grow sufficiently or even mature before the end of the summer season. In tropical climates, the annual rain cycle can also constrain reproduction timing.

large *Daphnia* and copepods eat small *Daphnia* (Zaret, 1980). If mortality due to predation is mainly due to fishes, then producing small young will be a more successful strategy than producing large young. On the other hand, if copepod predation predominates, producing large young will be the more successful strategy.

Age-Specific Energy Allocation Patterns—Reproduction, Growth, and Body Maintenance

Juveniles allocate energy to growth and maintenance. A common strategy for adults is to stop growing and to allocate energy to reproduction and maintenance. If energy is allocated to adult growth, the tradeoff is that initial reproduction may be lower, but after some time, the larger adults can produce many more offspring than can the small adults.

Rate of Population Growth versus Resting Egg Production

As described in the egg-bank model section, fast reproduction works well when environmental conditions are optimal. The fastest population growth occurs in asexual-reproducing organisms (such as branchiopods, rotifers, and plants), in which all offspring are females and in which energy is allocated to subitaneous egg production rather than resting eggs.

The rate of population growth also depends on the environment—temperature has the ability to either increase or inhibit growth. Typically, there is an optimal temperature for growth. The optimal temperature is not necessarily the highest temperature (Threlkeld, 1986). This is an environmental constraint on life-history strategy (see chapter 9).

Age of First Reproduction and Size- or Age-Selective Predation

Aquatic organisms often face a tradeoff between reproductive output and survival in the presence of size-selective predators. This tradeoff determines the duration and timing of the juvenile developmental period. Early maturation has the advantage of tending to increase reproductive rate (Cole, 1954). If predation falls most heavily on the largest (oldest) adults, the optimal strategy is to reproduce early at a small size. However, reproducing at a small size also results in smaller and fewer offspring, which can lower birthrate. On the other hand, if size-selective predation is intense on small offspring and small adults, the best strategy may be to delay age of first reproduction in order to produce a few, large offspring (Sommer et al., 2001; Zaret, 1980).

Possible reproductive strategies include reproducing only once to producing many batches of offspring. To some extent, the success of these different strategies depends on the probability of surviving to reproduce more than once, which can depend on the environment or on the timing of predation, as well as on the pattern of size-selective predation.

Life Span

Length of life (**life span**) is an aspect of life-history strategy that is closely related to other factors such as seasonal timing, growth rate, pattern of reproduction, and even locomotion behavior. Mathematical analysis of life-history strategy suggests that there is often little advantage to a long life, in terms of contributions to the next generation. One way to understand this is that offspring of an old adult are produced at the same time as the (naturally more numerous) grandchildren that are being produced by early-reproducing adults. On the other hand, many organisms grow as they age, and old, large adults can sometimes produce many more offspring at a time than can smaller younger adults.

The tradeoff between life span and locomotion behavior (see table 6.3 for an example) was discussed in chapter 4. Branchiopods, such as *Daphnia,* tend to allocate energy toward reproduction and have short lives. Calanoid copepods, such as *Diaptomus,* allocate energy toward escape response, enjoy a longer life than branchiopods, and produce relatively few offspring over their life span.

Life-History Adaptations

All of these life-history factors (and others) are under some degree of genetic control—they can be passed from one generation to the next. If these heritable characters increase reproductive success, relative to the average population rate of reproduction, the characters are considered beneficial **adaptations.** Some genotypes will be successful and some will not. Through time, the interaction between the genotype and the environment will produce a specific life-history strategy. Table 6.3 gives examples of adaptive life-history strategies and their environmental context.

It is important to understand that a specific life-history strategy will work best in a specific environment. A change in climate or a change in predator populations can have a large effect on what is the most successful strategy (Boersma et al., 1999).

Life history of a given genotype depends on environmental conditions or signals (Dodson et al., 1994; Larsson & Dodson, 1993). Conditions such as food concentration and temperature have a strong effect on life-history parameters such as growth rate and amount of reproduction. (Effects of food concentration and temperature will be discussed in chapter 9.) These are environmental parameters that have an immediate and direct effect on life histories of aquatic organisms. Less food often means fewer and smaller offspring that are more susceptible to starvation. Temperatures that are too cool or too hot result in suboptimal population growth.

Signals such as predator smell or **photoperiod** (day length) cause changes in behavior or development that have strong effects on life-history parameters. Chemical signals modify energy allocation, growth rates, and developmental patterns. A change in photoperiod can also modify life history. For example, day length is a major signal determining sex ratio in *Daphnia* (Hobaek & Larsson, 1990).

TRANSITION

The material in this chapter is focused on descriptions of populations and population dynamics. Natural next questions are, "What controls population size and rate of change?" and "How do multiple populations of different species live together?" The next four chapters present material that helps answer these questions. In chapter 7, we take a close look at the ecological interactions, especially competition and predation, that constrain the distribution and abundance of aquatic organisms. This is followed by chapter 8, which discuss how communities change over time. Community ecology is followed by two chapters (9 and 10) on ecosystem topics, in which we consider constraints imposed by physics and chemistry on aquatic organisms and their interactions, distribution, and abundance.

Study Guide

Chapter 6 Population Dynamics in Limnology: Population Size Changing with Time

Questions

1. What is *population dynamics?*
2. What is a rare species in an aquatic habitat?
3. How does the coefficient of variation assist interpretation of variability in populations with different size mean values?
4. Can an individual-based model also be age-specific?
5. How can different life-history strategies have the very same average result—contribution by each adult to the next generation of one successful offspring? What are the tradeoffs?
6. The value of *r* has units of "per time." For smaller organisms, this time unit is "per day," and for larger organisms, the time unit is more conveniently "per year." How is the value of *r* related to daily (or annual) percent change of the population?
7. Limnologists studying aquatic populations can easily calculate the rate at which a population is changing, as percent change per unit time. If a *Daphnia* population is growing at a rate of 15% per day, how many days will it take for the population to double in size?
8. What are major differences between the exponential and the egg bank models of population growth?
9. How does the concept of biological fitness relate to the concept of tradeoff in the context of life-history strategy for some groups of common aquatic organisms? Review the example of tradeoffs given in this and earlier chapters: chapter 2—the tradeoff faced by stream insects feeding in a current; chapter 4—the copepod tradeoff between photoprotection and vulnerability to predation.

Words Related to Aquatic Populations and Population Dynamics

adaptation
age class
age-specific model
average
biological fitness
biomass
bloom
coefficient of variation
constraint
cycle
density
density-dependent
dynamic
egg bank
exponential growth model
fecundity
genotype
individual-based
life-history strategies
life span
loss
mean
metapopulation
mortality
phenotype
photoperiod
population
propagule
recruitment
size
size-selective
stability
standard deviation
strategy
surrogate
survivorship
tradeoff
variance

Major Examples and Species Names to Know

Tommy Edmondson and the egg ratio technique

Don Hall's surprise

Steve Stearns and the general life-history problem

Additional Resources

Further Reading

Gotelli, N. J. 2001. *A primer of ecology.* 3rd ed. Sunderland, MA: Sinauer Press. 265 pages.

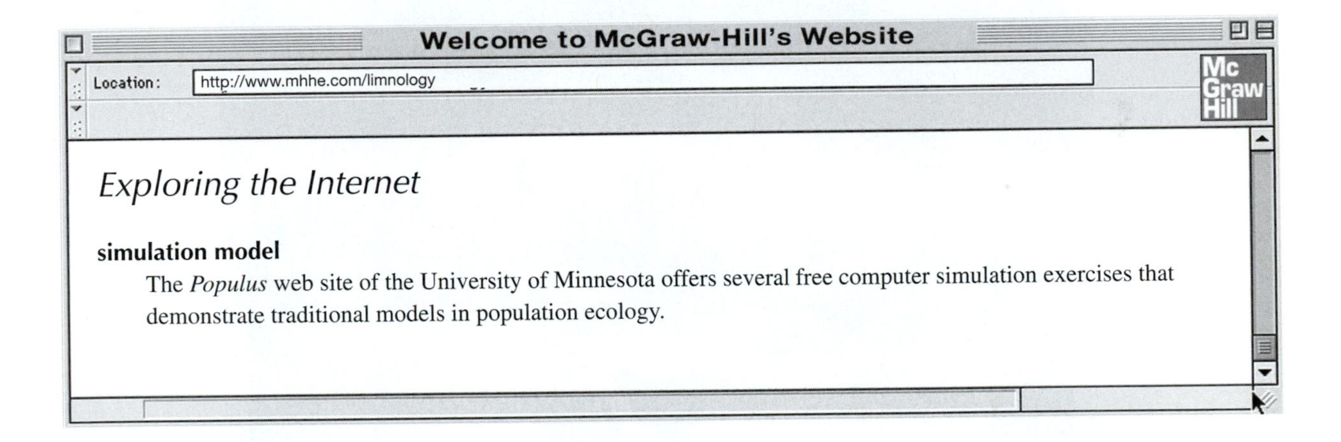

7

"In community ecology, it does not suffice to ignore all species except one or two, as in population biology."

T. F. H. Allen, 1998

outline

Community Ecology: Species Interactions and Community Structure

THE COMMUNITY CONCEPT

At this point in the introduction to limnology, we have described Hutchinson's ecological theater, and the players in the theater, as they stand about and come and go. Now it is time to consider interactions among the entire cast, or community, of players. This chapter discusses the components of aquatic communities, while chapter 8 focuses on how community structure changes over time. But first, it is necessary to define "community," the interactions among community components, and the nature of "community structure."

General Community Definitions

Ecologists use the concept of **community** in a variety of ways (Allen, 1998). At its very simplest, an ecological community is a *list* of species (or taxa) present in a specific area. Building on this concept, ecologists often think of a community as a group of *interacting* species, whether or not all the species happen to live in the same place at the same time. The community can be defined by ecological interactions, such as predation and competition, and these interactions need only occur for brief times, as, for example, during the annual cycle of events in lakes and streams.

The entities that interact are, of course, individuals. However, to simplify an already complex system, community ecologists typically assume that populations of similar organisms interact in the same way—they organize their science at the species level, which begs the question, "What species concept is used by community ecologists and limnologists?"

Community Building Blocks

Communities are made up of interacting entities, which can be categorized as **functional groups** (such as guilds) or species. Depending on the way organisms are categorized, community composition can be described using entities at a variety of scales. The following six categories (entities) are arranged according to a scale of generalization, from the most general (guild concept) to the most specific (clone concept).

Guild Concept

Guilds are groups of species that have a similar function or way of making a living. The species do not need to be closely related from a phylogenetic (evolutionary) perspective. An example of a guild is all the planktonic, algae-eating herbivores in a lake, including protists, rotifers, and microcrustaceans.

Most community ecologists would consider guilds a group of species rather than a species concept. However, ecosystem ecologists dealing with complex trophic systems find it useful to combine many species together into trophic guilds of organisms that have the same way of gathering energy—for example, the category *pelagic predators on zooplankton* might include both fish and invertebrate predators.

Functional Species Group Concept

Ecologists, especially those studying marine and large-lake systems, may be faced with the necessity to model or experiment with a system that includes scores or even hundreds of closely related species. When many species are morphologically and ecologically similar, and there is no particular need to distinguish among them, they are lumped together according to function. This classification scheme reduces the number of entities in the system and avoids concerns about careful (or obsessive?) taxonomy. A guild (for example, of planktonic herbivores) might be made up of several of these functional groups (such as cyclopoid copepods, calanoid copepods, cladocerans, and rotifers).

Morphological Species Concept

This concept focuses on phenotypic expression (the **phenotype** is a physical manifestation of the **genotype,** or genetic material). Organisms that appear similar, and different from other individuals, are considered the same morphological species (Mayr, 1963). "Appearance" can refer to external form, internal anatomy, behavior, or even molecular characters, such as similarity of enzymes or nucleic acids.

The morphological species is the concept that limnologists most often employ in models and descriptions. Just to make things interesting, limnologists must also take into account gender differences and polymorphic species, which have several morphologies in one species (as in Osenberg et al.'s [1988] study of sunfish). Related complications are **allometry** (Thompson, 1942), which causes morphological differences related to body size, and **phenotypic plasticity,** expressed by organisms such as *Daphnia* that have different individual morphologies depending on what environmental signals were received during development (Tollrian & Harvell, 1999).

Biological Species Concept

A species is composed of individuals that either interbreed, or can potentially interbreed, to produce fertile, viable offspring (Mayr, 1963). Because of the reproductive criterion, this concept relates species to the fields of genetics and evolutionary biology. This is the species concept that underlies the ecological understanding of **adaptation**—how genotypes meet environmental constraints. An individual (or species) is adapted to its environment if the individual possesses inherited traits that optimize the production of fertile and viable offspring in its specific environment. Note that different biological species can be morphologically identical. **Cryptic species** are morphologically similar but are different at the molecular level and are reproductively isolated. Mating trials often reveal cryptic species. For example, mating trials with the copepod *Acanthocyclops vernalis* revealed several cryptic species and suggested that individual species in this morphological group may have typical distribution ranges as small as a single shallow lake (Dodson et al., 2003).

Cryptic species are often difficult to recognize. Ulrich Einsle, who worked near Lake Constance in southern Germany, focused 20 years of research on distinguishing cryptic species of cyclopoid copepods using a variety of analyses, including morphological, electrophoretic, and chromosomal. His successful study of cyclopoids is reviewed by Wyngaard (2000). The bottom line of Einsle's research is that standard approaches failed to distinguish species, but species could be distinguished on the basis of the timing of decreases in chromosome number (chromosome diminution) during embryonic development. A subtle character, indeed!

In aquatic communities, there are (at least) tens of thousands of biological species. The time, resources, and knowledge needed to test the biological species concept make the reproductive criterion impractical, except in a few special cases. Therefore, the species concept used in the majority of studies of community ecology is the morphological species concept.

Molecular Species Concept

More recently, limnologists have been relying on molecular techniques to gauge the number of species in a lake or stream. Molecular diversity is morphological data that are perhaps closer to genetic information than the gross morphology used by traditional taxonomists. Also, molecular techniques are much easier to perform than mating trials. Using molecular techniques, limnologists often detect cryptic species. Excellent examples of this approach to the species question are found in papers such as those by Lee and Frost (2002) for the calanoid copepod genus *Eurytemora* and by Weider et al. (1999) for *Daphnia.* Studies of adjacent *Daphnia* populations in two lakes (DeMeester, 1996) reveal remarkable differences at the molecular level, suggesting that populations in each lake are adapted to the specific local conditions and may, in some cases, represent cryptic species.

Clone Concept

Many aquatic organisms exist as clones—a population derived asexually from a single ancestral female. This is the finest level of species concepts, and it is used by population ecologists and population geneticists. For example, Boersma et al. (1999) found that different clones of *Daphnia* had ecologically important differences in their behavior and life history that were adaptations to avoiding fish predation by escaping in time or space.

MAJOR AQUATIC COMMUNITIES

There are a number of major communities in lakes and streams (figure 7.1). These communities are to some degree distinct, but they are also interrelated (Schindler & Scheuerell, 2002). For example, fishes are large, mobile omnivores that move among communities, coupling different communities through their feeding, nutrient transport, and excretion. Despite this interrelation, the concept of distinct communities assists limnologists in understanding the complex aquatic systems known as lakes and streams.

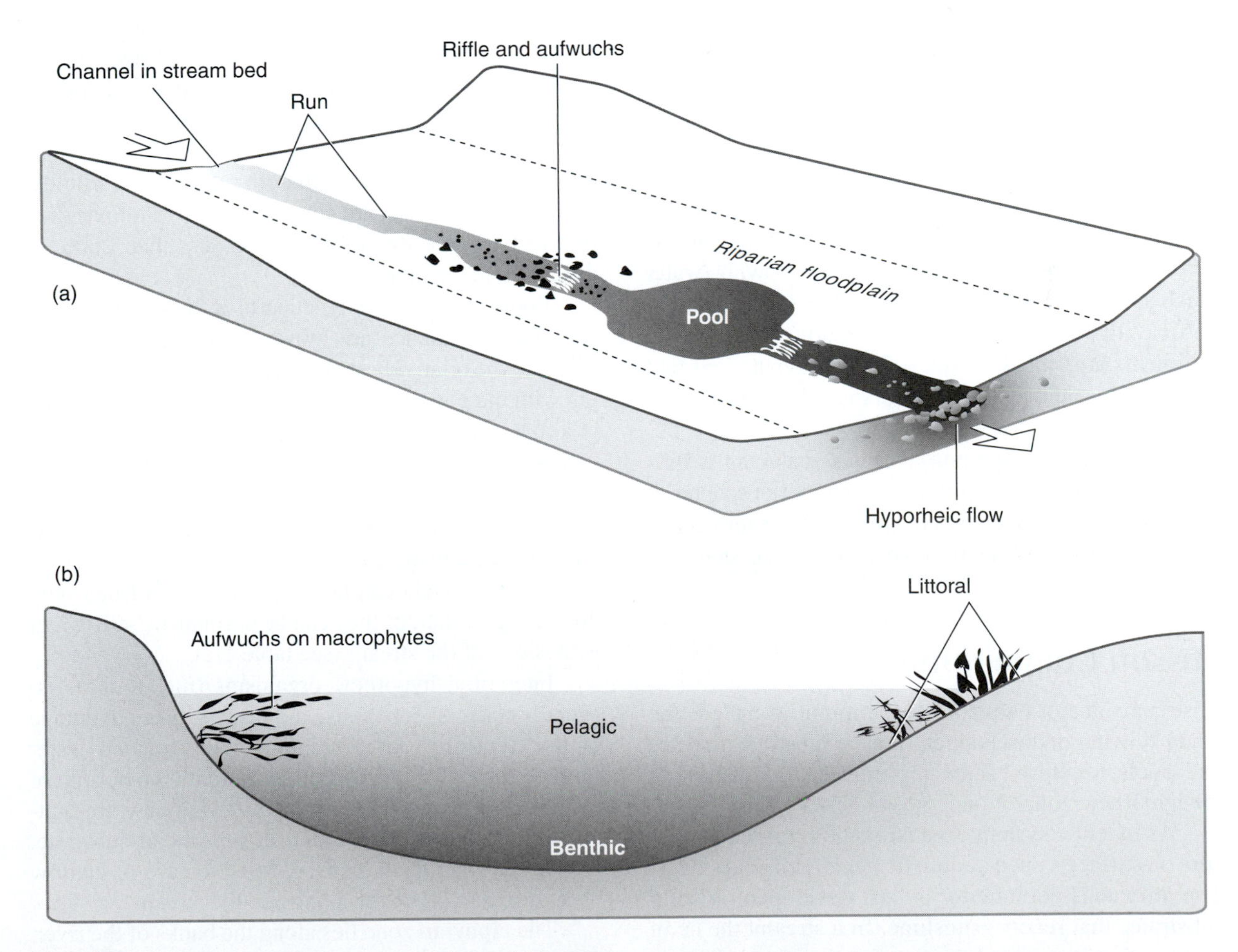

FIGURE 7.1 Major communities in (a) lakes and (b) streams.

Lakes (standing water) are called **lentic** environments, while streams (flowing water) are called **lotic.** Each of these communities can be represented by a food web. Each community is the focus of a major area of specialization of limnological research. Communities are often named according to their location in a lake or stream.

Lake Communities

In lakes (figure 7.1a), major habitats include the pelagic, littoral, benthic, and the aufwuchs. Algal mats are also provide structure for aquatic communities. The **pelagic** zone is the open water and the **littoral** zone is the shallow water near shore. Pelagic communities are characterized by plankton and plankton-eating fishes, while littoral communities are based on aquatic macrophytes and organisms that live on or among the plants.

The **benthic** community is located on the bottom of a lake (Brinkhurst, 1974). This community is characterized by heterotrophic organisms that live in or attached to the bottom, with typically fewer species (especially of fishes, insects, and crustaceans) at greater depths. Substrate in benthic habitats ranges from fine silt to large boulders, to bedrock, depending on the origin of the lake and the kind of streams coming into the lake. The farther the bottom is from flowing water (such as a stream flowing into the lake), the smaller the particles that will settle onto the benthos. Dark, organic sediments are called **gyttja,** a Scandinavian word that reflects the role of feces in producing the rich, organic sediment.

Aufwuchs (periphyton) species make up a millimeter-scale community that we first met in chapter 3. The community is seen as a mat of microorganisms living on surfaces of aquatic macrophytes and stones in the littoral zone.

Stream Communities

In streams, major habitats for communities include the water flowing in the channel, surface substrate in riffles and pools, shallow backwaters in the floodplain, and the hyporheic zone of underground flow (figure 7.1b).

As in lentic systems, the aufwuchs community occurs on surfaces, such as aquatic vegetation and stones. The aufwuchs community is best developed on solid substrates that receive sunshine. In a stream, the main body of flowing water is habitat for organisms that can swim against the current, such as fishes. The main watercourse of a stream with the highest water velocity, the **channel,** is often occupied by relatively few species. In contrast, coarse sediments, which provide some protection from the main current, are occupied by a diverse assemblage of species living near or in the substrate. **Riffles** are areas of fast-flowing water moving over coarse substrate (boulders, cobbles, and gravel). Size scales for coarse substrate range from meters for boulders, about 10 cm for cobbles, down to 1 cm for gravel. The water in riffles is turbulent because of fast flow over uneven substrate. The water flowing in riffles and in riffle substrate (**hyporheic** flow) is relatively high in oxygen because of the turbulence.

Fishes and benthic invertebrates can live in or near the coarse substrate associated with fast-flowing water by hiding in crevices, resting in still water behind large rocks, and having streamlined body forms. Invertebrates living on stones typically have flattened bodies that do not resist the water's flow (Peckarsky et al., 1990). Swimming animals that live in riffles, such as trout, find eddies behind stones or coarse, woody debris (logs) where they are protected from the power of the current.

Flowing water that is smooth rather than turbulent is called a **run.** Runs are characterized by laminar flow of water (with a Reynolds number less than 2000, as discussed in chapter 2).

Pools are relatively quiet parts of the stream, where flow and turbulence are minimal. Benthic organisms living in the relatively still water are adapted less to living with the power of water, and adapted more to living in a fine-grained substrate (silt, sand) that may be oxygen-poor. Silt particles are less than 0.1 mm in diameter, while sand particles are about 1 mm in diameter. Pools removed from the main channel are called **backwaters.** Organisms in backwaters resemble those in ponds. Backwaters can be temporary pools filled only during high water, or they can be permanent, still-water extensions of the stream (see table 1.2).

Interstitial hyporheic organisms (meiofauna), especially rotifers, copepods, and insects, take advantage of the (typically) large amount of oxygenated water flowing through coarse substrate beneath the bottom of the stream channel. The organisms living here are typically the same as those found in superficial substrate. Hyporheic organisms can be collected from shallow wells dug into substrate alongside the stream.

The **riparian** zone lies along the banks of the river. This region includes the **floodplain**—the area along the stream that floods during times of high runoff. Riparian, or floodplain, vegetation is adapted to occasional flooding on an annual basis. When the water recedes, ponds are often left behind.

Food Web Models

Food web models are characteristic of community ecology (see figure 1.13). The essence of a food web is a list of taxa, broken up into components representing groups of organisms that have the same predators and the same prey. Food web models are pictured as boxes (containing lists of taxa) connected by arrows indicating diet. Arrows point from food to consumer.

The food web diagram is similar to the trophic structure (figure 1.12), but the distinction is that the food web focuses on predator-prey and competitive interactions, while the trophic structure maps the flow of energy or a chemical, such as carbon or phosphate.

An example of a food web model component is a list of photosynthetic plant species (macrophytes) living in the shallow water of the littoral zone. Typically, the photosynthetic algae and associated bacteria and invertebrates (aufwuchs) living on the macrophytes would be grouped into a separate component because the algae have different predators. Food web models composed of few components in a linear sequence are called *food chains*. Webs are more complex systems, with multiple components per trophic level and multiple connections among components.

Food web models are maps of important ecological interactions among members of a community. Species that eat the same food (lower on the food web) are potential competitors, and species connected by an arrow are members of a predator-prey pair.

COMMUNITY INTERACTIONS

Limnologists are still trying to understand the factors that produce communities. One possibility is that communities are the result of random events of extinction and dispersal. This view is called the **neutral assembly** model of community assembly (Hubbell, 2001). An alternate hypothesis is the **niche assembly** model, in which a community is seen as the result of interactions among species that are adapted to a particular environment.

The niche assembly model has been the favorite model of limnologists, but the neutral model has not yet been tested. In this section, we will explore details of the niche assembly model, focusing on competition, predation, and mutualism.

"Traditionally, phytoplankton was considered to be a physically controlled community. Its seasonality was attributed to seasonal change in temperature or light . . ., or to the balance between sinking by gravity and resuspension by turbulence. . . In our data set, there are rather more examples of biological interactions, such as resource depletion followed by competition, grazing and predation, determining the temporal development of planktonic communities."

Sommer et al., 1986

Competition for Limited Resources

Competition and predation are two of the major interactions among members of communities.

General Concepts

Competition involves a struggle for one or more **limited resources,** such as food, energy, or specific chemicals like nitrogen or phosphorus. **Intraspecific** competition occurs between individuals of the same species. Males and females or young and adult organisms can experience intraspecific competition. **Interspecific** competition occurs between members of different species. For example, the scores of diatom species that occur in a lake are likely to be involved in interspecific competition for silica (Tilman, 1977). A successful competitor can reduce (deplete) availability of a limited resource to a level that allows the successful species to still flourish, but a level that causes other species to decline. Limited resources have often been reported in freshwater communities. For example, the abundance of primary producers (algae and plants) is positively correlated with inorganic nutrients, and herbivore abundance is correlated with abundance of primary producers (Leibold, 1999).

There are two main categories of the competitive interaction: **exploitative** and **interference.** Exploitative competition, also called "scramble" competition, occurs when different individuals or species struggle independently for nutrients. This is the kind of competition that is most often encountered—for example, competition between diatoms (see chapter 3) for nitrate and phosphate (Tilman, 1977). Competition among algae species for a limited resource is implicit during the annual sequence of algae species in a temperate lake (see the PEG model of annual species succession described in chapter 8). For fishes, competition for food among sunfish species affects their distribution within a lake, as well as their abundance, morphology, and life history (Mittelbach et al., 1999; Osenberg et al., 1988).

Interference competition occurs when one individual or species actively prevents another competitor from utilizing the resource—for example, *Daphnia* catching and even killing other smaller herbivores (rotifers—see chapter 4) that are eating the same food supply (Gilbert & Stemberger, 1985).

In the extreme playing-out of competition, only one species survives and its less successful competitors become extinct. This end point to competition for a limited resource is called **competitive exclusion.** Thus, the concept of competitive exclusion is critical to the niche assembly hypothesis of community composition.

Competition Avoidance

When ecologists look at nature, they see complex communities with many species. Thus, the interesting questions are less about competition and more about how organisms manage to avoid competition. Major ideas about how similar species avoid competition include environmental adaptations, which allow each species to be specialized in a particular environment; habitat partitioning, which allows organisms with similar requirements to gather resources in different parts of the lake; and resource partitioning, which allows organisms to divide up resources.

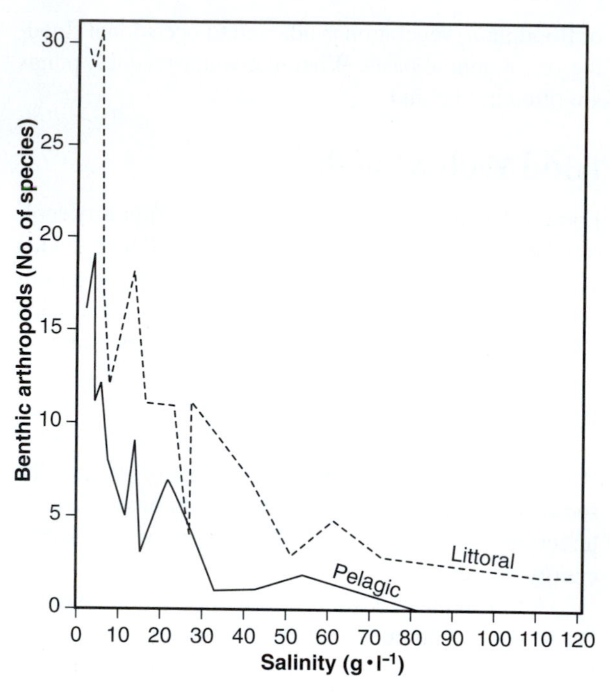

FIGURE 7.2 The number of species of benthic arthropods (macroinvertebrates) that occur in saline lakes in Saskatchewan, Canada. More saline lakes have fewer species. **Source:** Data from Hammer, 1995.

Environmental specialization According to the niche assembly hypothesis, each species probably has a unique set of environmental requirements. For example, many studies have shown that different *Daphnia* species have unique temperature optima and therefore occur in different depth strata of the lake at the same time (Threlkeld, 1986) or dominate the plankton at different times of the year (e.g., Spaak et al., 2000). Similarly, species of freshwater cladocerans show distinct optima for salinity (Boronat et al., 2001). In extreme environments, only a few specialists can survive. Hammer (1995) found fewer species of benthic macroinvertebrates in the more saline lakes in Sasketchewan (figure 7.2). Algal communities in lakes distributed over thousands of kilometers have changed during the last century, probably in response to industrial pollution (Vinebrooke et al., 2002). Acid deposition has lowered lake pH, shifting species composition toward filamentous green and phytoflagellate species.

Hutchinson (1961) asked why the pelagic community of phytoplankton is relatively diverse (100 or so species), when the seeming uniformity of the habitat suggests there should be only one or a few competitive winners—he called this situation the "**Paradox** of the Plankton." He reasoned that the constantly changing aquatic environment is responsible for the relatively high diversity. The changing conditions constantly alter the ability of any one species to be competitively successful, so that competitive exclusion (and reduction in diversity) is kept at bay. Another factor of possible importance is that many of the pelagic species might be present accidentally, having been swept out of the littoral zone by currents. One possibility that Hutchinson did not consider is expressed by the null assembly model. It is possible that all the species are equally good competitors and that communities are only the result of dispersal and extinctions (Hubbell, 2001). Although this possibility goes against the expectations of most aquatic ecologists, it has yet to be carefully tested. The niche assembly hypothesis expresses the expectation of most limnologists, including Hutchinson. It states that communities are composed of species that are adapted to the local environment and are able to outcompete other species. At least in some cases, adaptations might be more important than competition. For example, the occurrence and abundance of algal species can be predicted using physical and chemical data, without resorting to a competition effect (Reynolds, 1989; Smayda and Reynolds, 2001).

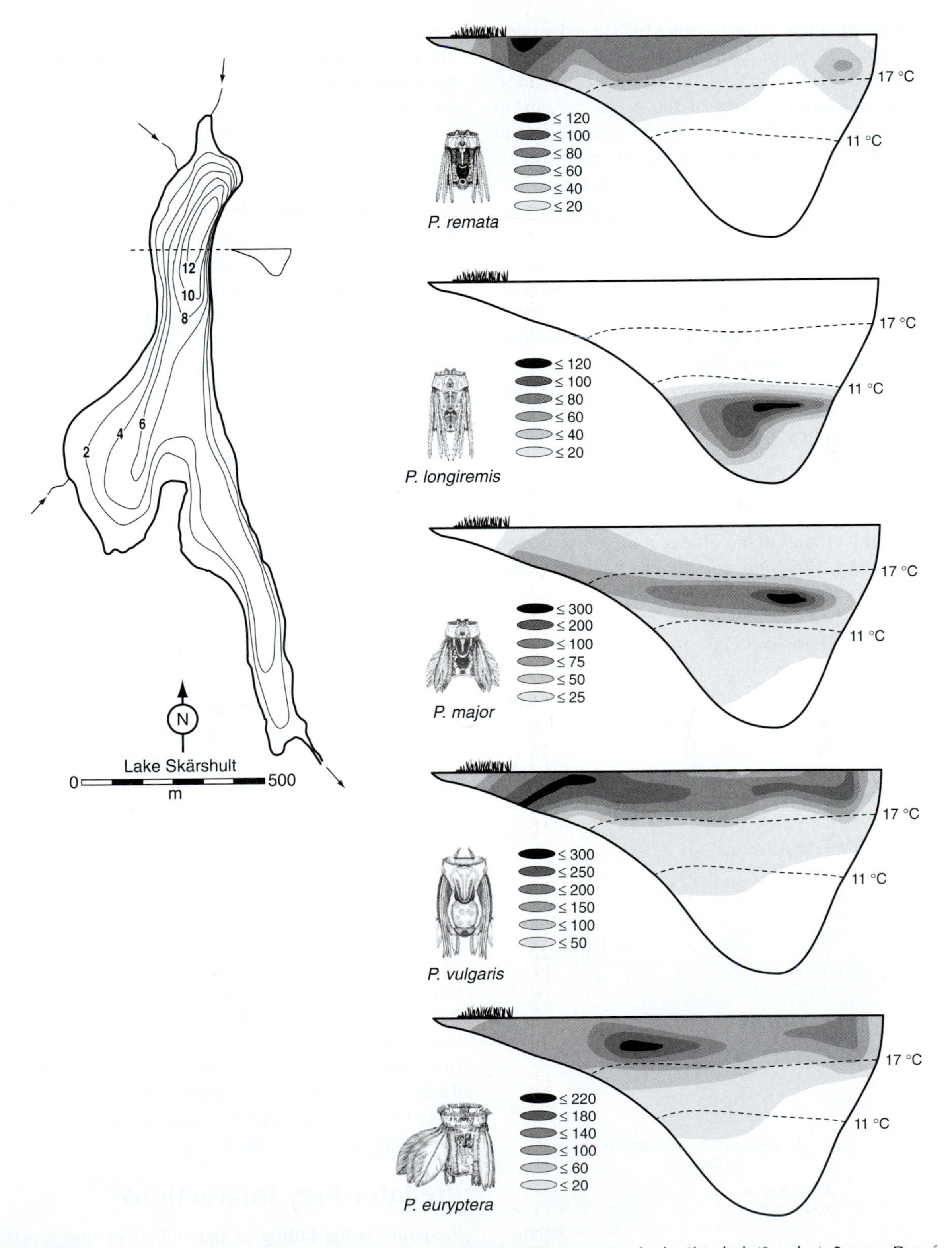

FIGURE 7.3 Different species of the rotifer *Polyarthra* are found in different parts of Lake Skärshult (Sweden). **Source:** Data from Berzins, 1958.

Habitat partitioning Hutchinson (1978) used a common association of rotifer species to illustrate habitat specialization (figure 7.3) within the context of local association. Rotifer species of the genus *Polyarthra* (see chapter 4) often co-occur in the same lake, but they show habitat partitioning in that different species tend to live in different parts of the lake. In Lake Skärshult in south Sweden (Berzins, 1958) *Polyarthra remata* is found in the littoral zone, *P. major* is found in the thermocline, and *P. longiremis* is hypolimnetic. *P. vulgaris* and *P. euryptera,* which are both abundant and epilimnetic zone, are of different sizes—the former about half the length of the latter. This difference in size may reflect different algal preferences, with the large rotifer eating large algae and the small rotifer eating small algae. Thus, the habitat partitioning allows several, very similar species to show an association at the scale of the lake.

As many as 20 to 30 species of water fleas (mostly in the branchiopod family Chydoridae, called *chydorids*) can be found in the littoral zone of an average temperate lake. Chydorid associations (a group of littoral water fleas) are a good example of adaptive radiation of related organisms living in the same environment. These different species cover a wide range of size and body shape (figure 7.4). Fryer (1968) stressed the importance of body shape and size over chemical factors in determining the distribution and evolution of chydorids. Differences in body size are associated with the ability to use the same overall habitat in different ways. For example, consider the amount of surface area available in the littoral zone, beneath 1 square meter of surface. For a person, there is about 1 square meter of surface area to stand on. However, for a large chydorid, there is much more surface area (such as macrophyte leaves) available to crawl over. A small chydorid would have even more surface area available for its activities because it fits into crevices not available to the large species (figure 7.4). Because species of different sizes are able to feed in different parts of the same habitat, body size differences allow many otherwise similar species to co-exist.

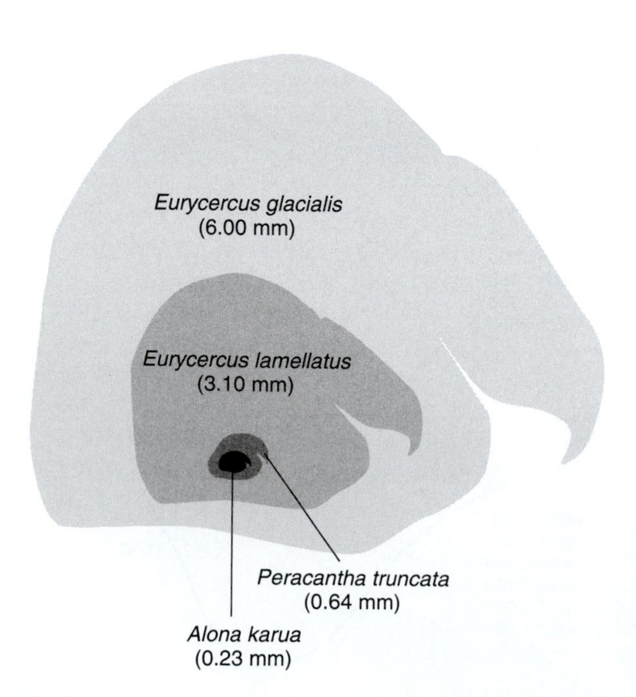

FIGURE 7.4 Range of morphological and size variation in species of chydorid water fleas. The ratio of length of the largest to the smallest is about 24; the ratio of volumes is about 24^3, or 15,625.

Selective feeding mechanisms and resource partitioning Many species of pelagic microcrustaceans and rotifers eat algae, which is often a limited resource. Competition is avoided if different kinds of zooplankton are specialized on slightly different aspects of the food resource (Richman & Dodson, 1983). There are two main ways that zooplankton gather algal and bacterial particles (as discussed in chapter 4). One strategy is to capture, taste, and ingest or reject individual particles. Protists, rotifers, and copepods use this method. The other strategy is to capture particles by filtration, collect the food into a ball (bolus), taste the bolus, and then ingest or reject the bolus. Cladocerans, such as *Daphnia,* use this method. The individual-capture method works well when algae are scarce (energy is conserved in capture because only desirable food is captured), or when palatable algae are present along with toxic algae (the herbivore can choose individual cells and reject toxic cells). The bolus-capture method works best when there is a high concentration of high-quality food. *Daphnia* spend a lot of energy to collect large quantities of food. If food density is low, the bolus-capture method costs more energy to run than it brings in, and if toxic algae are present, the bolus either has to be rejected (wasting the energy used to collect the bolus) or ingested (meaning the animal ingests some toxic algae along with the desirable food). The two feeding strategies are examples of how animals with the same *general,* limited resource can avoid competition.

Predator-Prey Interactions

Predation is the killing of individuals of one species (prey) by another species (predator). The prey become, in most cases, the food of the predator, and in lakes and streams, the prey can be a bacterium, a protist, or an invertebrate or vertebrate animal. Aquatic animals that consume plants are predators, but because of the nature of their plant diet, are often called *grazers* or *herbivores. Omnivores* eat both plants and animals.

Predation is a significant factor in the ecological struggle of aquatic organisms to persist from year to year. The summer is a time of especially intense predation, a time when populations of prey species can experience no increase in numbers or even a population decline due to predation. For example, grazing zooplankton can filter a volume equivalent to the entire lake every day (Porter, 1977). Because the algae are also reproducing rapidly, they do not appear to change in numbers. However, the species of algae rapidly shift from those preferred by zooplankton to those rejected by the herbivores. At a higher level in the food web, the presence or absence of carnivorous fishes in a lake can completely alter the plankton community and its dynamics (Brooks & Dodson, 1965).

Selective Predation Strategies

Predation is always **selective**—predators never feed at random. Different species of predators have unique preferences as to species and size of prey. There are two major patterns of selective predation: vertebrate predation and invertebrate predation.

Vertebrate predation A Bohemian limnologist, Jaroslav Hrbáček, noticed in the 1950s that different ponds were inhabited by different species of zooplankton. At the time, the standard explanation was that the ponds differed chemically or physically, and that these differences resulted in the dominance of different species. However, Hrbáček knew that all the ponds were in the same floodplain (of the Poltruba River); they had the same kind of bottom sediments; they were about the same size; they were, in most cases, right next to each other; they were filled with the same Poltruba River water; and they contained the same kind of algae. The critical difference Hrbáček saw was that some of the ponds contained fishes of the sort that eat zooplankton (planktivores) and some had only bottom-feeding fishes or low numbers of fishes. He made the creative and revolutionary suggestion that the fishes were having an effect on the kind of zooplankton that could live in a pond.

John Brooks (then at Yale University in Connecticut) heard Hrbáček's talk at an international meeting and remembered that in Connecticut there were numerous small lakes along the coast that were physically and chemically quite similar, but were inhabited by quite different kinds of zooplankton. Some of the lakes contained small species, such as rotifers and *Bosmina* (adults less than 0.5 mm long); some contained few rotifers and *Bosmina* but lots of larger zooplankton such as *Daphnia* (adults more than 1 mm long) (figure 7.5). Brooks wondered if the differences he was seeing in the Connecticut lakes could be due to fish predation and initiated a field study. Brooks and Dodson (1965) reported that lakes with the small zooplankton species were also occupied by a land-locked herring (the alewife fish), a fish specialized on eating large zooplankton (figure 7.6). In the lakes with larger zooplankton, the alewife was absent. Although there were other fish species that ate zooplankton, none was as focused on zooplankton as the alewife. A comparison of the species of Crystal Lake in 1942 (before alewife were introduced) with the zooplankton in 1964 (after the alewife introduction) showed, just as expected, that the association of species had changed from large to small species.

This phenomenon of *size-selective vertebrate predation* has come to be recognized as a major factor in determining the kind of species present in a lake. If there is a group of similar lakes that differ in the intensity of fish predation on the zooplankton, the lakes with the more intense predation will have the smaller zooplankton. The lakes with the more intense fish predation will also, as Hrbáček found, have more total phytoplankton and larger species of phytoplankton, as compared to lakes with less fish predation and larger species of zooplankton. This is because larger zooplankton, such as *Daphnia,* are more efficient at eating phytoplankton than are the small zooplankton, and they also can eat larger phytoplankton species. When fish predation is intense and zooplankton are small, large phytoplankton are relatively safe from being eaten.

Fishes typically take zooplankton larger than about 1 mm. This preference is mostly related to the feeding mode. Fishes typically feed by first seeing their prey and then capturing (ingesting) individual prey items. Apparently, prey smaller than about 1 mm are just too small to see or are not worth eating. The visual feeding mode also means that zooplankton can escape fish predation by living (at least during the day) in the deeper and darker parts of the lake.

Selective predation is also seen in herbivores. For example, zooplankton are selective as to size and quality of algae (Richman & Dodson, 1983). Larger herbivores are also selective. Moose require both aquatic and terrestrial vegetation (Pastor et al., 1988). When feeding on land, moose select deciduous plants over evergreen shrubs and trees. Beavers also show selectivity in

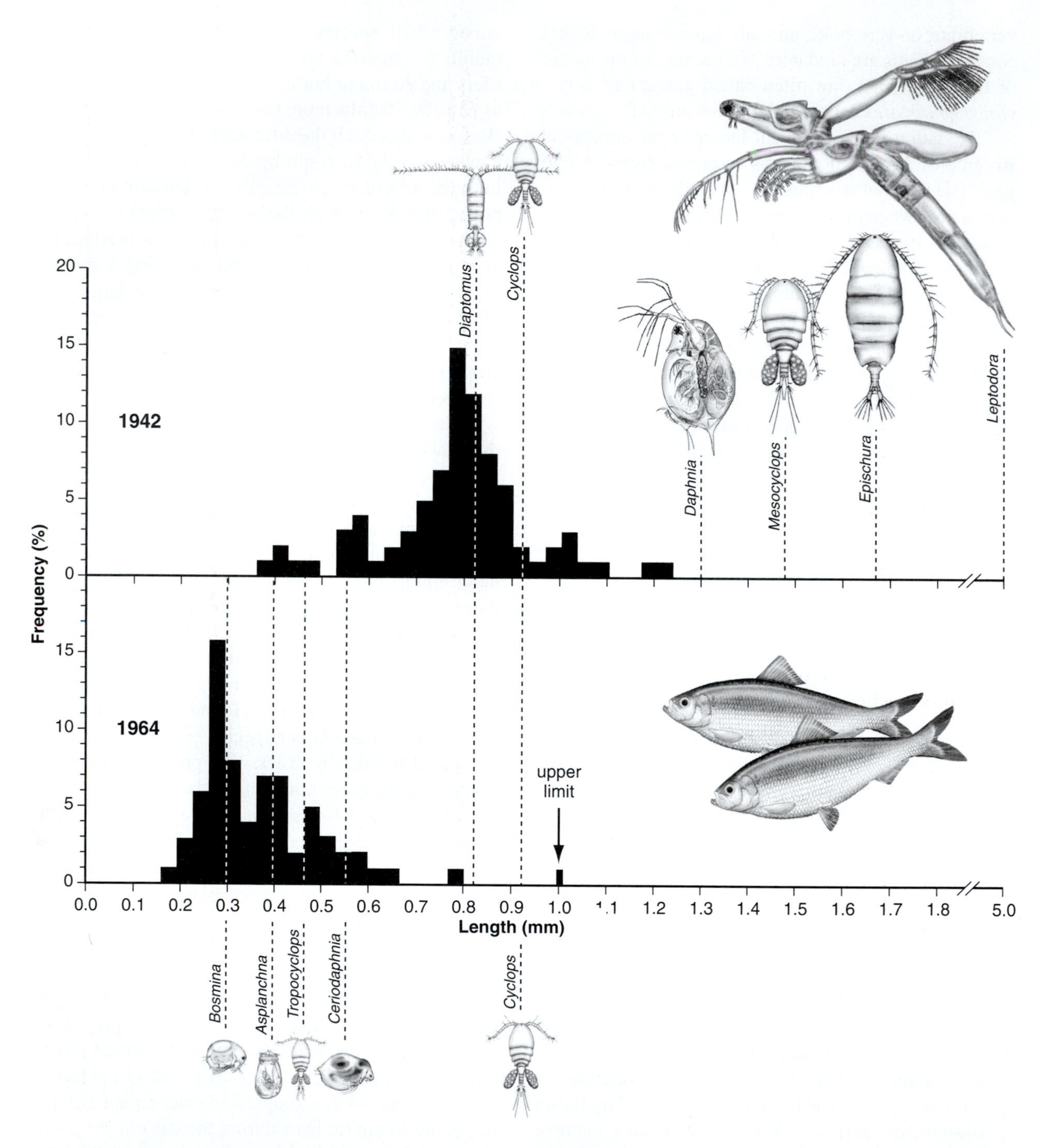

FIGURE 7.5 Histograms representing size-frequency distributions of zooplankton in Connecticut lakes. The histogram in the top graph represents the sizes of zooplankton in lakes lacking the alewife (*Alosa*), a plankton-eating fish (see figure 7.6). Pictures of plankton are placed at their average size—for example, the *Daphnia* is about 1.3 mm long. The Fish (alewives) are about 30 cm long. **Source:** Data from Brooks & Dodson, 1965.

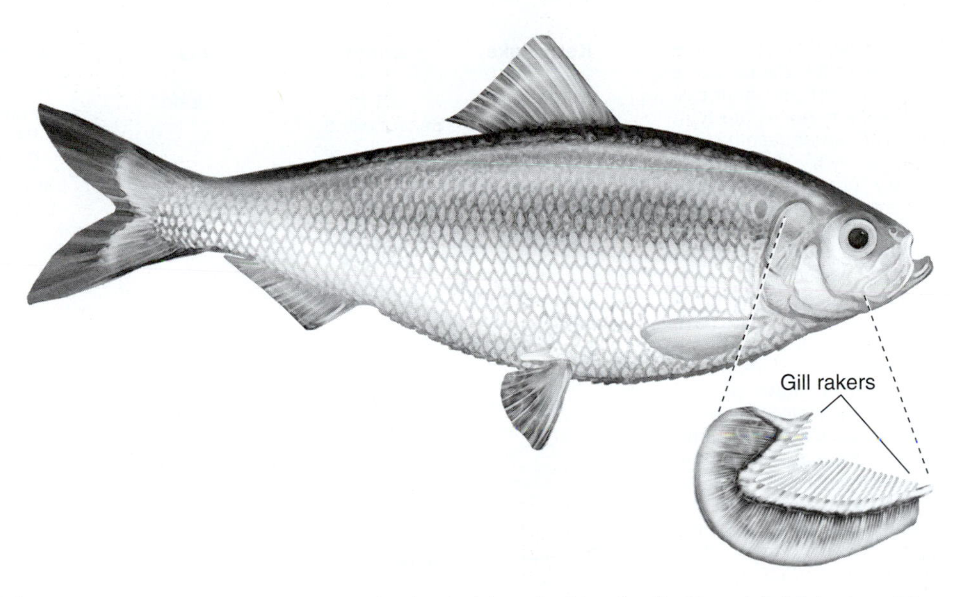

FIGURE 7.6 The alewife (*Alosa*) in side view, and a detail of the gill of the alewife. The adult fish is about 30 cm long.

their dietary choices among riparian wood plants (Naiman et al., 1988). This selective grazing has significant consequences for riparian plant community structure, just as zooplankton grazing selectivity is a driver of algal community diversity (Porter, 1977).

Invertebrate predation Brooks and Dodson (1965) hypothesized that small zooplankton are relatively uncommon in lakes with large zooplankton and low fish predation because the large zooplankton reduce the phytoplankton to a level at which the smaller zooplankton species cannot survive. That is, the larger species win the competition game for food. A few studies of grazing zooplankton have supported this hypothesis, but many have not, suggesting there is more to the absence of the small species than simply competition for food. For example, Tessier et al. (2001) found a trade-off between food quality and abundance that is not related to body size. *Daphnia,* and perhaps other water fleas, are able to either collect large amounts of food that are digested inefficiently or collect small amounts of food that are digested efficiently. The trade-off is regulated by whether gut passage time is fast or slow.

Relatively few small zooplankton live with large zooplankton species because the large species are predators as well as being competitors for algae. For example, *Daphnia* kill and sometimes consume small rotifers (Gilbert & Stemberger, 1985), and some of the larger copepods, such as *Epischura, Heterocope,* and large *Diaptomus,* eat smaller zooplankton (Dodson, 1984). Most of the larger cyclopoids, such as *Mesocyclops,* also eat smaller zooplankton (Kumar & Rao, 2001). Thus, when fish predation is weak, the larger zooplankton species dominate in a lake. These larger species, besides reducing the food supply, are an additional source of predation on small zooplankton such as *Bosmina.*

The larger predaceous zooplankton are size-selective, but, unlike fishes, larger zooplankton often specialize on prey smaller than about 1 mm. Although some of these invertebrate predators use vision to locate prey (such as the backswimmers and odonate larvae), most use mechanical cues. *Chaoborus* and copepod predators sense prey organisms by the vibrations the prey produce as they swim and eat. If the prey stop moving, they effectively disappear from these predators' sensors. For example, *Bosmina* attract copepod predators that sense vibrations produced by swimming and feeding (Kerfoot, 1978). When captured, *Bosmina* might escape and exhibit a "dead man's response"—ceasing all activity and slowly sinking out of range of the bemused copepod. The copepod, although unable to sense the nonvibrating prey, is occasionally successful by making a jump to where the *Bosmina* ought to be, by taking into account the time since the strike and the prey's sinking rate.

Lake classification and predation The kind of predators, vertebrate or invertebrate, that dominate a lake depend on the morphology of the lake. Thus, the planktonic

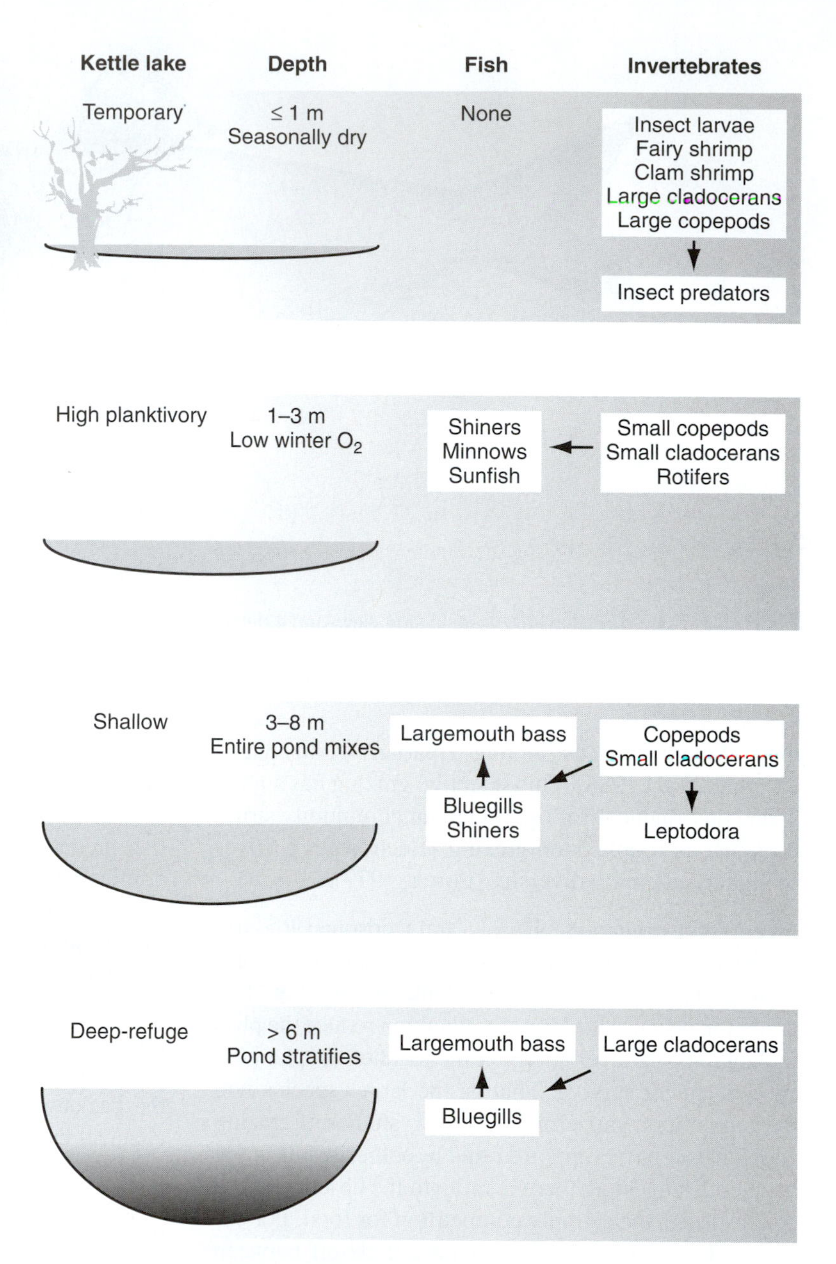

FIGURE 7.7 Lake depth gradient and plankton food web for small glacial kettle lakes located in the midwest and northeast United States. Small-volume ponds are typically dry in summer and, consequently, fishless, but contain a rich diversity of invertebrates that rely on dormant stages. Fisheries of permanent lakes are dominated by warm-water species, especially sunfish, with largemouth bass being the dominant piscivore. However, small, productive lakes often suffer from partial winterkill caused by low oxygen under ice cover. These lakes have no piscivores and, consequently, contain an abundance of highly effective planktivores. Large-volume lakes typically have bluegill sunfish as the dominant planktivore but differ greatly in zooplankton assemblage depending on whether the lake thermally stratifies in summer. Lakes that stratify provide a deep, cold-water, refuge in which zooplankton can hide from fish during the day. The zooplankton of such lakes contain large-bodied species of *Daphnia* as the dominant grazers. In lakes too shallow to stratify, these large-bodied species are replaced by smaller grazers which are better able to withstand the increased planktivory risk. **Source:** Modified from Tessier & Woodruff, 2002.

community in a lake depends indirectly on the lake's physical structure. For example, in temperate North America, small ponds tend to freeze solid in the winter, while larger lakes merely acquire a surface coating of ice. In the smaller lakes, fishes will "winter-kill" either from a lack of oxygen or from freezing. Similarly, predators that require a year or more of water are not likely to be found in temporary ponds. Deep, stratified lakes provide more habitats for prey refuge (the metalimnion or even hypolimnion) compared to shallow, unstratified lakes. Tessier and Woodruff (2002) conceptualized the interplay between lake morphology (size and shape) and predator assemblage to classify communities found in lakes of glaciated North America (figure 7.7).

Defenses Against Predation

Because predators are selective, it is possible for prey to have specialized defenses. Defenses fall generally into three categories: behavioral, morphological, and life history (see Larsson & Dodson, 1993).

Behavioral defenses The first line of defense is often behavioral, such as migrating to refuge habitats to avoid predators (including **diel** vertical and horizontal migrations), swimming in an inconspicuous manner, swarming, and escape behavior. Diel, or daily, vertical migration is the best-studied defense against predators. This behavioral strategy is discussed in Chapter 8.

Swarms are associations of large numbers of organisms in water. They are surprisingly difficult to study, requiring large number of samples in three dimensions (Folt, 1999; Malone & McQueen, Folt & Burns, 1999). Nevertheless, swarms appear to exist and to provide some protection from being eaten by predators, perhaps by saturating or confusing the predator.

Escape behaviors, which are behavioral defenses, were discussed in detail in chapters 3, 4, and 5.

Morphological defense and cyclomorphosis Morphological defenses include thick shells (turtles and some mollusks and, to a lesser extent, some rotifers and water fleas) and long spines (especially some rotifers, water fleas, and insects). These morphological defenses make it difficult for a predator to handle or swallow prey once captured. For example, Dodson (1984) studied *Daphnia* living in small arctic ponds. The hypothesis was that *Daphnia* species living with predaceous calanoid copepods would have thicker and stronger exoskeletons than species living in ponds lacking the copepods. The copepods feed by capturing prey, such as *Daphnia,* and chewing through the outer covering. Measurements of exoskeleton thickness and toughness supported the hypothesis as did feeding trials. Similar protection has been reported for the shell-like lorica of some rotifers (Hampton & Gilbert, 2001) and even the shelled amoebae (tintinnids—Capriulo et al., 1982).

Many small organisms, especially in the plankton community, show a seasonal cycle in morphological form, called **cyclomorphosis.** While the body form of individuals remains constant, shape does change from generation to generation. Cyclomorphosis often occurs in species of protists, rotifers, and branchiopods. Copepods rarely show even the slightest cyclomorphosis. Additional species, fishes and macroalgae, for example, show phenotypic plasticity, but not necessarily on an annual cycle. For a review of recent studies, see Kerfoot and Sih (1987) and Tollrian and Harvell (1999).

Cyclomorphosis was first described by Woltereck (1909), who showed that seasonal morphological variation often seen in *Daphnia* was due to phenotypic plasticity within a species (even a single clone), and not due to seasonal succession of different species (figures 7.8 and 7.9). The summer generations have long tail spines and high (often pointed) helmets, while the winter generations have short tail spines and low, rounded helmets (compare the helmet shape of the *Daphnias* in figure 4.8).

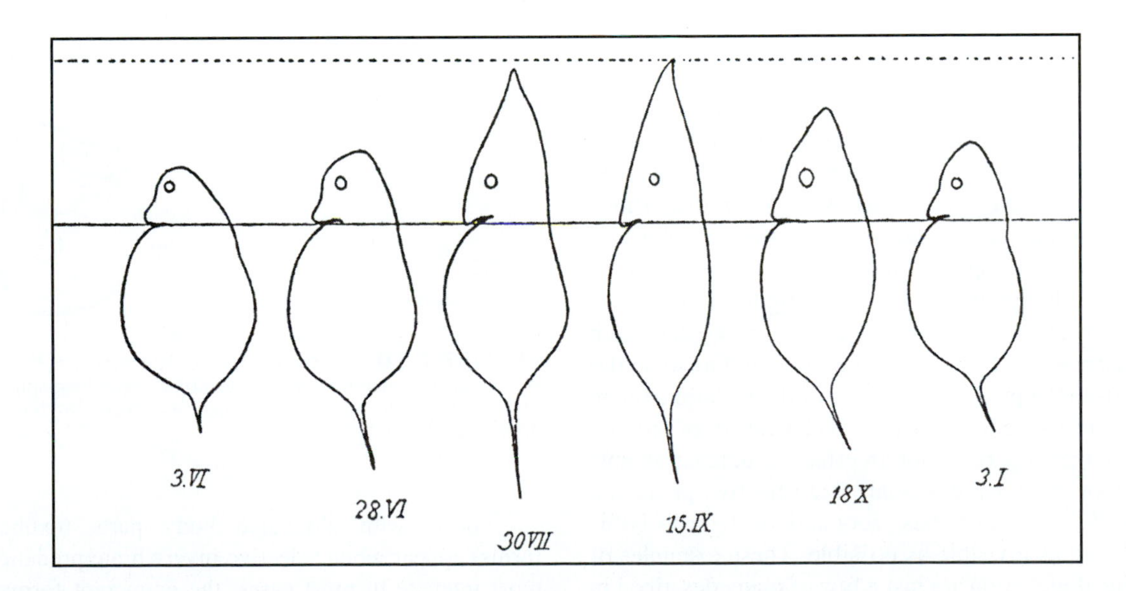

FIGURE 7.8 The first study that suggested there was an annual, among-generation change in the shape and size of *Daphnia*. **Source:** From Woltereck, 1909.

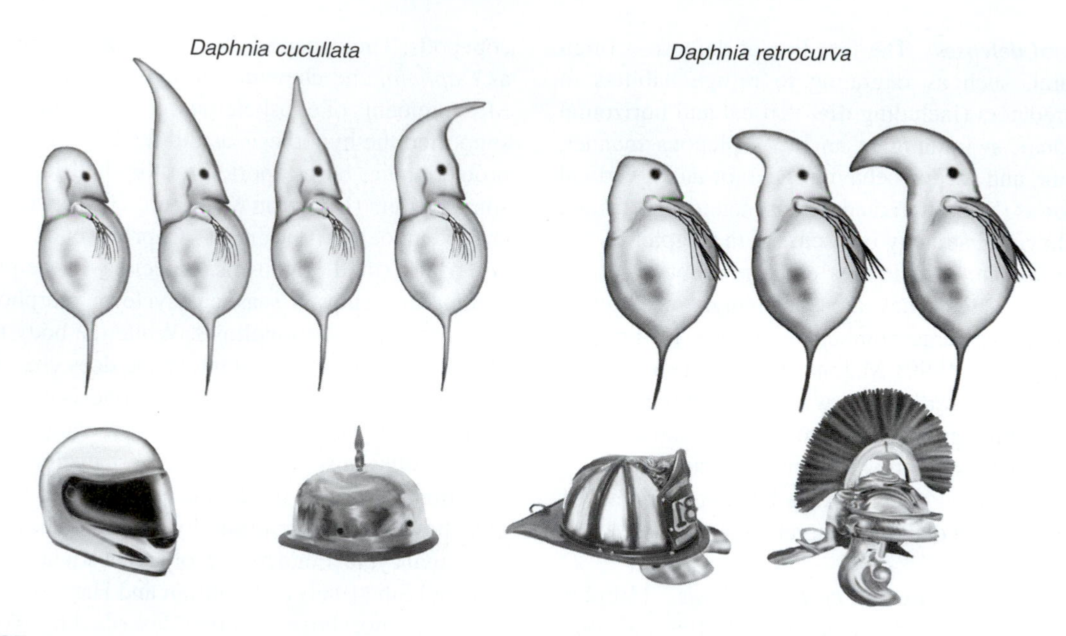

FIGURE 7.9 *Daphnia cucullata* (left) and *Daphnia retrocurva* (right) are two helmeted species that show annual cyclomorphosis. **Source:** The figure was inspired by Margalef (1983).

The phenomenon of cyclomorphosis provides a challenging opportunity to explore adaptive significance. For a phenomenon to persist in a population year after year, there must be a benefit, at least for part of the year. Many experiments with a number of predator-prey pairs have shown that prey individuals with elongated body spines or helmets (relative to individuals with short spines or helmets) have a lower mortality when specific predators are present (Tollrian & Harvell, 1999). The first well-studied example of cyclomorphosis as a predator defense was for the rotifers *Branchionus* and *Asplanchna*. Gilbert (1980) showed that long-spined *Brachionus* were safer than short-spined forms in the presence of the predator *Asplanchna.* The spines interfere with *Asplanchna*'s ability to ingest the *Brachionus* (figure 7.10). Similarly, juvenile *Daphnia* pulex that grow neck teeth (Krueger & Dodson, 1981) are protected to some degree from the predaceous larvae of the phantom midge *Chaoborus* (see figure 4.18b). The strategy of producing elongated parts is not an effective defense against fishes or other large, visual, size-selective predators. When living with fishes, zooplankton tend to be as small and as invisible as possible. These examples of chemical induction are just a few of many described in Kerfoot and Sih (1987).

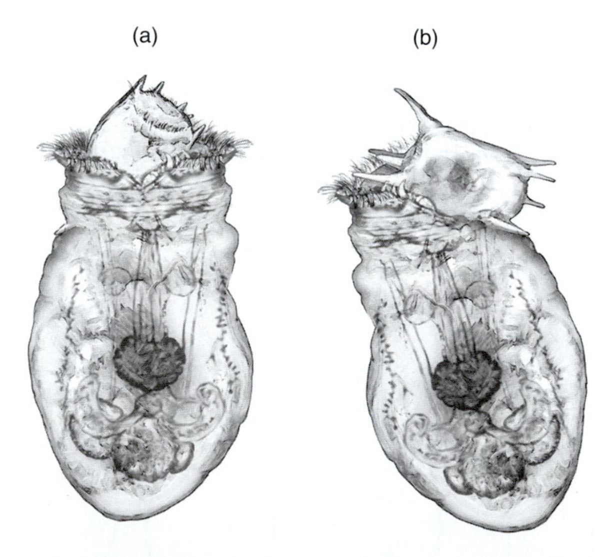

FIGURE 7.10 *Asplanchna* eating *Brachionus.* (a) The short-spined form of an adult female *Brachionus.* (b) A long-spined female *Branchionus* that had been exposed to *Asplanchna* smell during development.

Forms with elongated body parts (exuberant forms) appear when selective invertebrate predation is most intense. In most cases, the exuberant forms are seen in the summer. However, there are a few cases of "reverse cyclomorphosis," in which the exuberant forms are seen in the winter. For example, winter generations of *Bosmina* have long tail spines (mucrones) and antennules in the winter when the calanoid copepod *Epischura* is an active predator (Kerfoot, 1978).

There are two major lines of thought on why some morphological defenses are present only part of the time (Tollrian & Harvell, 1999). One idea is that there is a cost to expressing the morphological defense—a cost that reduces reproductive rate or competitive ability. The cost of cyclomorphosis may be related to energy allocation, which affects other aspects of life history. Another possibility is that there is no particular cost but different morphologies are the best defense at different times of year, depending on the annual change in predator regime.

Life-history defenses Timing or duration of life-history stages, adult body size, number of offspring, and size of offspring are all life-history characteristics that can be adjusted to defend against predation (Larsson & Dodson, 1993). Prey species typically have life histories adjusted so that the most vulnerable stages (juvenile or adult) are the shortest and occur when predation is as low as possible. For example, *Daphnia* living in the presence of fishes tend to reproduce rapidly and have many small offspring (Dodson, 1988), which is probably an adaptive response to fish predation.

Life history, morphology, and behavior are interrelated, and if the energy budget is limited, trade-offs are expected. For example, if an animal allocates energy from whole-body growth to growing long spines, mortality due to predation will be lower per unit time, but mortality during the longer life span may be greater. To be adaptive, the benefit of morphological defenses (the benefit is increased relative survivorship in the presence of certain specific predators) must be greater than the cost of the defenses (slower reproductive rate and longer time of vulnerability to all predators). Langergren et al. (2002) found that antennule length in *Bosmina* was correlated with density of the predator *Leptodora,* suggesting the longer antennules seen in the summer are a cyclomorphic defense. Longer antennule length was correlated with lower reproductive capacity—a life-history cost. Similarly, Barry (1994) found that a kairomone produced by backswimmers (Hemiptera) induces head crests in *Daphnia.* The crests protected *Daphnia* against notonectid predators, but the protection had costs such as longer developmental time and lower physiological vitality.

Chemical Induction of Predator Defenses

Changes in behavior, morphology, and life history are induced (caused to happen) by an environmental signal. Signals include a change in photoperiod, food concentration, or a chemical signal from a predator, such as a fish or *Chaoborus* (Larsson & Dodson, 1993). Chemical signals, produced by a predator, that affect behavior, life history, or development of prey are called **kairomones.**

The mechanisms of kairomone action are an area of active and exciting research. One great example of the mechanism at the cellular level has been reported for *Daphnia.* Exposure to a chemical signal causes a dramatic increase in cell nucleus size as the chromosomes are replicated several times (Beaton & Hebert, 1997). The larger number of chromosomes apparently stimulates cell growth, so an increase in helmet length is due to an increase in cell size, not an increase in the number of cells.

Until the 1960s, it was generally assumed that cyclomorphic organisms used environmental cues to control expression of phenotypes. Temperature, turbulence, food concentration, and perhaps day length had been proposed as possible environmental cues. Gilbert (1966) reported that a chemical (small protein) was the kairomone signal that induced development of long spines in the rotifer *Brachionus* (see figure 7.10). The kairomone was produced by *Asplanchna.* This observation was thought to be an unusual oddity, until examples of kairomones turned up in other predator-prey systems. Twenty years later, lab and field results reported by Krueger and Dodson (1981) suggested that a cyclomorphic system in *Daphnia pulex* was controlled by a kairomone produced by *Chaoborus.* Since then, many other prey species have been shown to undergo cyclomorphosis in response to chemical cues produced by a variety of predators (e.g., Dodson 1989; Havel, 1987; Stemberger & Gilbert, 1987; Tollrian & Harvell, 1999).

Kairomones can also influence life history via effects on yolk production. In a study of *Daphnia,* Stibor (2002) found that fish kairomones caused yolk production to start earlier, which speeded development and shortened the time to first reproduction. Fish-exposed *Daphnia* also produced more total yolk than did control animals. *Chaoborus* kairomones had no effect on timing of yolk production or amount, but caused yolk to be produced at a faster rate than seen in fish-exposed animals.

Aquatic organisms live in an **olfactory field** of dozens, if not hundreds of biologically significant chemical signals (Dodson et al., 1994; Larsson & Dodson, 1993). For example, *Daphnia* can probably sense at least a dozen different chemical signals. These signals provide information about whether to accept or reject food and about the presence of macrophytes, specific predators, competitors, congeneric individuals, and males. Production of males and haploid eggs is also induced chemically.

The existence of several different chemical signals, all of which could be in the water at the same time, suggests the possibility of hierarchical responses: Signals have an effect modified by the presence or absence of other signals. Black (1993) measured the response of a *Daphnia* clone to two predator kairomones from *Chaoborus* and *Notonecta.* Morphological and life-history responses to the two-kairomone-exposure were different from the response to either kairomone alone, but appeared to be a reasonable compromise. We still have a lot to learn in the area of signal processing and of the role of the internal state of the prey species. For example, how do learning and motivation affect reaction to a chemical signal?

Kairomones come in a bewildering variety of chemical forms (table 7.1 on p. 177).

Comparison of Ecological Importance of Competition and Predation

A number of studies (for example, Cooper & Smith, 1982; Gliwicz et al., 1981) have compared the effects of food limitation and predation on zooplankton. These studies were done by (1) correlating herbivore populations with food level, or (2) estimating the value of d using the b from the Edmondson (1965) egg ratio technique. Then, one looks for a correlation between the mortality rate, d, of the zooplankton species and either predator population size or algal concentration. In general, these studies have shown that zooplankton tend to be limited by food level rather than by predation, but at certain times of the year, particularly in the late summer, predation has a significant effect on prey abundance (see chapter 6 and Luecke et al., 1992).

Indirect Effects of Predation

Indirect effects are mediated through intermediate species. For example, two size-selective, plankton-eating predators, fishes (yellow perch, figure 5.2e) and larval insects (*Chaoborus*), have indirect effects on phytoplankton via their direct effects on zooplanktonis herbivores. The experimental addition of a top predator (northern pike; figure 5.2f) to Lake-221 (The Ontario Experimental Lake Area) resulted in a reduction of yellow perch and an increase in *Chaoborus,* a reduction in herbivorous zooplankton, a shift in algal dominance from green to cyanobacteria. During at least in the first 2 years, phytoplankton biomass and phosphorus (P) increased because of nutrient recycling and excretion by pike and zooplankton (Findlay et al., 1994). Vanni and Findlay (1990) used plankton enclosures to test experimentally for indirect effects of predators on phytoplankton community structure. The predators were yellow perch and *Chaoborus.* They found that phytoplankton increased relative to predator-free controls only in enclosures with fishes and not in enclosures with *Chaoborus,* suggesting an indirect effect of fishes. Fishes render phosphate (an important algal nutrient) available to phytoplankton in two ways: (1) directly by releasing phosphate at a high rate, and (2) indirectly by increasing rates of phosphate excretion by the zooplankton community (because small zooplankton have higher biomass-specific excretion rates than large zooplankton). *Chaoborus* have less effect on phosphate release. Phytoplankton taxa, which increased biomass in the presence of fishes, were those with high phosphate requirements, including green and blue-green algae and dinoflagellates.

Keystone Predation

A **keystone predator** is a selective predator, whose most preferred food, is the most successful competitor among a group of prey species. This selective predation can increase species diversity among prey species that would otherwise compete more strongly for limited resources. Robert Paine began using this term in the late 1960s to explain an observed correlation between the presence of a predaceous starfish and high diversity of the intertidal community of species that included prey for starfish (Paine, 1969). The same concept of keystone predation applies to freshwater habitats. For example, fish predation is size- and species-specific and tends to remove the largest *Daphnia* from the pelagic zone. The absence of large *Daphnia* (an efficient eater of algae) results in higher algal densities (often a limited resource) and the survival of cladoceran species (and their invertebrate predators) that do not typically co-occur or thrive with *Daphnia* (Carpenter et al., 1985).

PLATE 1 University of Wisconsin summer Limnology class, Lake Mendota, Wisconsin. **Source:** Photo by Jeffery Schell.

PLATE 2 Dove Lake, an oligotrophic lake below Cradle Mountain, Tasmania. **Source:** Photo by V. Dodson.

PLATE 3 Crater Lake, Oregon. A spectacular, ultraoligotrophic, 589-m-deep lake (surface area = 48 km^2) at 1883 m elevation, in the crater (caldera) of Mt. Mazama. The volcano last erupted about 7000 years ago. The photo was taken 15 March 1999, when the lake was not frozen (it freezes rarely), but the surrounding terrain was covered with about 4.5 m of snow. **Source:** Photo by Jim Phelan.

PLATE 4 Boiling Lake in an active caldera on the Caribbean island of Dominica, near the Valley of Desolation, 15 January 2001. The boiling, sulfur-laden lake is being observed by Calvin Anthony (left) and a French family from Martinique. **Source:** Photo by Dimitri Sokolenko.

PLATE 5 Lake Mendota, Wisconsin, with ice fishermen, in February 1989. **Source:** Photo by Steve Carpenter.

PLATE 6 Fly fisherman next to the dam of a beaver pond in Copper Creek, near the Rocky Mountain Biological Laboratory, in July 1969. The dam is made of willow and aspen sticks. The stream flows from left to right.

PLATE 7 A small sink hole pond in a karst region of the Owen's River Valley, east-central California.

PLATE 8 Dr. Laura Torrentera and research assistants, observing *Artemia* living in a saltern near Xtampu on the north coast of Yucatan, 1991. The saltern (basin and rockwork) was originally built by pre-Colombian Mayans.

PLATE 9 A kettle pond, "Page Creek South," in Marquette Country, Wisconsin, May 1998. **Source:** Photo by Dick Lillie, Wisconsin Department of Natural Resources.

PLATE 10 An artificial agricultural pond. Cows are standing on the earthen dam and in the pond. **Source:** Photo taken June 2001 near Conception Abbey, Missouri.

PLATE 11 A shallow playa pond near Amarillo in the Texas Panhandle, about 1992. **Source:** Photo by Wyman Meinzer.

PLATE 12 Lake Texoma, a reservoir on the border of Oklahoma and Texas. The dam is on the lower right of this picture. Note the muddy water entering the reservoir from the northern and western tributaries (north is at the top of the image.) This is a satellite image taken 9 October 1994. **Source:** U.S. NASA STS068-247-79.

PLATE 13 Amanda (Venable) Olson collecting aquatic insects in a wetland of Allequash Lake (Vilas County) in northern Wisconsin, an NSF Long-Term Ecological Research site. The student is a member of the University of Wisconsin 2000 summer Limnology class. The emergent plants are grasses and sedges. The floating plants are lily pads. **Source:** Photo by Amanda Olson.

PLATE 14 A large number of shallow rock pools (also called gnammas) along the summit of Warrdagga Rock, a large, granite inselberg located 360 km NNE of Perth, Western Australia. These pools are a meter or so in diameter. **Source:** Photo courtesy of Ian Bayly.

PLATE 15 Why-Not-Bog near Crystal Lake, Vilas County, Wisconsin in July 1998. The UW-Madison summer Limnology class is standing on a floating mat. The mat is composed of sedges, sphagnum moss, pitcher plants, bog rosemary, cranberry vines, and numerous other plants. *Utricularia* is growing, suspended in open water at the edge of the brown, tea-stained water. Water depth is about 2 m at the edge of the mat. **Source:** Photo by V. Dodson.

PLATE 16 Dr. Bobbi Peckarsky and students collecting stream macro-invertebrates in a Colorado stream, near the Rocky Mountain Biological Laboratory, in 1992. **Source:** Photo by Dr. Chester Anderson, Courtesy of Rocky Mountain Biological Laboratory.

PLATE 17 Electro-shocking for fish in Camp Creek, a small, second-order stream in Richland County in southern Wisconsin. This is a Wisconsin Department of Natural Resources stream-monitoring crew. **Source:** Photo by Dr. Brian Weigel.

PLATE 18 Dr. Everett Fee with vertical phytoplankton samples from a lake in the Experimental Lake Area, Ontario, July 1985. The samples demonstrate a hypolimnetic algal plate. The clear sample nearest Everett is from the hypolimnion, below the algal plate. The dark-green sample is from the algal plate, and the light-green sample was taken just 1 meter above. The next three clear samples are from the epilimnion.

PLATE 19 Dr. Peter Leavitt holding a Secchi disk above a thick surface bloom of the blue-green alga *Aphanizomenon* (a cyanobacterium) in Lake Mendota, Wisconsin. Normally, the disk would be suspended from a line and lowered into the water to measure water clarity. **Source:** Photo by Peter Leavitt.

PLATE 20 An aerial view of Clear Lake, California. Large patches of cyanobacteria can be seen near the surface, in the nearest part of the lake. **Source:** Photo by Don Mauser.

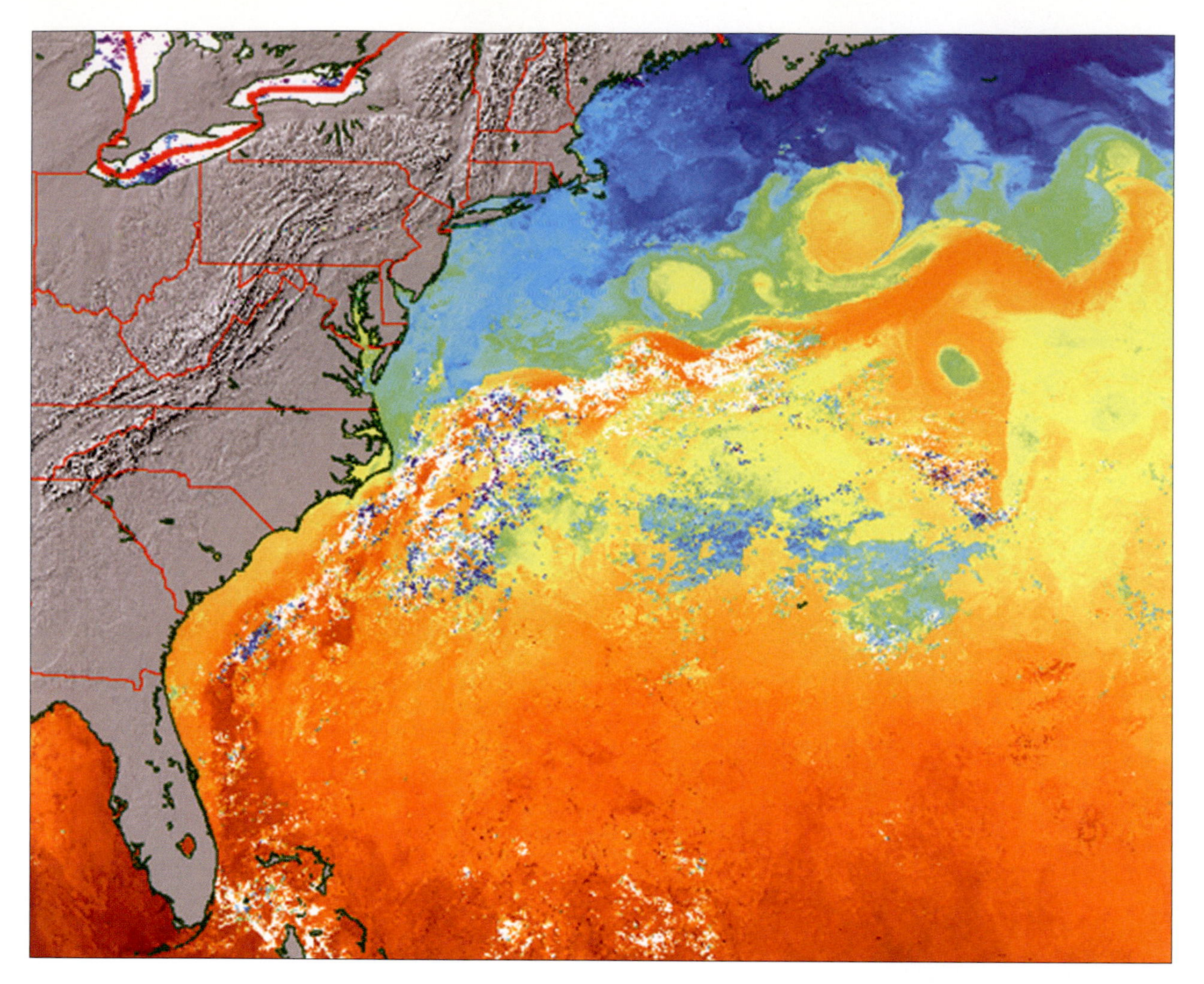

PLATE 21 Turbulence on a large scale in the North Atlantic. This is false-color satellite image of the Gulf Stream ocean current, from the Advanced Very-High-Resolution Radiometer (AVHRR), which is flown on the NOAA polar orbiting satellites. The image is a 3-day composite centered on 01 June 1997. Red indicates the warmest surface-water temperatures; red to orange—24° to 28°C; yellow-green—17° to 23°C; light blue—10° to 16°C, and dark-blue—2° to 9°C. Large-scale gyres (eddies) are indicated in the Gulf Stream by the circular masses of warmer or cooler water. The Gulf Stream flows in a northeast direction and rubs against cooler and relatively stationary coastal water. **Source: U.S.** NOAA AVHRR Archive #0556.

PLATE 22 Wind-induced waves breaking along the shore of Lake Mendota, Wisconsin, April 2001. The waves are about 0.75 meters high.

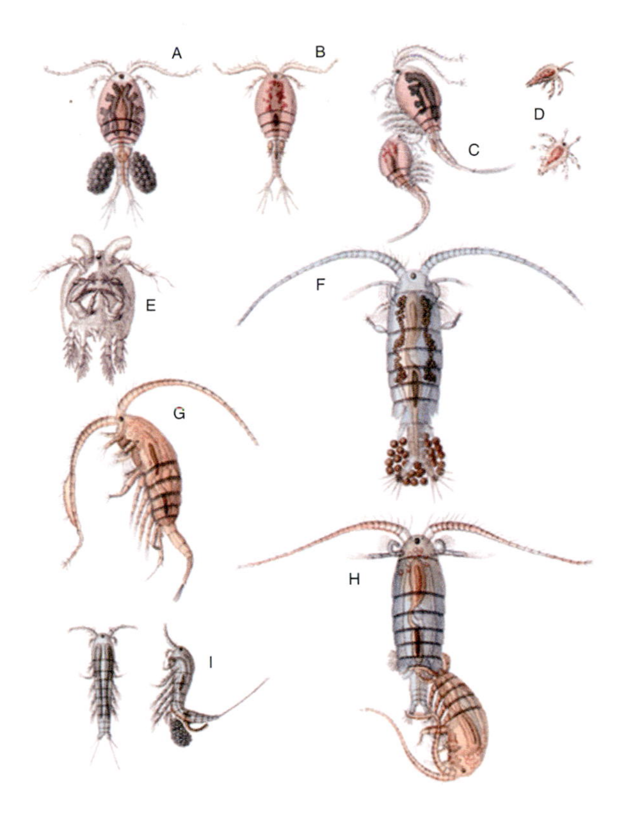

PLATE 23 Copepods from Jurine: (a) Female cyclopoid with two egg sacs; (b) male cyclopoid copepod with two bent first antennae; (c) mating cyclopoid copepods—the male is holding the female's fourth pair of legs with his first antennae; (d) two nauplii, in side and top views; (e) head appendages of a cyclopoid copepod—first antennae (cut off), second antennae, mandible, first maxilla, second maxilla, and maxilliped; (f) adult female calanoid copepod with one egg sac; (g) male calanoid copepod with right first antenna modified for grasping a female; (h) calanoid copepods mating—the male is holding the female's urosome (tail segments) with his fifth pair of thoracic legs and the bent right first antenna; (i) two harpacticoid copepods. **Source:** Modified from Jurine, 1820.

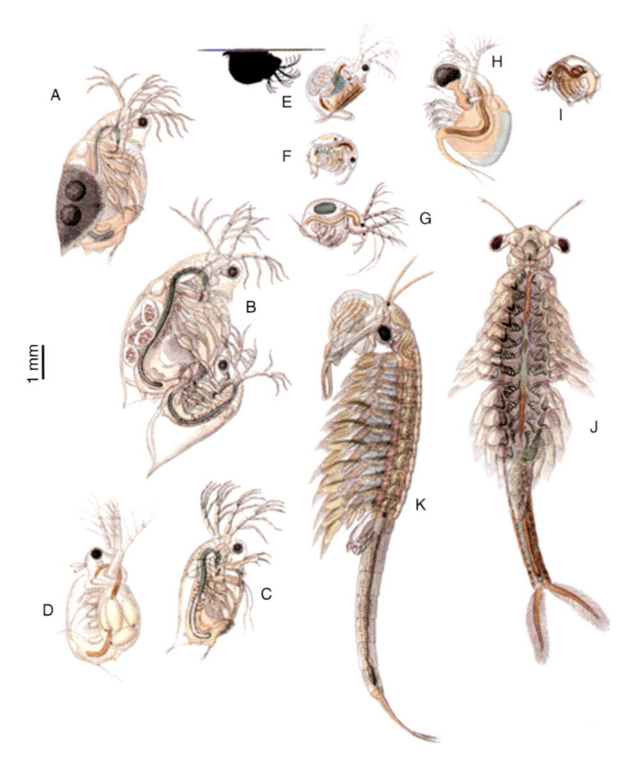

PLATE 24 Cladocera from Jurine (a) An adult female *Daphnia,* with modified carapace (ephippium) holding two resting eggs; (b) mating *Daphnia;* (c) an adult male *Daphnia,* (d) an adult female *Moina,* carrying four embryos in the brood chamber; (e) female *Scapholeberis* showing the darkened ventral carapace margin, and a silhouette showing a *Scapholeberis* with its straight ventral margin up against the water surface; (f) female *Bosmina* carrying one embryo; (g) female *Macrothrix* carrying one embryo; (h) female *Polyphemus,* a predaceous water flea; (i) female *Chydorus;* (j) adult female fairy shrimp, showing the filtering setae on the thoracic legs; (k) adult male fairy shrimp, with enlarged second antennae used to grasp his mate. **Source:** Modified from Jurine, 1820.

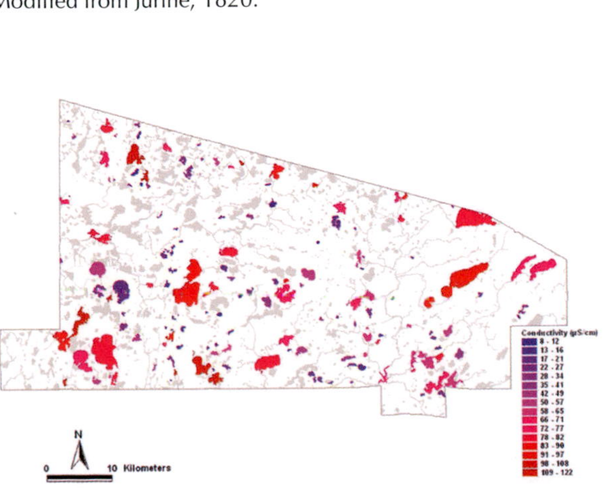

PLATE 25 Conductivity of lake waters, in Vilas Co., WI. The conductivity of each lake is color-coded, with the darkest blue indicating 8-12 $\mu S\ cm^{-1}$ and the darkest red indicating 109-122 $\mu S\ cm^{-1}$. (Map prepared by Joan Riera, for the NSF Long Term Ecological Research program).

PLATE 26 Lake District, Northeast Wisconsin, Vilas County, site of NSF LTER North Temperate Lake Research. **Source:** NSF LTER NTL archives.

PLATE 27 Tourists admiring the Goose Necks (deeply cut meanders) in the San Juan River, Utah. The river was much larger when mountain glaciers were melting at its source. This is "The River" of Ann Walka's poem—see the beginning of chapter 12. **Source:** Photo by V. Dodson.

PLATE 28 Manicougan crater and lake (reservoir) in northern Québec, Canada. The circular lake is about 70 km in diameter and the water level has been increased by a dam. About 214 million years ago, a meteorite exploded over the land surface, creating a round pool of melted rock surrounded by a ring of fractured rock. The original crater was covered with as much as 1 km of sediment, which has since been eroded away, leaving a central mass of the once-melted rock and shocked rock, surrounded by a circular channel where less-resistant fractured rock has been removed by glaciation. This image (Landsat_20010601.jpg) shows the full Landsat-7 scene (185 by 185 km), acquired on 1 June 2001. The image is made up of Landsat bands 5, 4, and 3 in red, green, and blue, respectively. Most of the area is forested (appears dark-green). The pink/brown areas are clear-cuts (logging areas), fire scars, and/or non-forested wetlands (the large one just south of the crater is a logging area). (Thanks to Jonathan Chipman of the University of Wisconsin Environmental Remote Sensing Center.) **Source:** U.S. NASA LandSat-7.

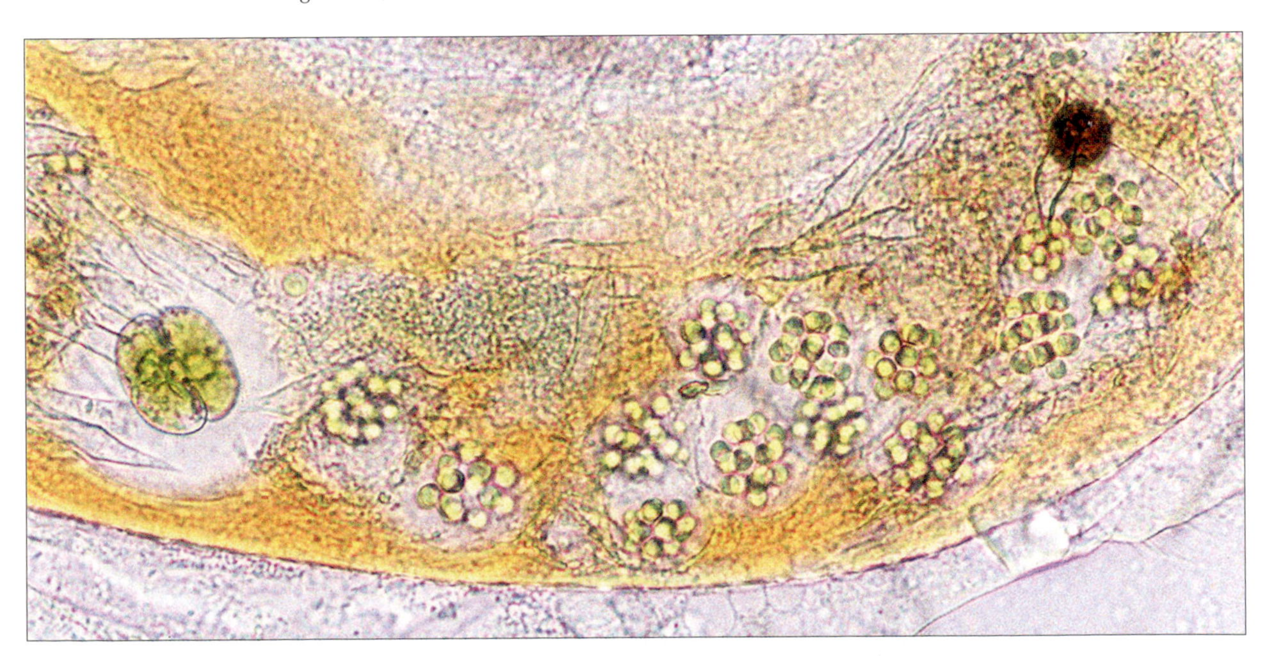

PLATE 29 Small colonies of green algae, each surrounded by a clear, gelatinous sheath, can be seen in this close-up photo of the contents of the posterior end of a *Daphnia* gut. **Source:** Photo by Pieter Johnson.

PLATE 30 Iceberg Lake, Glacier National Park, Montana. The lake water is a distinctive opaque, blue-white color due to "glacial milk"—finely ground rock released by the glacier melting in the cirque. **Source:** Photo by V. Dodson.

PLATE 31 Lake Mendota, Wisconsin, summer anglers and motor boats. **Source:** Photo by Stephen Carpenter.

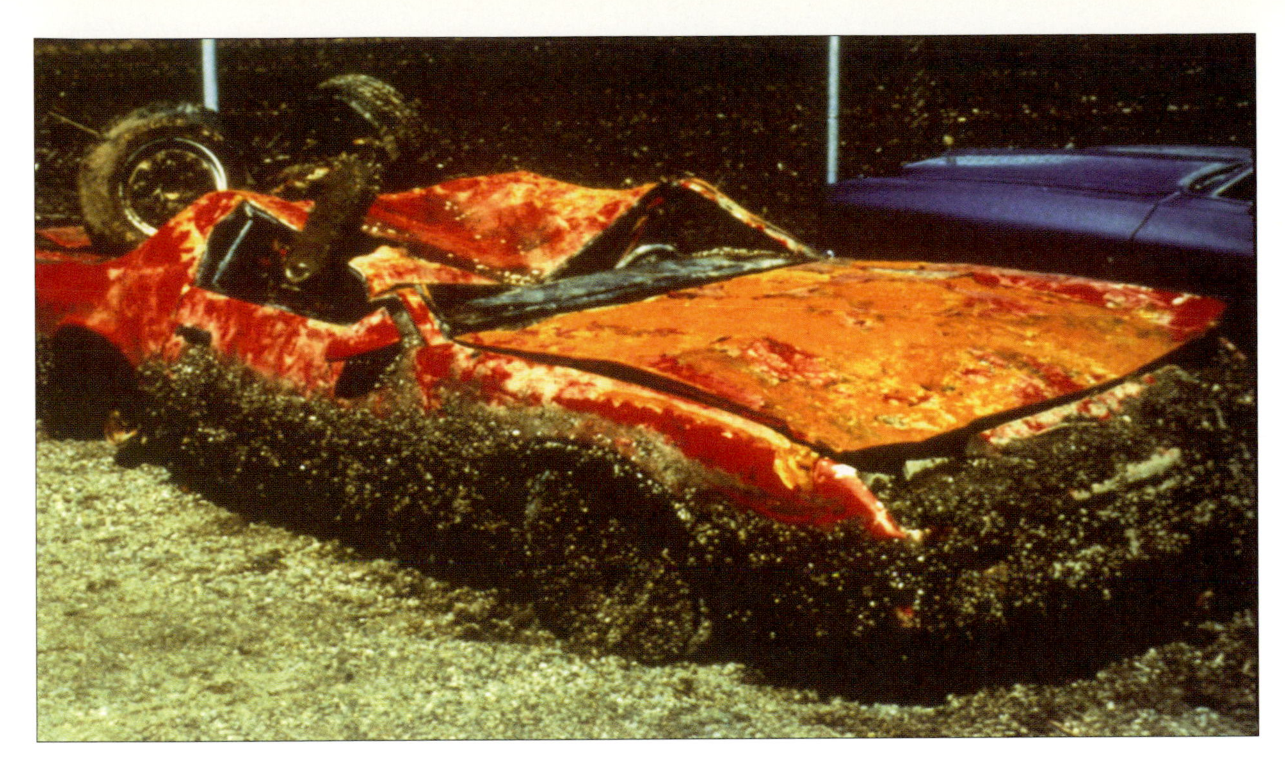

PLATE 32 An automobile body just after it was pulled from Lake Erie, in the spring of 1990. The car was lying upside down in the water, which is why the roof has no mussels on it. The lower panels are coated with attached masses of Zebra mussels. A mussel car! **Source:** Photo by Ron Griffiths, Ontario Ministry of the Environment.

PLATE 33 Dr. Roberto Rico Martínez, an unidentified fisherman, and Dr. Marcelo Silva Briano at Lake Chapala. Mats of attached, wave-resistant green algae are visible on the rocks, in the wave splash zone.

PLATE 34 Fish market in Jinja, Uganda, 1992. The large fishes in the cart are Nile perch. These fish were caught in Lake Tanganyika with a gill net. **Source:** Photo by Jim Kitchell.

PLATE 36 Professor Angel M. Nieves collecting invertebrates from limestone cave pool in Camuy, Puerto Rico. **Source:** Photo by A. Nieves.

PLATE 35 Student near the shore of Lake Mendota, Wisconsin, holding up a mass of the invasive, nuisance aquatic plant, Eurasian *Myriophyllum.* **Source:** Photo by Don Chandler.

PLATE 37 Dr. Carlos Santos Flores sampling a subtropical wetland in Cartagena Lagoon, in Lajas, Puerto Rico. He is using a plankton net to collect microcrustaceans. The aquatic macropyhytes are dominated by cattails (*Typha*), a grass and two floating plants, *Eichhornia* (water hyacinth, the larger plant with purple flowers) and *Pistia* (water lettuce, the low, light-green plant). **Source:** Photo by Carlos Santos Flores.

PLATE 38 Dr. José De Santiago using a large pipette (turkey baster) to sample aquatic organisms in a tree hole pool, in a Puerto Rican forest tree. **Source:** Photo by Dr. Carlos Santos Flores.

PLATE 39 Lake Missoula, Montana. Ancient shorelines are visible as horizontal rows of trees and shrubs on the hillsides near Missoula. **Source:** Photo by V. Dodson.

PLATE 40 Professors Edmondson and Hrbacek using a Birge plankton net to collect zooplankton from Lake Lenore, Washington. This is the same long lake shown in aerial view in figure 11.14.

PLATE 41 On 31 October 1999, algal blooms occurred in Lake Mendota and Lake Monona, both located in Madison, Wisconsin. These areas are depicted as the bright-green features in the largest lake and second-largest lake shown in this photograph. This image is a false-color composite resulting from combining three dates of Landsat-7 Enhanced Thematic Mapper Plus satellite data. Red and near infrared spectral data were used to compute a normalized difference vegetation index (NDVI) for each of the three dates. The NDVI values for the image corresponding to the date of bloom are shown in green in the composite and depict clearly the spatial distribution of the blooms. (A more detailed description of the production of this image can be found in *Remote Sensing and Image Interpretation,* 5th edition, by Thomas M. Lillesand, Ralph W. Kiefer, and Jonathan W. Chipman, John Wiley & Sons, 2004). **Source:** Image courtesy of the Environmental Remote Sensing Center, Gaylord Nelson Institute for Environmental Studies, University Wisconsin-Madison.

Table 7.1 Specific Chemicals Used in Induction of Defenses Against Predators (Including Herbivores)*

Species Producing Chemical Signal	Species Receiving Chemical Signal	What is Known About the Chemical Structure of the Signal	Kairomone Effects on Prey
Cyanobacteria (Brittain et al., 2000)	Herbivorous zooplankton	Toxins, including a heptapeptide	Direct contact with toxic algae might cause rejection of food, but the toxin can be a different chemical than the "taste" chemical.
Predatory ciliate—*Lembadion* (Peters-Regehr et al., 1997)	Prey ciliate—*Euplotes*	Part or all of a surface protein of *Lembadion*	Reception induces *Euplotes* to develop an elongated shape that is difficult for *Lembadion* to ingest.
Asplanchna (Gilbert, 1966)	*Brachionus*	A small peptide	Low concentrations in water induce spine elongation during embryonic development.
Tetrahymena vorax, large cannibalistic form (Butzel & Fischer,1983)	*Tetrahymena vorax,* small form, starving	Natural catabolic excretory products from nucleic acids	Starving animals release nucleic acid components that induce development of cannibals.
Brachionus (Gilbert, 1980)	*Asplanchna*	Vitamin E	Presence of vitamin E, a dietary requirement in prey, induces increased size and changed morphology in offspring.
Brachionus female (Snell, 1998)	*Brachionus* male	β-amyloid glycoprotein on surface	Contact of female surface guides male mate choice behavior.
Zooplankton grazers (*Daphnia*) (Wiltshire & Lampert, 1999)	Green algae (*Scenedesmus*)	Urea?	Low concentrations of a kairomone induce clumping of algae, which reduces desirability for grazers.
Chaoborus larvae (Parejko & Dodson, 1990)	*Daphnia pulex*	Organic molecule < 500 MW, somewhat polar and ionizes (to anion) at about pH 6	Chemical produced in *Chaoborus* hind gut dissolves in water; low concentration in water causes morphological change during embryonic development and changes life history.
Fishes (Lass et al., 2001)	*Daphnia* species	Trimethylamine?	Chemical produced by fish dissolves in water; may induce zooplankton DVM.
Adult oysters (Decho et al., 1998) (marine species, given for comparison)	Oyster larvae	Tripeptide (glycyl-glycyl-L-arginine)	Induces larval oysters to settle (swim downward, attach the foot) in oyster beds. The signal is active down to about 10^{-10} molar concentration.
Humans (Kashian & Dodson, 2003)	*Daphnia* species	Pesticides of various chemical structures	Low concentrations in water mimic natural kairomones to induce morphological and life-history changes.

*"?" = controversial

Mutualisms and Symbiosis

A **mutualism** is an interaction among species in which both species benefit. Mutualistic interactions may have evolved from competitive or predatory interactions. For example, predator-prey relationships are not always as simple as they seem. Until recently, it was assumed that zooplankton, such as *Daphnia,* ate phytoplankton, and that the phytoplankton lost out as a result—the only responses open to phytoplankton species were resting stages or defenses such as toxicity, large size, spines, or hard coverings. Dr. Karen Porter (1977) tested this assumption using 1-m^3 plastic bags suspended in a lake. The bags were filled with lake water (including algae) and various numbers of *Daphnia.* Dr. Porter found that small, undefended algae decreased in number, defended algae showed little or no change in abundance, and a group of green algae actually increased. The more grazing, the more of these algae appeared in the enclosures. The algae were special in having a **gelatinous** covering, so that the small, green algae were encased in a clear matrix (plate 39). Laboratory observations confirmed that these algae were being ingested, and some were being digested, but most survived the gut passage. As the gelatinous green algae passed through the gut, they absorbed enough phosphate to stimulate several cell divisions after leaving the *Daphnia.* In this case, being consumed by a predator was the best thing that could happen to the gelatinous green algae!

COMMUNITY STRUCTURE

Communities have *structure,* which is often described using measures of species **diversity, composition,** and **size-spectrum.**

Diversity

We have already been introduced to the concept of taxonomic diversity of aquatic organisms in chapters 3 to 5. Community ecologists use the concept of diversity (or **biodiversity**) in a different way: Ecological diversity is related to the number of different kinds of organisms. It can be measured as species **richness** (the number of species or other taxa). If a gill net sample of a lake produces 10 different kinds of fishes, then the lake has a fish species richness of 10. Richness is often measured at the level of the individual lake (Dodson et al., 2000). However, it can also be measured for a lake district or for a zoogeographic region (Shurin et al., 2000). Comparison of species richness at different scales allowed Shurin et al. (2000) to conclude that ecological interactions at the local level are important in limiting the importance of dispersal in determining species associations.

Diversity is also measured using various special indices. A long-time favorite index of ecologists for expressing species diversity is the Shannon-Weaver (1949) index:

$$H' = -\text{sum}\ (p_i * ln(p_i))$$

where p_i is the frequency (percent abundance) of the ith species, and sum means to add up, for all the species in a community, the products of p_i and the natural logarithm of p_i. The Shannon-Weaver index incorporates the concept of **evenness.** A community has the greatest degree of evenness (and the highest value of H' for a given number of species) if all the species are equally abundant. If a community does not show much evenness, it will have **dominant** (the most abundant) species and **rare** species.

Species Area Curves

The question of how many different kinds of organisms live in a lake or stream is a difficult but important question whose best answer is based on a wide range of ecological knowledge, tools, and spatial-temporal scales (Cottingham, 2002). For a lake, the answer depends on the taxonomic group in question, the size and productivity of the lake, and the nearness of other lakes. The answer also depends on how the lake (or stream) was sampled.

Taxonomic group In plankton communities, diversity measured as species number is inversely related to the size scale of the organisms. In the pelagic zone of a typical temperate lake, there may be hundreds of algae species, scores of protozoans, a few dozen species each of rotifers and crustaceans (usually more rotifer species), only a few species of insects and the larger crustaceans, and 20 or so species each of macrophytes and fishes. The number of bacteria species is unknown but is at least in the hundreds.

Lake size Pennak (1957) gathered species-richness data from 27 Colorado lakes and 42 other lakes of worldwide distribution. He found that small-to-medium-sized lakes are characterized by an association of one to three copepods, two to four water fleas, and three to seven rotifers: the freshwater zooplankton is characterized by low diversity.

Dodson (1992) looked at species richness of crustacean plankton in 66 North American lakes and ponds. In this study, he found the smallest temporary ponds to have only a couple of species of crustacean zooplankton, while the largest (Laurentian Great Lakes) had up to 34 species. The study found that the most important factor for predicting the number of species in a lake was the size (surface area) of the lake (figure 7.11).

The use of two other factors slightly improved the prediction based on lake size: the rate of primary productivity and the nearness of other lakes. For lakes of the same size and annual primary productivity, those that are near many other lakes, as in a lake district, tend to have more species than do isolated lakes. For example, the isolated Crater Lake, Oregon, and ponds isolated in the Utah desert have fewer species than would be predicted by their size and productivity.

Lake productivity Lakes with moderate productivity tend to have more species than either ultraoligotrophic or hypereutrophic lakes of the same size (figure 7.12; see Dodson et al., 2000). For example, clear mountain lakes, sewage lagoons, and temple tanks have fewer species than do lakes of the same size that are mesotrophic, such as moderately productive Lake Wingra and Lake Mendota, Wisconsin.

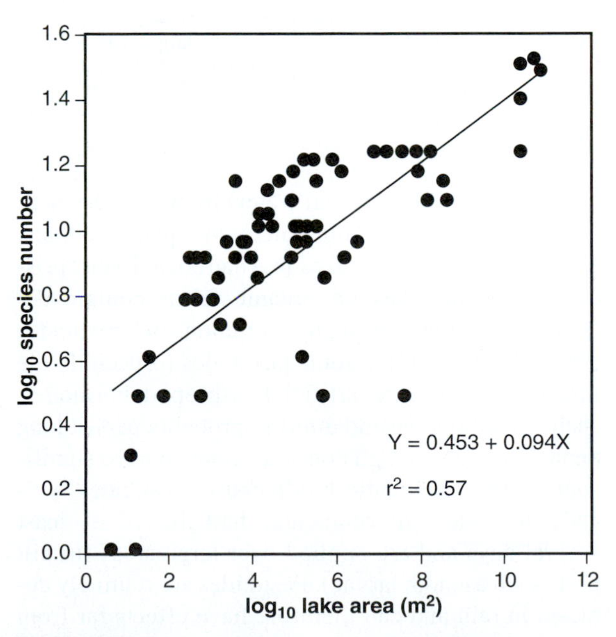

FIGURE 7.11 Species area curve for crustacean zooplankton of 66 North American ponds and lakes. The surface area of Lake Mendota, Wisconsin is about 4×10^7 m^2. **Source:** Data from Dodson, 1992.

Primary productivity varies with latitude, being lowest at high latitudes, higher in temperate latitudes, and perhaps slightly lower in tropical latitudes. It is possible that there is a latitudinal gradient in species diversity. An old story is that tropical lake communities are less diverse than temperate zone lakes, despite reasonably high productivity. However, the results of several recent studies suggest that zooplankton species richness is not affected by latitude (Santos Flores, 2001). The reason for the perception that there are fewer species in tropical lakes is probably that tropical lakes have been poorly studied and there are still a lot of new tropical species to be described.

Sampling technique The species list for a lake depends strongly on how the lake is sampled. In the area of sampling technique, limnologists lag far behind terrestrial ecologists (Mittelbach et al., 2001). Cooper et al. (1998) recommended that limnologists sample and do experiments at multiple rather than single scales, to better understand pattern and process in aquatic systems. (For an example of a graphical model of **biocomplexity** at multiple scales in aquatic habitats, see figure 9.3.)

Limnologists who study lakes are generally satisfied to generate a species list by taking one or a few vertical plankton net tows in the center of the lake, regardless of the size of the lake. This approach tends to miss species in large lakes. Also, there is a question of when and how often to sample aquatic habitats. Species lists from lakes are typically not the same from year to year (Arnott et al., 1999), which is a good argument for long-term ecological research!

The U.S. National Science Foundation **(NSF)** supports a long-term ecological research program, often called **LTER.** NSF supports basic research and trains and educates citizens in science and mathematics (for example, check out the WOW Internet site listed at the end of chapter 2).

Pesticides and Diversity

Pesticides are chemicals designed to kill animals, often arthropods, in agricultural and urban settings. Chemical contamination by industrial and agricultural chemicals, if taken to the extreme, results in a drastic decrease in biodiversity. However, even moderate contamination of aquatic habitats by pesticides or inorganic fertilizers may lead to changes in composition and, in some cases, an increase in diversity. For example,

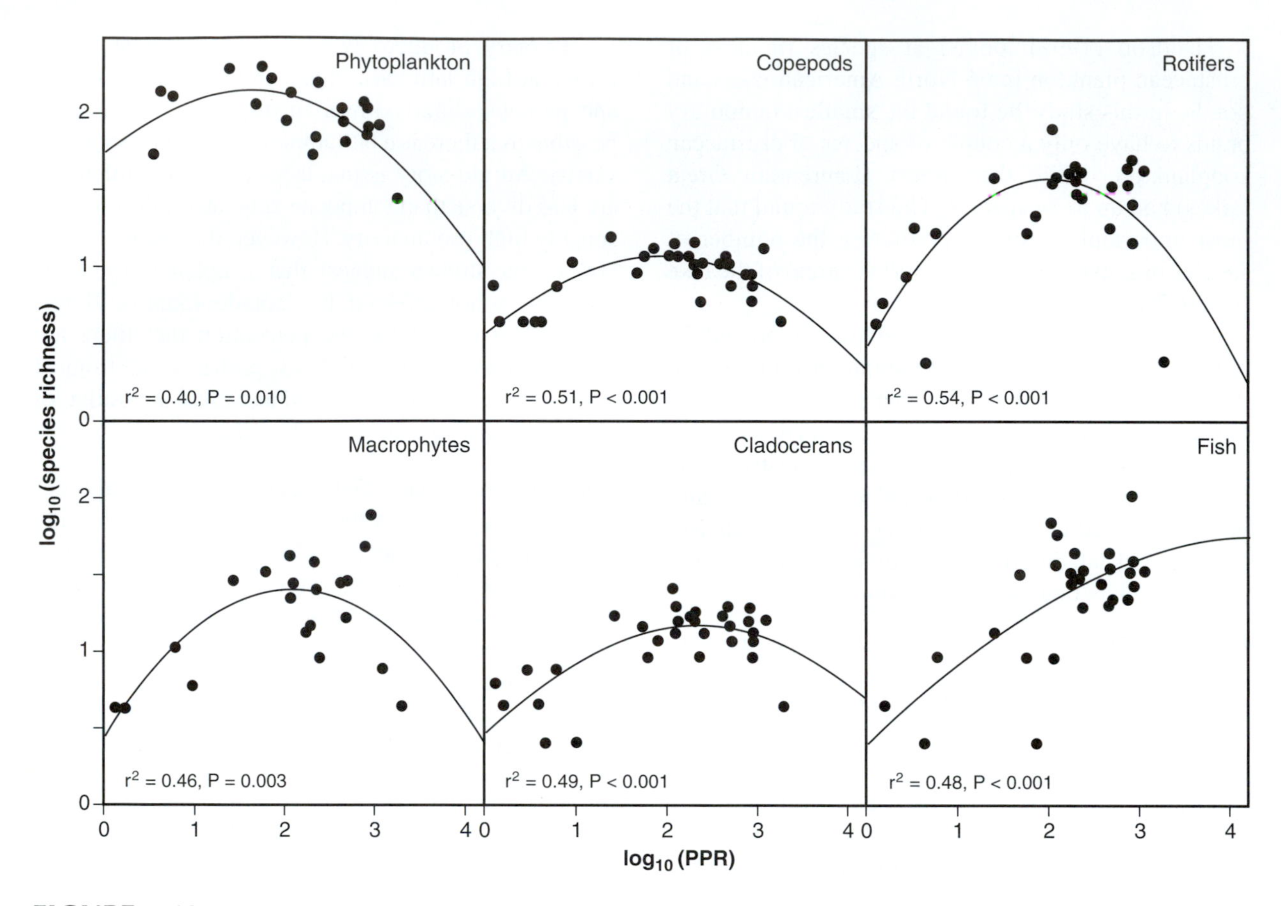

FIGURE 7.12 The relationship between species richness and primary productivity (PPR). These graphs are based on data from 33 well-studied lakes and ponds. The smallest pond (0.5 ha) is a sewage treatment pond in Madison, Wisconsin; the largest lake (67,000 ha) is Lake Biwa in Japan. The lowest primary productivity (1.3 g C m^{-2} yr^{-1}) is a high-altitude pond in Colorado, and the most productive (1300 g C m^{-2} yr^{-1}) is the Madison sewage treatment pond. Lake Mendota, Wisconsin has a surface area of 3937 ha and a productivity of 342 g C m^{-2} yr^{-1}. **Source:** From Dodson et al., 2000.

moderate contamination with the insecticide carbaryl in small ponds was followed by a decrease in crustacean diversity, but an increase in the number of less-sensitive rotifer species (Hanazato, 2001). Because increased productivity is often associated with agricultural activity in the watershed of a lake, it is difficult to know if the reduced biodiversity associated with high productivity (figure 7.12; Dodson et al., 2000) is due to increased plant nutrients (nitrogen and phosphorus) or increased agricultural contamination due to pesticides.

Pesticide use is not systematically reported, so limnologists do not know the extent to which their aquatic communities have been exposed to toxic chemicals. Pesticides and their carriers are broadcast widely in agriculture and silviculture. Many of these pesticides volatilize into the atmosphere, are distributed by large-scale weather patterns, and are then brought back to Earth (and water) by wet or dry deposition. Pesticides are ubiquitous in aquatic habitats, except perhaps frozen lakes in Antarctica. Even "pristine" lakes or lakes on organic farms contain detectable amounts of many common-use pesticides (Donald et al., 2001). Some pesticides (especially the insecticides that are artificial arthropod hormones, such as methoprene and dimilin) probably persist long enough at high enough concentrations to have significant effects on aquatic biodiversity. Pesticides typically have an environmental half-life of at least several days and are applied over large areas (to kill pests such as moth larvae). Pesticides are routinely detected in rain and can therefore have effects far from their point of application (Hayes et al., 2002). Thus, even in seemingly "pristine" watersheds, aerial application of pesticides holds the possibility of causing large ecological changes in aquatic habitats.

Spatial Heterogeneity and Diversity

In limnology, **spatial heterogeneity** is measured as patchiness on surfaces or as variability in substrate composition. **Patches** are areas of distinct habitat that fit together to make up the two-dimensional pattern of aquatic habitats as seen from above. At a smaller scale, patches are heterogeneous when composed of different-sized particles, such as sand, gravel, and cobblestones. Understanding how aquatic systems work depends on an appreciation of the role of environmental heterogeneity (Cooper et al., 1997).

In wetlands or littoral zones, species diversity increases with increasing habitat diversity. For example, in a New York state swamp, Anderson and Leopold (2002) found that wetland plant diversity was highest in areas that were partly shaded. Similarly, Knutson et al. (1999) reported that amphibian diversity is positively correlated with patch diversity in wetlands.

The number of species in a lake or stream is intimately related to the diversity of habitat in that system. Compared to smaller lakes, larger lakes have additional habitats, such as hypolimnia (see figure 7.7), that provide places for additional species to live. The increase in the number of habitats is one of the main mechanisms that results in the observed species area curve for lakes (Dodson, 1992).

In streams, environmental diversity (heterogeneity) affects species diversity directly and ecosystem processes indirectly (Cooper et al., 1997).

Community Composition

Community composition is defined as a list of species. This list is often restricted to a single ecological guild. The diversity of a community (measured as either richness, or H′) can remain constant, while composition changes over time. Imagine you are sampling a lake for fish species. At each sampling date, you might find 10 fish species, but not the same species each time (richness remains the same, but composition changes).

Diversity has so far been defined as the number of species, as species richness or evenness. However, diversity also depends on community composition. Food web structure also contributes to diversity. Allen et al. (1999) distinguish between **complicatedness** (the number of species in a functional group) and **complexity** (the number and relationships of connections of functional groups). The implication is that communities with simple organization are less diverse than communities with many systems parts. A community with three functional groups has a more complex hierarchy (and is, in that sense, more diverse) than a community with two functional groups.

Species Associations

Species found living in the same lake or stream are associating. The human animal loves to see pattern in Nature, and pattern identification is an integral part of community ecology. One aspect of pattern in Nature is the association of species.

In some cases, especially if there are extreme physical or ecological differences between sites, some species tend to be found together (Dodson, 1979, 1987). If the environment is not extreme, when similar sites (lakes) are analyzed, aquatic species (e.g., zooplankton) may appear to be distributed at random (Dodson & Lillie, 2003).

Species can occur together in aquatic habitats because of history and local adaptation:

1. *History:* Two or more species may just happen to be in the same part of the world. On the global scale, species associations can be the result of evolutionary history and the constraints of zoogeography.
2. *Local adaptation:* Species may have requirements that are met only in a specific type of aquatic habitat. On the local scale, species associations are the result of adaptation to local conditions and to the interplay of predator-prey or competitive relationships.

Local conditions tend to change, and species adapt to these changes. For example, changes in algae due to cultural eutrophication (Hairston et al., 1999) or in lake pH due to acidification (Fischer et al., 2001) result in rapid, intraspecific changes in zooplankton physiology.

An important question is whether species occur in associations at random or if there are patterns of association. Objective techniques of pattern identification depend on statistical analysis of data for species distribution, which in turn depend on the null hypothesis. That is, the statistical analysis is based on a model of how often species would be observed to occur together if species were distributed at random among sites. The exact nature of the null model is a matter of on-going discussion, but it is critical to the identification of species associations (see Gotelli, 2000; Sanderson, 2000). A URL address for a simulation model for analysis of species associations is given at the end of this chapter. Bell (2001) gives a particularly clear description of the importance of using the correct null hypothesis (in his case, the Neutral Community Model) when interpreting ecological data.

Global associations At the global scale, there are clear differences in the distribution of aquatic species between the two major groups of continents. All the continents were combined into one large landmass, called Pangea, about 260 million years ago. As the Earth's tectonic plates drifted about, the continents split into two super continents—Laurasia and Gondwanaland—during the Triassic about 200 million years ago. The super continents continued to split apart throughout the Cretaceous and Quaternary, forming into the continents we recognize today:

Laurasia: North America, Europe, and most of Asia

Gondwanaland: South America, Africa, Australia, and Antarctica

Zooplankton of the Laurasian continents resemble each other more than they do those of the Gondwanaland continents, and vice versa. For example, the copepods of South America are more similar to those of Australia than to the copepods of North America. The taxonomic differences between aquatic organisms of the two groups of continents are at the family and genus levels.

Zooplankton within the Laurasian continents are different at the species level, with the most diverse clusters of species in the glaciated area of these continents. For example, recent analyses using molecular techniques show many species that were recently thought to be identical in Europe and North America are actually genetically different. These differences suggest a burst of speciation associated with the glacial period that started about 1.5 million years ago.

Local associations At the scale of the local habitat, species associations can be found for all the major groups of aquatic organisms. The mechanisms that produce associations help limnologists to understand how lakes work. For example, bacteria often form symbiotic associations in which the metabolic activities of two or more species are in some way linked. Because one species produces a chemical that another requires, and vice versa, the two species are almost always found living together.

The products of bacterial metabolism (inorganic and organic molecules, including vitamins and energy sources) are essential to the life of all other organisms. Large organisms, such as worms, insects, or fishes, cannot survive for long without an intact bacterial fauna in their gut. Hence, the multicellular organism and the bacteria species tend to co-occur. For example, the gut of stream-dwelling medicinal leeches (*Hirudo medicinalis*) naturally contains a nearly pure culture of the bacterium *Aeromonas veronii* biovar *sobria* (Graf, 1999).

There are usually two to three common water flea species in the plankton of moderate-sized lakes (Pennak, 1957). Pelagic species are often associated according to the predator regime (see the section on Size-Selective Predation). In the presence of size-selective fish predation, it is common to find several small water flea species living together in the same lake. *Ceriodaphnia* and *Bosmina* frequently co-occur in the presence of the intensely predaceous alewife fish (Brooks & Dodson, 1965). If fish predation is low and invertebrate predation is intense, one finds dominance of relatively large species, such as *Daphnia pulicaria* and *Daphnia galeata mendotae.*

Two or more calanoid copepod species (*Diaptomus*) are often found living together in moderate-sized lakes (Pennak, 1957). Typically, the species will not be the same size and may have different seasonal patterns. The difference in size of co-occurring diaptomid species was first thought to be due to competition for food (Cole, 1961). The idea was that animals that differed in body mass by a factor of 2 (or length by a factor of 1.26, the cube root of 2) might have different enough food preferences to allow coexistence. The different body sizes meant that the copepods were eating different species of phytoplankton and were thereby able to live together. However, more recent research has shown that the larger of the two diaptomid species is often predaceous, so the species pairs may also be predator-prey pairs (Dodson, 1984).

Cyclopoid copepods show a less striking pattern of species association than do calanoid copepods. When two or more cyclopoid species are found in the same pond, there is a clear difference in life history and food preferences, and usually the species show little spatial overlap. Closer analysis of cyclopoid species may reveal coexistence of many cryptic species. For example, in Wisconsin, the *Acanthocyclops vernalis* group includes four very similar species, and it is common to find two of them in the pelagic zone of a lake: *A. brevispinosus* and *A. limnetica,* with *A. robustus* sometimes occurring in a single lake with one or the other or both. True *A. vernalis* is more typical of temporary ponds. Co-occurrence of *Acanthocyclops* species is probably an example of species that can live together, despite being potential competitors for food, because they are in the same geographic area and they have the same basic environmental tolerances and requirements.

Size-Spectrum

The **size-spectrum** of a community is the relative abundance or biomass of organisms or particles across a range of sizes. Typically, there are, of course, many more small organisms than there are large organisms, but biomass is more consistent across a wide size range. Sprules et al. (1991) performed a spectral analysis of Lake Michigan for the size range from picoplankton (see chapter 3) to salmon. The spectrum provided insight into trophic interactions across the food web and gave insight as to which ecological factors were controlling fish production.

Brooks and Dodson (1965) described plankton communities as size-spectrums (see figure 7.5). This ground-breaking study of size-selective predation set the stage for understanding the role of biological interactions in community dynamics, the topic of the chapter 8.

Study Guide

Chapter 7 Community Ecology: Species Interactions and Community Structure

Questions

1. What is a *community?*
2. What are the components of "community structure"?
3. What are the major communities found in still and flowing water?
4. How important are predation and competition in determining which species make up a community?
5. This chapter presented two major lines of thought as possible explanations of the seasonal aspect of cyclomorphosis. Can you design an experiment to distinguish between the two alternatives? What test organism would you use?
6. What is the "Paradox of the Plankton" and how is the paradox resolved?
7. Look at the organisms living in a lake or stream. Can you find evidence to support either the neutral assembly model or the niche assembly model?
8. Use Dr. Karen Porter's story of *Daphnia* and gelatinous green algae as an example of mutualism. How is this mutualism different from simple predation?

Words Related to Freshwater Communities

adaptation
allometry
backwater
benthic
biocomplexity
biodiversity
channel
community
competitive exclusion
complexity
complicatedness
composition
cryptic species
cyclomorphosis
diel
diversity
dominant
evenness
exploitative
floodplain
functional group
gelatinous
genotype
guild
gyttja
hyporheic
interference
interspecific
intraspecific
kairomone
keystone predator
lentic
limited resource
littoral
lotic
LTER
mutualism
neutral assembly
niche assembly
NSF
olfactory field
paradox
patch
pelagic
pesticide
phenotype
phenotypic plasticity
pool
rare
richness
riffle
riparian
run
selective
size-spectrum
spatial heterogeneity
spectrum

Major Examples and Species Names to Know

Locate Pangea, Laurasia, and Gondwanaland on a map

Hutchinson's example of *Polyarthra* in lakes, and habitat specialization

The pattern of diversity of waterfleas and copepods in lakes of different sizes

The species area curve for lakes in North America; be able to draw the graph and label the axes

Relate the feeding behavior of cladocerans and copepods to their optimal habitat and to competition for limited nutrients

Describe Dr. Gerald Cole's example of pairs of *Diaptomus* (calanoid copepod) species in Arizona desert ponds, and explain how the copepod distribution pattern is probably due to predation, not competition

Describe Dr. John Brooks's study on the effect of alewives on community structure in Connecticut lakes; draw the Brooks-Dodson graph, including all labels, and explain what it means

For Professor Woltereck's example of cyclomorphosis in *Daphnia,* explain costs and benefits of cyclomorphosis

What Was a Limnological Contribution of These People?

G. Evelyn Hutchinson

Jaroslav Hrbáček

John Brooks

Karen Porter

Exercise

Species Area Curve and Diversity Indices

1. For the six lakes listed in the following table, enter the number of microcrustacean zooplankton species and the lake surface area, using the web site. *http://www.limnology.wisc.edu/*

   Click on "Data" and then "Biodiversity/Species Lists." Click on "MS Excel Datafile for Dodson, Arnott, and Cottingham" (the first choice).

   Use the data to fill in the table.

Lake or Pond	Number of Crustacean Zooplankton Species	log Species Number	Surface area (ha)	log Surface Area
MadMetro Pond 2B				
Mexcut Pond L1				
Mirror Lake				
Crystal Lake				
Trout Lake				
Lake Mendota				

2. Graph log species number vs. log area. This graph is a *species area curve.*
3. Use this graph to predict the number of crustacean zooplankton in Lake Michigan. ___________
4. How good is the richness prediction for Lake Michigan? Check the report at *http://www.epa.gov/glnpo/monitoring/plankton/mich83-92/michplankton83-92-02x.pdf*
5. Give an explanation of the difference between what you predicted and what is reported.

6. How large a lake area would support 30 zooplankton species? ____________
7. Go to *http://www.limnology.wisc.edu/*

Click on "Data," then on "Online Datasets" (do not click on "Data Catalog" one line above).

Click on #3, "Plankton."

Click on "Dataset" at #4, "Zooplankton."

Enter your user name and your password (whatever you like, but record it somewhere).

When the error message appears, click on "New User."

Fill out the user information and click on "Submit."

At the bottom of the next page, write your purpose—such as "Summer Limnology Z315 class project" and click on "Submit."

Check: LAKEID, SAMPLE_DATE, SPECIES, AND NUMBER_PER_LITER, Allequash Lake, and set the beginning sample date to 2000.

Use only the data for 24 February 2000.

The species richness for Allequash Lake on this date was ____________.

Fill out the following table, to find the value of the index of species diversity, H'.

i	n_i (number per liter)	p_i fraction of the total	$ln\ (p_i)$	$p_i * ln\ (p_i)$
1				
2				
3				
4				
5				
6				
7				
8				
Total		—	—	

Note: n_i = the density of the ith taxon (number per liter). Using the total number of organisms per liter, you can calculate p_i = the decimal fraction of the total density for each taxon. The value of p_i ranges from just above zero to 1.0. You are calculating an index of zooplankton diversity.

Sum of $p_i * ln\ (p_i)$ times –1 = H'= ____________

Additional Resources

Further Reading

Tolrian, R., and C. D. Harvell. 1999. *The ecology and evolution of inducible defenses.* Princeton, NJ: Princeton University Press. 383 pp.

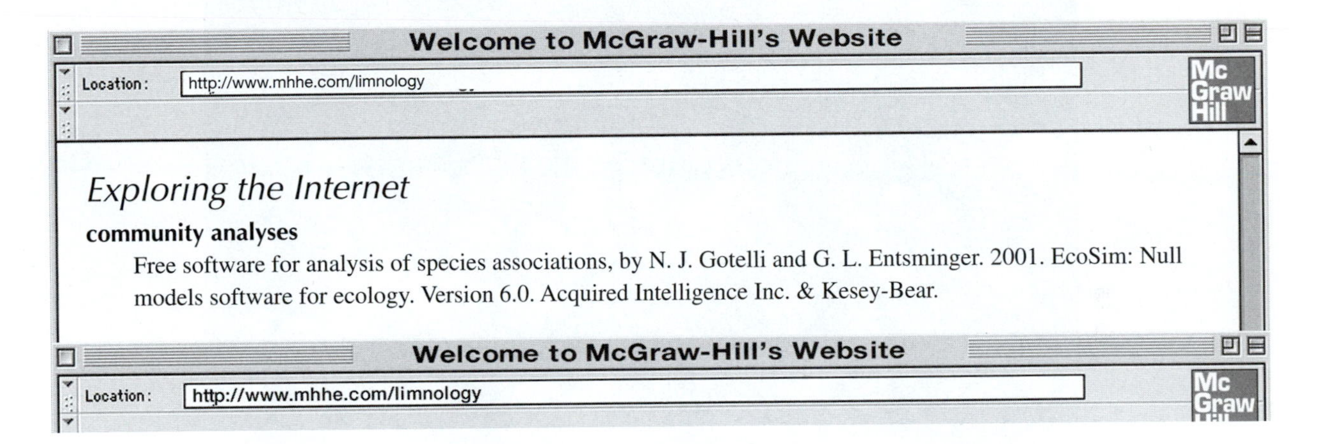

Exploring the Internet

community analyses

Free software for analysis of species associations, by N. J. Gotelli and G. L. Entsminger. 2001. EcoSim: Null models software for ecology. Version 6.0. Acquired Intelligence Inc. & Kesey-Bear.

8

"Differences (among animals) are manifested in modes of subsistence, in habits, in actions performed. . . And of creatures that live in the water some live in the sea, some in rivers, some in lakes, and some in marshes, as the frog and the newt."

Aristotle, *Historia Animalium* (probably class notes), ca. 350 B.C.

Community Ecology: Freshwater Communities Changing Through Time

COMMUNITY ORIGINS

Imagine you are standing on the shore of a newly formed lake. Perhaps you are at the edge of a slowly melting glacier, or maybe you are watching a beaver complete its dam, or you are on the dam of a new reservoir as the construction equipment disappears down the road. The lake will eventually be home for a group of organisms that will form a community. At the moment, however, the lake is only beginning to be filled with life. A limnological question is, "How are communities established and how do they change over time?"

Species that arrive first in a new lake are termed **pioneer** species. They are typically well-adapted for **dispersal** (movement) from one freshwater habitat to another. A quick review of chapters 3 to 5 provides examples of dispersal adaptations, such as asexual (vegetative) reproduction, resistant resting eggs, and the ability of dispersal stages (such as spores, eggs, and seeds) to ride on wind or water currents and to be carried by predators.

Any lake receives a constant rain of dispersal stages. The first pioneer organisms to arrive at the lake will be the smallest—bacteria and protists—organisms that travel on the wind as spores and are normal components of soil (Finlay, 2002). These spores are present as soon as liquid water is formed into a new lake. Resting eggs of rotifers and microcrustaceans typically reach a new lake within a few weeks, and aquatic macrophytes, mollusks, insects, fishes, and other aquatic vertebrates will all appear within the first year or two.

As described in chapter 7, these species will begin to interact (niche assembly model) or not (neutral assembly model). In either case, as populations wax and wane, some will occasionally become locally extinct and newly dispersed species have the opportunity to become established in the lake.

Several lines of evidence suggest that once a community is established, newcomers find it difficult to establish a new colony in the lake. For example, lakes within a few meters of each other might contain the same (morphological) *Daphnia* species, but the populations are genetically distinct (DeMeester, 1996), suggesting little or no long-term successful dispersal after the initial pioneers. Using artificial communities established in microcosm experiments,

Shurin (2000) showed that zooplankton communities are surprisingly resistant to colonization by new species.

This chapter is about community dynamics—how communities change in terms of composition and diversity over time. These changes can be seen at the small scale of migrations occurring over a period of hours or days, as well as seasonal and long-term changes related to extinctions and invasions.

Limnologists have used the ability of communities to change to improve water quality (biomanipulation) and for purposes of restoration and conservation (alternate stable states).

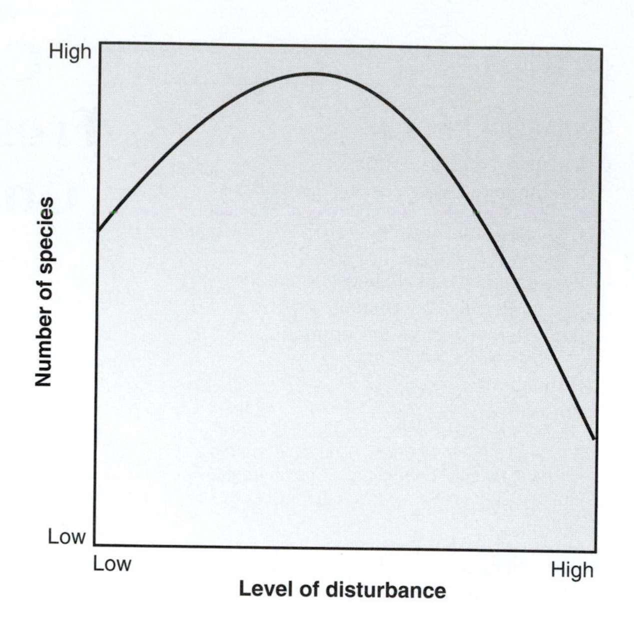

FIGURE 8.1 The general graphical model showing the intermediate disturbance hypothesis.

COMMUNITY DYNAMICS

Communities are typically in a state of flux, as populations change in size and species come and go due to extinctions, invasions, or migration. Limnologists often have the advantage of a good paleorecord of past changes in their communities. Microfossils from lake sediments can tell amazingly complete stories of past community dynamics, as discussed in the section on **paleolimnology** in chapter 11.

At less than geological time scales, disturbances often cause important community changes. An ecological **disturbance** has been defined as "any relatively discrete event in time that disrupts ecosystem, community, or population structure and changes resources, substrate availability, or the physical environment" (Pickett & White, 1985). In a community context, a disturbance could change community structure. Disturbances can be at the scale of a day (diel vertical migration), a year (annual succession), or longer (climate change, response to rare and extreme events such as the extinction of a species or the invasion of a new species).

It is generally thought that a moderate amount of disturbance is beneficial for communities, in that disturbance can increase biological diversity and guard against ecological stagnation. Connel's (1978) *intermediate disturbance hypothesis* (figure 8.1) is a cornerstone of ecological theory that limnologists have adopted enthusiastically. The effect of disturbance on biological diversity can be investigated using relatively large-scale natural or manipulative field experiments. For example, Cardinale and Palmer (2002) moved stones around in the bed of a warm-water, mesotrophic stream in Virginia to learn how disturbance affected primary production and respiration of the aufwuchs (**biofilm**) community. They found that moderate disturbance of stream substrate increased caddisfly species diversity, which in turn significantly affected several important ecosystem processes, such as increasing nutrient recycling and stream productivity.

Community Dynamics on a Daily Basis

Populations in aquatic communities show large-scale migration behaviors. Two examples are the daily vertical migration of plankton and fishes in open water, and the drift of insects downstream in moving waters. Migratory movements remind us that a community includes species that are present only part of the time.

Diel Vertical Migration (DVM) of Zooplankton

Limnologists have long known that populations migrate on a daily basis:

> "In the morning and evening, and even during the day when the sky is overcast, the daphnids usually stay at the surface. But during very hot weather, and when the sun beats fiercely down on the pools of stagnant water where they live, they sink down in

the water, and stay at a depth of six or eight feet or more; often not a single one can be seen at the surface."

Bayly's (1986) English translation of Cuvier's (1817) Le Regne Animal, vol. 17, Les Crustaces (Masson:Paris), p. 235

Many species of aquatic organisms show a behavior called **diel** (daily) **vertical migration (DVM)**—the behavior is particularly characteristic of pelagic branchiopods, copepods, and mysids. Populations are often observed to migrate, on a daily cycle, over great distances vertically, in both fresh and marine waters. Amplitudes reported in the literature vary from about a meter to hundreds of meters in oceans (the record is held by a calanoid copepod *Calanus* species in the North Sea; Marshall & Orr, 1972). Examples and a review of studies on *Daphnia* migration are given in Dodson (1990). For zooplankton, the diel migration is typically downward at dawn and upward at dusk.

DVM takes the animals into very different habitats, in terms of temperature, illumination, food density, turbulence, oxygen and solute concentrations, and predator densities. The diel pattern seems to often have the greatest amplitude during the summer and fall months. The larger developmental stages typically have the greater amplitude.

Aquatic ecologists have long wondered what organisms gain from vertical migration. There are obviously some costs to migration, such as energy used in swimming and being in a suboptimal habitat. There are a number of hypotheses about the adaptive nature of DVM.

Most theories about the benefits of DVM have three components: a light response, a physiological advantage, and a predation avoidance component. It is easily shown in the lab that zooplankton are very sensitive to changes in light level. This sensitivity allows organisms to migrate at the correct time of day, using light as a cue. However, the light response by itself is probably not the ultimate reason for migration.

Vertical migration takes organisms from warm water to cold water and back again. This migration, which costs very little energy, can easily be paid for by a small advantage of either warm or cold water. Stich and Lampert (1984) used a flow-through lab culture technique to test the hypothesis of metabolic advantage. In Lake Constance, Germany, *Daphnia hyalina* exhibits a pronounced diurnal vertical migration, while *D. galeata* stays near the surface. In the lab experiments, both species grew better and produced more offspring under conditions of constant high temperature and high concentration of food. Under the extreme "migration" conditions of fluctuating temperature and food, *D. hyalina* was more successful than *D. galeata,* even though *D. galeata* grew better under the constant conditions. Stich and Lampert's results did not support the hypothesis that daphniids gain some metabolic advantage from vertical migration. Their results are in agreement with an earlier suggestion that the larger *D. hyalina* is more susceptible to predation, and therefore shows the vertical migration, while *D. galeata* is more specialized physiologically and can thereby afford some predation. Stich and Lampert hypothesize that zooplankton, based on the metabolic disadvantages of migrating, should exhibit diel vertical migration only when food is abundant.

Zaret and Suffern (1976) studied vertical migration in Lake Gatun (a tropical lake in the Panama Canal) and Fuller Pond (a temperate lake in Connecticut). In both cases, they found that the zooplankton assumed distributions that lessened mortality from the dominant planktivores. Laboratory estimates of feeding rates suggest that the fish predators can exert intense selective pressures against those zooplankton that do not show vertical migration.

Winfried Lampert (1993a) synthesized the physiological and predator avoidance components of DVM into the current, and most satisfying, theory of the cost and benefits of DVM. This theory states that the **benefit** of DVM is reduced predation (ultimately to visual predators) during the day and warm water at night for maximum possible growth rate. The **cost** of DVM is the reduced growth rate that occurs during the daylight because of the cold water into which organisms migrate to avoid predators.

Experiments, in the laboratory or in cages in lakes, have clearly demonstrated that zooplankton migrate only when a chemical signal is present (Loose, 1993). This signal is a smell produced by predators, such as fishes. In the absence of the signal, zooplankton tend to seek the most food in the warmest water. Although marine zooplankton and fishes often show extreme DVM, they do not appear to respond to chemical signals from predators (Bollens et al., 1994).

Other Groups that Show Diel Vertical Migration

Dinoflagellates Vertical migration of dinoflagellates can contribute significantly to observed diurnal periodicity in stratification and photosynthesis of phytoplankton. For example, in high clear mountain lakes the dominant dinoflagellate (*Gymnodinium*) migrates downward with increasing light intensities (Ilmavirta, 1975). The maximum rate is approximately 1 m/hr^{-1}. In clear oligotrophic lakes, photosynthesis is reduced near the surface during the midday hours as a result of light saturation, photoinhibition, and downward migration. In turbid highly productive lakes, *Gymnodinium* has a competitive advantage compared to other phytoplankton, because it migrates down into deep nutrient-rich water at night and rises to near the surface for photosynthesis during the day (Lieberman et al., 1994).

Cyanobacteria Some species that produce gas vacuoles can migrate upward as photosynthesis progresses and makes gas available to fill the vacuoles. On a sunny summer morning, it is possible for a lake to change in less than an hour from a clear lake with a few flecks of algal colonies to a lake covered with solid scum.

Rotifers These organisms show little DVM. When present, the amplitude is only 1–2 m (Magnien & Gilbert, 1983). Rotifers probably have no evolutionary incentive to escape the size-selective predation of fishes, because they are too small for fishes to find effectively, but they may avoid migrating invertebrate predators. For example, a *Polyarthra* population shows an inverse migration (rising during the day), as its predator *Tropocyclops* sinks during the day (Gilbert and Hampton, 2001). The cyclopoid sinks during daylight hours to escape visually oriented notonectid predators.

Insects The predaceous phantom midge larvae (*Chaoborus;* see figure 4.18b) migrate vertically, especially if planktivorous fishes are present (Berendonk & Bonsall, 2002). *Chaoborus* species migrate to avoid fish, but they also are following their migrating prey. The variety in DVM shown by different species of *Chaoborus* provides material for natural experiments to explore direct and indirect effects on zooplankton behavior and of multiple predators in different habitats.

Fishes Fishes, especially juveniles, may show diel horizontal and vertical migrations. The migration is thought to optimize predator avoidance (avoiding larger fishes), prey availability (such as zooplankton), and perhaps the metabolic advantages of warmer water (Levy, 1990). Daily migration of fishes (and other voracious predators, such as *Chaoborus*) provides a complex challenge to zooplankton that are also migrating.

Stream Drift

One of the easiest ways to sample the insects of moving water is to leave a net in the current overnight. One typically catches large numbers of organisms (usually insects living in the substrate) that are drifting in the current. This downstream **drift** begs the question of why there are any insects remaining in the stream.

Although the number of insects caught in a drift net can be impressive, they represent only a small fraction of the total number of insects living in the bottom material of the stream. Thus, even if all the drifting insects were washed downstream, it would have a small effect on the remaining populations. In addition, insects living in stream bottoms tend to face into the current and to walk upstream. This difficult-to-measure upstream movement may compensate for drift. Even a small average movement upstream by the majority of the population would balance downstream drift by a small portion of the population. Also, the adult stages of most drifting insects fly upstream to mate and deposit eggs.

Small, fast-flowing cold-water trout streams provide a most congenial location for studying drift of stream insects (also called **macroinvertebrates**). Allan (1987) studied drift in a small trout stream, Cement Creek, Colorado. The amount of drifting macroinvertebrates in this stream was greatest in June and July—about 10 times as much at night as during the day—and the total amount was proportional to the number of insects in the stream bed. Professor Peckarsky and her students also studied the ecology of stream insects in Colorado trout streams (plate 16). Results of a series of carefully designed artificial stream and whole-stream experiments (McIntosh et al., 2002) show clearly that stream insects (mayfly larvae) drift more at night only if brook trout are present—macroinvertebrate behavior apparently depends on the smell of the water. In addition to the diel drifting patterns in these Colorado streams (artificial and natural), Peckarsky et al. (2002) found that the mayflies exposed to trout smell developed faster and matured at a smaller size, compared to the same species in fishless streams. This life-history response, which occurred in a single year, is thought to be an adaptive response in development to size-selective trout predation. Thus, both zooplankton in lakes and macroinvertebrates in streams appear sensitive to chemical signals from their predators—signals that influence their behavior and development.

Seasonal Ecological Succession in the Pelagic (Peg Model)

All communities are dynamic. The species present in a lake or stream change constantly, as does the pattern of which species are dominant or rare. Ecological **succession** often focuses on eukaryotes, but bacteria also show successional changes in biofilms (Jackson et al., 2001). Annual cycles of population change are observed in all freshwater habitats—here we give an example featuring eukaryotes from lakes.

Most of the species in a lake are present at all times but are unnoticed because they are in the form of dormant resting stages, such as spores or eggs. Understanding the annual flux of lake species is one of the main goals of aquatic ecology.

One attempt to describe and understand the annual flux of species is the model produced by the Plankton Ecology Group (**PEG**), a committee of limnologists who are members of the International Society for Limnology (Sommer et al., 1986). They gathered to design a description of the *annual succession* pattern of pelagic species in a typical northern temperate zone lake. This descriptive model of the generalized annual succession of species in the pelagic zone applies to many north temperate lakes. The following 5 steps are a simplified version of the original PEG model typical of a stratified eutrophic lake in Northern Europe (figure 8.2):

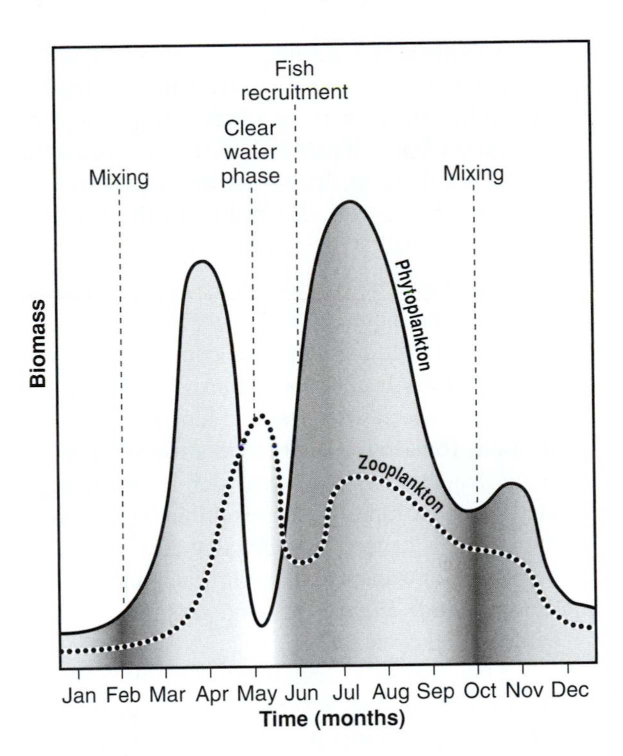

FIGURE 8.2 The PEG (Plankton Ecology Group) model for annual succession of species in a typical eutrophic Northern European stratified lake. Compare this model to Don Hall's Surprise (figure 6.6).

1. Toward the end of winter, longer days facilitate phytoplankton growth. As the ice melts and wind stirs the water, nutrients mix throughout the water column. Small, fast-growing algae such as cryptomonads and small centric diatoms increase in abundance. Planktonic herbivores with short generation times graze on this crop of small algae. The herbivores have either over-wintered as adults or hatched from resting eggs. Small species (such as *Bosmina* and rotifers) increase their populations first and are followed by slower-growing species (such as copepods and large *Daphnia* species). During this time, nutrients are moving from the water into algae and zooplankton.
2. As the water warms, the herbivore populations increase exponentially up to the point at which their density is high enough to produce an intense community filtration rate that exceeds the reproduction rate of the phytoplankton. As a consequence of intense herbivore grazing, the phytoplankton biomass decreases rapidly to very low levels. There then follows, usually in late spring, a **clear-water phase** that persists until inedible algal species increase to cloud the water. Nutrients are recycled by the grazing process and may accumulate in the water during the clear-water phase. During the clear-water phase, herbivorous zooplanktonic species become food-limited and both their body weight per unit length and their fecundity decline. This results in a decrease in their population densities and biomasses. (Remember that the relative importance of food limitation and predation were discussed in chapter 6 under the topic, "Don Hall's Surprise." Refer back to figure 6.6.)

At about the time of the clear-water phase, many species of fishes reproduce, and juvenile fishes (which are typically eating plankton) move out into the pelagic zone. Fish predation accelerates the decline of herbivorous planktonic populations to very low levels. Because the fishes tend to eat the largest zooplankton, there is a shift toward a smaller average body size (*Bosmina,* small *Daphnia* species) among the surviving crustaceans.

3. With warmer summer conditions, reduced grazing pressure, and increased nutrients, the phytoplankton crops start to build up. The composition of the phytoplankton becomes complex due to both the increase in species richness and functional diversification. To use an analogy with a forest, there are small "undergrowth" species such as green algae that reproduce rapidly and are optimal zooplankton food. Large "canopy" species that are avoided by herbivores and adapted to warm water and low nutrients include *Dinobryon,* large diatoms, dinoflagellates, and cyanobacteria.

   Algae quickly deplete the soluble reactive phosphorus and silica to nearly undetectable levels, and algal growth becomes nutrient-limited. Nitrogen depletion favors a shift to nitrogen-fixing species of filamentous cyanobacteria. Lack of nutrients prevents an explosive growth of "edible" algae, although herbivores recycle some nutrients and allow some algal growth. The population densities and species composition of the zooplankton fluctuate throughout the summer.

4. In the fall, days become shorter, the water cools, and the lake mixes again. Mixing replenishes dissolved nutrients, which allow algae to grow faster. An algal community develops: species adapted to turbulence, moderate nutrient concentrations, and grazing. Large, unicellular, filamentous algal forms appear, especially large or filamentous diatoms. This association of low-food-quality algae is accompanied by a variable biomass of small, edible algae. This algal composition, together with some reduction in fish predation pressure, leads to an autumnal maximum of zooplankton that includes larger forms such as large *Daphnia* species.

5. As the days shorten into winter, light and temperature begin to limit photosynthesis. Primary production drops to near zero and algae decreases in abundance to the winter minimum. Herbivore biomass decreases as a result of reduced fecundity due to both lower food concentrations and decreasing temperature. Some species in the zooplankton produce resting stages at this time. However, many species of zooplankton can survive winter as juveniles or adults. As the lake cools, some cyclopoid species awaken from their juvenile diapause in the sediments and contribute to over-wintering populations in the zooplankton.

The PEG model is focused on phytoplankton and zooplankton. Seasonal or annual changes are most accessible to limnological study because of time and space constraints typical of limnological research (Magnuson, 1990). However, these annual changes occur in the context of longer-term community dynamics that happen over centuries or millennia.

Long-Term Changes in Communities

The species list for any terrestrial or freshwater community depends on a number of factors, such as the size of an island or lake, the level of primary productivity, and the nearness of other similar habitats (this material was discussed in chapter 7.)

"Island" Biogeography

MacArthur and Wilson (1963) developed a simple model of community dynamics called *Island Biogeography.* This model is dynamic in that it deals with interacting processes (immigration of new species, extinctions) and it is an example of neutral assembly, in that all species in the community are assumed to be similar in terms of dispersal and competitive ability. Unfortunately, this is also an "equilibrium model," which implies a system that does not change in response to disturbance or a changing environment. Equilibrium theories were popular in the 1960s but have fallen out of favor as ecologists have decided to acknowledge that change is ubiquitous and eternal. Even though this model assumes equilibrium conditions, it has proven very useful in understanding community dynamics.

The Island Biogeography Model was originally developed for terrestrial animals on islands, but it works equally well for freshwater organisms in aquatic habitats separated by dry land. The model depends on two ecological processes: **immigration** (rate of dispersal of new species into the lake) and the **extinction** rate of species already established in the lake (figure 8.3).

Imagine a lake that continuously receives individuals (spores, eggs, or whatever) dispersing to the lake

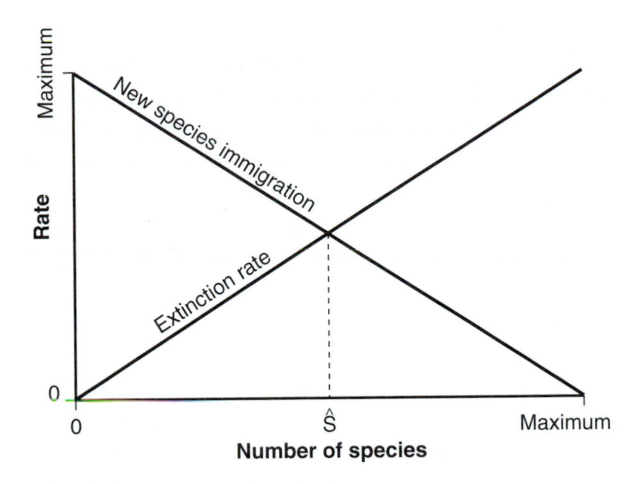

FIGURE 8.3 The Island Biogeography Model. The Ŝ indicates the equilibrium number of species predicted when the extinction rate equals the rate of immigration of new species.

from the surrounding landscape. Some of the species in the lake will, for whatever reason, become locally extinct. There are two important things to know about the processes:

- The rate of immigration is for *new species* not already in the lake. If the rate of immigration of individuals is constant, the rate of establishment of new species will decline as more and more species become established in the lake. In other words, the immigration rate of new species (new species per year) is highest at first (when no species are living in the lake) and declines as species richness increases in the lake. If the lake contained all the species present in the surrounding habitat, the rate of immigration of new species would have to be zero, even if new individuals were arriving every day.
- The rate of extinction increases with diversity. This is partly because with more species, the population of each might be smaller and therefore more likely to be extirpated by a chance event, and partly because with a constant extinction rate per species—the greater the species richness—the more species will become extinct each year.

The exact shapes of the lines in the Island Biogeography Model (figure 8.3) are probably not important—they are usually drawn as straight lines. The exciting aspect of the model is that the two lines cross. The species richness below the crossing point is the equilibrium number of species (species richness) predicted by the model. Notice that the model predicts that if the community contains more than the equilibrium number, extinction rate is greater than immigration rate, and the number of species will decline. Conversely, for a species richness less than the equilibrium value, immigration rate is greater than extinction rate, and species richness is predicted to increase toward the equilibrium value.

An important implication of the Island Biogeography Model is that while the number of species in a habitat tends to be near the equilibrium value, species composition is likely to change over time. When species go extinct, they won't necessarily be replaced by immigrants of the same species.

Extinctions

According to Island Biogeography theory, the number of species living in a lake or stream is the result of the rate at which new species are established and the rate at which species become locally extinct (**extirpated**). When a species disappears from all possible habitats, it is extinct.

Little is known about the process of extirpation, especially for smaller organisms such as invertebrates and protists. Hutchinson (1965) pointed out that it is difficult to know if a species is absent or just rare in a lake. For example, it would be difficult to find a species of copepod present at an average **density** (number per unit volume) of one individual per cubic meter—one might conclude the species was not in the lake. Nevertheless, such an animal would have a population size (in Lake Mendota, Wisconsin) of 480 million individuals. In addition to the problem of rareness, species can persist as resting stages for decades or centuries—they are not present as active forms but they are also not extinct.

Many aquatic vertebrates are officially recognized as being at risk or in danger of becoming extinct because of their relatively small population sizes and the global explosion of habitat destruction by humans. Aquatic vertebrates are large enough to be easily seen. There are far fewer officially endangered invertebrate species. Some examples include freshwater clams, aquatic insects, and fairy shrimp and tadpole shrimp—all species large enough to see easily. There are no officially endangered microscopic invertebrate species.

Invasion

Invading (**exotic**) species modify the composition of many aquatic communities. For example, global warming is facilitating the invasion of cold lakes by smallmouth bass. As the lakes warm, the bass become established in lakes previously dominated by trout, changing the lake's community composition and pathways of energy flow (Vander Zanden et al., 1999).

Invaders are species newly arrived from a distant lake region. In the long-term context, all aquatic species were originally invaders in the sense that they were the pioneer species that first occupied newly formed lakes and streams. After the original communities are established by pioneer species, new species can become established in the habitat, as long as several hurdles are cleared (figure 8.4; for a more detailed analysis, see Kolar & Lodge, 2001 and Sakai et al., 2001). A potential invasive species always has an origin—it is **endemic** in its native habitat. The first step of an invasion is for a species to disperse from its native habitat and to gain a foothold in a new region away from the native habitat (figure 8.4a).

Human activity is by far the most efficient mechanism for moving aquatic organisms from one place to another. Water pumped into and out of a ship to adjust its position in the water (**ballast** water), discards from the aquarium trade, and releases by fish managers and anglers all contribute to the transport of freshwater species from one continent to another. Dumont and Negrea (2002) tell a nice story about the probable transport of an *Alona* (chydorid branchiopod) species from South America to the crater lake Rano Raraku on isolated Easter Island. They hypothesize that Captain Cook unknowingly transported the *Alona* in water casks about 1774. The construction of transportation, and especially *canal building,* connects previously isolated watersheds, providing corridors for rapid dispersal of aquatic organisms into new parts of a continent (Kraft et al., 2001).

If the species successfully disperses and colonizes a new location, it must become established (figure 8.4b). Successfully established species are often called *exotic species.*

To be considered an invader, an exotic species must increase and spread, to become common in its new location (figure 8.4c). The extent of spread can sometimes be predicted by studying the distribution of the species in its native homeland. For example, the predatory spiny water flea *Bythotrephes longimanus* (see figure 4.7h) has invaded the North American Great Lakes and some surrounding lakes. Based on the kind of lakes inhabited by *Bythotrephes* in Europe, MacIsaac et al. (2000) predict that it will continue to spread and eventually occupy additional hundreds or thousands of lakes in North America. Similarly, Karatayev et al. (1998) analyzed existing former Soviet Union data on zebra mussel (*Dreissena polymorpha*) physiology and habitat preference, providing a basis to predict the eventual distribution and abundance of this invasive molluscan species in North America.

If an invasive species has an adverse ecological or economic impact on communities or people in the new

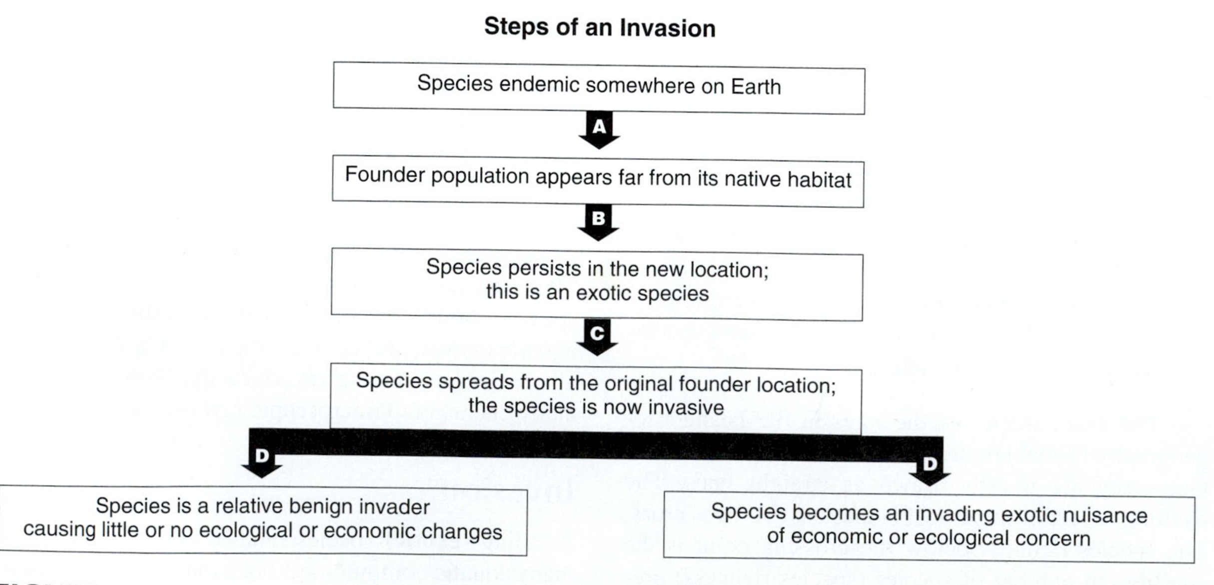

FIGURE 8.4 A model of the invasion process.

location, it is considered a **nuisance,** or a detrimental invader. If the invader has little or no effect on humans or freshwater communities, it is a **neutral** invader. An invading species that is useful to people or that benefits the freshwater environment would be considered beneficial. The USGS has a great web site that provides more information on invaders of North American lakes (see the address at the end of the chapter).

Examples of nuisance or detrimental invaders of aquatic habitats are well known the world over. For example, the Nile perch (plate 34) has been introduced into many habitats, including the great African rift lakes (such as Lake Tanganyika), where these fishes are causing catastrophic changes in community structure and fish abundance (Kitchell et al., 1997). A fish parasite such as the sea lamprey, which has invaded the Great Lakes and taken a huge toll on salmonid production, is an example of a detrimental invader; so is the spiny water flea *Bythotrephes,* which causes a precipitous decrease in zooplankton diversity when it occupies a lake (Yan & Pawson, 1997). An example of a neutral invader is the calanoid copepod *Eurytemora affinis,* which has been in the Great Lakes since the 1960s, with no discernable effect. The African *Daphnia lumholtzi* is also a neutral invader (Havel et al., 2000) that has been in North American lakes and rivers since 1989, with no discernable effect on zooplankton. Some would consider exotic salmonids stocked into the Great Lakes as beneficial invaders.

Look for examples of the three types of invaders (detrimental, neutral, or beneficial) in the following three histories.

Fish Invasions of the Laurentian Great Lakes

Perhaps the most complete story of invading species is for the Laurentian Great Lakes (Mills et al., 1993). At least 146 animal species (vertebrates and invertebrates) from somewhere else (mostly Europe) have invaded the Great Lakes in recorded history. These invasions were facilitated by the building of the Welland Canal (completed in 1919), which connected the lower and upper Great Lakes, allowing some species (such as lampreys and alewives) to swim from Lake Ontario to the upper lakes. Increasing ship traffic has been responsible for introducing the majority of species from other continents. Ships often take on water as ballast, steam to a Great Lakes port, dump the water, and take on cargo. The dumped ballast water often contains living plants and animals from the home port. Invading fish species have had dramatic effects on all communities of the Great Lakes. For examples of the effects of invading fish species, see table 8.1, which lists the history of economically important, invading animal (mostly fish) species in the upper Great Lakes.

A Molluscan Invader in Aquatic Communities

Mussels are bivalve mollusks that attach themselves to surfaces using strong, adhesive threads. One species that lives in fresh water, the zebra mussel (*Dreissena polymorpha;* see figure 5.1c and plate 32), is an invasive species that is a significant economic nuisance in North America (Ludyanskiy et al., 1993). Two hundred years ago, this mussel was restricted to living in and near the Caspian and Black seas. As canals were built and shipping increased with the developing industrial revolution, the mussel began to appear throughout Europe. Sometimes ships that have unloaded cargo fill their holds with water, so that the ship rides at the correct level. When the ship takes on more cargo, the water is dumped. Any animals that were in the ballast water are therefore transported to new ports. This apparently happened about 1987 in the Great Lakes. In 1988, zebra mussels were discovered in Lake St. Claire near Detroit. It is probable that these mussels traveled as larvae in ballast water of a large ship from a European port. The mussel has since spread throughout the lower Great Lakes and into the Ohio, Mississippi, and Hudson river watersheds. Zebra mussels can travel from lake to lake as larvae (in water transported in boats or bait buckets) or as adults attached to boats (the adults can live for days out of the water and also live in bait wells).

The zebra mussel invasion is of some ecological and economic concern because it can be very abundant, to the extent of covering suitable substrate in lakes and streams (plate 32; also see chapter 5). *Dreissena* qualifies as a nuisance invader because it has a multitude of direct and indirect ecological effects (figure 8.5), some of which are undesirable. While it is generally agreed that improving water clarity is beneficial, the mussel also costs society in several ways. Zebra mussels live on the insides of water intake pipes, thus greatly reducing the amount of water taken into water supplies and cooling plants. Removing zebra mussels, and designing new intake systems, are expensive endeavors. Native North American bivalves do not form thick beds on surfaces and, therefore, North American (unlike European) engineers have not previously designed intake pipes to accommodate this "bio-fouling." Accommodation to zebra

Table 8.1 A History of Major Invading Animal Species the Upper Great Lakes, Especially Lake Michigan

This history focuses on exotic fishes that have become common or nuisances; many other organisms have invaded the lake but have not become abundant (Mills et al., 1993). Drawings of the fish are shown in figure 5.2; also see the web sites listed at the end of chapter 5.

Date	Event
1880s	*Carp* are introduced from Europe by the federal government, as a food source.
1899	Commercial fish harvest of mostly native Great Lakes fishes peaks at 119 million lbs and remains high through the 1930s.
1912	*Rainbow smelt,* stocked into Crystal Lake, Michigan, escape into the Great Lakes, to become a dominant planktivore in the upper Great Lakes over the next decades, competing with the native white fish species.
1919	Welland Canal is built between Lake Ontario and Lake Erie, facilitating movement of coastal fishes such as the *sea lamprey* and *white perch* into Lake Michigan.
1921	*Sea lamprey* (native of Lake Ontario) appear in Lake Erie, but are scarce because of lack of suitable spawning habitat. The lampreys later appear in Lake Huron (1932), Lake Michigan (1936), and Lake Superior (1946).
1920s	The beginning of declines of native planktivores, including lake herring (1920s) and other white fish species through the 1930s and 1940s. Several species are extinct by the 1940s.
1939	The first *alewife* appears in Lake Huron (1939) and later in Lake Michigan (1949). Planktivorous alewife and smelt overlap in diet with 20 or so native planktivore species. Alewives and smelt lay eggs on the lake bottom, a refuge from their predation. By eating pelagic eggs and larvae, the exotic species hastened extinction of about half the native planktivores—those that spawn in open water.
1940s through 1950s	Native lake trout catches plummet. Most commercial fishing ceases. The lamprey did not become abundant until the late 1940s and early 1950s. Causes of the crash include the increasing lamprey population and an almost exponentially increasing commercial fishing industry.
1958	In Lake Michigan, more than a million pounds of alewife are caught commercially, large zooplankton species decrease dramatically, victims of increased size-selective predation—*Daphnia* are replaced by the smaller *Bosmina* and *Diaptomus.*
1960s	Several salmonid fish species are stocked, in increasing numbers, with the goal of rebuilding the sport and commercial fishery. The salmonids have, at first, plentiful prey (alewives), producing extraordinarily successful fishing. There is a spectacular alewife die-off in 1967, a result of the high population. Salmonid stocking continues today, although after salmonids reduced the alewife population, fishing success declined.
1965	Forbidden levels of DDT and dieldrin are found in fatty fish: chubs, lake trout, carp, and salmon; results in warnings to avoid frequent fish meals.
1986	*Ruffe,* a fish from the Baltic Sea, appears and becomes very common. Ruffe, which feed on eggs laid on the bottom, are suspected of reducing the native walleye population.
1988	*Zebra mussels* are discovered in Lake St. Claire near Detroit and spread rapidly to other Great Lakes and adjacent inland lakes and streams. Mussels clear the water, displace native clams, and compete with zooplankton for food.
1988	The invading predaceous spiny water flea *Bythotrepes* becomes abundant in Lake Michigan, raising concern of direct adverse effects on zooplankton and indirectly adverse effects on salmonids and other fish, such as yellow perch.
1990	Round and tube nose *gobies* (natives of the Caspian Sea) appear in the lakes and are suspected of competing with native species. Since 1990, exotic species have continued to invade the Great Lakes (see the Sea Grant web site cited at the end of this chapter).
2001	A review by Hewitt and Servos (2001) describes many agricultural and industrial chemical contaminants (including the lamprey larvicide) that act as endocrine disruptors, potentially affecting normal growth and development of all species of aquatic organisms.

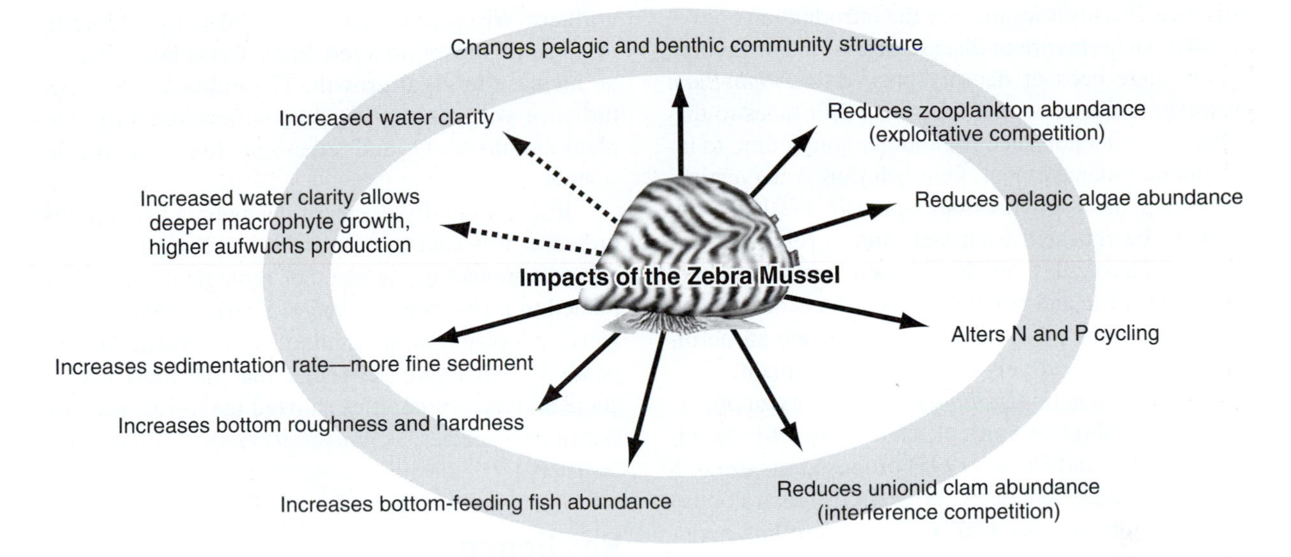

FIGURE 8.5 Zebra mussels, which are invading lakes and streams in North America and northern Europe, affect their habitat in many ways.

mussels will require refitting many water supplies, industries, and power plants.

Zebra mussels also displace or prevent native bottom organisms (fishes and mussels) from feeding. This is especially a concern for species of native unionid clams. Zebra mussels grow on and cover the larger clams and threaten the extinction of many native species (Burkalova et al., 2000), an excellent example of interference competition for space.

Zebra mussels have had a major effect on the plankton of North American lakes. In shallow water (such as Green Bay in Lake Michigan and the western end of Lake Erie, and any lake water less than a few meters deep), zebra mussels filter enough of the entire water volume each day to cause a noticeable clearing of the water (Culver, 1992). This means that the lakes become clearer (a benefit), but the primary productivity is going into the inedible (by humans) mussels rather than into zooplankton and fishes.

By covering the bottom of a lake, zebra mussels become the main food supply for bottom-feeding animals (such as some species of fishes and ducks). Some of these large animals can eat zebra mussels and some can't, so there will probably be changes in the species of vertebrates found in and on lakes invaded by the mussels.

This invasion is still very much in progress. We can expect that the mussel will have a large effect on aquatic communities. However, it is also likely that this invader, like other species in the past, will become less of a pest as other species learn to live with the zebra mussel.

A Plant Invader in Aquatic Communities

Eurasian water milfoil (*Myriophyllum spicatum;* see figure 5.7d) is a widespread nuisance (plate 35) waterweed (aquatic macrophyte to limnologists). This is a long, stringy plant with feathery leaves that roots in the bottom of a lake, grows to form dense beds, and pokes its small, white flowers out of the water. Each plant has a single stem that can be several meters long. Although the individual plants are weak, beds are dense enough to entangle and drown swimmers.

The plant was probably introduced into North America from Europe in ballast water or perhaps by the aquarium trade. It was first reported in the Potomac River and Chesapeake Bay in the late 1800s. Since then it has spread rapidly throughout the United States and Canada, often becoming a nuisance because it grows so thickly that it shades out native plants and because it is not a good food for fish, insects, or mollusks.

As with many invading species, *Myriophyllum spicatum* has been a nuisance for only a couple of decades. It appeared in Lake Mendota, Wisconsin in the early 1960s, where it displaced other species and became the dominant aquatic macrophyte for about 25 years. It then began to decline somewhat in abundance. We do not know what happened to reduce the nuisance. Possibilities include the introduction or evolution of an herbivore or disease that would be favored by the huge beds of densely packed *Myriophyllum*. However, a nuisance that takes 2 to 3 decades to disappear is still a nuisance for a long enough time to invite human management. One solution is harvesting. Removing macrophytes does open up the beds, but needs to be repeated each year and is relatively expensive. Herbicides are less expensive but are not species-specific and may have side effects on fish and humans. A third possibility is to introduce an herbivore that will selectively eat the nuisance plant.

The invasion of *Myriophyllum* presents an opportunity for **biological control,** using a specific herbivore. Sheldon and Creed (1995) proposed the use of a North American aquatic weevil (beetle) as a control agent. They have since had success controlling populations of *Myriophyllum* in New England and Wisconsin by broadcasting laboratory-reared beetles. The challenge was to find a native (North American) insect that would reduce the survivorship and reproduction of the Eurasian water milfoil and not adversely affect other aquatic macrophytes. Relatively few species of insects (or other herbivores like snails) eat milfoil, perhaps because of a chemical defense mechanism. Dr. Sheldon experimented with at least 12 different aquatic insect species that were found actually eating the milfoil. One of the insects, a small (2-mm long), native aquatic weevil (a kind of beetle with a long, thin mouth for sucking plant juice) fit her criteria. This weevil is now being cultured in large numbers and introduced to lakes with dense beds of milfoil. The beds tend to disappear in 2 to 3 years, replaced by a diverse association of many species of native plants and a few Eurasian milfoil plants.

Resistance

Community stability is a concept used in different ways by different scientists. In order to clarify the concept, Holling (1973) proposed two related concepts: **resistance** and **resilience.** Resistance is a measure of how much a community changes in response to a disturbance. This magnitude of response is also called "inertia"—the inverse of what is called "sensitivity." Resistance can be measured in terms of change in diversity and composition, or a loss of species following a given disturbance. For example, Cottingham and Carpenter (1998) disturbed three northern Wisconsin oligotrophic lakes by adding nitrogen and phosphorus fertilizer. These lakes showed an increase in algal growth. The eutrophication disturbance was accompanied by a decrease in phytoplankton diversity and evenness (but not species richness).

If a community does not change readily to a disturbance, it is said to be *resistant* to change. A major question in ecology is whether biological diversity is related to resistance to change. Shurin (2000) investigated zooplankton communities of small, fishless ponds in Michigan. He found that the most diverse zooplankton communities showed the lowest invasion rate of new species, as if *diversity confers resistance to invasion.*

Resilience

Resilience is the rate at which communities recover from a disturbance; it is also called "rate of return" or "rate of recovery." Resilience can be measured as the rate at which species richness returns to predisturbance levels. For example, Dodson and Lillie (2001) wanted to know if the millions of dollars spent on wetland restoration in Wisconsin had been well spent. They looked at the community of microcrustacean and insect plankton taxa in small ponds and wetlands in three kinds of watersheds: least-impacted (or reference), agricultural, and restored. A restored watershed is one that has been taken out of agriculture and managed for conservation purposes. Results of the study showed that agricultural ponds had about half the taxa of least-impacted ponds, and that ponds in restored areas showed a significant linear correlation between taxon richness and time since restoration. It appeared that the restored communities took about 6 years to return to the average richness seen in least-impacted ponds: The resilience was on the order of one taxon per year. Thus, the millions spent on wetland restoration appear to be having a beneficial effect.

BIOMANIPULATION—TROPHIC CASCADES

Community structure is determined by a large number of direct and indirect interactions among aquatic organisms, as influenced by land use by the human population.

Cultural Eutrophication

Humans living in the watersheds of lakes perform land-use activities, such as agriculture and sewage disposal, that result in large-scale addition of inorganic chemicals, especially nitrate and phosphate, to the water of lakes and streams. These chemicals act as fertilizers for phytoplankton, contributing to decreasing water quality. Two major aspects of water quality, clarity and odor, are strongly influenced by the amount and kinds of algae that grow in lakes. Large phytoplankton species, and especially large colonies of cyanobacteria (such as *Microcystis;* see figure 3.2a), are associated with poor water quality. As the human population grows, nutrient-related water quality has been generally decreasing as lake productivity increases (see table 2.5 for characteristics of eutrophic and oligotrophic lakes). Eutrophication resulting from human activity has been called **cultural eutrophication** (Hasler, 1947).

Society often desires to reverse cultural eutrophication—to improve water quality. Interest is particularly high when lakes become dominated by evil-smelling, turbid blooms of cyanobacteria. Several approaches have been used to manage nuisance algae resulting from cultural eutrophication.

The earliest approach was to simply poison the algae. Water quality can be improved using in-lake poisons such as copper sulfate (or more recently, organic herbicide). For example, Lake Mendota, Wisconsin was treated with copper sulfate from 1912 to 1958, and with various herbicides (diquat, endothall, and 2,4-D) since 1976 (Andrews et al., 1986). These chemicals work but are expensive to apply (especially in large lakes) and their use raises concerns about human and wildlife health. Limnologists have long been interested in more effective, safer, and less-expensive alternatives. In some cases, whole lakes can be treated to remove phosphate by adding chemicals (such as alum) to the lake that combine with phosphate (Robertson et al., 2000).

In many cases, the water quality of a lake or stream can be improved by reducing the inputs of nutrients from **point sources** (piped inputs, such as sewage treatment plants) or **nonpoint sources** (such as diffuse inputs draining from agricultural fields and feedlots). Reduction in nitrogen and phosphate nutrients is a bottom-up approach to improving water quality. Edmondson (1994) used this bottom-up approach to improve water quality (reduce cyanobacteria abundance) in Lake Washington. One of the great success stories in limnology is improvement of Lake Washington water quality by diverting the sewage of Seattle from the lake to Puget Sound (figure 8.6; see also Edmondson, 1994). Lake Washington, which had been receiving sewage since the 1940s, suddenly changed from clear water to murky water because of a cyanobacteria bloom of *Oscillatoria rubescens* in the summer of 1955. There was a public outcry and an immediate desire to do something. Laying his reputation on the line, Edmondson identified phosphate as the problem and strongly recommended sewage diversion as the best solution. He made the recommendation in an atmosphere of controversy, when the detergent industry was funding research showing that other chemicals (not phosphate detergents!), such as carbon dioxide, were the problem. Edmondson's recommendation was

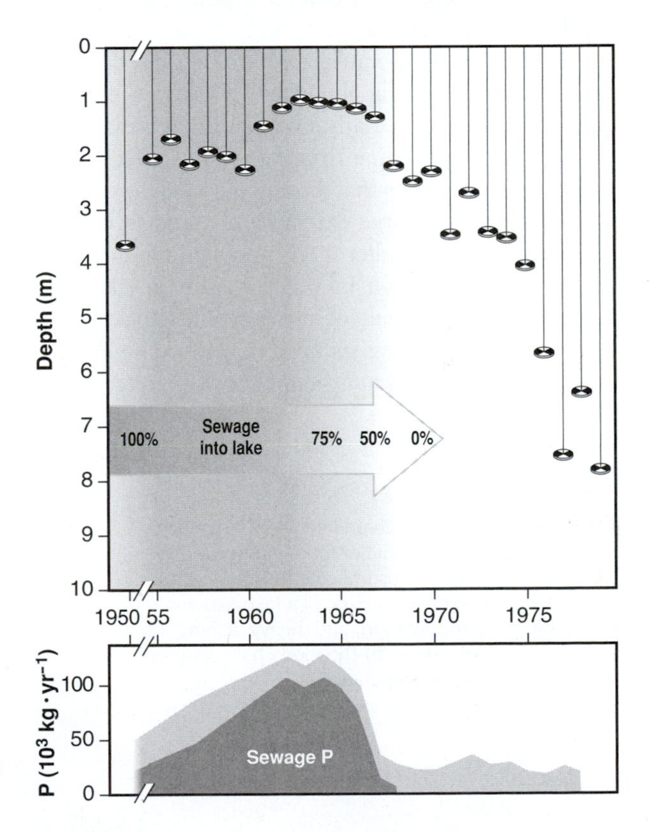

FIGURE 8.6 A success story in cultural eutrophication. Lake Washington, which had been a clear, oligotrophic lake, was polluted by sewage in the 1940s and 1950s. As a result of increasing phosphate concentration, cyanobacteria increased in abundance, clouding the water, and Secchi disk depth decreased. As sewage was diverted from the lake in the early 1960s, the lake cleared and is once again a clear, oligotrophic lake. **Source:** Data from Edmondson & Litt, 1982.

based mainly on a small number of laboratory experiments and a few natural experiments such as Hasler's (1947). The well-designed, whole-lake experiments that confirmed phosphate's role in cultural eutrophication were still in the future (e.g., Schindler, 1977). This was a gutsy action for the professor, speaking to politicians nervous of the cost of a diversion opposed by well-paid lobbyists. Local governments followed Edmondson's suggestion and spent many millions of dollars to divert sewage from the lake. After a partial diversion in 1963, deterioration in water clarity slowed. Water clarity improved as the diversion process continued, much to the relief of the local limnologists. Lake Washington is now as clear as it was before becoming an urban sewage receptacle, and Edmondson is credited with using limnological knowledge to good effect in a complex, politically charged and large-scale system.

Bottom-up algal nutrient control is safer than poisons for reducing algal abundance. However, it can be quite expensive, especially in agricultural areas where the algal nutrients come from nonpoint sources. Also, while it is well known that the amount of algae in a lake is determined to a large extent by the amount of inorganic nutrients Schindler (1977) pointed out that there is significant variation in water quality in lakes that contain the same amount of nitrogen or phosphorus (see figure 10.5). Limnologists have developed an alternative to the bottom-up management option, a technique called *biomanipulation.*

Biomanipulation

Hrbáček (1962) had the responsibility of managing fish production in small riparian ponds along the River Poltruba in Czechoslovakia. He observed a relationship between the amount of zooplankton and fishes in these ponds—more large zooplankton occurred at the beginning of the summer; more fishes were produced at the end of the season. At the time, this was a daring hypothesis because limnologists thought aquatic organisms had little or no effect on each other. Research inspired by Hrbáček resulted in a shift in the limnological understanding of how communities worked, from a water chemistry perspective to an ecological perspective. One of the consequences of this paradigm shift was a new way of managing water quality, based on the indirect effect of fishes.

Joe Shapiro et al. (1975) suggested that ecological relationships could be used to improve the water quality in lakes that have experienced cultural eutrophication and overfishing of the large piscivorous (trophy) fishes. For example, some of the variation in algal abundance reported by Schindler (1977) might be due to community-wide interactions, and Shapiro called the technique of harnessing predator-prey interactions **biomanipulation.** Building on Shapiro's concept, limnologists at the University of Wisconsin provided an important conceptual analysis of biomanipulation (Carpenter et al., 1985). Biomanipulation is a top-down technique that uses food-chain interactions to reduce algal abundance (figure 8.7).

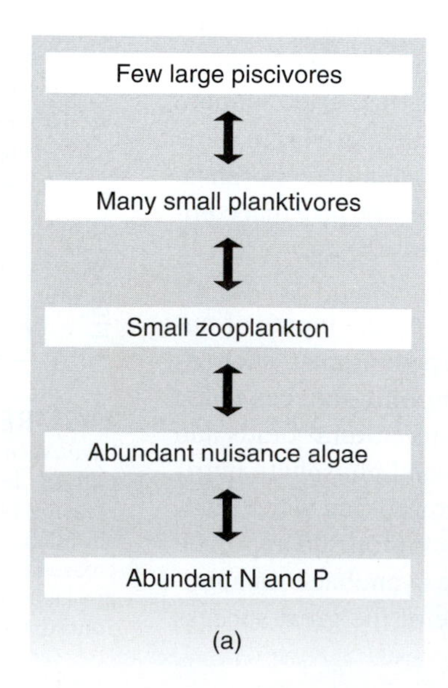

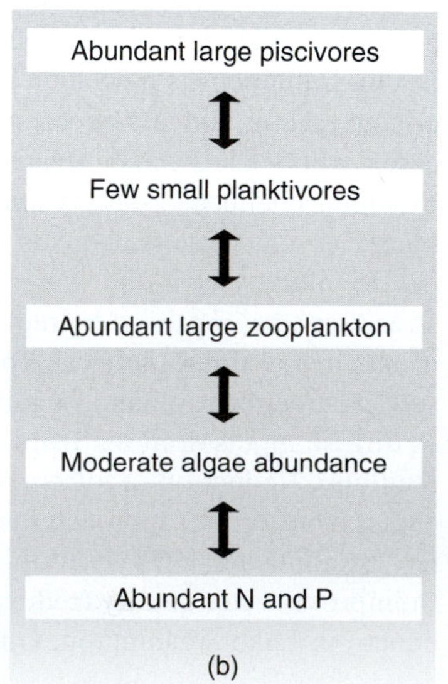

FIGURE 8.7 Food chain interactions of pelagic organisms in a mesotrophic or moderately eutrophic lake. The arrow directions indicate the flow of food. (a) The food chain characteristic of a lake experiencing cultural eutrophication and heavy fishing for large, piscivorous fishes. (b) The food chain resulting from successful biomanipulation.

The basic idea of biomanipulation is that by increasing the density of effective planktonic herbivores (grazers) such as large species of *Daphnia,* nuisance phytoplankton (figure 8.7a) can be reduced in abundance and the algal community can be altered to an association of small and inoffensive species. It is not feasible to stock sufficient large *Daphnia* to directly reduce algal abundance. However, *Daphnia* abundance can be increased by decreasing the abundance of plankton-eating fishes. The most intense planktivores are usually small fishes such as alewives, ciscoes, yellow perch, and sunfish. It is not feasible to directly remove the small, plankton-eating fishes, but they can be reduced in number by increasing the abundance of large, fish-eating (piscivore) fishes such as walleye, bass, pike, or large salmonids, to produce the food chain shown in figure 8.7b. Thus, the biomanipulation hypothesis is that water quality can be improved indirectly by increasing the density of large, predaceous fishes in a lake, a top-down strategy.

Wisconsin limnologists performed a daring (expensive and controversial) long-term and large-scale whole-lake experiment in Lake Mendota (Lathrop et al., 2002). Assisted by a fortuitous die-off of a major planktivore (the cisco), stocking of the lake with predaceous walleye did appear to result in an improvement of water clarity. The Lake Mendota biomanipulation experiment revealed the importance of the human factor. A major practical problem is that it is difficult, if not impossible, to keep people from overfishing the stocked populations of large, predaceous (game or trophy) fishes. When huge numbers of walleye were added to Lake Mendota, fishing and bait shops as far away as Chicago instantly received the alert and notified their patrons. Lake Mendota then experienced an unusually heavy amount of angling, which removed a significant portion of the walleye.

Biomanipulation has had some successes, especially in mesotrophic lakes with only moderate water-quality problems.

ALTERNATE STABLE STATES

This chapter began with a discussion of the different communities that can be found in specific aquatic habitats. Each distinct habitat was described as having a characteristic community.

When we look at nature, we find that each distinct habitat can actually be occupied by one of several communities. For example, if we collect aquatic vegetation from a number of shallow lakes in southern Wisconsin, we might find that some lakes are dominated by aquatic macrophytes, and some lakes have few macrophytes but dense phytoplankton.

A survey of primary producers in 74 shallow lakes in southern Wisconsin (Dodson & Lillie, 2002) revealed the presence of two alternate communities (figure 8.8). In this example, the lakes appear to belong to one of two groups: either open-water lakes dominated by algae, or lakes with little open water dominated (covered) by aquatic macrophytes, such as duckweed, pond lilies, and emergent sedges and cattails. These differences in primary producers will have effects on all other aspects of the pond. The water in macrophyte-dominated ponds tends to be clear and to have low oxygen concentrations. Herbivores are presented with very different kinds of food, which will in turn affect the kind and abundance of predators in the pond.

This phenomenon of distinct alternate communities (alternate stable states) has been observed in shallow lakes in many parts of the world (Scheffer, 1998).

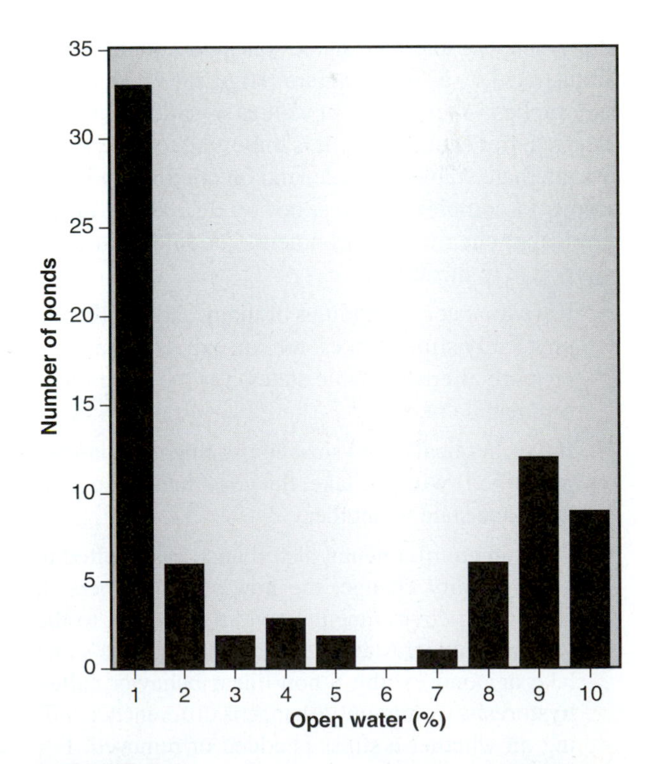

FIGURE 8.8 Results of a survey of 74 shallow lakes in southeastern Wisconsin. In this bimodal distribution, the lakes tend to either have mostly open water (algae-dominated) or very little open water (macrophyte-dominated).

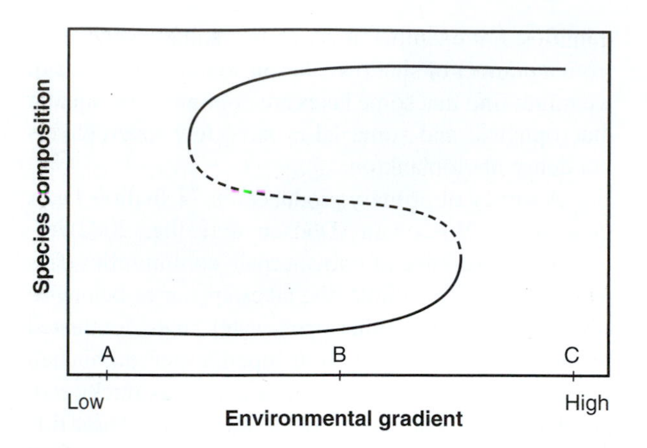

FIGURE 8.9 A fold diagram, used to express the concept of alternate stable states in community ecology. In this diagram, the extremes (A and C) in the environmental gradient (which might be phosphorus concentration in a lake) are associated with a single kind of community (represented by % abundance of one or more dominant species). At intermediate levels (B) of the environmental gradient, either community is possible, and the community that actually occupies a site will depend on the past history of the site. The dotted line represents ephemeral intermediate communities that are not stable over time.

The concept of alternate stable states (or communities) is illustrated by a "fold diagram" (figure 8.9). This concept applies to terrestrial as well as aquatic communities (Scheffer et al., 2001). It is important to understand that alternate stable states depend on community interactions in complex systems, not on differences in the physical or chemical environment. The fold diagram is interpreted to mean that:

- If we look at communities of many chemically and physically similar lakes, we can expect to find one or more alternate stable states, for the same environmental conditions.
- If the chemical or physical environment is changed (disturbed) within a lake, the community can shift from one state to another.
- Once an environmental disturbance has resulted in a community change, the new state can persist, even when environmental conditions return to the original values. Alternate stable states in the same lake or pond exhibit a non-linear behavior called **hysteresis.** Community change is different depending on whether a stress is added or removed. For example, the "environmental gradient" in figure 8.9 might refer to phosphorous concentration. Early culture eutrophication can add phosphorous to a lake (as from concentration A to concentration B), with little or no effect on species composition. At some critical threshold (between concentration B and C), species composition suddenly shifts. If this new second community state is undesirable, then the obvious management strategy is to decrease the amount of phosphorous in the lake. However, even if this is achieved, the graph shows that a reduction may have no effect on species composition. Phosphate can be reduced to a concentration considerably lower than the level that caused the shift, and the community will remain in the second state. To get the community to shift back to the first state, phosphate must be reduced to a low level (between A and B).

Scheffer (1998) discusses the biological interactions that lead to alternate stable states in shallow lakes. In short, the photosynthetic community can be dominated by either pelagic algae or by rooted macrophytes, but the two groups of species do not co-exist because of light and nutrient limitation. Algae shade macrophytes and inhibit growth from the sediments. Rooted macrophytes can store enough phosphorus to inhibit growth of algae in open water. At very low total phosphorus concentrations, a shallow lake will be always be dominated by rooted macrophytes and the water will be very clear. At very high total phosphorus concentrations, a shallow lake will always be dominated by algae, especially cyanobacteria. However, at intermediate phosphorus levels, either community is possible. A disturbance, such as a pulse of phosphorus or herbicide treatment of the macrophytes, can change the community. Once changed, the new community will tend to persist until the next disturbance comes along. For example, cultural eutrophication can shift a shallow lake from a mesotrophic, clear-water state to a eutrophic state dominated by dense cyanobacteria populations. Even if the input of phosphorus is reduced, the lake can remain eutrophic because of:

- The shading effect of the cyanobacteria, which inhibits macrophyte growth;
- The phosphorus is dissolved in the water or residing in algae, instead of being stored in macrophytes;
- The hypolimnion is more anoxic, which helps keep phosphate in solution rather than in the sediments; or
- The presence of bottom-feeding fish, such as carp, which stir up sediments and regenerate phosphorus available for algal growth.

This conversion to a condition of eutrophy (high productivity) has been observed in the majority of temperate lakes, largely because the majority of temperate lakes have watersheds with agricultural or urban land use.

Indian Lake, Wisconsin is a eutrophic, shallow lake in an agricultural watershed, dominated by cyanobacteria (Lathrop, 1988). The whole lake was fertilized with nitrate to discourage (nitrogen-fixing) cyanobacteria that occurred at nuisance levels. In the spring following the fertilization (which can be considered a disturbance), the water was clear and the shallow lake quickly filled with macrophytes. However, the change was not stable—cyanobacteria reappeared later in the summer and shaded out the macrophytes. Thus, the environment appears to support the algae-dominated community, with the possibility of an ephemeral macrophyte community developing after a disturbance. The condition of Indian Lake is represented in figure 8.9 by a position on the environmental gradient between points B and C, probably closer to point C.

Although ecological interactions resist change from one state to another, restoration of desirable water quality can be achieved if enough environmental and biological changes are made in a lake. For example, Delevan Lake, in southeastern Wisconsin, developed a serious nuisance algae problem in the latter part of the 20th century, as agriculture and population density increased in the watershed and input of phosphate increased. The lake lost clear water and macrophyte beds and became dominated by dense algae. The lake became less eutrophic in response to several, rather heroic, management measures (Robertson et al., 2000):

1. Best-management agricultural practices were adopted to reduce phosphate input into the lake;
2. Phosphate was removed from the lake water using an alum treatment (see chapter 9);
3. Bottom-feeding fishes (which re-suspend phosphate from sediments) were removed by poisoning; and
4. Predatory fishes (which eat small fishes, which eat zooplankton)were re-introduced to the lake, resulting in more zooplankton to eat algae (the process of biomanipulation).

The lake no longer has problems with nuisance algal blooms, the lake water is clearer, and there are now extensive beds of macrophytes.

The Delevan Lake example maps onto the fold diagram (see figure 8.9). The lake started at point B (on the X-axis) with moderate phosphate and macrophytes dominating. As phosphate was added to the lake (moving along the X-axis toward point C), the community composition suddenly changed to algae dominance. Moderate reduction of phosphate input (alone) would have moved the environmental conditions back to point B, but the community would have remained dominated by algae. Only an intense and multifaceted management effort resulted in enough environmental change (toward point A) to cause the lake to shift back to macrophyte dominance.

SUMMARY

The examples of complex interactions involved in biomanipulation and alternate stable states make frequent reference to energy and chemicals in freshwater systems. In many ways, the chemistry and energy flow in aquatic systems are influenced, modified, or determined by living organisms. This is why the organisms were introduced first, followed by their interactions. The next two ecosystem chapters, chapters 9 and 10, focus on how organisms and their interactions affect energy flow and chemical cycles in aquatic habitats.

Study Guide

Chapter 8 Community Ecology: Freshwater Communities Changing Through Time

Questions

1. In aquatic systems in your area, do you see evidence of the intermediate disturbance hypothesis? Where are the lakes in your area on the disturbance continuum?
2. How well does the PEG model fit lakes in your area? How does the climate and geography of your area compare to that of Northern Europe? What is the annual succession of littoral species in your region, and how does this succession depend on species interactions and annual changes in the climate?
3. What does the Island Biogeography Model predict for community structure and community dynamics of a lake or river? What would you expect to see when looking at 20 years of species lists for the same body of water? How does the concept of "rare species" affect your expectations? What factors could increase immigration rate to a lake? If the immigration rate increases and the extinction rate remains the same as before, what will happen to species richness in the lake? In your area of the world, are the extinction or immigration rates changing?
4. What are the economically and ecologically important freshwater invader species in your locale? Are these species nuisances, beneficial, or neutral? Is it possible for an exotic species to be both a nuisance and a benefit?
5. In lakes and streams in your area, is there evidence for cultural eutrophication? What is driving the eutrophication? Are there reasons to reverse the process, and what would be necessary to do so? What is the desired goal?
6. What did we learn from the biomanipulation experiment in Lake Mendota, Wisconsin?
7. Why was Dr. Edmondson elated when he saw the Secchi disk depth increase in Lake Mendota (see figure 8.6)?
8. Explain the concepts in the hypothesis: "Diversity confers resistance to invasion."
9. Can you summarize ecological succession, as in the annual cycle for the open water of a temperate lake, described by the PEG model?
10. Explain the Island Biogeography Model and describe its relationship to community diversity, dynamics, and invasions.
11. Apply the invasion model to an exotic species in your region.
12. Explain the adaptive benefits and costs of diel vertical migration to zooplankton.
13. What species interactions result in alternate stable states of communities in aquatic habitats?

Words Related to Freshwater Communities

ballast
beneficial
benefit
biofilm
biological control
biomanipulation
clear-water phase
cost
cultural eutrophication
density
diel vertical migration (DVM)
dispersal
disturbance
drift
endemic
exotic
extinction
extirpation
hysteresis

immigration
invader
macroinvertebrate
neutral
nonpoint source
nuisance
paleolimnology
PEG
pioneer
point source
resilience
resistance
succession

Major Examples and Species Names to Know

Know the general pattern of the PEG Model for temperate zone lakes. Does the PEG model apply to pelagic communities in your area?

Be able to describe three examples of invasions given in this chapter: fishes in the Great Lakes, zebra mussels in North America, and *Myriophyllum* in Wisconsin lakes. In which example is biological control most effective?

What Was a Limnological Contribution of These People?

Ulrich Sommer

Tommy Edmondson

Martin Scheffer

Arthur Hasler

Bobbi Peckarsky

Additional Resources

Further Reading

Edmondson, W. T. 1991. *The uses of ecology: Lake Washington and beyond.* Seattle: University of Washington Press. 329 pp.

Lampert, W., and U. Sommer. 1997. *Limnoecology: The ecology of lakes and streams.* Translated by J.F. Haney. New York: Oxford University Press. 382 pp.

Scheffer, M. 1998. *Ecology of shallow lakes.* London: Chapman & Hall.

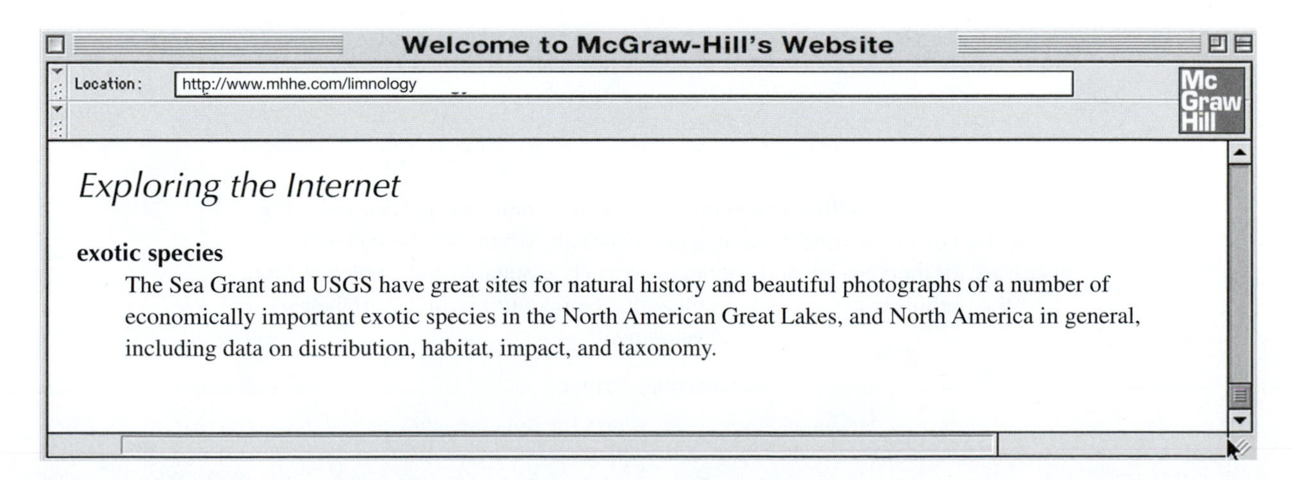

Exploring the Internet

exotic species

The Sea Grant and USGS have great sites for natural history and beautiful photographs of a number of economically important exotic species in the North American Great Lakes, and North America in general, including data on distribution, habitat, impact, and taxonomy.

9

"Every oyster-bed is thus, to a certain degree, a community of living beings, a collection of species and a massing of individuals, which find here everything necessary for their growth and continuance, such as suitable soil, sufficient food, the perquisite percentage of salt, and a temperature favorable to their development"

Karl Möbius, 1883

outline

Aquatic Ecosystems and Physiology: Energy Flow

THE ECOSYSTEM CONCEPT

An **ecosystem** is essentially a model of how energy and chemicals flow from one component (reservoir for an energy or a chemical) to another. The components typically include mixtures of both living organisms and nonliving material. Just as species are the units of community ecology, the units of ecosystem ecology are calories (for energy flow) or moles (for chemical cycling). These units can be expressed as biomass, the wet or dry weight per unit area or volume.

It is difficult to draw a boundary around an ecosystem because the various components can have quite different boundaries. For example, in the carbon cycle of a lake, one component is the atmosphere, which is global, and another is the bottom sediment, which is inside the lake. Because ecosystems have complex boundaries, the components and rates characteristic of an ecosystem model are often presented as averages per unit of volume or area, rather than per habitat. For example, *gross primary* **productivity,** the rate at which carbon is fixed by photosynthesis in a lake, is often expressed in terms of amount per unit of lake surface area: $mg\ C\ m^{-2}\ day^{-1}$.

A. G. Tansley (1935), a terrestrial plant ecologist, first used the word ecosystem to mean:

> ". . . the more fundamental conception is, as it seems to me, the whole system (in the sense of physics), including not only the organism-complex, but also the whole complex of physical factors forming what we call the environment of the biome [a group of organisms]—the habitat factors in the widest sense." . . . "The fundamental concept appropriate to the biome [a group of organisms] considered together with all the effective inorganic factors of its environment is the *ecosystem.*"

Models of Energy Flow

The ecosystem concept was picked up and refined by Ray Lindeman, a student of G. Evelyn Hutchinson. Lindeman (1942) used the ecosystem concept to describe how species and the environment of Cedar Creek Bog, Minnesota were related by patterns of energy flow. Lindeman drew the first trophic structure diagram (see figure 1.14a), a graphical model that could be generalized to any aquatic ecosystem. This bold action—of creating a general model using data from a single lake—was controversial, arousing the ire of two influential U.S. limnologists of the time: Paul Welch in Michigan and Chancy Juday in Wisconsin (Cook, 1977). Lindeman's trophic model was innovative and controversial because ecologists of that time were more accustomed to seeing lakes characterized by lists of species or chemicals. Limnologists such as Birge and Juday had been collecting data on species distributions and water chemistry for decades (there are still bales of these data sheets in storage at the University of Wisconsin State Historical Society). Welch and Juday were outraged that a young upstart was proposing a new way of conceptualizing lakes on the basis of data on energy flow from a single lake, but the intervention of Hutchinson, and a disclaimer by the editor, allowed the paper to be published in the journal *Ecology.*

Lindeman organized his ecosystem model by using the *trophic level* concept, an idea first used by the British ecologist Charles Elton in his 1927 book, to mean groups of organisms that had a common energy or nutrient source. Trophic levels include *primary producers* (*autotrophs,* usually photosynthetic), *primary consumers* (plant-eating herbivores), *secondary consumers* (primary carnivores), and so on to *top carnivores* (figure 9.1). The ecosystem, or *trophic structure model,* is a system made up of components and rates of transfer. If the ecosystem model is focused on energy, the components are called *trophic levels* (figure 9.1). Lindeman's (1942) ecosystem model has become a mainstay of limnology. Hutchinson developed large-scale nutrient cycle models based on the same ecosystem concept (for example, Hutchinson, 1944). These global-scale *biogeochemical* ecosystem models are discussed in chapter 11. This chapter focuses on energy; chapter 10 is focused on the chemical cycling of four chemicals of major limnological importance.

While the concept of discrete trophic levels is a useful generalization for summarizing ecosystem structure, organisms sometimes do not fit neatly into the categories. For example, many organisms change diets as they develop, and some species are omnivores, taking food from more than one trophic level. Young trout eat zooplankton, while a large adult eats primarily small fish. Medium-sized trout eat both zooplankton and small fish that eat zooplankton. Developmental changes or an omnivorous diet make it difficult to assign an organism to a specific level. Vander Zanden and Rasmussen (1996) suggested that dietary data or stable isotope data be used to assign a trophic position to a species, instead of a trophic level. Trophic position can be intermediate between two levels, to reflect developmental changes or an omnivorous diet. For example, the dominant pelagic "herbivores" such as *Daphnia* and *Diaptomus* are actually omnivores. Pace and Vaque (1994) used short-term feeding trials in which they fed *Daphnia* plankton from three lakes. They found that *Daphnia* were a significant predator on protozoans and rotifers in two eutrophic lakes in New York and one oligotrophic lake in Michigan, USA.

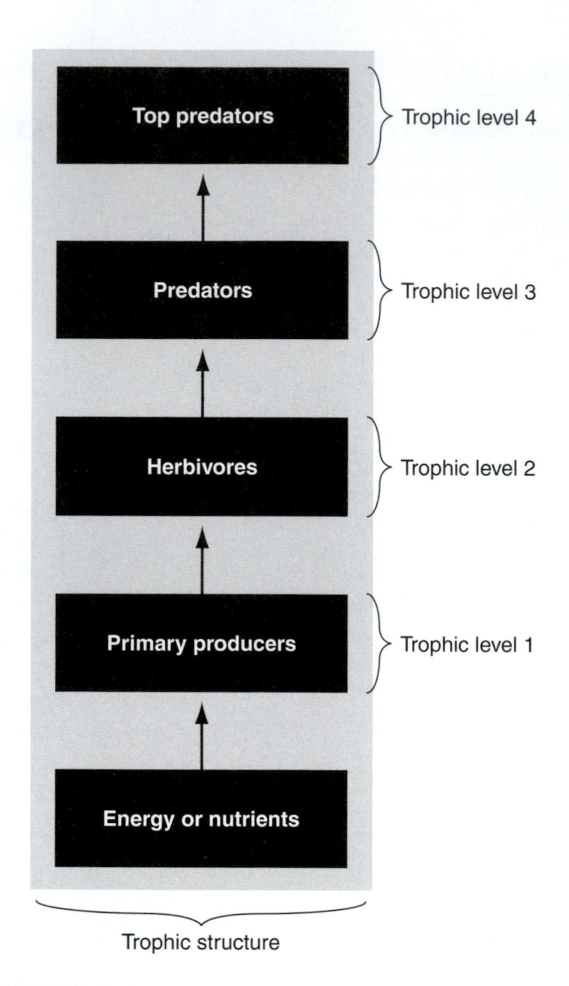

FIGURE 9.1 A simple hypothetical example of a trophic structure with four levels.

Eugene Odum's (1953) ecology text synthesized and popularized the general ecosystem concepts of earlier ecologists. Odum's text became the standard for a generation of ecologists from the 1950s through the

1970s. In 1957, Eugene's brother, Howard Odum, drew a compelling trophic structure diagram modeling energy flow from a thermodynamic perspective (see figure 1.14b). H. Odum's basic model is still used over 40 years later.

The models of Lindeman and Odum describe *trophic dynamics,* or the rates of flow of energy. Trophic dynamic models have an energy source, a path of energy flow, and an eventual sink for the energy. The energy source can be based on photosynthesis taking place within the boundaries of the lake or stream (**autochthonous** production), or on inputs of organic material imported from the outside the lake or stream (**allochthonous** production). Autochthonous production is the result of bacterial, algal, or macrophyte photosynthesis that occurs within a lake or stream. Allochthonous production occurs when leaves or other organic materials from the watershed fall or are swept into water or (as in H. Odum's example from Silver Spring, Florida of tourists throwing bread into water to feed birds or fish). Energy flows from one component to another due to biological processes (chiefly predation and decomposition) or physical processes (such as sedimentation and water currents). The ultimate energy sink is outer space, to which degraded energy (heat) ultimately flows. Notice that energy flows, but unlike chemicals, does not cycle through the ecosystem.

The graphical model for an ecosystem is called a *trophic structure* or a *chemical cycle* (see chapter 10). The ecosystem model (see figure 1.14b) often superficially resembles a food web (see figure 1.13). However, there are some distinctions that are usually observed between the two concepts (table 9.1).

Rates of transfer of energy from one trophic level to the next are due to nonliving chemical and physical processes and to the actions of living organisms. For example, sunlight is the most important energy input into ecosystems. Photosynthesis converts light energy to biologically useful chemical energy in the form of high-energy bonds in organic chemicals such as carbohydrates, proteins, fats, and the energy-rich carrier molecule adenosine triphosphate (ATP). Although of much less ecological importance, chemical and thermal energy in hot springs can also be converted, by bacteria, into energy-rich organic compounds.

The ecosystem concept applies at a wide range of *scales* (the size at which a system is observed) (figures 9.2 and 9.3). The smallest ecosystems, at the micrometer (μm) scale, are individual bacteria or cells that are actively gaining and losing both energy and chemicals. A slightly larger ecosystem is at the millimeter (mm) scale of bacteria populations at the mud-water interface. Examples of large-scale ecosystems are lakes, watersheds, and even the entire Earth.

Space-Time Diagrams

Ecosystem ecologists often use a space-time diagram to organize their thoughts about entities and processes. An example is given in figure 9.3. The diagram (graph) typically has logarithmic x and y scales in order to squeeze wide ranges of size and duration onto a single image. The domain of each entity or process is defined by its minimum and maximum extents in space and time. When comparing two entities or processes, the one with the longer duration or larger spatial scale provides context. For example, ponds and lakes exist in a context of larger geographic entities such as watersheds and lake districts. Entities interact most intensely when they have a common time or space scale.

Table 9.1 Distinctions Between the Graphical Models Associated with Ecosystem Ecology (the "trophic structure"—figure 1.14b) and Community Ecology (the "food web"—figure 1.13).

	Ecosystem Model	Food Web
Boxes contain	Reservoirs of living and/or nonliving energy or material	Species names
Arrows represent	Movement of energy or chemicals	Predator-prey relationships
Boundary	Each reservoir has its own extent or range	Often fairly well defined and the same for all the species involved

(a) **Bacteria ecosystem**

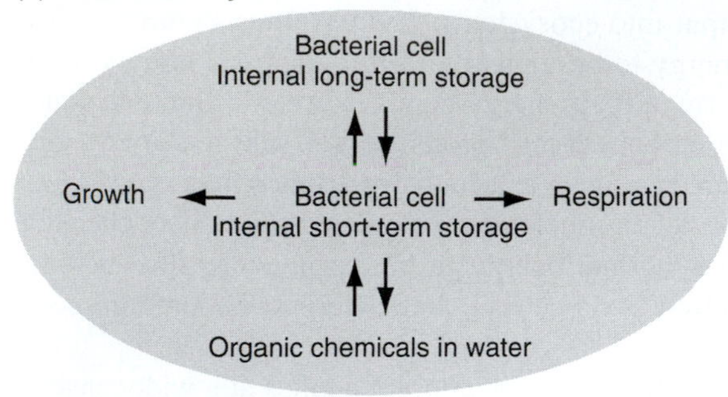

(b) **Lake ecosystem**

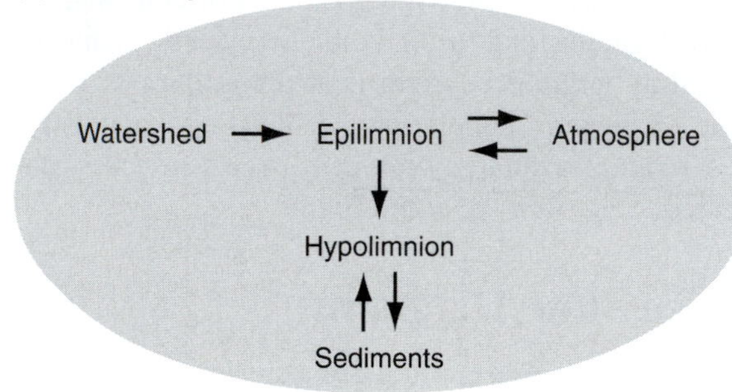

(c) **Earth ecosystem**

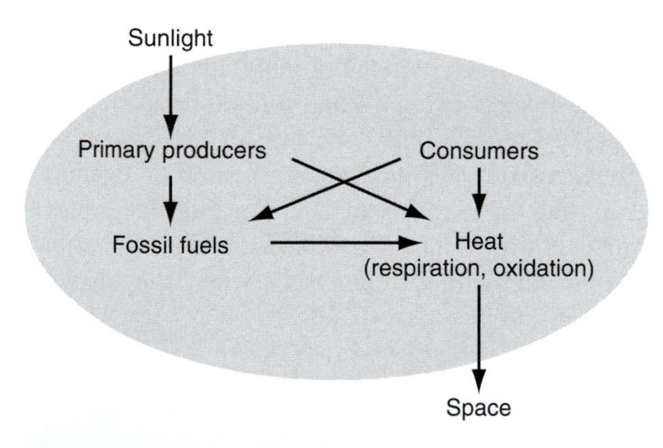

FIGURE 9.2 Examples of ecosystems at different scales: (a) bacteria; (b) lake, (c) the Earth.

PRIMARY PRODUCTIVITY

Production in ecosystem terms relates to fixation of carbon, most often via photosynthesis. Photosynthesis is a major ecosystem process, the source of energy entering most ecosystems. Each year, more than 150 billion metric tons of carbohydrate (sugar) are produced worldwide by the photosynthetic process. Photosynthesis is the physiological process by which energy from sunlight is captured and fixed as chemical energy.

The total rate of energy intake at a trophic level is called *gross productivity.* This is *gross primary productivity* (**GPP**) for autotrophs, *gross secondary productivity* for herbivores, and so on. Gross primary productivity is, in most cases, the result of photosynthesis, a chemical reaction that changes light energy into chemical bond energy (figure 9.4).

Light energy absorbed by the photosynthetic pigments (colored, light-absorbing chemicals) and associated proteins splits the water to produce oxygen and high-energy electrons. The electrons are used to produce energy-rich chemicals (the light reaction), which then participate in the energy-using reaction (dark reaction) that reduces CO_2 to carbohydrate. The carbohydrate contains useful chemical energy for driving other metabolic reactions resulting in growth, reproduction, and repair of tissues.

Not all the energy fixed by gross primary productivity is available for storage or metabolic processes. Roughly half of the total amount of energy transferred from light to chemical bonds is quickly lost during the chemical reactions associated with metabolism. This loss of energy is called *respiration,* the rate at which energy is consumed by an organism. Both plants and animals respire. Respiration is basically the photosynthesis equation running backwards. At night, respiration of all organisms in the lake uses oxygen and produces carbon dioxide. The absence of photosynthesis at night means that respiration can cause a decline in dissolved oxygen in a lake, whereas photosynthesis generally exceeds respiration during the day.

The amount of energy left over after accounting for respiration is called *net primary productivity* (**NPP**). The NPP is productivity stored as tissue or offspring and potentially available as food to the next trophic level:

$$\text{NPP} = \text{GPP} - \text{Respiration}$$

Photosynthetic Organisms

Most photosynthetic organisms, from cyanobacteria to higher plants, contain chlorophyll *a.* Photosynthetic organisms also possess several other accessory pigments including the yellow and orange **carotenoids** and the reddish **phycobilins.** Macrophytes, euglenoids, and green algae all contain the pigment chlorophyll *b.* Other photosynthetic protists (e.g., diatoms, dinoflagellates,

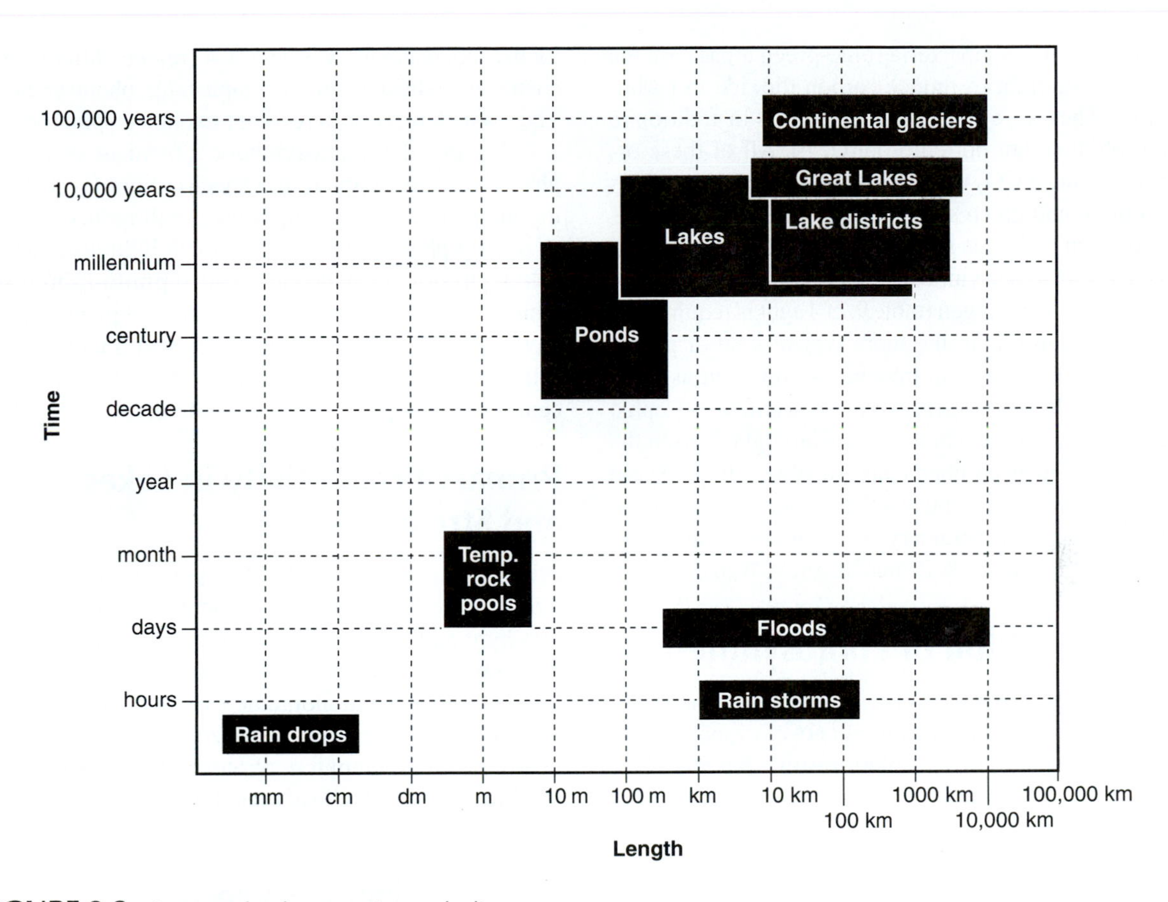

FIGURE 9.3 An example of a space-time scale diagram.

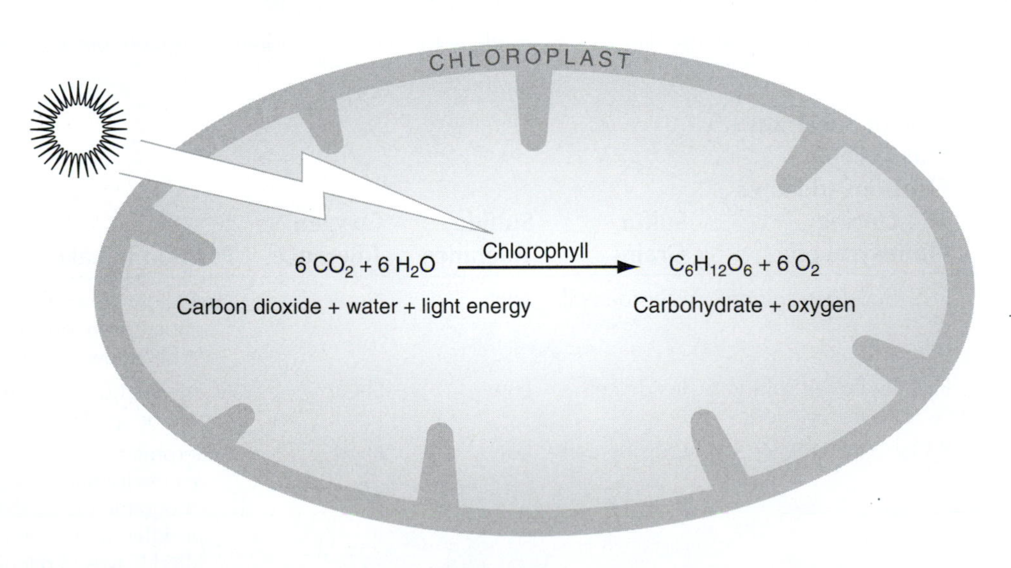

FIGURE 9.4 A simplified form of the photosynthetic reaction.

red algae) and cyanobacteria (blue-green algae) use the energy of sunlight to reduce carbon dioxide to carbohydrate. They have chlorophyll pigments in addition to chlorophyll *a* (but not chlorophyll *b*). All of these organisms reduce CO_2 using water as an electron donor.

Purple and green sulfur bacteria use hydrogen sulfide (or even hydrogen gas) as the electron donor, instead of water. The photosynthetic reaction produces sulfur or sulfate instead of oxygen (table 9.2). Light is required for the reaction, just as in the more typical kind of photosynthesis. Sulfur bacteria may be as important as other primary producers in meromictic or stratified eutrophic lakes. The energy can come either from light or from the oxidation of reduced chemicals, such as sulfide. These microorganisms also typically form thin, dense plates at or below the depth boundary between anoxic and oxygenated water (Culver & Brunskill, 1969; Wetzel, 1973).

Light Inhibition of Photosynthesis

At high concentrations of oxygen, the photosynthetic rate of most algae and higher plants, but not of cyanobacteria, is inhibited or reversed. This **photorespiration** resembles the response of terrestrial plants that use the **C3** pathway for the dark reaction. "C3" and "C4" refer to different biochemical pathways for accomplishing photosynthesis. The **C4** dark reaction is resistant to photorespiration.

A particular photosynthetic organism is usually physiologically adapted to a specific light level. Moving the plant to a lower light level will reduce the rate of or stop photosynthesis. Too much light also can result in physiological changes, called **photo-inhibition,** which reduce the photosynthetic rate. C3 plants tend to have the highest photosynthetic rate at about 10% of full sunlight, and have a lower rate at higher light concentrations. C4 plants are less inhibited by high light.

Primary Productivity in Lakes and Streams

Primary productivity (PP) takes place in a number of places in freshwater habitats. The main requirement is that light and liquid water are present, along with the necessary organic nutrients, especially sufficient nitrate, phosphate, and inorganic carbon or sulfide.

In the pelagic open-water zone, productivity depends on phytoplankton photosynthesis. The least-productive lakes fix less than 1 milligram of carbon

Table 9.2 A Summary of the Physiology and Ecology of Photosynthetic Bacteria that Use Light Energy but Use a Form of Sulfur Instead of Water as an Electron Acceptor, in the Reduction of Carbon Dioxide.

Many of these organisms can also fix carbon dioxide in the dark, by oxidizing reduced forms of sulfur.

Sulfur Bacteria	Type of Substrate that Accepts Electrons (donates H^+) During Photosynthesis	Sulfur Grains	Sulfide Tolerance	Oxygen Tolerance	Position in Lake
Green	H_2S (hydrogen sulfide)	Outside cell	High	None	As shallow as possible to be in the photic zone, but deep enough to be below all oxygen
Purple	H_2S (hydrogen sulfide) (or even H_2 alone)	Inside cell	Low	Slight	At intersection of oxygen and sulfide gradients
Purple, Nonsulfur	$S_2O_3^{2-}$ (thiosulfate)	Outside cell	Low	High	Anaerobic mud where there is access to high concentrations of organic material; also bogs, puddles, and the hollow, rain-filled leaves of pitcher plants (Hepburn et al., 1927)

Source: Based on material from Brock, 1999.

per day per square meter (annual average), while the most productive lakes average about 1 gram of carbon fixed per day per square meter. The highest rates of primary productivity for the pelagic zone occurs in sewage treatment lagoons and similar enriched lakes (table 9.3). For example, a Madison, Wisconsin sewage lagoon (not aerated) produced about 1300 grams of carbon per square meter per year (Dodson et al., 2000). By way of comparison, the average U.S. corn (= maize) production, for the entire plant, amounts to about 1500 grams of carbon (total plant) per square meter per year. The highest rates of pelagic production can only occur when nonlimiting levels of nutrients are present in streams or lakes. These conditions occur in watersheds associated with agriculture, industrial waste, or sewage treatment. If nutrients or light are limiting, productivity can be much lower (table 9.3).

A general rule of thumb is that 1 square meter of the Earth's surface will have about the same net primary productivity, whether the surface is terrestrial or aquatic. Lakes fix about as much carbon per unit surface area as would a grassland or forest in the same place, with the same climate.

The euphotic zone is that part of the lake or stream receiving at least 1% of incident surface light. This turns out to be about three times the Secchi depth. The depth at which photosynthesis just balances respiration is called the **compensation depth.** As discussed in chapter 2, the relative depths of the compensation depth and the thermocline depth have far-reaching effects on the vertical stratification of primary producers, herbivores, and chemical characters such as the vertical oxygen profile.

In the mixed layer (epilimnion) of lakes and in flowing water, the algae tend to be circulated from

Table 9.3 Some Examples of Net Primary Productivity in the Open Water of a Range of Lakes

Name	Biome (in United States unless otherwise noted)	Trophic Classification	Watershed	Limiting Factors	g C/m^{-2}/year^{-1}
Mexican Cut Pond	Colorado alpine, high elevation	Ultra-oligotrophic	"Pristine"	Nutrients, temperature	1.3
Char Lake	Canadian high arctic	Ultra-oligotrophic	"Pristine"	Nutrients, temperature	4.1
Lake 239	Canadian boreal forest	Oligotrophic	"pristine," logging	Nutrients, temperature	21.3
Lake Washington	Pacific Northwest mixed forest	Mesotrophic	Urban, residential, agricultural	Nutrients, light	96
Lake Thonotosassa	Subtropical Florida	Eutrophic	Urban, residential, agricultural	Nutrients, light	166
Plussee	Northern Germany mixed forest	Eutrophic	Residential, agricultural	Nutrients, light	186
Lake Mendota	Temperate deciduous forest, Wisconsin	Eutrophic	Urban, industrial and agricultural	Light	342
Lake Valencia	Tropical deciduous forest, Venezuela	Eutrophic	Urban, industrial, and agricultural	Light	821
Madison Sewage Lagoon	Temperate deciduous forest, Wisconsin	Hyper-eutrophic	Urban, industrial and agricultural	Light	1300

Source: Data from Dodson et al., 2000

light to dark. Algae that are able to deal with the intense light-to-dark regime persist. The denser the algae, the more likely an individual cell is to be in the dark because of the shade produced by other algae. Thus, very dense algae populations favor floating or motile algae.

In very clear lakes (plates 3 and 18), the photic zone extends down into the hypolimnion where algae are often concentrated as a very thin plate, as discussed in chapter 2. There is no mixing of the water in the hypolimnion, so the algal plate is stable. The plate is also quite distinct, often about 1 meter thick. The plate's upper and lower boundaries are sharp because of insufficient light below the plate and insufficient nutrients above the plate. The upper boundary is set by nutrient limitation and the lower boundary is set by light limitation.

In the littoral zone of lakes, pools, and streams, macrophytes accomplish much of the total photosynthesis. A significant amount of primary production is due to the aufwuchs community of protists and bacteria growing on the surface of macrophytes or rocks. The littoral zone has the potential for the greatest annual level of primary productivity per unit surface area in aquatic habitats because it has the highest light, nutrient, and temperature conditions. The maximum rate observed is about 25 times that of phytoplankton (Wetzel, 2001). However, in most lakes, the littoral zone is a linear habitat, restricted to a narrow band around the edge of a much larger, open portion. Geometry dictates that the ratio between pelagic and benthic productivity is large for large lakes (Vadeboncoeur et al., 2002). A noted exception is Neusiedlersee in Austria, a large but very shallow lake that is almost all littoral zone; the shallow lake is nearly covered with extensive reed beds and other emergent vegetation. Cultural eutrophication of a lake can shift the majority of primary productivity from the littoral and benthic zones to the pelagic zone (Vadeboncoeur et al., 2003). As nutrient concentrations increase, phytoplankton increasingly shade attached macrophytes and aufwuchs.

In streams, backwaters often have extensive littoral zones, and riffles have well-developed aufwuchs communities. At the terrestrial margin of the littoral zone, the riparian zone provides large inputs of organic material as deciduous leaves and coarse woody debris.

The benthic zone (the bottom of the lake or stream below the macrophyte-dominated littoral zone) supports few photosynthetic plants because of low light levels. Clear-water oligotrophic lakes typically have benthic zones covered with a thick mat of algae or moss, growing slowly in very dim light.

Measuring Primary Productivity

The *oxygen light-dark bottle* method for measuring algal productivity is based on changes in oxygen concentration. Lake water incubated for a few hours in clear, glass bottles (suspended in the lake so as to receive sunlight) generates oxygen as the algae photosynthesize. This oxygen production can easily be used to calculate net primary productivity. However, the algae are also respiring, using some of the produced oxygen for metabolic processes. The total amount of photosynthesis observed in the bottle is therefore indicated by the observed increase in oxygen plus the amount of oxygen used in respiration. Respiration is measured by incubating lake water in dark (covered) bottles at the same depth as the light bottles. The rate of oxygen loss in dark bottles is the respiration rate. The rate of GPP, then, is the sum of the NPP rate and the respiration rate.

The problem with this technique is that it takes several hours of incubation, during which time there are significant changes in the bacteria and phytoplankton in the bottles. A curious observation made by alpine limnologists is that in very oligotrophic water, light bottles tend to gain less oxygen than do dark bottles (Pennak, 1978).

The **^{14}C tracer** method is based on the radioactivity of ^{14}C, which has a half-life of 5760 years. A small amount (trace) of ^{14}C in the form of $NaH^{14}CO_3$ is added to a sample of lake water. The lake water with tracer is incubated in a glass bottle, either in the lake or lab, for an hour or so. The total amount of nonradioactive carbon dioxide is measured in the lake water. After the incubation, the algae are filtered out of the water. The radioactivity of the algae indicates how much tracer has been fixed as organic carbon. The total amount of carbon fixed is assumed to be proportional to the amount of tracer fixed. The assumption is that the ratio between tracer and nonradioactive CO_2 is the same inside the algae as it is in the water. For example, there might be 1000 times as much normal carbon as radioactive carbon in the spiked lake water. We can measure the amount of radioactive carbon in the algae and multiply by 1000 to estimate the total amount of carbon that was fixed during incubation.

The ^{14}C method is about 100 times more sensitive than the light-and-dark bottle method. The ^{14}C method measures something between GPP and NPP, closer to NPP if short incubation times are used. The problem is that some of the fixed ^{14}C is immediately respired by the algae back into the water. The ^{14}C method has been used extensively all over the world, so it is good for comparative purposes.

Primary production can also be estimated by *weight change* per unit area due to accumulation of organic material during a time interval, such as a summer growing season. This technique is borrowed from agricultural science and is sometimes used for aquatic macrophytes and fishes (which have a monthly or even annual life cycle), but the weight change technique is not often used for phytoplankton (which have a life cycle on the scale of days).

Tropical Productivity

You might think annual NPP would be highest in tropical lakes because the warm water allows rapid photosynthesis all year long. However, there are several factors that tend to reduce NPP in tropical lakes. Just being located in the lowland tropics is not enough to ensure a high rate of annual production.

Although low-elevation, tropical lakes are warm all year long, the effective growing season is often quite short, compared to that of temperate lakes. The GPP usually shows a peak for a month or so, associated with the beginning of the rainy season when nutrients are washed into the lake. In addition, days are never longer than about 13 hours in the tropics, compared to a maximum of about 15 hours for temperate-zone lakes. The shorter tropical day length means that during the growing season, tropical lakes have less photosynthesis during the day and more respiration during the longer nights. Finally, winds are often strong in the tropics and able to mix sediments into the water, especially because many tropical lakes are shallow. For example, Lake Chapala in Central Mexico (plate 33) is a warm and shallow polymictic lake enriched with sewage and agricultural runoff. However, Lake Chapala is so murky from suspended silt due to daily winds that it has low primary productivity.

Energy Inputs Other than Sunlight

Nonphotosynthetic Primary Productivity

Several kinds of bacteria can use hydrogen sulfide (H_2S) to reduce carbon dioxide to carbohydrate in the absence of light. Bacterial production by nonphotosynthetic colorless bacteria is called "chemosynthetic," and occurs in sulfur springs. This nonphotosynthetic primary production is a minor component in most freshwater systems, but it supports a large and diverse community of animals living near hot, deep-sea, mid-oceanic vents.

Wave Energy and Productivity

When waves break on shore, they provide energy that can result in erosion of the shoreline, oxygenation of water, and rapid circulation of water supplying nutrients to the organisms growing in what is typically a high-light environment. Organisms adapted to live with breaking waves actually take advantage of the energy provided by the waves (the organisms need not spend their own energy to circulate water). For example, several species of large algae live attached to rocks in the breakwater zone of both freshwater and marine systems. Studying a marine system, Leigh et al. (1987) found that despite extreme wave action, some rocky shores of the Pacific Northwest have high algal productivity. On Tatoosh Island (which faces the open ocean and experiences intense waves), the tough macroalga (kelp) called the sea palm (*Postelcia*) produces more than10 kg dry matter/ m^2/year. This is a much higher level of productivity than that observed in typical freshwater examples, in which maximum open-water rates of production are about 1.3 kg dry matter/m^2/year (see table 9.3). The most productive species are restricted to areas of intense wave action. Waves and the associated turbulence increase the capacity of kelps to acquire nutrients by increasing the flow of water over the algae. The intense wave action also bashes floating logs against the intertidal rocks, which removes less-resistant predators and competitors. In a sense, the kelps are benefiting from the wave energy, just as they benefit from the sun's energy.

The same supplement of wave energy might also be true for attached algae along lake shores and in streams. For example, *Cladophora* (rock hair) is a tough, filamentous green alga found attached to hard surfaces in the shallow breaking-wave zone (see plate 33). In cases of extreme wave action (as shown in plate 22), there are encrusting blue-green algae (cyanobacteria) or lichens on rocks at the shoreline. Like the kelps, these algae benefit from high-energy wave action, which brings nutrients and removes competitors for light. Similarly, in streams, cyanobacteria attached to rocks in riffles show increased productivity with increased current speed (Dodds, 1989).

SECONDARY PRODUCTION

Whereas primary production typically begins with carbon dioxide and sunlight, secondary production is based on the oxidation of high-energy organic compounds produced by the primary producers. Secondary producers include most bacteria, many protists, and all animals.

Ecological Efficiencies

One of Lindeman's goals was to measure the ratio of the amount of energy entering one trophic level to the amount of energy entering the previous level. For example figure 1.14a shows that energy from the sun flows into the phytoplankton as primary production. Some of that fixed energy then flows from phytoplankton to zooplankton. The ratio of the two rates of energy flow is the ecological efficiency of phytoplankton. Lindeman considered this efficiency one of the unifying principles of trophic dynamics and offered the concept as a useful way of understanding how energy flows through lake ecosystems. In general, about 10% of the energy coming into a trophic level flows to the next level.

Fishes and Primary Productivity

The total yield of fishes from a lake depends on both the shape (depth) of the lake and the amount of primary productivity (Ryder et al., 1973). There is a strong, positive correlation between fish production and primary production (Pauly and Christensen, 1995). Because primary productivity is correlated with the amount of available phosphate (as discussed in chapter 9), there is also a correlation between phosphorus input to a lake and the fish yield (figure 9.5). For a given lake, fish yield increases with primary production, if for example, the lake experiences cultural eutrophication. Although total fish yield increases with nutrient enrichment, the dominant fish species change from more desirable (such as trout and perch) to less desirable (such as carp). At extremely high levels of productivity, total fish yield is reduced in response to the low oxygen conditions associated with hypereutrophication.

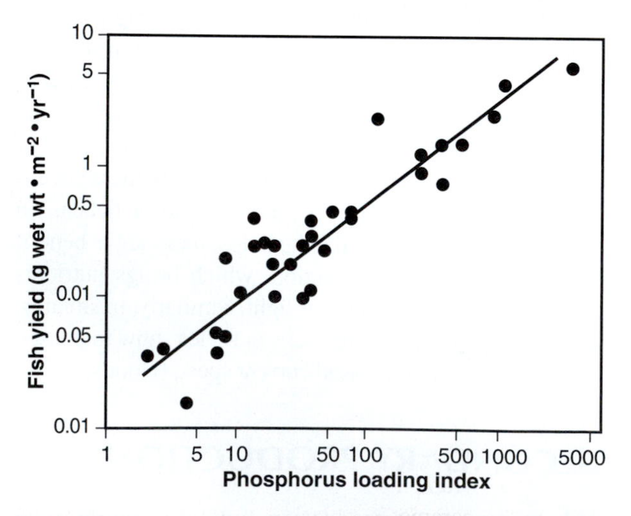

FIGURE 9.5 The relationship between annual fish production and the input of phosphorus to a lake (phosphorus loading). This graph has logarithmic scales. **Source:** Data from Jones & Lee, 1991.

Microbial Loop

The *microbial loop* provides an example of the system-scale consequences of ecological efficiency (Stockner & Porter, 1988). The microbial loop is a subsection of the pelagic trophic structure. It is composed of small pelagic species consisting of bacteria, protists, and rotifers (figure 9.6). The microbial loop tends to be an important energy flow path when inorganic nutrients (nitrate and phosphate) are scarce and allochthonous dissolved organic compounds are present to stimulate the growth of bacteria and small protists (Meyer, 1994).

The trophic levels of the microbial loop are: (1) small autotrophs (picoplankton—see chapter 3), (2) protistan

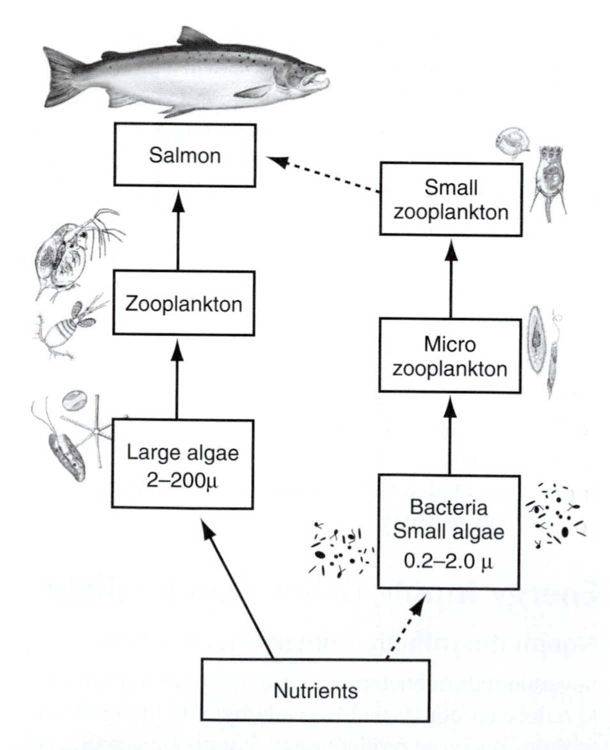

FIGURE 9.6 The microbial loop model. Compared to a standard model (see figure 9.1), this trophic structure has an extra level at the microbial level, resulting in less energy being available for the top trophic level.

micrograzers, (3) small invertebrate predators and grazers, (4) larger invertebrate predators such as copepods and rotifers, and (5) fishes. In this example, the microbial loop has three trophic levels between autotrophs and fishes. The alternate food web is: (1) autotrophs (nanoplankton), (2) large grazers such as *Daphnia* and copepods, and (3) fishes. In this nanoplankton-based food chain, there is one trophic level between the autotrophs and the fishes. The ecological efficiency of each trophic level is about 10%. Thus, lakes with a strong microbial loop will provide significantly less energy to fishes, compared to lakes dominated by the three-level food chain. Lakes with food webs dominated by the microbial loop have lower standing stocks of smaller juvenile sockeye salmon, compared to lakes dominated by the nanoplankton-based food web.

Allochthonous Inputs

In small streams, a major source of energy is organic material, such as fallen tree leaves from outside the stream—the allochthonous energy source. As the plant material begins to decay (serve as a substrate for bacteria), it provides food for many species of herbivores, such as stream insects that shred and scrape decaying organic material (see figures 4.18c and g).

INTRODUCTION TO PHYSIOLOGICAL ECOLOGY

Physiology describes how organisms work—it is a mechanistic view of biology at the level of the organism. Physiological ecology, also often called *organismal ecology,* takes the perspective of the interaction of individual organisms with their environment. The interaction of most importance is metabolism, which is the energy-flow system, on an individual basis, for a specific organism and its environment. A common physiological ecology question is, "How does this organism work, in terms of energy requirements, body size and form, and tolerances to its environment?"

Surface-to-Volume Ratios

The size of an organism has a profound effect on its physiology. A general model of an organism is the sphere (physiologists love this kind of simple model!). For a sphere with radius r, the surface area (S) depends on the square of the radius, while the volume (V) of the sphere depends on the cube of the radius:

(Equation 9.1)

$$S = 4\pi r^2$$

(Equation 9.2)

$$V = (4/3)\pi r^3$$

The surface-to-volume (S/V) ratio, therefore, is proportional to $1/r$, so larger spheres have a smaller surface-to-volume ratio. For example, if you are comparing one sphere with another that has 10 times the radius, the larger sphere has a surface-to-volume (S/V) ratio that is 1/10 (one-tenth) that of the smaller sphere.

For the more spherical forms (single-celled protists including algae and bacteria), the spherical model can easily be used to understand the relationship between form (size scale) and function. For example, the efficiency of gas and other chemical transport is related to the surface area; requirements for oxygen and other chemicals and storage are related to volume. Therefore, small size is an automatic (structural) adaptation to low oxygen concentration because small organisms have a favorable S/V ratio for taking up oxygen and a low demand relative to the amount taken up. Small animals, less than 2 millimeters in at least one dimension, need little or no specialized gas-transport systems because of the favorable S/V ratio, and because diffusion is an effective means of gas transport over small distances (Alexander, 1979).

The surface-to-volume ratio is most useful for comparing organisms of the same *shape.* Objects have the same shape if they have the same linear proportions among their parts. For example, large and small algae might be roughly spherical, and therefore have the same shape but a different size. An algal cell and a macrophyte have different shapes and different sizes.

External similarities can be confusing. For example, a *Paramecium* (1 mm long) and a fish (1 m long) have superficially similar shapes (figure 9.7). However, the organisms have very different shapes if internal organs are considered. The *Paramecium* is more or less a simple football shape. The fish, which might appear to also be football-shaped, is actually a much more complicated shape because it has its high-surface-area organs folded up inside its body (figure 9.7). These organs include the gills, the gut (and associated villi), and the kidneys. If we considered paramecia and fishes to have the same shape, then clearly the *Paramecium* would have the larger S/V ratio. When the total surface area of the fish is taken into

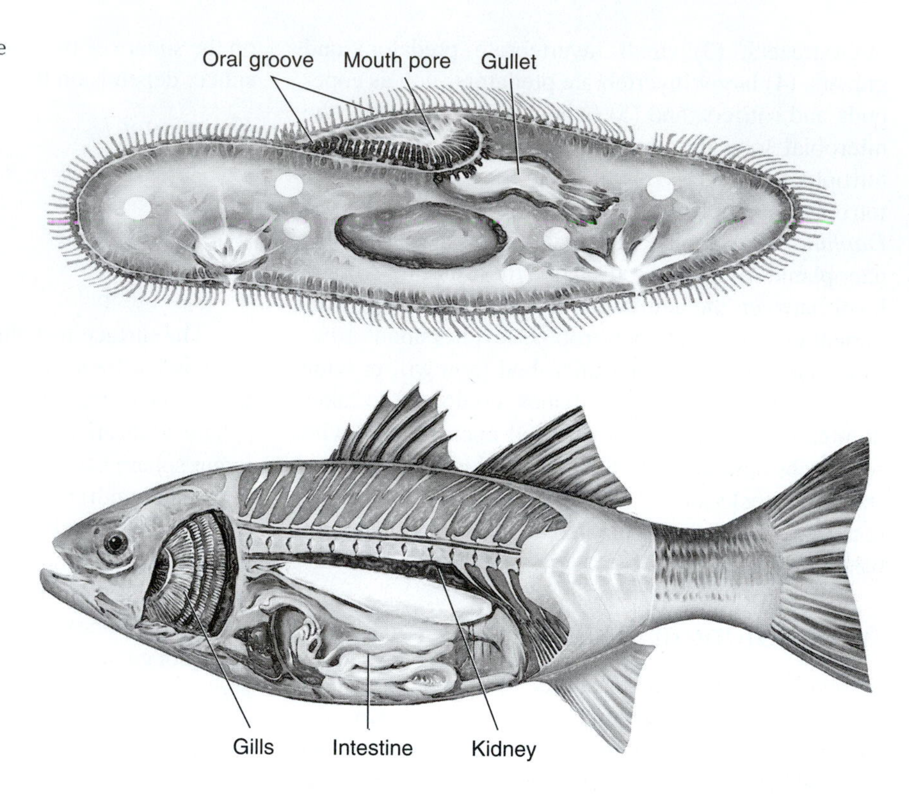

FIGURE 9.7 Surface-to-volume ratio for a paramecium and a fish.

account (gill, gut, and kidney surface), the *Paramecium* and the fish actually have approximately the same S/V ratio.

Metabolic Rate and Body Size

Metabolic rate scales with several factors, including body size (Peters, 1983), temperature, and feeding rate. First, we will look at the effects of body size.

All organisms use energy-rich chemicals as a fuel source for metabolic activities. Respiration is the rate at which chemical fuel is oxidized to produce energy-rich organic chemicals that can be used in growth and metabolic maintenance. Respiration is measured by either the amount of energy consumed per unit time or by the number of calories of organic compounds consumed per unit time.

Not surprisingly, smaller organisms use less oxygen than larger organisms (figure 9.8). What is surprising about respiration rate is that smaller organisms are less efficient than large animals—there is a general tendency for smaller organisms to have higher metabolic rates per unit of body mass (table 9.4).

As some of the smallest organisms, bacteria have some of the highest (weight-specific) respiration rates. (Weight-specific rate means the amount of calories or

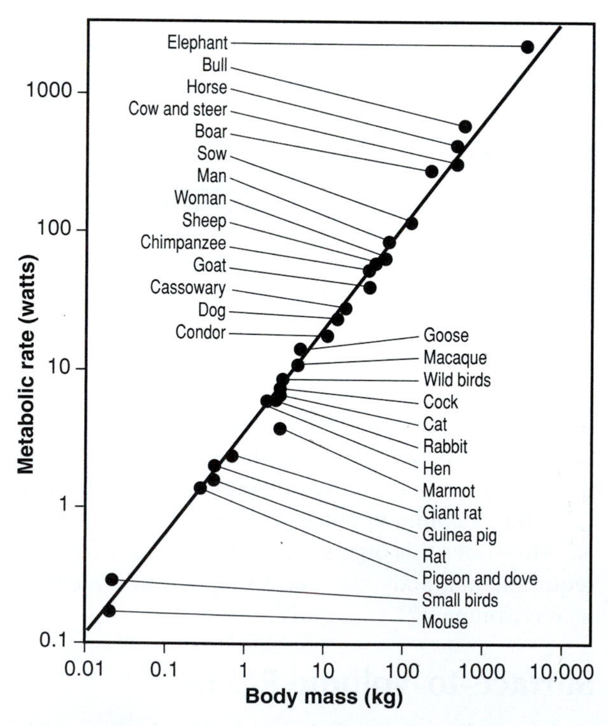

FIGURE 9.8 Mouse-elephant metabolic rate curve per unit weight. This graph presents data for terrestrial vertebrate animals. Gillooly et al. (2001) claim that the same relationship exists for aquatic invertebrates. **Source:** Data from Schmidt-Nielsen, 1984.

oxygen used per gram of weight.) The high weight-specific metabolic rate of bacteria has important consequences for ecosystem function—one aspect is the rate at which organic chemicals are oxidized in various components of the ecosystem. The total bacteria biomass (weight per unit area) in a lake or stream is often similar to the biomass of algae, zooplankton, or fishes (see table 3.2). However, the oxygen requirement of the bacteria is 10 to 100 times as much as that of zooplankton or fishes (table 9.4). Thus, ecosystem respiration depends largely on bacterial respiration, which is determined by temperature and the amount of organic chemicals dissolved in the water. The more organic material, the more bacteria can grow, and the more oxygen consumed per unit time. With enough organic material, water can become anoxic, to the detriment of oxygen-requiring organisms such as arthropods and fishes. A measure of the potential ecosystem respiration rate is the biological oxygen demand (**BOD**). BOD is measured by incubating water containing bacteria and organic material, and measuring the amount of oxygen consumed in a specific time. Bottles are usually incubated for 5 days at 20°C, to produce the standard BOD_5 (APHA, 1998).

Although a maximum of about 14 milligrams of oxygen can be dissolved in 1 liter of (cold) water, it is possible that the BOD will greatly exceed this saturation value. For example, raw sewage routinely has a BOD of about 100 milligrams per liter. A BOD value greater than the oxygen saturation level suggests that the water source is in danger of becoming anoxic.

Metabolic Rate and Temperature

Metabolic rates, such as respiration, increase exponentially over a reasonable range of temperatures. Below this range, there is little or no metabolic activity, and at higher temperatures, the organism dies as heat denatures metabolic enzymes. The temperature range of exponential increase in metabolic rate can be as much as 20°C.

Table 9.4 Metabolic Rates of Aquatic Organisms of Various Sizes, at 20°C.
Conversions used: wet weight = 10 × dry weight; 1 ml O_2 = 4.8 cal = 0.0048 Cal.

Aquatic Species or Group	Individual Metabolic Rate (calories/hour^{-1})	Body Weight, Wet (gm)	Metabolic Efficiency (calories gm^{-1} body weight)
Bacteria[1]	10^{-13}	10^{-12}	100
Flagellates[1]	10^{-12}	5×10^{-11}	20
Infusoria[1]	10^{-9}	10^{-7}	10
Amoebae (*Chaos*)[1]	10^{-8}	10^{-5}	1
Daphnia pulex[3]	1.4×10^{-6}	0.35×10^{-3}	1.4
Daphnia magna[2]	4.8×10^{-6}	1.3×10^{-3}	3.7
Daphnia magna[3]	3.4×10^{-6}	1.5×10^{-3}	2.3
Chaoborus (Diptera)[3]	1.5×10^{-5}	8×10^{-3}	0.18
Stenonema (Ephemeroptera)[3]	1.6×10^{-5}	20×10^{-3}	0.08
Pisidium (Mollusca)[3]	4.4×10^{-6}	50×10^{-3}	0.088
Trout fry[4]	3.7×10^{-5}	2.2	0.017
Brook trout, juvenile[4]	1.6×10^{-5}	23	0.00067
Carp[4]	2.7×10^{-4}	602	0.00044
Sturgeon[4]	2.1×10^{-3}	6500	0.00033
Sturgeon[4]	2.2×10^{-3}	10,400	0.00028

[1] Hemmingsen, 1960
[2] Porter et al., 1982
[3] Lampert, 1984
[4] Winberg, 1966 (tables corrected—"ml" in the translation taken as μl)

Within this range, metabolic rates approximately double for each 10°C-increase in temperature. This rate of increase in 10°C-increments is called the $\mathbf{Q_{10}}$.

The temperature preferences of an organism are related to its enzyme kinetics. An enzyme is a protein catalyst that assists in the conversion of one chemical to another. The rate of the chemical reaction depends on temperature (figure 9.9a). At low temperatures, the enzyme acts slowly. The reaction rate increases to some maximum rate at the **optimal temperature.** Above the optimal temperature, the enzyme becomes less efficient as thermal energy tears it apart. An organism has a preferred, optimal temperature and this temperature is closely tied to the optimal temperatures of all its enzymes (although the individual typically has a lower preferred temperature than its individual enzymes). Natural selection works to coordinate the optimal temperatures of all the enzymes within an individual.

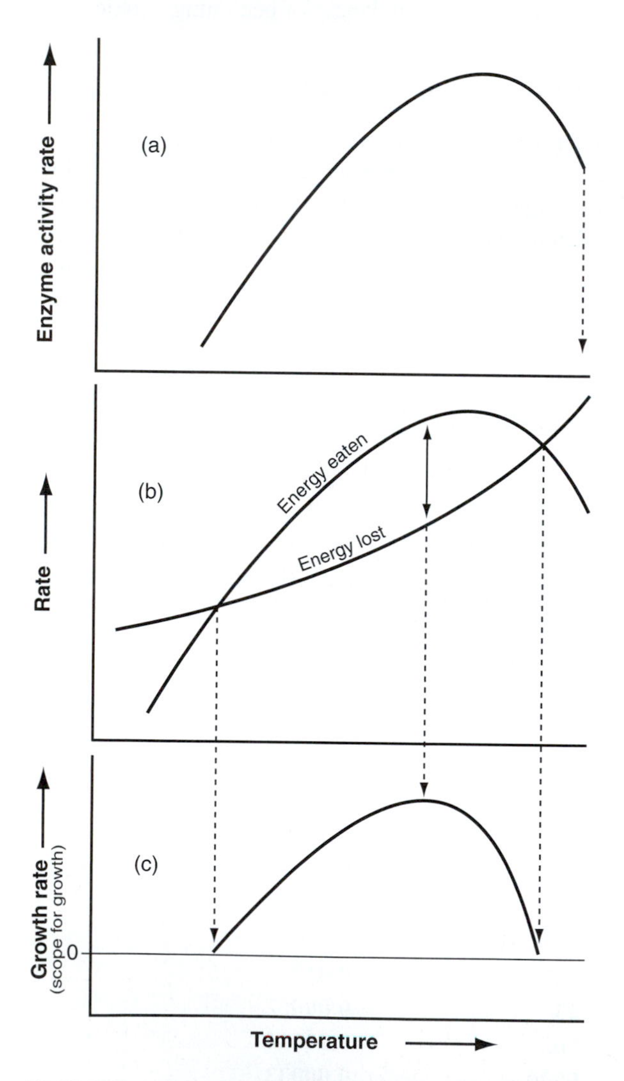

FIGURE 9.9 The effect of temperature on physiological rates. (a) A typical enzyme kinetics curve. (b) The relationship between temperature and the rates of feeding and respiration. The arrow between the two curves represents the difference between the two curves at that temperature. The arrow is at the temperature associated with the maximum difference and is the scope for growth at that temperature. Where the two curves cross, the scope for growth is zero. (c) The scope for growth curve. **Source:** All graphs redrawn from Kitchell, 1998.

A specific organism has an optimal temperature range, around its preferred temperature, in which it is active and can grow, reproduce, and maintain itself. At cooler temperatures, some aspect of acquiring or using energy will be too slow to support life. At warmer temperatures, the respiration energy requirement will be too high to allow the life cycle to be completed.

The suitable temperature range can be only a few degrees or, perhaps, 20 degrees. A narrow temperature range of tolerance is associated with highly specialized metabolic enzymes that function well only in the narrow temperature range. Organisms with a broad temperature range are probably not able to use energy as efficiently at any one temperature. Antarctic fishes are adapted to a very narrow range of temperatures, between about −2°C and 0°C. An increase of 10°C will kill these fishes. This extreme temperature specialization is made possible by environmental constancy. The temperature of the Antarctic Ocean, near the permanent ice shelves, varies annually by less than 1°C.

In temperate zone lakes and streams, freshwater organisms experience a habitat that varies from near freezing to about 30°C. For example, *Daphnia* have multiple generations in lakes. The summer generations live out their lives in a few weeks at the higher temperatures, while a winter generation might last several months at near-freezing temperatures. This is not to say that all *Daphnia* are equally well adapted to a wide range of temperatures. For example, the two common species of *Daphnia* in Wintergreen Lake, Michigan have different temperature requirements—one preferring cooler water in the spring and one prospering in the lake in warmer summer water (Threlkeld, 1986).

Scope for Growth and Temperature Preferences

Many physiological rates, such as feeding rates of fishes (figure 9.9b), resemble the enzyme kinetic curve because the physiological behavior is the result of many enzymes acting in concert.

Respiration is, in general, the rate at which metabolism proceeds, often measured as the rate at which an organism consumes oxygen. This rate increases exponentially with temperature (figure 9.9b) to a peak, then crashes. *Scope for growth* is the difference between the rate at which energy is acquired (consumed) and the rate at which energy is used in metabolism (figure 9.9b). Because of the shapes of the feeding and the respiration curves, the greatest difference between the two curves is always to the left of the optimal feeding temperature—the greatest growth rate is at a temperature lower than the temperature at which the most food is eaten!

The scope for growth is a **unimodal curve** (figure 9.9c). At some optimal temperature, growth will be greatest; growth rate will decline and even become negative at higher or lower temperatures. When measured in the laboratory, different species of fishes have different scope-for-growth optima (figure 9.10a), the result of their different genotypes. When these same fish species are caught in the wild, they are found to occur in parts of the lake or stream that correspond to their laboratory-determined maximum scope for growth (figure 9.10b).

Interaction of Temperature and Body Size

After surveying the literature on metabolic rate for all organisms, Gillooly et al. (2001) were able to make a remarkable generalization:

> *"Mass- and temperature-compensated resting metabolic rates of all organisms are similar. . ."*

In their model, "all organisms" included microbes, invertebrates, vertebrates, and plants, with temperature optima from 0°C to 40°C. They found that if body size and temperature were taken into account, there is only one metabolic rate. Dr. Gillooly is fond of saying that once temperature and body size are taken into account, the apple in your lunch and the fish in your aquarium have similar (resting) metabolic rates!

Metabolic Rate and Other Factors

Body size and temperature are two key factors that influence metabolic rate. Other factors include feeding rate, activity level, stress, and diapause.

The amount of food available to an organism strongly affects respiration and growth rates. Plants provided with an excess of CO_2 and optimal light show extraordinarily fast growth. Animals have much lower metabolic rates when they are starving than when they are well-fed. However, too much food is also not optimal. When food is abundant, assimilation becomes less efficient as food moves quickly through the gut. Also, there are indirect effects of abundant food. For example, Porter et al. (1982) reported that dense algae suspensions were adverse for *Daphnia*. The *Daphnia* spent so much energy cleaning their filters

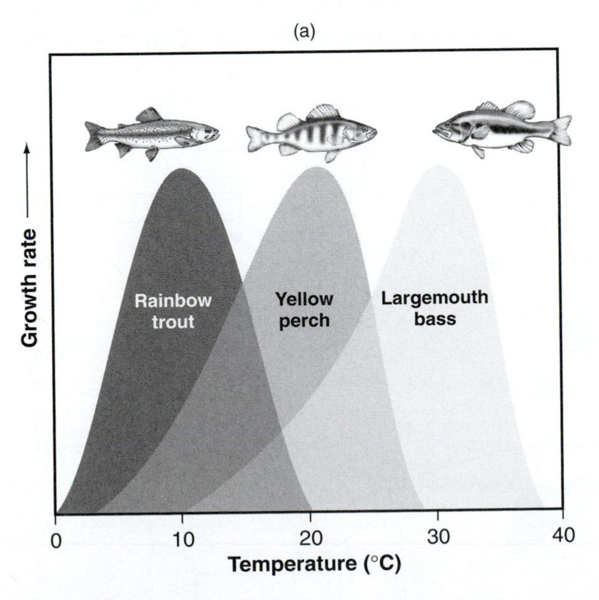

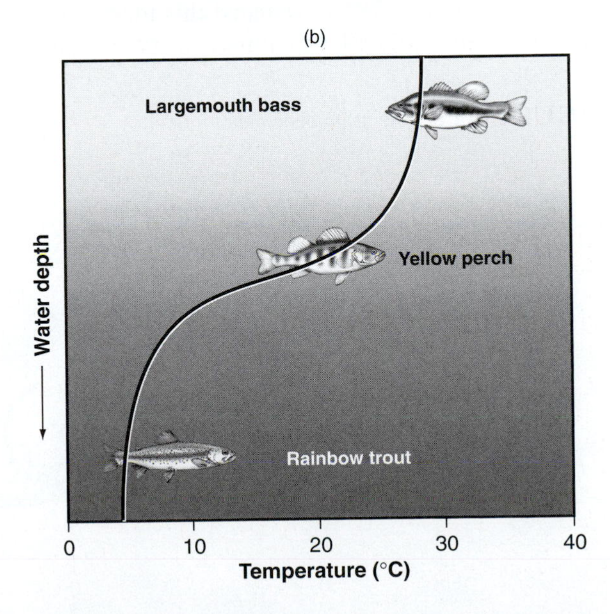

FIGURE 9.10 (a) Fish temperature preferences determined in the laboratory. (b) Preferred habitats of these same fish species in a summer-stratified lake. **Source:** From Kitchell, 1998.

that their respiration rate exceeded the assimilation rate—the animals were starving to death in the midst of abundant food.

An increase in activity level increases metabolic rate. The metabolic rates presented in table 9.4 are for resting, or quiescent, organisms. Movement, especially escape responses, can increase metabolic rate by a factor of 10 or more.

Stress is a physiological response to the environment, in which the animal's metabolic rate is elevated when it receives a signal. In aquatic systems, the signal is often chemical in nature and can be the "smell" of predators or competitors. The signal often induces a change in behavior, which increases activity level. This change in activity level requires energy that could otherwise be used for growth or reproduction.

Diapause is an induced metabolic rate of near zero. As described in chapters 3, 4, and 5, many organisms have a diapausing life stage. This diapausing stage (resting egg, seed) has the advantage of requiring little from the environment and being resistant to extreme environmental variation.

BALANCED BIOENERGETIC MODEL

Eastern European limnologists, especially in the former Soviet Union and Poland, provided leadership in the study of ecological energetics. The International Biological Project (**IBP**) expanded this theme in the late 1960s and early 1970s—limnologists were measuring energy contents, primary productivity, and respiration rates of all imaginable aquatic organisms. With the assistance of IBP staff, Jim Kitchell and his colleagues (1977) developed a **bioenergetic** model as part of an IBP study of fishes in Lake Wingra, Wisconsin. The model has developed and proven very useful in understanding the ecology of fishes and their prey (Hanson et al., 1997; Jobling, 1994). Although developed to explore fish ecology, the model has also been applied to other organisms such as zebra mussels and *Daphnia*.

The purpose of the model is to use physiological knowledge, at an individual or population level, to predict individual growth, reproduction, or feeding rates. The bioenergetic model is an energy budget based on energy reservoirs and transfers at the individual level (figure 9.11). In other words, it is like an ecosystem model at an individual level.

The beauty of this model is that energy is neither created nor destroyed in the organism—energy input is balanced by energy loss (a fish, in the case of figure 9.11). Energy flow in this model is measured as grams of carbon or calories per day. Energy that goes into the fish is either stored as growth or reproduction, or it leaves the fish as heat (respiration) or waste food. In the simplest form of the model, energy enters a fish as consumed food. Some of this food cannot be assimilated (taken into the bloodstream) and is released from the body as soluble and indigestible solid wastes. The waste material contains a portion of the energy that was consumed. The food that is assimilated contains energy that can be used in metabolism

FIGURE 9.11 The fish bioenergetics model.

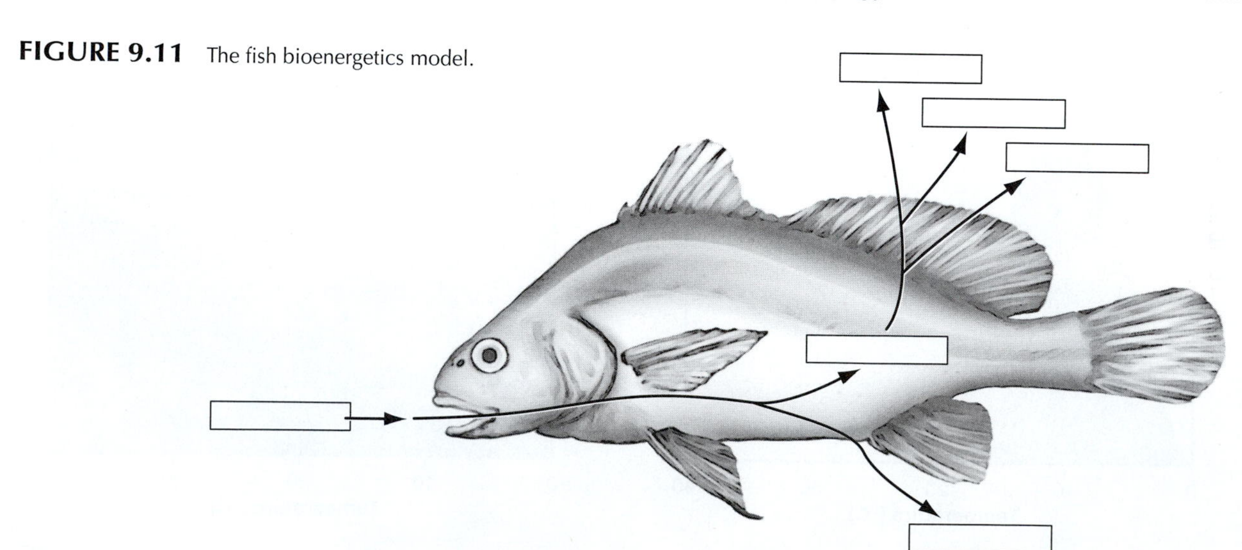

(that is, respiration), growth, and reproduction. In equation form, the model is:

(Equation 9.3)

where:

$$C = W + R + G + X$$

C = energy consumed

W = energy lost as waste

R = energy lost as respiration (the result of metabolism)

G = energy stored in chemical bonds as growth

X = energy allocated toward reproduction

The equation for the bioenergetic model is also sometimes written to include assimilation:

(Equation 9.4)

$$A = R + G + X = C - W$$

where:

A = assimilated energy—energy taken into the bloodstream

The four components of this model depend on four main variables: body size (and developmental stage), water temperature, the quality of food, and the quantity of food. The units of the four components are rates, either calories per day per organism, or the weight-specific rate: calories per day per gram of organism. The rates can also be expressed as grams of wet or dry weight, or grams of carbon in the numerator.

The model can be used to make predictions on a daily or annual basis. The actual equations (that predict the components from body size, temperature, and food parameters) are complex but based on standard physiological relationships. The model software is available (Hanson et al., 1997). The bioenergetics model was used to evaluate and predict fish population dynamics in Lake Michigan (Kitchell & Crowder, 1986). The model correctly forecast a decline in forage fishes (alewives) as increasing numbers of salmonids were stocked in the lake.

The model unit is either the average fish in a population or the members of a population. The simplest approach is to model the consumption or growth of a single average individual from hatching to its eventual death. More complex, individual-based models simulate the development of thousands of individuals, taking into account variation in the environment and in physiological relationships.

Because this model is a closed system, it is possible to predict any one of the components from values of the other three. The model has been used to predict growth rates and to test our understanding of what controls growth rates in a wide variety of organisms, including fishes, zooplankton, and mollusks. For example, the model has been used to predict the growth rate of salmon, lampreys, and zebra mussels in Lake Michigan. In the early 1980s, the model successfully predicted that Lake Michigan was being stocked at a level that could not be supported by the alewife population. As predicted, the salmon population has declined in the 1990s, partly because the alewife population, the forage species, has also declined (see table 8.1).

By simply rearranging the equation, the model has also been used to estimate feeding rates of fishes, based on observed growth rates. For example, the model indicated that the fishes of Lake Skadar, Montenegro (Balkans), were growing too fast, given the supply of zooplankton and benthic invertebrates. This suggested the presence of an additional food source, which was subsequently found to be filamentous algae.

SUMMARY

The material in this chapter is concerned with movement of energy through ecosystems and individuals. Quantitative and graphical models are important tools for the understanding of energy flow. In chapter 10, a similar approach will be used to understand the movement of chemicals from one ecosystem compartment to another.

Study Guide

Chapter 9 Aquatic Ecosystems and Physiology: Energy Flow

Questions

1. How does the microbial loop affect fish production? Do fish managers desire a microbial loop? If a fish manager has a microbial loop and wants to get rid of it, what would you advise as the best policy?
2. How is the concept of "optimal" related to the concept of "the greatest"?
3. How are enzyme kinetics related to distribution patterns of fishes in lakes?
4. Would a fish manager be able to use the bioenergetic model if energy (food) is super-abundant?
5. What is the difference between gross primary productivity and net primary productivity? How does ecological efficiency relate to this model? (Remember that NPP = GPP − respiration.)
6. Animals cannot perform photosynthesis. Do green plants perform both photosynthesis and respiration?

Words Related to Energy Flow and Physiology

^{14}C tracer	C3	GPP	photorespiration
allochthonous	C4	IBP	phycobilins
autochthonous	carotenoids	NPP	productivity
bioenergetic	compensation depth	optimal temperature	Q_{10}
BOD	ecosystem	photo-inhibition	unimodal curve

Major Examples and Species Names to Know

Lake Chapala, Mexico, and the reasons for low tropical NPP

The Lake Skadar fish growth example of the missing item in the diet

How the bioenergetics model and cesium balance models work, and the calculations associated with these models

What Was a Limnological Contribution of These People?

Ray Lindeman

G. Evelyn Hutchinson

Jim Kitchell

Exercises

Fish Bioenergetics

Fill in the blanks of this illustration of a fish representing the balanced bioenergetic model.

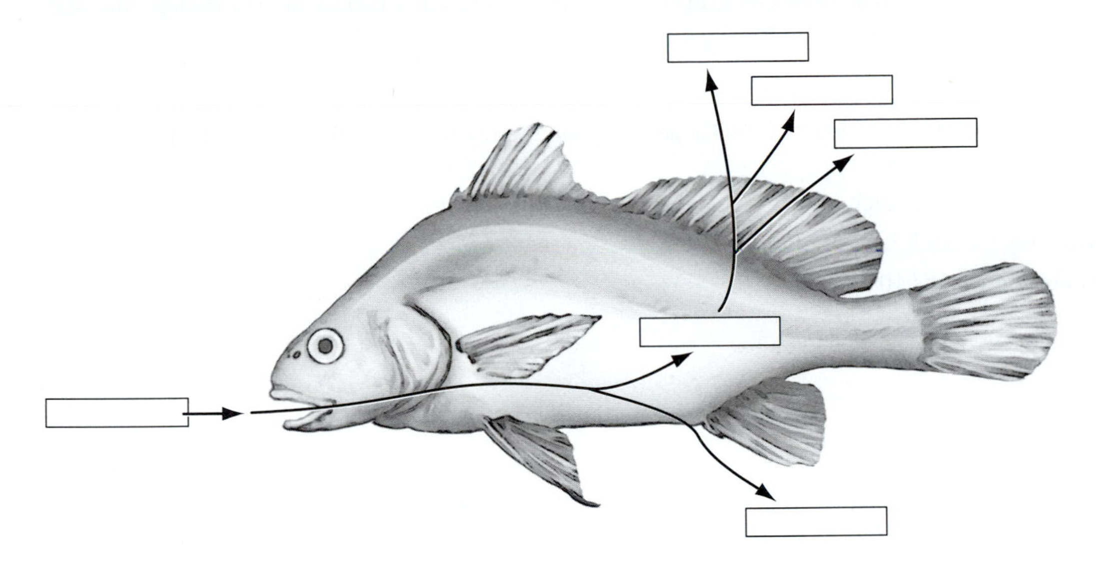

Check your answer with figure 9.11. Then, using the following information, answer the following questions:

From laboratory observations, you find that a certain *juvenile* fish has these characteristics:

- Feeding rate of 100 grams of carbon per day
- Half the food that is eaten (ingested) is excreted
- Respiration rate is half of assimilation rate

1. What are the units of the growth rate?
2. Do you have enough information to estimate the fish's growth rate, using a simple bioenergetics model? If not, what additional information do you need?
3. If you have enough information, what is the growth rate of the fish? Include the proper units!
4. In one sentence, give an example of an ecological trade-off in the context of the bioenergetic model.

Calculate Feeding Rate—Cesium Method

What is the feeding rate of a 10-kg yellowfin tuna? This exercise is an example of physiological modeling. Because the feeding rate of free-swimming tuna cannot be measured directly in the ocean, it is necessary to use a bioenergetics-type model to estimate the feeding rate. This model will use the input and loss rates of cesium, a cation that is relatively inactive in fish metabolism and can be easily measured in fish tissue. For background, see Olson and Boggs (1986).

What We Know:

1. Concentration of cesium [Cs] in food = 1 mg Cs per kilogram of prey fish. Note that square brackets are used to indicate "concentration."
2. Prey fishes weigh 100 grams each.
3. [Cs] in tuna is 2 mg Cs per kilogram of tuna (= 2 ppm)
4. The tuna in question weighs 10 kg.

The Model:

$$B = M \times [Cs] \times a$$

where:

$$M = \text{mass of the tuna in kg (units are kg/tuna)}$$

$$[Cs] = \text{mg per kg of Cs in the tuna}$$

$$a = \text{constant with units of day}^{-1}$$

The amount of Cs lost per tuna per day depends on the mass of the tuna and the concentration of Cs in the tuna. This is a standard physiological assumption.

The constant "$a = 0.01$" is determined by observing tuna swimming in a tank and measuring their feeding rate under controlled, but small-scale, conditions. The variable B is the answer to the question, "How much Cs does a 10-kg tuna lose per day?

units =

$B =$

How much Cs is contained in each 100-g prey fish? ___________ units = ________

How many prey fish does a 10-kg tuna need to eat per day just to balance the loss of Cs? __________

Additional Resources

Further Reading

Colborn, T., D. Dumanoski, and J. P. Myers. 1996. *Our stolen future.* New York: Dutton Press. 306 pp.

Hutchinson, G. E. 1957. *A treatise on limnology. Volume I: Geography, physics, and chemistry.* London: John Wiley & Sons. 1015 pp.

Peters, R. H. 1983. *The ecological implications of body size.* Cambridge, UK: Cambridge University Press.

Thompson, D. 1942. *On growth and form.* 2nd ed. Cambridge, UK: Cambridge University Press. Especially Volume I, Chapter 2: "On Magnitude."

Wetzel, R. G. 2001. *Limnology: Lake and river ecosystems.* 3rd ed. New York: Academic Press.

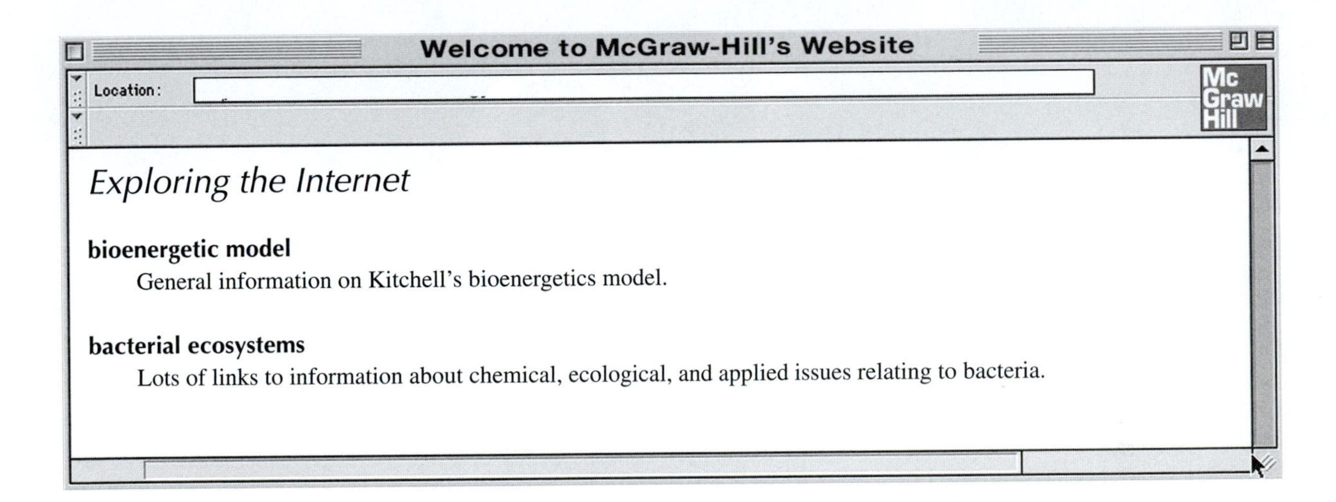

Exploring the Internet

bioenergetic model
General information on Kitchell's bioenergetics model.

bacterial ecosystems
Lots of links to information about chemical, ecological, and applied issues relating to bacteria.

10

“. . .we find no vestige of a beginning, no prospect of an end.”

James Hutton, 1788

Aquatic Ecosystems: Chemical Cycles

outline

CHEMICAL CYCLES IN WATER AND BIOGEOCHEMICAL CYCLES

Ecosystem models of chemicals, like the energy models, are made up of components (reservoirs) and rates of transfer. A major difference between energy systems and chemical systems is that energy is thought of as flowing through the system (from sun to organisms to heat), whereas chemicals cycle from one compartment to another and back again (especially in lakes). There is no significant flow of energy from the top carnivores back to the primary producers, but chemicals frequently make the return trip many times over.

Chemicals are quantified as mass (such as grams), mass concentrations (grams per liter), moles (grams molecular weight), molar concentrations (moles per liter), and molar equivalents for ions (moles per liter multiplied by the ionic charge). **Loading** refers to the annual total amount of a chemical added to a lake or passing a point in a stream. Loading is often expressed as the total annual input to a lake divided by the lake's surface area.

In ecosystem models, the chemical of interest (such as carbon, nitrogen, or phosphorus) can take on different forms (oxidized or reduced) in different reservoirs, but it typically moves cyclically among the various reservoirs. In lakes, chemicals cycle among various reservoirs and only slowly leave the lake. Stream systems display a more linear or spiraling pattern of movement of nutrients—chemicals are washed into the stream, caught by primary producers, and eventually released back into the stream by herbivores and predators (Newbold et al., 1982).

Long-term ecosystem studies in the Hubbard Brook, New Hampshire watershed have documented important interactions among vegetation, soil and bedrock, and water chemistry. Clear-cutting of a subwatershed greatly increased stream concentrations of many ions, compared to uncut subwatersheds (Bormann & Likens, 1970). Pictures and details of water chemistry are available at the website listed at the end of this chapter.

Hutchinson (1944) developed the concept of **biogeochemistry:** large- (even global-) scale chemical cycles that include living organisms and large-scale abiotic reservoirs. (For more details on biogeochemistry, see Schlesinger [1997].)

Sediments, particulate organic matter, and the pelagic zone are examples of ecosystem reservoirs of chemicals. Each reservoir can include living and nonliving members. For example, the reservoir "sediments" might include inorganic material (sand or clay); organic, nonliving material (partly decayed remains of organisms and feces); and organisms (including all the microbes, protists, plants, and animals) living in the sediments.

The transfer of nitrogen from the atmospheric reservoir (as nitrogen gas) to the soil reservoir (as nitrate) each time lightning occurs is an example of a nonliving process. An example of a living process is bacterial nitrogen fixation, in which atmospheric nitrogen is reduced to ammonia.

Chemical cycles include the watershed and ground water as well as the visible surface water in a stream or lake. This seems obvious, but was a little-studied part of limnology until the 1960s, when scientists started worrying about where algal nutrients in lakes and streams came from (Vallentyne, 1974; Vollenweider, 1976). This modern view of lakes and chemical cycles is the opposite of the "lake as a microcosm" concept proposed by Forbes (1887).

All chemical cycles are fascinating to the ecosystem ecologist, but we will give only five examples here. The *carbon cycle* is presented first because carbon compounds have a profound effect on water chemistry, are a component of photosynthesis and acid deposition, and can even be a dramatic threat to human health. The *oxygen cycle,* which is placed second, is intimately related to the carbon cycle via photosynthesis. Oxygen production and use, at both the individual and whole-lake scales, is probably the most acutely important process in aquatic habitats. The chapter continues with a discussion of three important limiting nutrients. The *phosphate cycle* is important, because phosphate is most often the chemical nutrient that limits plant and algal growth and, when in excess, like nitrogen compounds, can pollute water. The relatively complex *nitrogen cycle* follows. Nitrogen occurs in several different forms and plays a major role in acid deposition, organismal metabolism, and cultural eutrophication. When phosphate is not the limiting nutrient for algal growth, the next most likely candidate is nitrate. The *silicon cycle* is relatively minor in terms of the overall water chemistry of aquatic habitats. However, silicon is included here because its abundance can limit the growth of one major group of algae, the diatoms.

THE CARBON CYCLE IN WATER

Forms of Carbon

Carbon exists in a number of chemical forms in water (figure 10.1), including:

Carbon dioxide (CO_2) A small component of the Earth's atmosphere. It readily dissolves in water. In aquatic systems, CO_2 is consumed by photosynthesis and produced by metabolism (respiration).

Dissolved inorganic carbon (**DIC**): Includes carbon dioxide, carbonic acid, bicarbonate, and carbonate. Bicarbonate ion (HCO_3^-) is formed when CO_2 dissolves in water to form carbonic acid (H_2CO_3), which **dissociates** (breaks down) into bicarbonate and hydrogen ions. At low pH, most of the DIC is in the form of carbon dioxide or bicarbonate (figure 10.2). At high pH, bicarbonate breaks down to form a carbonate ion (CO_3^{-2}) and a hydrogen ion. Carbonate is poorly soluble and tends to precipitate readily.

Dissolved organic carbon (**DOC**): Methane, the simplest form of organic carbon (CH_4) is produced by anaerobic decomposition. This gas is dissolved in water and forms bubbles when decomposition produces a supersaturated gas solution. Under conditions of low temperature just above 0°C and high pressure, methane combines with water to form a semi-solid **hydrate.** Methane forms this hydrate in permafrost or at great depths in ocean sediments or deep in sediments of Lake Baikal, Siberia (Kuzmin et al., 2000).

Dissolved organic material is mostly in the form of stable organic acids (**humic acids**) that have resisted bacterial decomposition (Hutchinson, 1957). DOC can be produced in the watershed (especially in boreal evergreen forests), in peaty margins of bogs, or from deciduous leaf litter (Meyer et al., 1998). In water with low rates of bacterial decomposition (anoxic swamps or cold oligotrophic bogs), DOC can stain water, producing a tea-colored, acidic solution. DOC absorbs a broad range of wavelengths of sunlight, so DOC reduces photosynthesis by absorbing light energy (PAR), and at the same time protects organisms from harmful UV radiation (see chapter 4). Klug and Cottingham (2001) found that DOC had important synergistic interactions with other nutrients in determining algal community structure, as well as aspects of lake stratification and water chemistry.

Particulate organic carbon (**POC**): All living and dead organisms are based on organic compounds. Thus, any living or dead organism, feces, or discarded parts of organisms (such as dead leaves) contribute to POC.

Carbon Reservoirs

Like all chemicals, carbon is stored in several major carbon cycle reservoirs in the environment and in living organisms (figure 10.1). These reservoirs include:

Rocks: Carbon is a component of many different minerals. It is often present as a carbonate, especially in **limestone** (calcium carbonate) and **dolomite** (calcium and magnesium carbonate), but also in some igneous rocks.

Atmosphere: Carbon dioxide makes up about 0.035% of the atmosphere at the present time. As the human population burns fossil fuels and forests, the concentration is steadily increasing. A website address is given at the end of this chapter for the prime CO_2 monitoring station, on the island of Hawaii.

Before the evolution of organisms capable of performing photosynthesis (before about 3.8 billion years ago), the Earth's atmosphere was perhaps 20% carbon dioxide, with no oxygen. Photosynthesis transformed this primeval atmosphere—the patient photosynthesis of cyanobacteria over millions of years produced an atmosphere similar to the current one, which has very little CO_2 and about 20% oxygen.

Most scientists agree that as atmospheric CO_2 increases in concentration, the average world temperature also increases due to a greenhouse effect because CO_2 is very efficient at absorbing heat (Crowley, 2000; Kerr, 2001; Vitousek et al., 1997). More information on this issue is available at the NASA website address given at the end of this chapter.

Oceans: When carbon dioxide dissolves in marine or fresh water, some portion of the CO_2 undergoes a chemical reaction with water to produce carbonic acid and bicarbonate, which can further dissociate into carbonate ions. After chloride and sulfate, inorganic carbon compounds are the most abundant anions in the oceans. Because the interaction of inorganic carbon and water involves hydrogen ions, inorganic carbon can act as a buffering system that tends to stabilize the acidity of water (the carbonate buffer is discussed next in the CO_2, Bicarbonate, Carbonate Equilibrium section of this chapter).

Seawater contains about 26 milligrams of inorganic carbon per liter. This small amount per

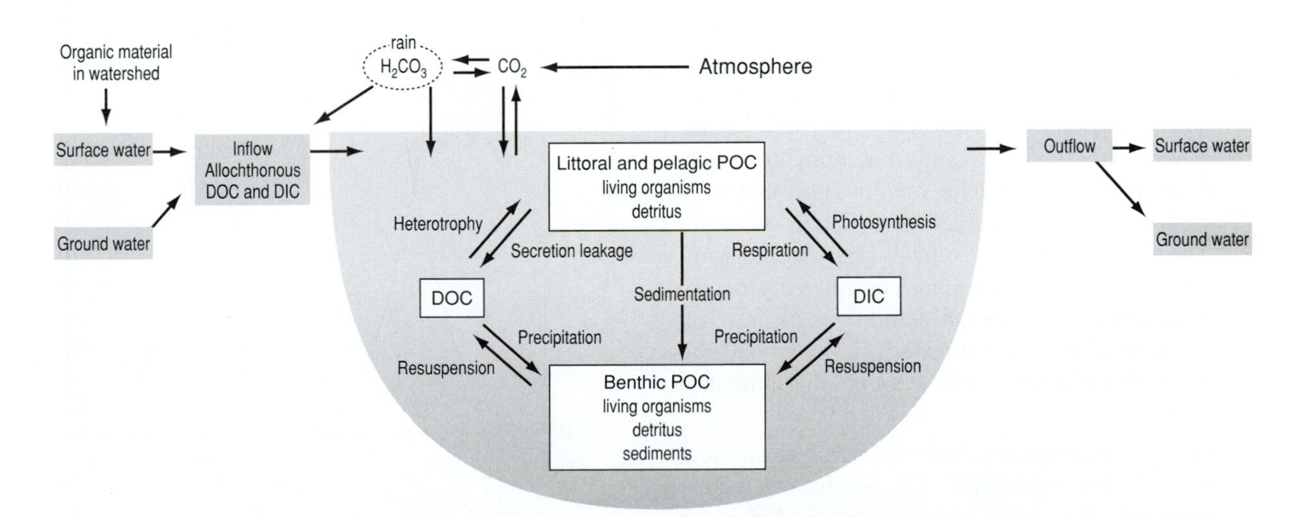

FIGURE 10.1 The carbon cycle for Lawrence Lake, Michigan. POC = particulate organic carbon; DIC = dissolved inorganic carbon; DOC = dissolved organic carbon. The boxes represent various carbon reservoirs in lakes.

liter implies a huge reservoir of inorganic carbon for the world-ocean. The inorganic carbon is approximately in equilibrium with the atmosphere, and the equilibrium depends on temperature. Thus, there is concern that global warming (partly due to increased atmospheric CO_2) will accelerate as the warming climate causes CO_2 to move from the oceans into the atmosphere (Fedorov & Philander, 2000).

Fresh water: Carbonates wash from the soils through streams, ground water, and lakes toward the oceans. Bicarbonate is the major anion in fresh waters (except in salt lakes). As in the oceans, the interaction of inorganic carbon with water supports the major buffering system in fresh water.

Detritus: This is the reservoir of disintegrated organic debris. In limnology, **detritus** refers to organic material in the form of fine particles mixed with bacteria. Detritus is often formed as the accumulation of bacteria, decaying plant material, and feces—it accumulates on the bottom of lakes and pools in streams. Green-to-black sediments rich in feces are termed **gyttja;** brown organic sediments rich in decayed plant materials are called *peat* (with plant parts still recognizable) or **dy** (if the plant material is broken down to humic acid colloids). A **colloid** is a homogeneous or slightly granular gelatinous substance, often in microscopic particles. In lake water, colloids are made up of organic material such as mucus or dy or of clay or iron (ferric) hydroxide.

Primary producers: Carbon (mostly as organic compounds) is a major constituent of living organisms, about 40 to 50% of dry mass. Photosynthetic algae and macrophytes absorb dissolved carbon dioxide and, in many cases, bicarbonate as the raw material for photosynthesis. The process of photosynthesis converts inorganic carbon to organic carbon. The intensity of primary productivity in lakes and streams, in terms of grams of dry weight per square meter, is similar to that of terrestrial systems in the same general area.

Consumers: In terms of biomass (weight per unit area or volume), organisms high on the food chain make up only a small carbon reservoir. This is because of the ecological inefficiencies discussed in chapter 8.

The CO_2, Bicarbonate, Carbonate Equilibrium

When carbon dioxide dissolves in water, it can participate in a number of reactions, depending on the hydrogen ion concentration (acidity—see figure 10.2). If hydrogen ions are abundant (as in acidic water), some of the carbon dioxide will react with water to form carbonic acid. If, however, hydrogen ions are scarce (as in alkaline water), carbon dioxide dissolving in water combines with water to form bicarbonate ion and a hydrogen ion. The carbonic acid dissociates (breaks up) readily and is difficult to measure in alkaline water. If hydrogen ions are particularly scarce, the bicarbonate ion will dissociate into another hydrogen ion and a carbonate ion. All of these chemical species are in **equilibrium**—rates of dissociation are equal to rates of recombination for all chemicals in the system. In a state of equilibrium, the inorganic carbon reactions are not one-way, but are reversible. Concentrations of the various forms depend on temperature and the various compounds and ions that participate in the

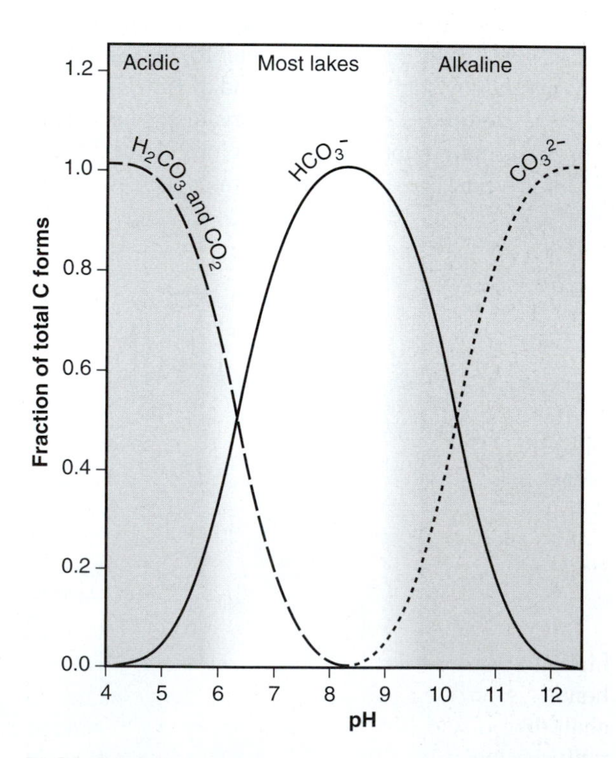

FIGURE 10.2 Dissolved organic carbon and pH in water—the relative proportions of inorganic carbon forms at different pH values.

reactions. If the concentration of any one of the chemical forms is changed, reactions will occur that minimize the change. For example, as plants remove carbon dioxide from water, bicarbonate combines with hydrogen ion to form carbonic acid, which dissociates into carbon dioxide and water (the limnological implications of this particular reaction are discussed next). In chemical shorthand, the interactions between inorganic carbon and water can be represented as:

$$CO_2 + H_20 \rightleftharpoons H_2CO_3 \rightleftharpoons H^+ + HCO_3^{-1} \rightleftharpoons 2H^+ + CO_3^{-2}$$

Limnologists use this equation as a model to understand and predict aspects of water chemistry. For example, pure water in contact with the atmosphere becomes a weak solution of carbonic acid as it takes up CO_2 from the air. Most of the carbonic acid breaks down into bicarbonate and H^+. Because the H^+ concentration rises, the solution becomes acidic (the pH drops). Because H^+ ions are present in excess, the bicarbonate tends to recombine with H^+ to make carbonic acid, rather than breaking down further to form more H^+ and carbonate. Thus, for pure water in contact with air, there is little or no carbonate present.

Pure water in contact with carbonate (such as limestone rock) will dissolve some of the carbonate. Some of the carbonate combines with H^+ in the water to form bicarbonate. The solution is then alkaline because the bicarbonate removed H^+ from the water. Because H^+ ions are in short supply, the bicarbonate will tend to dissociate back into H^+ and carbonate.

pH, Inorganic Carbon Equilibrium, and Photosynthesis

As photosynthesis removes CO_2 from water, the carbon equilibrium tends to replace the CO_2 and, in doing so, removes H^+ from the water. Thus, photosynthesis, by removing DIC, increases the pH. At night, when the plants (and animals and bacteria) are respiring, the reverse happens: There is a diel pH cycle associated with the diel photosynthesis cycle.

When photosynthesis is intense (high temperature, high nutrients, dense algae), the pH is so high that carbonate is no longer soluble, and it precipitates as a chalk-like material called **marl.** This precipitation, or **whiting** (figure 10.3) of carbonate, is assisted by warming of water because calcium carbonate is less soluble in warm water than in cold water.

Most algae (protists) can use only CO_2 as a carbon source for photosynthesis. As photosynthesis proceeds and more and more CO_2 is removed from the water, the water becomes more alkaline and the inorganic carbon is mostly in the form of bicarbonate and carbonate. However, cyanobacteria can use both CO_2 and bicarbonate as carbon sources. Thus, cyanobacteria have an advantage of being able to continue photosynthesis as the pH rises during the day in eutrophic water.

Acid Deposition

Acidic compounds come into watersheds in a number of forms: acidic dust, snow, fog, and rain. The main sources of acids in liquid water acid deposition (rainwater) are:

Carbonic acid (H_2CO_3): From CO_2 dissolved in rainwater

Nitric acid (HNO_3): Principally from oxidation of nitrogen in internal combustion engines, but also from natural sources that produce nitrogen oxides, such as lightning

Sulfuric acid (H_2SO_4): Principally from sulfur oxides produced when coal is burned; volcano fumes are a natural source of sulfur oxides

Rain in the northeast United States ranges from pH 3.5 (summer) to 4.2 (winter). This is extremely acidic—similar to bog lake water. The preindustrial pH of rain was about 5.6, which is the pH produced by atmospheric CO_2 dissolving in rainwater. The most acidic rain falling in the United States has a pH of about 4.3 (see the U.S. Environmental Protection Agency web page cited at the end of this chapter).

Buffers

Chemicals that reduce the effects of acids can have significant effects on water chemistry (review chapter 2). For example, as acid (hydrogen ion) is added to a carbonate solution, some of the hydrogen ions combine with carbonate, so the concentration of hydrogen ion changes less than would be predicted on the basis of the amount added to the solution. The major **buffers** in lake water are the inorganic carbon buffer and the organic acid buffer (found in bogs).

Titration

Titration is the process of adding acid until a measurable change occurs or until the pH is reduced to a certain value. Human activity has long been adding mineral

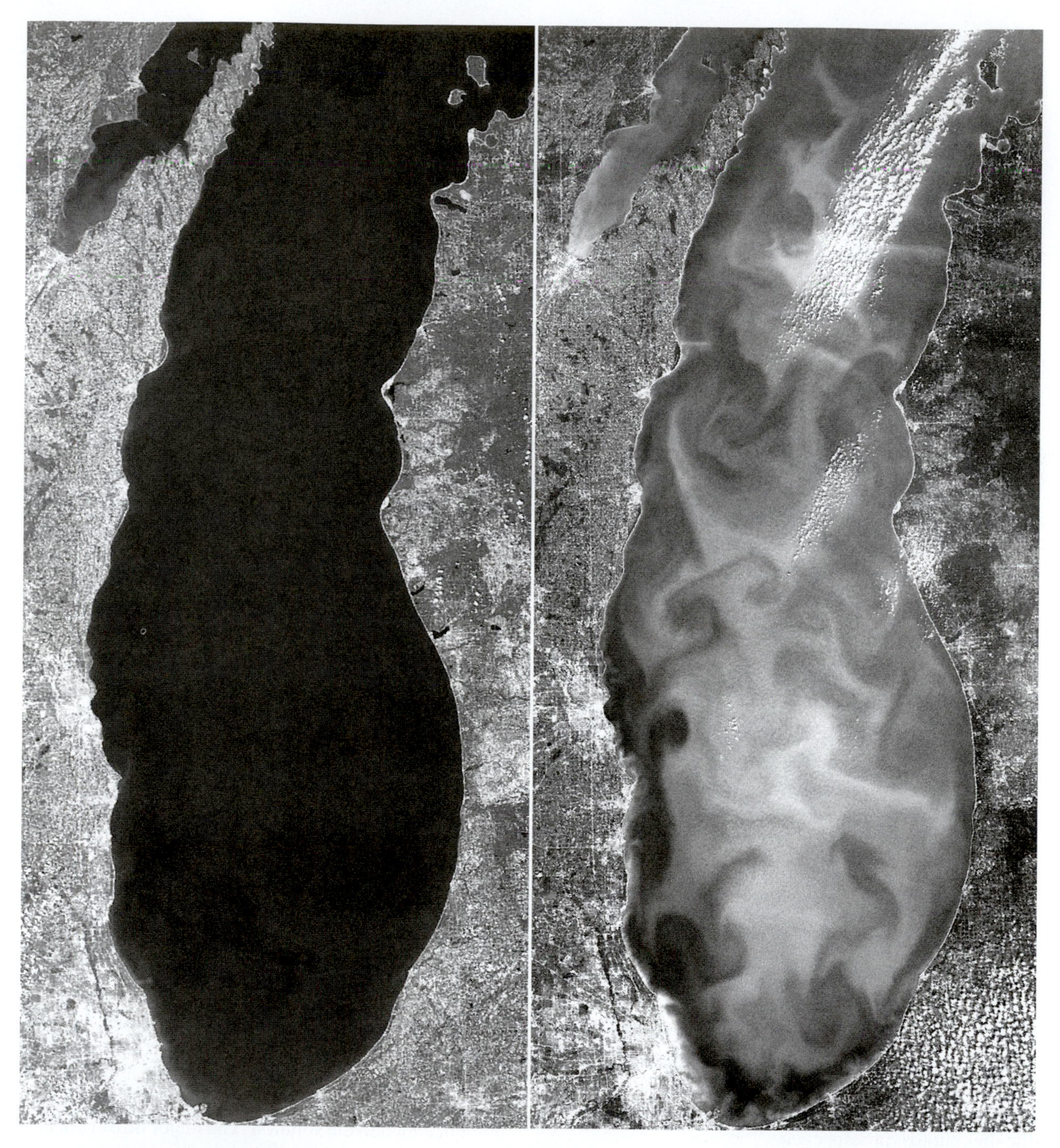

FIGURE 10.3 Lake Michigan before and after "whiting" caused by precipitation of carbonates. On July 13 2001(MODIS image 20010713) Lake Michigan appears dark as it absorbs radiation, but on September 2, 2001 (MODIS image 20010902) during a "whiting event" the masses of fine carbonate particles reflect a significant amount of light. Both images were acquired by Terra/MODIS (Terra is one of the satellites in NASA's "Earth Observing System" series; MODIS is one of the five sensors on Terra). The data shown in these images are spatially and radiometrically enhanced versions of MODIS band 4 (wavelengths 545-565 nm; in the "green" part of the visible spectrum). For questions relating to these images, please contact Jonathan Chipman (jchipman@wisc.edu; 608-263-3266) at the University of Wisconsin Environmental Remote Sensing Center.

acids (nitric and sulfuric) to aquatic systems. Major sources include mining of sulfide ores, which produces sulfuric acid, and burning of fossil fuels to produce both sulfuric and nitric acid in the atmosphere. Acids associated with mining typically affect only the watershed that contains the mine, while atmospheric acids can travel long distances via wind (Ohio to Ontario, Czech Republic to Norway) to affect water chemistry (Baker et al., 1991; Hedin et al., 1994).

The acids in rain **titrate** lakes—acid can be added for a period of time before any significant change occurs in pH. When a lake's buffering capacity is exhausted, pH can drop rapidly. The process of titration can make lakes too acidic for most aquatic organisms, but it can be reversed by adding carbonate (ground limestone or chalk) to lakes. This is effective but expensive, requires roads (or large planes) to transport the carbonate, and causes precipitation of algae and a decline in zooplankton for a few weeks.

The Dark Side of CO_2

Carbonated water (soda) is used as an anesthetic for zooplankton—it immobilizes the animals. The anesthetic properties of CO_2 have occasionally resulted in large-scale limnological and terrestrial tragedy.

The anesthetic nature of carbon dioxide is responsible for at least a couple of tragedies involving "killer lakes" in the African country of Cameroon (Sigurdsson, 1987). This small country in the bight of western Africa lies across a range of active volcanoes that produce, among other things, large amounts of CO_2. On 5 August 1984, 34 people were killed by mysterious causes near Lake Nyos, Cameroon, a crater lake whose eastern part contains a deep hole, about 1200 feet wide and 300 feet deep. The lake's deep waters presently contain large concentrations of CO_2. On 21 August 1986, an explosive fountain of water was observed in Lake Nyos. CO_2 released by this event killed 1746 people. The CO_2 might have been released directly into the crater as one large bubble, but was probably the result of a positive feedback process in which a landslide or earth movement stirred some of the deeper water upward. It is supposed that this water was saturated with CO_2 held in solution at the high pressure of the deep water. When the deep water came toward the surface, the CO_2 was released as gas bubbles, which created a rising current, which brought more CO_2-saturated water toward the surface until an eruption occurred that removed about 75% of the CO_2 from the water, or about 2 million tons of CO_2, or 35 billion cubic feet at 1 atm. The lake dropped 3 feet, probably due to loss of CO_2, not water. Calculations suggest that enough CO_2 was released to cover the affected 24-square-mile area 50 feet deep: 40% CO_2 produces instant death; 10% produces coma and death in 10 to 15 minutes. CO_2 is 1.5 times heavier than air, so it formed a ground-hugging cloud that flowed down a couple of stream valleys from the lake, killing most animals (but not plants!) for several miles. Thus, while carbon dioxide is an indispensable component of fresh water, it does have its dark side. (Look for pictures of Lake Nyos on the Internet.)

THE OXYGEN CYCLE IN WATER

Forms of Oxygen

Like carbon, oxygen occurs in many forms in aquatic systems, including:

Oxygen gas (O_2): The principal form of significance for organisms in the environment is the diatomic gas. Oxygen is a powerful oxidizing agent (electron acceptor). When other chemicals, such as carbohydrates, combine with oxygen, a great deal of energy is released. This energy can be used in metabolism or, if the reaction is intense enough, it produces fire. Oxygen tends to damage living tissue, resulting in a need for repair or regeneration of tissue and eventual aging and death. Powerful oxidation can also be accomplished by partially combined (reduced) oxygen. Many bacteria use nitrate (NO_3^{-1}) or sulfate (SO_4^{-2}) as oxidizing agents.

Ozone (O_3): An atmospheric gas that is indirectly of great importance to aquatic systems. Ozone is formed in the upper atmosphere when ultraviolet light interacts with oxygen, and it breaks down in chemical reactions with other atmospheric components. Ozone is not found in even trace amounts in water because ozone is so reactive. However, in the upper atmosphere, ozone does absorb high-energy ultraviolet (UV) radiation coming from the sun. UV is active chemically—it readily destroys complex organic molecules such as nucleic acids. UV radiation is dangerous for organisms such as plankton and fish fry that lack photoprotective screening pigments (discussed in chapter 4), and this radiation can penetrate clear water for several meters. Williamson et al. (2001) found that the interaction between dark-colored DOC and ultraviolet radiation influenced development of plankton communities in new lakes produced as an Alaskan glacier melted. Lakes with clear water had fewer zooplankton species than similar lakes stained with photoprotective DOC.

Inorganic: Oxygen is a common constituent of important inorganic molecules in aquatic systems. Examples are water, nitrate, phosphate, silicate, and oxides of metals, such as iron oxide.

Organic: Oxygen is an important component of organic molecules. While carbon forms the backbone of these molecules, oxygen (and hydrogen) flesh out most of the structure of the typical organic molecule. Except for hydrocarbons, essentially all organic molecules contain at least one oxygen atom, in, for example, alcohols, lipids, sugars, starches, and organic acids (including amino acids, fatty acids, and nucleic acids).

Oxygen Reservoirs

The major reservoirs of oxygen include:

Rocks: Most of the oxygen on (and in) the Earth is combined with silicon and aluminum to form aluminosilicate minerals such as clay, feldspar and basalt. Because it is so tightly bound to other atoms in rock, this oxygen is essentially unavailable for living organisms.

Atmosphere: The diatomic gas (O_2) makes up about 21% of our atmosphere. Oxygen gas is soluble in water, with highest solubility (about 14 milligrams per liter), or 0.0012% at 0°C.

Water: The oxygen cycle is intimately linked to water, the carbon cycle, and to biological activity, such as photosynthesis and respiration. Chapter 2 includes a discussion of the sources of oxygen in water (mixing and photosynthesis) and the vertical profiles of oxygen in stratified water.

Oxygen is a waste product of photosynthesis. Before photosynthesis evolved, the Earth's atmosphere lacked oxygen. Cyanobacteria began performing photosynthesis, producing oxygen from water and carbon dioxide, about 2.5 billion years ago.

Most animals require at least a couple of milligrams of oxygen per liter to support respiration (see chapter 4). However, many surface waters are deficient in oxygen, due to bacterial decomposition or organic compounds. Oxygen concentrations below about 2 milligrams per liter are stressful or lethal, but many organisms have morphological, physiological, and behavioral adaptations to cope with low oxygen concentrations.

Metabolic Adaptations to Low Oxygen Concentrations

Some organisms do not require any oxygen for their metabolic functions. Many bacteria and some protists can live in anoxic environments where there is little or no free oxygen. These organisms use an anaerobic metabolism that substitutes chemicals such as nitrate or sulfate for oxygen. These oxidizers are less efficient than oxygen, but work well enough in anoxic environments.

Most multicellular organisms can use anaerobic metabolism for short periods of time but eventually need a supply of free oxygen. For example, vertebrates can use carbohydrates as electron acceptors, producing lactic acid. Anaerobic respiration produces very little energy compared to aerobic respiration in the presence of oxygen.

Breathing in Water

Organisms living in water often have a problem getting enough oxygen. Compared to air, water has a low concentration of oxygen. One liter of air weighs about 1.29 grams and is 21% oxygen, so there is about 0.25 gram of oxygen in 1 liter of air. One liter of water weighs 1000 grams and contains about 0.01 gram of oxygen. Thus, there is *25 times* as much oxygen in 1 liter of air, compared to 1 liter of water. This means that oxygen is much less available to aquatic organisms, compared to terrestrial organisms.

In addition to the problem of low oxygen concentration, there is the problem of the density of water. Water has a much greater mass (1000 grams per liter) compared to air (1.29 grams per liter at 0°C, 1.16 grams per liter at 30°C). This means that breathing (moving) water in and out of lungs would be about 800 times as difficult (energetically costly) as breathing air. To extract enough oxygen from water to support life, animals, especially large animals, use energy to pump water over large surface areas called **gills.** Note that aquatic animals do not use lung-like organs. This is probably because lungs work by drawing air into the body. Thus, lungs work best when handling a fluid like air with low mass (and therefore low inertia) and low viscosity. Compared to air, water has significantly higher mass and viscosity. It appears that it simply takes too much energy to use lungs with water.

Circulatory Systems

Organisms larger than a couple of millimeters cannot depend on diffusion to supply oxygen to their tissues (Alexander, 1979). Oxygen is moved by a circulatory system (a fluid-filled system of tubes through which the fluid moves). We are familiar with circulatory systems in animals, but plants also use circulatory systems to supply oxygen to tissues. Plants have difficulty keeping enough oxygen in roots growing in lake sediments because the sediments are often anoxic. To meet this environmental challenge, aquatic plants have shallow roots (or do not put roots into the sediments), and some plants have a tube or spongy tissue system for pumping oxygen to roots from the leaves and stems.

Behavioral Adaptations to Low Oxygen Concentrations

A number of aquatic insects, such as diving beetles, back swimmers, and water boatmen, use air bubbles as oxygen sources. The air bubble sticks to small hairs on the surface of the insect and provides a larger surface area for getting rid of CO_2 and a reservoir for oxygen. The insect periodically swims to the surface to renew the oxygen in its air bubble. Mud minnows (Magnuson et al., 1983) survive under the ice in anoxic "winter-kill" lakes by sucking on air bubbles trapped under the ice. (Lakes are said to **winter-kill** when they run out of oxygen in the winter, killing most of the fishes.)

Hemoglobin

This iron-containing pigment is found in many animal groups, from arthropods to vertebrates. The chemical is often present at lower concentrations when oxygen is sufficient, especially for invertebrates, and higher concentrations are induced by low oxygen stress. Increases in this chemical are used as an adaptation to low oxygen concentration, allowing animals to live in water with only 2 to 4 milligrams per liter. Hemoglobin combines with oxygen at high pH and releases the oxygen at low pH. Respiration in animals reduces the pH (the carbon dioxide produced by respiration combines with water to produce bicarbonate and hydrogen ions). Thus, the pigment can combine with oxygen on the outer surface of the animal and then be transported into the body (by a circulatory system) where the respired carbon dioxide reduces the pH and causes the oxygen to be released.

Some plankton and some benthic organisms develop hemoglobin **facultatively** (when they need it). For example, *Daphnia* can produce or lose a pink color. The color (due to increased hemoglobin content) of *Daphnia* is an example of a **homeostatic** physiological mechanism. (A homeostatic mechanism is like the thermostat in your house that keeps the house at a constant temperature.) If the *Daphnia* is not getting enough oxygen, it manufactures more hemoglobin and becomes redder. If the *Daphnia* is getting sufficient oxygen, it stops production of hemoglobin and becomes clearer. The energy that would have been used to make hemoglobin can be diverted to growth or reproduction.

THE PHOSPHORUS CYCLE IN LAKES

Forms of Phosphorus

Phosphorus is most often the limiting nutrient for plant growth in fresh water. It is present in several major forms in lakes (figure 10.4):

Total phosphorus (TP): All the phosphorus in a sample of water. The total phosphorus in a water sample is determined by converting all the phosphorus (whether soluble or from organisms and detritus) in an unfiltered sample into inorganic orthophosphate. The following categories are components (not necessarily independent components) of TP. In many cases, total phosphorus is the preferred indicator of a lake's nutrient status, because TP levels remain more stable than other forms over the annual cycle.

Typical concentrations of TP in lakes are 10 to 80 μg per liter, depending mainly on the land-use characteristics of the watershed (Wetzel, 2001). Very oligotrophic water might have only 1 μg per liter, and hypereutrophic water (e.g., water associated with sewage treatment plants or intense agriculture) might have as much as 200 μg per liter TP.

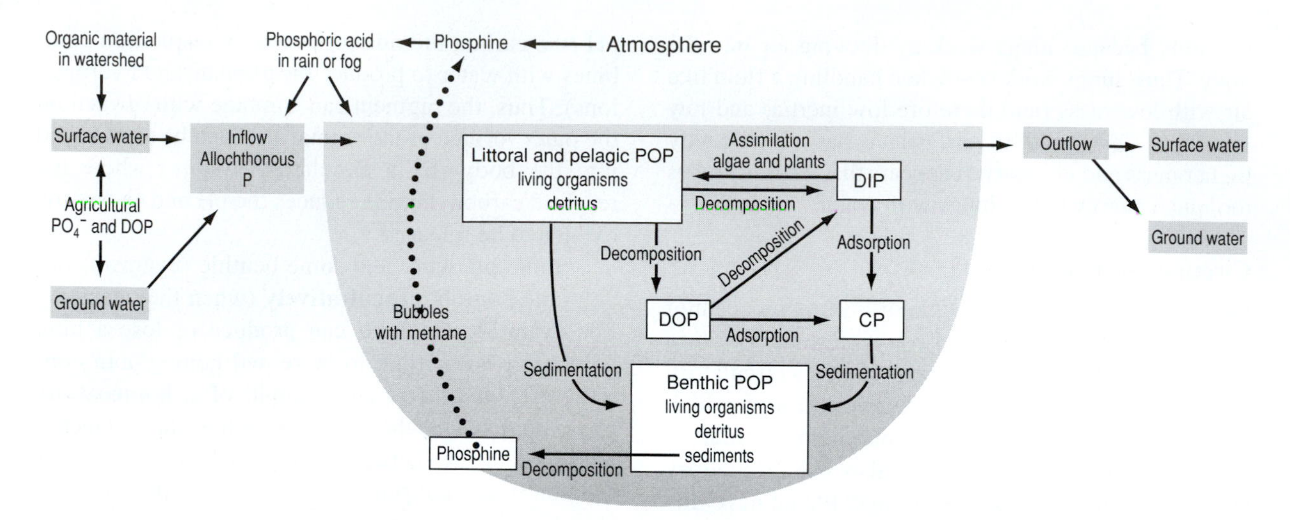

FIGURE 10.4 Major reservoirs of phosphorus in a stratified lake. TDP = total dissolved phosphorus; SRP = soluble reactive phosphorus; POP = particulate organic phosphorus; DIP = dissolved inorganic phosphorus; CP = colloidal phosphorus; DOP = dissolved organic phosphorus.

Soluble reactive phosphorus (SRP): Includes forms of phosphorus (both inorganic and organic) that are dissolved in water and are readily available for uptake by algae and macrophytes. This reservoir, which is in the ballpark of 10% of TP, is variable on an annual and even daily scale, depending on location in the lake, time of year, and intensity of primary production. SRP is highest during mixing events and in the spring epilimnion before algal growth takes off, and lowest in the summer epilimnion when algae are abundant.

Total soluble phosphorus (TSP): The phosphorus (whether organic or inorganic) that passes through a filter. TSP includes both SRP and any soluble, unreactive phosphorus. For example, fish farm effluent contributes phosphorus to the River Bush in northeast Ireland (Foy & Rosell, 1991). The total phosphorus loading was reported as 60% SRP, 10% unreactive phosphorus, and 30% phosphorus bound to particles.

Orthophosphate: The inorganic form of phosphate, ionic PO_4^{3-}, is a major component of SRP. Dissolved orthophosphate is also called dissolved inorganic phosphorus (DIP). The orthophosphate group can be bonded to a carbon atom (in organic chemicals) or to another phosphate ion, as in ATP (adenosine triphosphate).

The phosphate ion, like the carbonate ion, forms a *buffering system* in water. The simple phosphate ion, monohydrogen phosphate ion, and the dihydrogen phosphate ions are in equilibrium with hydrogen ions in water. If hydrogen ions are, for example, added by acid rain, the equilibrium shifts to absorb some of the excess hydrogen ions: Some phosphate ions (PO_4^{3-}) combine with hydrogen ions to become monohydrogen phosphate ions (HPO_4^{2-}), and monohydrogen phosphate ions take up hydrogen ions to become the dihydrogen phosphate ions ($H_2PO_4^-$). Because of the very low concentration of DIP, compared to bicarbonate, phosphate is rarely as significant a buffer as bicarbonate.

Dissolved organic phosphorus (DOP): Produced by living or decomposing organisms. Many breakdown products of living organisms are soluble in water, including creatine phosphate (a waste product for some animals), some detergents and pesticides, and nucleotides (such as nucleic acids and adenosine triphosphate).

Particulate organic phosphorus (POP): Includes phosphorus in organic chemicals in living or dead organisms, dead leaves, and feces.

Particulate inorganic phosphorus (PIP): Orthophosphate attached (adsorbed) to particles. Orthophosphate is nearly insoluble, especially in alkaline water and in the presence of the common divalent cations, calcium and magnesium. Also, the orthophosphate ion has a strong electrostatic attraction to other ions and polar chemicals, such as clay particles or the delicate flakes of lake snow (ferric hydroxide colloids).

Phosphine gas: The poisonous phosphine gas (PH_3) can be produced by anaerobic bacterial decomposition of organic matter (in the absence of oxygen). Phosphine has been detected in sewage treatment plants and in bacterial cultures (Jenkins et al., 2000), and may be associated with **will-o-the-wisps,** a mysterious phenomenon that may be due to burning swamp gas (which is mostly methane) possibly igniting phosphine as it spontaneously and rapidly oxidizes in air.

Reservoirs of Phosphorus

The major phosphate cycle reservoirs include (figure 10.4):

Rocks: Phosphorus is a component of many different minerals. As phosphate, it is often present as the mineral apatite (a calcium phosphate component of igneous rocks), as a minor component of limestone, and as a variety of relatively rare minerals, often in combination with arsenic and vanadium.

Soil: Watersheds hold large amounts of phosphorus, especially in agricultural areas, such as the 686-km^2 Lake Mendota, Wisconsin watershed. Bennett et al. (1999) developed a mass-balance model for phosphorus loading and discharge. In this system, the watershed annually receives 1307 metric tons of phosphorus, mostly as agricultural fertilizer and feed supplements for dairy cattle. About 700 tons are exported from the watershed in crops and animal products, and about 34 tons wash into the lake from the soil. This means that 575 tons are stored in the soil each year! This stored phosphorus is available to maintain the 34-ton input to Lake Mendota for many years, whether or not additional phosphorus is imported into the watershed.

Guano: Accumulated bird or bat droppings, rich in insoluble organic phosphates such as creatine phosphate. This material, which is neither quite soil nor a mineral, sometimes accumulates in masses large enough to be exploited as an agricultural fertilizer (Hutchinson, 1952; Schlesinger, 1997).

Atmosphere: Phosphate occurs in the atmosphere in small amounts as a constituent of particulate matter (wind-blown dust), and perhaps fleetingly as the gas phosphine (which oxidizes rapidly). Atmospheric deposition of phosphorus can be an important component of phosphorus loading to oligotrophic lakes. For example, atmospheric deposition is a component of the phosphorus budget of oligotrophic Lake Tahoe, California (Jassby et al., 1995).

Oceans: Phosphates are poorly soluble in water, especially in ocean water, which is slightly alkaline. Calcium and magnesium phosphates precipitate out, forming sediments that eventually become phosphate-rich rock. After millions of years, the phosphate sediments may be exposed by **tectonic** forces (slow crustal movements of the earth), and the global biogeochemical phosphate cycle continues.

Fresh water: Phosphates wash slowly from the soils through streams, ground water, and lakes toward the oceans. Phosphate tends to get caught in freshwater systems because fresh water is often rich in calcium and magnesium, which combine with phosphate to form insoluble precipitates. Also, phosphate ions readily stick to rocks and minerals (especially clays), to any organic material suspended in the water, and to mineral colloids (such as ferric hydroxide) that form in water.

Primary producers: All organisms require phosphate for all aspects of life, including growth, reproduction, and metabolism. Orthophosphate is the major intracellular anion. Those organisms at the bottom of the food chain often have great difficulty accumulating enough phosphate. For algae and macrophytes, phosphate is often the nutrient that most limits growth and primary production.

Consumers: Heterotrophs digest (break down) foods (plants or animals) and, in the process, obtain the phosphate they need to satisfy their metabolic requirements. When animals obtain an excess of phosphate in their diet, the phosphate is excreted, often as part of a large organic molecule (such as creatine phosphate).

Phosphorus can be a limiting factor for zooplankton growth and reproduction (DeMott et al., 2001). Of course, phosphate limitation of herbivores is especially important when algae are also experiencing a phosphate deficiency.

Animals higher up the food chain (omnivores and carnivores) seldom have a phosphate deficiency because their animal prey are relatively rich in the nutrient. Vertebrates even make their bones out of calcium phosphate! Thus, animals high on the food chain appear to have an excess of phosphate—the nutrient that is so rare in primary producers tends to accumulate at the top of the food chain. Kitchell et al. (1979) estimated that over half the TP in Lake Wingra, Wisconsin resides in fish bones.

Phosphate as a Limiting Nutrient

Lakes are dependent on atmospheric deposition, watershed sources, and resuspension of lake sediments as phosphate sources. Vollenweider (1976) showed that the average annual chlorophyll concentration in a lake (a rough indicator of the average algal biomass and primary productivity) was proportional to the amount of phosphorus entering the lake (figure 10.5). Schindler (1977) reasoned that phosphate would be a limiting factor more often than nitrogen because if nitrogen is limiting, then nitrogen-fixing organisms (such as cyanobacteria) will make up the deficit, and nitrogen will no longer be limiting. However, if phosphate is limited, there is no obvious biological mechanism for increasing the concentration. Schindler supported his argument with data from a lake survey, which showed algae abundance was strongly correlated with average annual total phosphorus concentration (figure 10.5). (For a larger data set, Carpenter found the same relationship, if a little more variable—see the website address given at the end of chapter 11.)

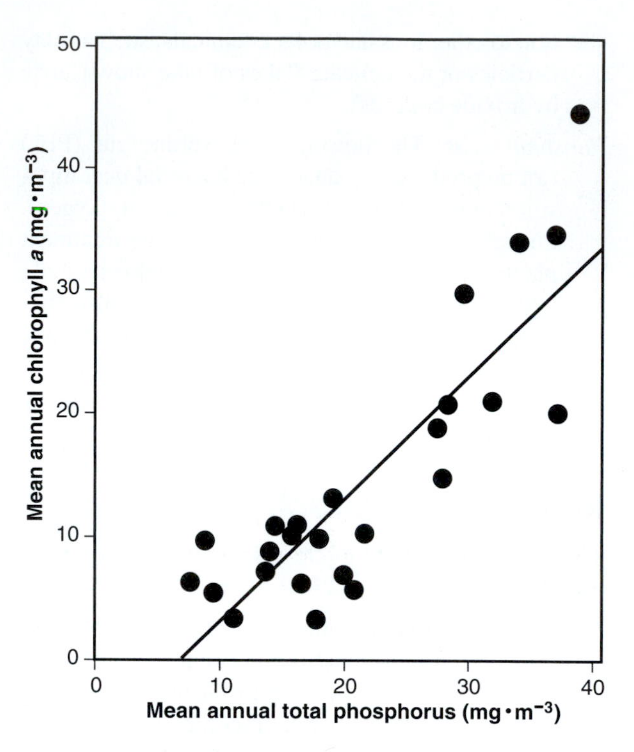

FIGURE 10.5 The relationship between average annual concentrations of total phosphorus and chlorophyll *a* (a measure of algal abundance) in lakes of the Experimental Lakes Area, Ontario, Canada. See figure 8.4 for the relationship between fish production and phosphorus concentration. The wide range of phosphorus concentrations in this figure was made possible by including data from fertilized and unmodified lakes. The correlation explains about 74% of the variation in algal abundance. **Source:** data from Schindler, 1977.

Stratification

Thermal stratification has a strong influence on the amount of available phosphate in lake water. The onset of summer stratification marks the beginning of removal of phosphate from the epilimnion. A number of mechanisms transport phosphate from the epilimnion to the hypolimnion, where it is trapped until the lake destratifies and mixes.

Organic particles such as dead algae, zooplankton, fishes, and plant tissue all contain phosphate and tend to sink out of the epilimnion. In addition, phosphate is sticky and tends to form weak electrochemical associations with fine clay particles or ferric hydroxide colloids. Ferric hydroxide, $Fe(OH)_3$, is very insoluble, forming snowflake-like, colloidal particles in water (MacIntyre et al., 1995). Ferric iron is formed in oxygenated water (oxidation is the loss of electrons, which have a negative charge). Ferrous (Fe^{++}) iron is oxidized to the ferric (Fe^{+++}) form if oxygen is present, and especially if the pH is between 5 and 8. After spring mixing, ferric hydroxide colloidal particles sink out of the epilimnion, sweeping up phosphate as they fall. If the hypolimnion is anoxic, the ferric form is reduced back to the ferrous form, which is soluble in lake water. The colloidal particles disperse, and the phosphorus is also released into hypolimnetic water in soluble form. The fall overturn in productive dimictic lakes often produces a bloom of algae because the phosphate-rich hypolimnetic water is mixed back into the upper euphotic zone.

Water quality managers sometimes desire to remove phosphate from water to reduce primary productivity or control nuisance algae. They can add alum (aluminum sulfate) to a lake to mimic the action of colloidal iron particles. The alum forms insoluble colloids to which phosphate sticks. Like ferric hydroxide colloids, the alum colloids slowly sink, adsorb phosphate ions, and deposit the phosphate on the lake bottom. It takes tons of alum to treat even a small lake. Welch and Schrieve (1994) report that alum treatment markedly improves water quality, but needs to be repeated about every 5 years.

Phosphate stored in sediments can be returned to the epilimnion during stratification by macrophytes, which have roots in the sediments. Also, bottom-feeding fishes such as carp stir and resuspend bottom sediments during normal feeding activity, returning phosphate to the epilimnion.

Luxury Consumption

When phosphate is available, algae take it up at a high rate and store it for future generations. Each time an algal cell divides, its phosphate store is partitioned among the daughter cells. The smallest algae are best at taking up phosphate because of surface-to-volume considerations.

Under certain conditions, bacteria take up large amounts of phosphate. For example, in sewage treatment plants, bacteria suspended in water release some of their phosphate if the water is anoxic. When the water is then oxygenated, the bacteria take up large amounts of phosphate as they grow and divide rapidly. The bacteria are then settled out of suspension, to form sludge. This luxury consumption can help reduce the phosphate concentration of sewage effluent from about 5 to 10 milligrams per liter to 2 to 4 milligrams per liter. The phosphate trapped in the sludge can then be used as agricultural fertilizer.

Recycling

Zooplankton and fishes play a major role in recycling phosphate in lake water. Zooplankton eat algae and excrete the organic phosphorus back into the water. Fishes eat zooplankton and excrete phosphorus into the water. Thus, the phosphate is taken from a particulate form (algae or zooplankton that could sink into the hypolimnion) to a soluble form, which is more likely to stay in the epilimnion.

Limiting Nutrient

In the 1960s, there was an exciting limnological controversy about what was causing the increasingly serious pollution of streams and rivers. Lakes were becoming choked with blue-green algae and streams were stinking with slimy algal growths. Limnologists suggested three possible nutrients that might be causing this unwanted productivity: carbon dioxide, nitrate, and phosphate. Heated arguments occurred among limnologists as to which was the *limiting nutrient.* The argument was not just among scientists; there were large corporate interests as well. Chemical companies were successfully selling huge amounts of phosphate-rich detergents and were resisting the possibility that phosphate could be an important limiting nutrient. Chemical companies tended to favor the possibility that carbon dioxide was the major limiting nutrient, which would let them off the hook.

This controversy led to myriad, clever laboratory and field experiments and observations. In the end, phosphate was determined to be, in many cases, the major limiting factor for algal growth in lakes and streams (leading to the development of phosphate-free detergents). Evidence for the dominant role of phosphate includes (Schindler, 1977; Vallentyne, 1970):

1. The concentrations of CO_2 and bicarbonate in lakes with algal problems, such as Lake Erie, are immaterial because of the large total amount of inorganic carbon and the buffering response of the system.
2. Carbon concentrations (100–120 mg l) have stayed the same in Lake Erie for the last 100 years, whereas the incidence of algal blooms has increased dramatically as phosphate concentrations increased.
3. Adding carbon, usually as organic compounds, has little or no effect on fish productivity in aquaculture situations.
4. Much research shows that addition of N and P fertilizers leads to very marked growth and high standing crops of algae in fish ponds.
5. Sewage treatment plants produce effluent that is low in organic chemicals and high in N and P. Over the last 90 years, the amount of carbon added to natural systems via sewage has decreased, N and P have increased, and algal nuisances have become worse.
6. A comparison of primary productivity and nutrient concentrations, across many lakes, shows a strong correlation between production and phosphate concentration.
7. Nitrate and inorganic carbon are much more abundant than phosphorus in lake water and can enter

FIGURE 10.6 Lake 226 at the Experimental Lake Area (ELA), western Ontario, Canada. **Source:** From *Science*.

the lake from the atmosphere. Thus, biological activities (such as nitrogen fixation and respiration) tend to restore deficiencies of nitrate and carbon dioxide in water.

8. In a most famous whole-lake manipulation, David Schindler (1977) fertilized half of Lake 226 (Experimental Lake area, western Ontario) with phosphorus (figure 10.6). The results were spectacular. The fertilized half of the lake turned bright green as algal populations exploded. The conclusion was that phosphate, not nitrate, was the limiting nutrient.

Cultural Eutrophication

There is a strong positive relationship between the amount of phosphate (total phosphate, or TP) measured in a lake and algae abundance (see figure 10.5).

According to Hasler (1947), *cultural eutrophication* is an increase in productivity (and accompanying decrease in water quality) due to **anthropogenic** (manmade, or "domestic") inputs of nitrogen and phosphorus to lakes and streams. Anthropogenic sources of phosphate and nitrate include sewage, agricultural runoff, lawn fertilizer, pet wastes, and atmospheric pollution. As the human population has grown, cultural eutrophication has become widespread in freshwater and marine habitats (Smith et al., 1999). Some well-studied examples include:

Gull Lake, Michigan: Septic tank sewage and lawn fertilizer from vacation homes and estates are main sources of fertilizer for the algae in this mesotrophic, rural lake. Lake productivity increased significantly between the 1960s and 1974 (Moss et al., 1980), as water clarity decreased and an anoxic summer hypolimnion developed.

Lake Wingra, Wisconsin: Dog feces and other urban runoff (from decaying leaves in the gutter, lawn fertilizer) are major sources of fertilizer for this small, urban lake in a watershed surrounded by homes (on the city sewer system) and the University of Wisconsin Arboretum (which contributes very little of the total fertilizer) (IES, 1999).

Lake Mendota, Wisconsin: This large, urban lake once received all of the sewage from the city on its shores. Now that the sewage is diverted, the lake receives fertilizer stored in the sediments, fertilizer from the same sources as Lake Wingra. Because Lake Mendota has a large watershed in rich farmland, most of the fertilizer input into the lake comes from agricultural sources: the natural nutrients in the soil (from limestone bedrock) and especially artificial nutrients (Bennet et al., 1999).

Lake Thonotosassa, Florida: This shallow, naturally eutrophic lake is surrounded by agricultural fields and moderate suburban development. Prior to 1970, the lake was heavily polluted by agricultural runoff, organic material such as citrus waste, and sewage effluent. Cowall et al. (1974) stated that, "During the dry winter months when citrus processing reached a peak, the flow of the [inlet] creek was composed primarily of effluents from these plants. Citrus pulp and the smell of oranges were evident several kilometers downstream. By the time the effluent reached Lake Thonotosassa . . . organic materials were in an advanced state of decay and the creek was usually anaerobic."

Reversing cultural eutrophication is a complex process. It is not simply a matter of stopping the input of phosphates or other nutrients into a lake because sediments store a relatively huge amount of the nutrient. For example, each time rain falls on the watershed and each time the lake mixes, nutrients such as phosphate are returned to the lake water from the sediments. (Bennett et al., 1999). The hysteresis behavior associated with alternate stable states (chapter 8) also complicates reversal of cultural eutrophication. Management of eutrophic lakes is considered further in chapter 11, as a part of Restoration Ecology.

Environmental Protection

Identification of phosphorus and nitrogen as major factors in algal growth had far-reaching consequences. Today, in the United States, streams and lakes typically lack unpleasant smells, toxic chemicals that kill fishes, a coating of foam, and unnatural colors. This is no accident. People had been complaining about the bad smells, scums, and fish kills of urban lakes for centuries. These problems intensified as the population grew. Society knew enough to sometimes filter solid matter out of sewage and to boil water if people were getting sick, but we really didn't know enough about water pollution to fix it. The 1960s was a time when many urban, suburban, and even rural streams and lakes resembled disgusting sewers. The water smelled and grew masses of slimy algae. Soap suds from phosphate-rich detergents formed huge mounds below dams and in sewage treatment plants. In Madison, Wisconsin, suds were occasionally blown a half-mile from the sewage plant, drifting over a major highway and endangering traffic. In the late 1960s, newspaper headlines declared Lake Erie "dead" because of fish kills and floating scum (actually, the productivity of the lake was at its peak—but the lake was dead for many human purposes). At this same time, limnologists finally had enough knowledge about toxic chemicals and algal nutrients to make a difference. The U.S. Clean Water Act of 1972 provided rules and guidelines for cleaning up our water surface and ground waters, and established the U.S. Environmental Protection Agency (**EPA**) to implement the act.

THE NITROGEN CYCLE IN WATER

Forms of Nitrogen

After phosphate, nitrate is the second most likely inorganic nutrient to limit plant growth in freshwater habitats. Nitrogen occurs in several different chemical forms in water (figure 10.7).

Total nitrogen (**TN**): All the nitrogen in a water sample. The total nitrogen in a reported sample does not typically include nitrogen gas dissolved in the lake water, but does include all inorganic and organic forms. Typical concentrations of TN in lakes are 0.4 to 2.7 μg per liter, depending mainly on the land-use characteristics of the watershed (Wetzel, 2001). Unlike phosphate, all the nitrogen compounds are very soluble in water. These forms include:

Nitrogen gas (N_2): A relatively inert gas that makes up most of Earth's atmosphere and dissolves readily in water. However, it does not react with water or

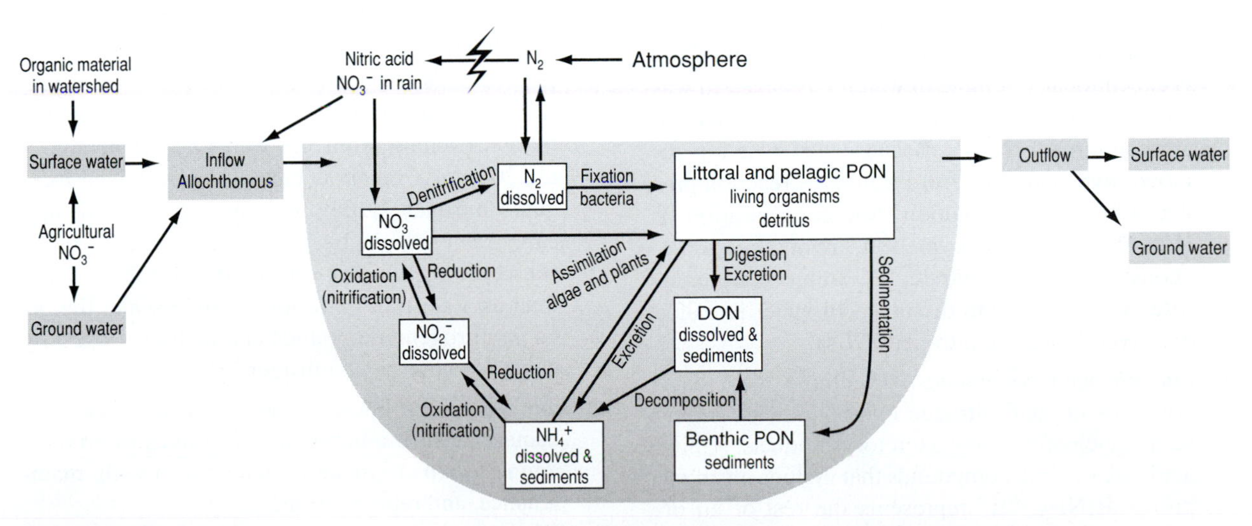

FIGURE 10.7 The nitrogen cycle. PON = particulate organic nitrogen; DON = dissolved organic nitrogen.

anything else, except in unusual situations. Lightning and certain bacteria can convert nitrogen gas into nitrogen oxides (NO_x) or ammonia (NH_4). Nitric acid (HNO_3), a major component in acid rain, is formed when gaseous nitrogen oxides dissolve in water. Large amounts of nitrate are created with each lightning flash. The intense heat of the lightning provides the energy needed by nitrogen gas to combine with atmospheric oxygen. This oxidized nitrogen combines with water to form nitric acid—rain from a thunderstorm is a very dilute solution of nitric acid and carbonic acid (and sulfuric acid if a coal-burning power plant is downwind).

Oddly enough, the reaction of nitrogen gas with oxygen releases a great deal of energy, so we might expect that all atmospheric nitrogen would be oxidized. This is not the case because the reaction requires a large amount of activation energy before the oxidation can occur. This activation energy is supplied by lightning or bacteria, or in the industrial process, by burning fossil fuels.

Nitrate and nitrite (NO_3^-) and (NO_2^-): Two ions that are formed when oxides of nitrogen interact with water. Nitrate is a relatively benign chemical that is used by many plants, protists, and microbes as a source of nitrogen for building proteins and other organic nitrogen compounds. Nitrite, on the other hand, can be quite toxic.

Ammonia (NH_3): The reduced form of nitrogen that is made by nitrogen fixation, either by bacteria or in an industrial process. Ammonia is also a waste product, used by many organisms to get rid of excess nitrogen or hydrogen ions, or both. Ammonia is exceptionally soluble in water. Dissolved in water, it combines with hydrogen ions to produce the ammonium ion NH_4^+. Because this reaction removes hydrogen ions from water, the reaction produces an alkaline solution (low concentration of H^+). When oxygen is present, ammonia spontaneously oxidizes to nitrate. The sum of nitrate, nitrite, and ammonium dissolved in water is called dissolved inorganic nitrogen (DIN).

Organic nitrogen compounds: The three main categories of organic nitrogen compounds are amino acids, nucleotides, and excretory products. Amino acids are organic compounds that include an amino group: R-NH_3 ("R" represents the rest of an organic molecule). Amino acids are the building blocks of proteins used in nearly all aspects of metabolism and structure. Organic nitrogen is also a major component of nucleotides such as ATP and nucleic acids. Animal excretory products that are organic include urea and uric acid.

Most organic nitrogen compounds are soluble in water. Larger organic molecules are less soluble, but they tend to be quickly broken down by bacterial digestion into small soluble molecules, such as amino acids. Small organic molecules are an important food source for bacteria and protists.

Reservoirs of Nitrogen

The nitrogen cycle in lakes is complex (figure 10.7) because of the many different oxidized and reduced forms of inorganic and organic nitrogen. Major nitrogen reservoirs include:

Rocks: Nitrogen exists in igneous rocks and the interior of the Earth as ammonia dissolved in fluid inclusions and as atoms included as irregularities within molten or crystalline rock. In sedimentary rocks, nitrogen is usually present as salts of nitrate.

Atmosphere: Nitrogen gas makes up about 78% of the atmosphere.

Oceans: Nitrogen gas and nitrates are very soluble in water. In low oxygen situations, such as in deep trenches or in productive estuaries at night, the nitrogen will be in the form of nitrites or ammonium ion.

Fresh water: Nitrogen gas dissolves in water, but this form of nitrogen has little interaction with organisms, except for nitrogen fixation. Nitrates are very soluble and easily leach out (wash out) from watershed soils toward the oceans. However, headwater streams rapidly take up ammonium and nitrate leaching from the watershed (Peterson et al., 2001). As much as half of the dissolved inorganic nitrogen (DIN) washing into a headwater stream is captured by biological activity—in this sense, the stream may act as a **sink** for DIN, rather than as a **source.** On a longer time scale, the organic nitrogen compounds can be transformed by decay into inorganic nitrogen.

Primary producers: Photosynthetic plants, protists, and bacteria absorb ammonium or nitrates to produce amino groups to make proteins for growth, maintenance, and reproduction.

Consumers: Animals and fungi digest (break down) proteins into amino acids, which are used to build up animal proteins. Because primary producers are high in organic nitrogen compounds (especially proteins), and because animals burn protein for energy, animals also are faced with the need to excrete nitrogen compounds, as discussed in the next section, Nitrogen Fixation.

Five major biological processes involving nitrogen in fresh waters are:

Nitrogen fixation: N_2 gas and chemical energy are converted to ammonium.

Nitrification: Reduced forms such as ammonium are converted to nitrite or nitrate.

Denitrification: Nitrate (partial reduction) is converted to nitrogen gas.

Assimilation: DIN or organic nitrogen is incorporated into organic compounds.

Excretion: Animals typically ingest too much nitrogen (as amino groups) and need to excrete the nitrogen as ammonium ion, urea, or uric acid.

Nitrogen Fixation

In the biological world, only bacteria can fix nitrogen—converting dissolved nitrogen gas into ammonia. The bacteria can live alone or be symbionts (for example, with legumes or alder trees), but only bacteria can perform the metabolic magic that combines nitrogen gas with hydrogen. Cyanobacteria (blue-green algae) fix nitrogen, using enzymes that must be protected from oxygen. Cyanobacteria have a variety of morphological adaptations that keep oxygen from the nitrogen-fixing enzymes. Some species have specialized, thick-walled cells called **heterocysts** that are dedicated to nitrogen fixation. Common (often nuisance) algae that fix nitrogen include *Aphanizomenon* and *Gleotrichia.*

Aphanizomenon (see figure 3.2a) makes thick colonies that look like grass clippings or, in extreme cases, green paint (plates 19 and 20) and has long, parallel filaments of cells set in a gelatinous matrix (most cyanobacteria cells produce mucus). The thickness of the heterocyst, the thickness of the colony, and the presence of the mucus (which may slow diffusion of oxygen into the colony) all combine to provide an anoxic environment inside the colony to support nitrogen fixation.

Gleotrichia looks like a 1 mm hairball (see figure 3.2b), with many filaments radiating out from a central area, all embedded in a gelatinous matrix. The heterocysts are always basal on each filament (at the center of the colony), where the oxygen would be at a low concentration, because of respiration of the surrounding cells. The cells at the tips of the filaments are much thinner than the basal cells, and so would be most efficient at taking up nutrients (favorable surface-to-volume ratio) and least affected by the gelatinous matrix.

Trace nutrients needed for biological nitrogen fixation include *iron* and *molybdenum.* Often, increasing the concentration of either element in the water will increase the rate of nitrogen fixation. In the open oceans, it is possible that iron, rather than nitrogen or phosphate, is the main limiting factor. Adding just a little soluble iron can greatly increase primary productivity (Monastersky, 1995).

Measurement of nitrogen fixation is accomplished using the reduction of acetylene gas to ethylene. This method works because the acetylene has the same general shape and size as the nitrogen gas molecule, and it is particularly sensitive because neither acetylene nor ethylene is present naturally in lake water.

Fixation is often a significant source of fixed nitrogen input to a lake ecosystem. As much as half can be contributed by nitrogen-fixing organisms; the rest comes from atmospheric deposition (mostly due to lightning) or runoff from the watershed. In temperate and boreal oligotrophic lake and stream systems, migrating salmon can make a significant contribution to the productivity (Schindler et al., 2003). Salmon feed at sea but return to small, freshwater streams to spawn, typically in areas of cold water and low primary productivity. Before spawning, many salmon are eaten by bears, which then enrich forest soil with their feces. After spawning, the remaining fishes die, and many are eaten by eagles, which add their feces to the watershed. Fishes also are decomposed by microbes in water. The question is, how significant is the nitrogen in the bodies of these salmon for primary producers in the large-scale ecosystem streams, lakes, and the watershed forest? The question can be answered using sophisticated, stable isotope tracing techniques because salmon nitrogen (from the sea) includes more of the heavy isotope than does nitrogen in freshwater habitats. Research has shown that in some cases, a significant fraction of freshwater and even watershed nitrogen comes from salmon (Helfield & Naiman, 2001). Alders are trees that can grow in nutrient-poor, riparian zones because they host nitrogen bacteria in the roots. If alders are present in a salmon-spawning watershed, the alders are likely to contribute even more fixed nitrogen to the watershed and streams than the salmon (O'Keefe & Edwards, 2002).

Nitrification

Bacteria can use ammonium as a place to dump oxygen from CO_2 when they are fixing chemical energy in organic molecules (chemotrophy). The CO_2 is reduced as the ammonium is oxidized—first to nitrite, then to nitrate. The reduced carbon (CHO, carbohydrate) can then be used in metabolic processes as a source of energy or as structural material. Nitrite is quickly oxidized to nitrate, either by bacteria or abiotically in a spontaneous reaction with oxygen in the atmosphere. In habitats in contact with the air, such as flowing river water or a lake epilimnion, there is plenty of oxygen, so nitrite is usually present only in very low concentrations (<100 μg l). Nitrite may be more abundant in low-oxygen habitats such as a summer hypolimnion or water experiencing organic pollution.

An unusual but medically important aspect of nitrification is the syndrome called **methemoglobinemia,** a condition affecting humans, especially infants. Intensive agriculture requires heavy fertilization of crops using ammonia. The ammonia is oxidized by bacteria and by the oxygen in the air (nitrification). Water running off the fields as surface water or ground water then has high concentrations of nitrate, which appear in drinking water from shallow wells. Humans ingest the nitrate in well water either directly or via human milk. This nitrate is not particularly dangerous, except for nursing infants. Before being weaned, the gut of human infants is at a pH of 4.5 to 5.5 (see table 2.1). Under these conditions, denitrifying bacteria reduce nitrate to nitrite. The gut of a weaned human has a pH of 2 or less, which is so acidic that it inhibits growth of the denitrifying bacteria, and the nitrate is not reduced to nitrite in the weaned human. Nitrite (but not nitrate) combines tightly with hemoglobin, producing *methemoglobin,* which does not break down easily and does not carry oxygen. (Carbon monoxide has the same effect—of poisoning hemoglobin.) The more methemoglobin, the more the body is starved for oxygen. Obvious signs of asphyxiation, such as blue lips, are associated with this potentially fatal syndrome. Methemoglobinemia is likely to be a problem if water contains more than about 45 mg NO_3^{-1} per liter. (Note: 10 mg l NO_3-N is the same as 50 mg l of NO_3.) Concentrations higher than the 45 milligrams of nitrate per liter threshold are characteristic of intensely farmed areas underlain by fractured limestone bedrock near the surface: The fractures quickly funnel the contaminated water from fields to shallow wells.

Denitrification

NO_3^- and NO_2^- can act as oxidizers (terminal electron acceptors) just like O_2. In the process of accepting electrons, the nitrogen compounds are reduced to nitrogen gas (N_2). Thus, in low-oxygen or anoxic environments, bacteria can use nitrate or nitrite in place of oxygen in the oxidation of organic compounds. For example, in the hypolimnion of eutrophic lakes, when oxygen is low, nitrate is also lower in concentration than in the epilimnion (figure 10.8).

Denitrification by bacteria converts nitrogen-oxygen compounds into nitrogen gas. This process is exploited in sewage treatment plants to remove nitrogen compounds from sewage effluent (the nitrogen compounds would otherwise act as fertilizers, encouraging algal growth: cultural eutrophication). If sewage is incubated with the correct bacteria, under warm, low-oxygen conditions, a significant portion of the nitrate is converted into nitrogen gas. The nitrogen leaves the treatment plant as a gas, helping the plant manager meet requirements for the maximum allowable amount of DIN in the sewage **effluent** (the water that leaves the plant).

Assimilation

Algae and bacteria can take up either ammonium or nitrate. Higher plants can take up nitrate. The nitrogen is used to make amino acids. Then, amino acids are polymerized or further metabolized to make proteins, nucleic acids, ATP, and many other organic nitrogen compounds.

Animals that consume primary producers digest (break down) proteins into amino acids. Amino acids are assimilated (absorbed through the gut wall) and used to build up animal proteins. Because primary producers are high in organic nitrogen compounds (especially proteins), and because animals burn protein for energy, animals also are faced with the need to excrete nitrogen compounds.

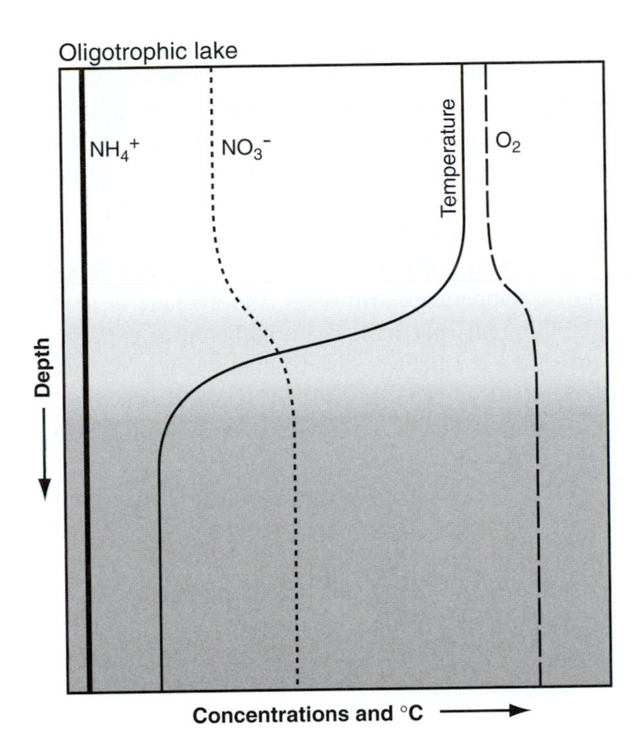

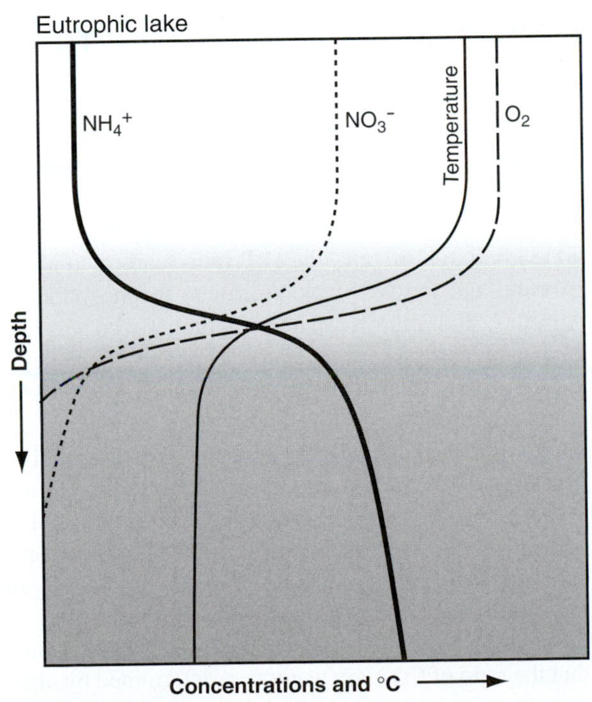

FIGURE 10.8 Models of the vertical profile of oxygen, temperature, and relative concentrations of ammonium and nitrate ions in oligotrophic and eutrophic lakes.

Nitrogen Excretion

Proteins and other organic chemicals that contain nitrogen are metabolized for energy by all organisms. Ammonia is a common waste product of metabolism. If oxygen is present in the water, the ammonia is quickly oxidized to nitrite or nitrate. However, if oxygen is absent, ammonia will accumulate (figure 10.8). Unless it is diluted rapidly and washed away, ammonia can be lethal.

When proteins are used for energy, the nitrogen from amino groups is excreted in one of several forms, constrained by the amount of energy needed and the toxicity of the excreted compound.

Animals employ a variety of chemicals as nitrogen excretory products (figure 10.9). For small aquatic snails, nitrogen can be excreted by diffusion of (highly toxic) ammonia into the water. Terrestrial snails of the same size are constrained to excrete nitrogen as cyclic carbon-nitrogen compounds (uric acid, discussed next) because ammonia cannot be easily washed away from terrestrial animals. If the terrestrial snails excreted ammonia, the liquid in their lungs would become poisonous.

Urea, $HN(CH_3)^2$ is the preferred (adaptive) form of excretion if the animal is too large to excrete ammonium and if water is readily available. Urea is much less toxic than ammonium. For example, many fishes excrete nitrogen in the form of urea. The disadvantage to urea is that it contains some reduced carbon that could otherwise be used as an energy source rather than being excreted.

Ammonium ion

Urea

Uric acid

FIGURE 10.9 Chemicals used by animals to excrete nitrogen.

Uric acid (or similar large molecules) are used to excrete nitrogen in arid environments. These compounds are more or less insoluble in water and are very nontoxic. Thus, they require no water for their excretion. However, they are large organic molecules that represent a lot of stored energy.

Ecosystem Manipulations and the Nitrogen Cycle

Bormann and Likens (1970) studied the ecological consequences of clear-cutting the 38-acre Hubbard Brook watershed in an experimental forest in New Hampshire. (Gene Likens is currently the Director of the Institute for Ecosystem Studies, an ecological field station near Millbrook, New York.) The question concerning logging was whether a clear-cut harvest has an effect on biogeochemical cycles, such as the nitrogen cycle. To explore this question, the forest vegetation was cut and left in place. As a result of the logging:

1. Water output increased 40% in amount and flood maxima increased in the streams, due to decreased ability of the soil to hold water and to decreased tree transpiration (plants release water during photosynthesis and respiration).
2. The nitrogen cycle was fundamentally altered. Before cutting, the forest accumulated nitrogen. After cutting, organic nitrogen from vegetation (and roots and soil) was oxidized into nitrate and H^+ (that is, to nitric acid). This loss of organic nitrogen decreased soil fertility, increased the nitrate content of the water, and decreased the pH. Instead of accumulating 2 kilograms per hectare per year of nitrate nitrogen, the soil lost 120 kilograms per hectare per year.
3. Associated with the production of H^+ by oxidation of the nitrogen compounds, metallic ions were released from the soil. Compared to undisturbed watersheds, the increases in net losses were calcium (10×), magnesium (7×), sodium (3×), potassium (21×), and aluminum (9×).

SILICON IN WATER

Dissolved silicon (Si) or silica (SiO_2), from aluminosilicate clays, is present in water in the form of the acid, silicic acid (H_4SiO_4). Silicate dissolved in water readily forms short polymers and even large, amorphous particles of colloidal material. Because silicic acid has four hydrogen atoms that can be lost sequentially, this acid forms the basis of a buffering system in water. Like phosphate, silicate is of minor importance in buffering water because carbonate is usually so much more abundant.

Dissolved silica can be present in water at concentrations of up to about 100 milligrams per liter, depending on temperature. However, the average is around 13 milligrams per liter. Diatoms begin to experience stress (at a competitive disadvantage) when concentrations fall below 5 milligrams per liter, and growth of diatoms stops when the concentration is in the range of .5 to .1 milligram per liter.

A major particulate form of silica is found in the shells of diatoms. The rapid diatom decline observed in the spring is due, at least in part, to a reduction of the available silicate (and other nutrients, such as phosphate). In a typical lake, the annual decline of diatoms is accompanied by the transport of tons of silicate to the hypolimnion and sediments.

ECOLOGICAL STOICHIOMETRY

Stoichiometry relates to the ratios of atomic weights of different chemical elements. Sterner and Elser (2002) provide an excellent review of the ecological role of stoichiometry in aquatic systems.

Common stoichiometric ratios include the **C:N** ratio (carbon to nitrogen), the **C:P** ratio (carbon to phosphorus), and the **N:P** ratio (nitrogen to phosphorus). Alfred C. Redfield was an oceanographer who was fascinated by patterns of the relative abundances of chemicals in marine water and living organisms. Redfield (1934) reported that the ratios of C, N, and P are remarkably constant, especially for the open ocean. The average relative proportions of C, N, and P are 106:16:1. That is, for every atom of phosphorus dissolved in deep seawater, there are 16 nitrogen atoms and 106 carbon atoms. The same average ratio is seen in **seston** (suspended particles). Most of this seston is living and dead algae and microbes. Redfield reasoned that the ratio of C:N:P in water was determined by algal growth requirements—that ecological interactions were actually causing the observed ratios in water (Redfield, 1958; Redfield et al., 1963). C, N, and P are micronutrients that potentially limit the growth rates of organisms. If one nutrient is in relatively lower supply, the other two nutrients will be taken up at relatively lower rates, and the balance will tend to be restored to the Redfield ratio.

The three nutrients of the Redfield ratio affect growth in different ways:

Carbon—Stored as carbohydrate or fat and therefore can increase body bulk even if there is insufficient N and P for reproduction.

Nitrogen—Determines the protein component of new material, and therefore growth in mass and condition of the tissue; can be limited by nitrogen even if there is sufficient C and P.

Phosphorus—Determines the amount of cell membrane, DNA (for cell division) and especially RNA (for protein synthesis). Thus, even if C and N are available, growth by cell division and production of new protein can be limited by P.

Algae are biochemically flexible and can grow in ways that depart greatly from the Redfield ratio. For example, in strong light, as photosynthesis proceeds rapidly, carbon is fixed in the form of carbon-rich carbohydrates. As carbon content increases, the algae automatically become more and more N and P limited (Sterner et al., 1997). The limitation is that there isn't sufficient N and P to balance the increasing carbon stores. As food, these carbon-rich and relatively phosphorus-poor algae (or seston) are essentially junk food—rich in carbohydrate (calories) but low in the P and N zooplankton need for growth and reproduction.

Water washing into a stream or lake will have a specific composition of different elements, depending on factors such as watershed, soil, and bedrock chemistry. Chemical and biological processes in water affect stoichiometric ratios of chemicals dissolved in water and the chemical make-up of organisms. The C:N ratio in lake water varies greatly depending on watershed chemistry, loading rates, stratification within the lake, and recycling of nutrients by animals eating autotrophs (Sterner et al., 1995). Although the Redfield ratio is a baseline, ratios reported for lake studies, for aquatic autotrophs, and invertebrate herbivores, actually vary over at least an order of magnitude from the ideal (Elser et al., 2000a). Departures from the Redfield ratio give an indication of which nutrients are limiting growth of algae and zooplankton and allow a deeper understanding of competition for limited nutrients. For example, nitrogen-fixing cyanobacteria, such as *Aphanizomenon,* have the competitive advantage when lake water has a low N:P ratio. In the epilimnion of eutrophic lakes, nitrogen-fixing species come to dominate the phytoplankton in the summer, when both nitrogen and phosphorus are in low concentrations (Smith, 1983).

Different species have different stoichiometric preferences. The diatom *Asterionella formosa* (see figure 3.8n) requires higher silicate concentrations, while another diatom, *Cyclotella meneghiniana* (see figure 3.8l), requires relatively higher phosphate concentrations (Tilman, 1977). *Asterionella* specializes on a higher Si:P supply ratio than does *Cyclotella.* In the laboratory, when the Si:P ratio was high, *Asterionella* became the dominant diatom, and when Si:P was lower, *Cyclotella* became dominant.

Among zooplankton, compared to *Bosmina, Daphnia* are not very efficient in absorbing phosphorus, and the *Daphnia* population growth rate can be limited by the amount of phosphate in their algal food, while *Bosmina* population growth rate is probably independent of the food phosphate concentration (Schulz & Sterner, 1999).

Stoichiometric effects can interact with expected top-down effects of predators on zooplankton and algae. Especially in oligotrophic lakes, the C:P ratio can be high enough so that even when zooplankton are released from fish predation, the zooplankton populations (if they are living on carbon-rich junk food) cannot increase enough to reduce algae abundances (Elser et al., 1998). Thus, lake elemental stoichiometry can constrain biomanipulation.

When top-down biomanipulation effects are observed in eutrophic lakes, the resultant community changes can have large effects on lake stoichiometry. Pike (a piscivore) were added to eutrophic ELA Lake 227 (Elser et al., 2000b). After the addition of pike, populations of smaller fishes declined and *Daphnia* populations increased. As the *Daphnia* population increased and reached peak abundance, the concentration of phosphorus in the *Daphnia* population increased, and concentrations of dissolved nitrogen and phosphorus increased in the lake water (due to increased grazing and the associated excretion of nitrogen and phosphorus). At the time of the peak *Daphnia* abundance, more than 30% of the phosphorus in the epilimnion was stored in the *Daphnia* population! As the nitrogen concentration in the lake water increased, nitrogen-fixing cyanobacteria became less abundant. Thus, there were several stoichiometry-related consequences to the top-down manipulation of adding a top predator.

TOXIC CHEMICALS

Since the 1800s, humans have been adding significant amounts of toxic chemicals through industrial and agricultural activities (Colborn & Smolen, 2002). Rates of contamination have increased dramatically since about 1945, as an exponential number of common-use chemicals such as plastics and pesticides entered the marketplace. Many of these new organic chemicals end up in freshwater systems. Heavy metal contamination also is becoming more widespread as the human population increases. Anthropogenic (man-made) contaminants are found in organisms all over the Earth, and many of them can be detected in ground and surface waters. For example, common-use pesticides are detectable in lakes and streams in agricultural and urban watersheds and even in remote and protected watersheds (Donald et al., 2001; Graymore et al., 2001). Several U.S. agencies keep tabs on freshwater contamination and make their data available on websites. Relevant agencies include the U.S. Environmental Protection Agency (**EPA**) and the U.S. Geological Service (**USGS**).

The study of the distribution and adverse effects of chemical contaminants is called **environmental toxicology** (Rand & Petrocelli, 1985; Trautmann et al., 2001), or **ecotoxicology.** Environmental toxicology favors studies of molecular mechanisms, while ecotoxicology focuses on the ecological aspects of anthropogenic contaminants.

Many anthropogenic chemicals and heavy metals are powerful, biologically active chemicals with significant effects on the organisms of freshwater systems. For example, modern agriculture relies on chemicals (**pesticides**) designed specifically to kill organisms such as nuisance plants, insects, nematodes, and fungi. Even when diluted by rain or surface water, these chemicals often have the potential to cause ecologically important, adverse effects at concentrations far below the lethal concentration.

The diversity of toxic chemicals presents a significant challenge in identifying cause-and-effect relationships. There are literally thousands of toxic anthropogenic chemicals present at low concentrations in fresh water. While there are only a handful of toxic heavy metals, they often have complex chemistries, with several forms (such as different oxides) for each metal. Metals often exist in soluble and insoluble forms, depending on the redox environment. Mercury and a few other metals exist in both inorganic and organic forms.

Acutely toxic chemicals act in a short time, killing (**lethal** effect) or damaging organisms. **Adverse** sublethal effects include reduced growth rate and/or undesirable changes in development or behavior. In many parts of the world, acute toxicity is a relatively small problem in aquatic systems. Toxic chemicals can also have **chronic** (long-term) effects. For example, many contaminants are recognized as being **carcinogenic** (causing cancer), **teratogenic** (causing adverse developmental effects), **mutagenic** (causing mutations in DNA), neurotoxic (adversely affecting nerve or cognitive function), or as being **endocrine disruptors** (acting like or interfering with hormones). In addition, some chemicals cause adverse immunological effects, damaging animals' immune systems.

Heavy Metals in Freshwater Systems

Several heavy metals occur in fresh water and are of ecological and limnological consequence (table 10.1). These metals are typically present at low concentrations, in the parts per million (**ppm**) or parts per billion (**ppb**) range, so they can be difficult and expensive to detect.[1] Nevertheless, heavy metals may have significant effects on aquatic biology (Rand & Petrocelli, 1985).

Heavy metals come from a variety of sources, both natural and artificial. Natural sources are metal-rich minerals from exposed bedrock or volcanoes. Human-related sources include principally mine refuse, industrial sources, and transportation and energy-related sources.

Several heavy metals are essential for proper ecosystem functioning. For example, iron, copper, zinc, and molybdenum are all necessary as enzyme components in plants and animals. Especially in the sea, but also in fresh water, iron concentration sometimes limits the rate of primary productivity, while molybdenum concentration may limit the rate of nitrogen fixation (Anbar & Knoll, 2002). These metals must be present at low concentrations (for proper biochemical functioning), but they can be toxic if present

[1] *In U.S. usage, 1,000,000 = 1 million, or 10^6;*
one thousand millions = 1 billion, or 10^9;
one thousand billions = 1 trillion, or 10^{12}.
1 billion = a million millions, or 10^{12}.

Table 10.1 Examples of Heavy Metals Important in Limnology

This is not a complete list, but these metals are some of the most often detected in fresh water.

Metal	Major Sources of Environmental Contamination from Human Activity	Required Nutrient?	Major Kinds of Biological Activity	Does the Metal Biomagnify?
Iron	Scrap metal, mining waste	Yes	Nitrogen fixation, oxygen transport, and other proteins and enzymes	Very little, but difficult to excrete
Molybdenum	Mining waste	Yes	Nitrogen fixation	No
Copper	Mining waste, electronics, plumbing	Yes	Oxygen transport, relatively toxic to algae	No
Arsenic	Well water (especially with dropping water tables)	No	Teratogen, mutagen, and carcinogen	Minor biomagnification
Lead	Mining waste, leaded gasoline, paint pigments, plumbing, ammunition	No	Neurotoxin; adverse effects on kidney, heart and reproductive organs; teratogen, probable carcinogen	In general no, but difficult to excrete
Cadmium	Metal plating procedures	No	Kidney failure, bone decalcification, carcinogen, teratogen	Yes
Mercury	Burning coal, electronic applications, gold mining	No	Neurotoxin affecting brain and nerve function; causes immune and reproductive system dysfunctions, damages enzymes	Yes

at slightly higher concentrations. Other heavy metals, such as arsenic, cadmium, lead, and mercury are simply toxic and are not required for normal biological activity (table 10.1).

With the exception of arsenic, molybdenum, and selenium, these heavy metals are poorly soluble in alkaline water and they easily attach to organic particles. Thus, exposure to toxic concentrations is possible in what may appear to be pure, clear water in pristine environments, such as clear oligotrophic mountain streams. Concentrations of several toxic metals can be especially high in soft water flowing from sulfide ore deposits and mine tailings.

Assessing average metal concentrations can be difficult if contaminated water rich in metals is released episodically—at night, for instance. Macroinvertebrates living in lakes and streams accumulate heavy metals in their exoskeletons, over periods of weeks or months. Colborn (1985) found that large stonefly nymphs were an especially good monitor of average metal concentrations, such as cadmium and molybdenum, in alpine streams affected by periodic releases of water from mines.

Anthropogenic Organic Chemicals in Freshwater Systems

Humans have recently added a huge number and a bewildering variety of toxic organic chemicals to aquatic habitats (Rand & Petrocelli, 1985). General classes include many different kinds of chemicals (table 10.2). Figure 10.10 displays a rogue's gallery of these chemicals.

Ecological Effects of Toxic Chemicals

Contaminants that rapidly degrade into harmless products are of most concern when they occur as a short-term input pulse, usually after a sudden rainfall

Table 10.2 Examples of the Great Variety of Organic Anthropogenic Toxic Chemicals That Can Be Detected in Aquatic Habitats

See Rand & Petrocelli (1985) for more details. Additional information is available on the Internet. For example, search using "aquatic dioxin" to find sites including the U.S. EPA site. Examples of the major classes are given in figure 10.10.

General Kind of Chemical	Major Sources of Exposure	Examples of Biological Activity in Aquatic Habitats
Dioxins	Products of combustion, also contaminants in herbicides	Developmental effects in eggs and juveniles of birds and fishes
PCBs = polychlorobiphenyls	Electrical industry (liquid resistors); fire retardants in construction material (wood, plastics, even concrete)	Endocrine disruption, carcinogen, mutagen, teratogen
Personal care products (PCPs) (especially musks, a diverse group of chemicals)	Use and disposal by humans	Endocrine disruption, immune responses
Pesticides, such as the herbicide atrazine and the insecticide DDD	Agriculture, home maintenance	Lower growth rates, carcinogenic, neurotoxins, endocrine disruption
Pharmaceuticals, such as erythromycin and chlofibric acid	Human and livestock excretion	Inhibit growth, affect microbial populations
Plasticizers (make plastics flexible; include phthalates and nonlyphenol)	Industry—cosmetics, perfumes, glues	Endocrine disruption, carcinogens
Products of combustion especially PAHs (polycyclic aromatic hydrocarbons) such as anthracine	Power plants, transportation, wood burning	Carcinogenic, mutagenic, teratogenic

transports an agricultural application or industrial release into a freshwater system. Persistent chemicals that break down slowly affect aquatic life for months, years, or even centuries. For example, PCBs were long used by industries in the Great Lakes watershed. After manufacture of PCBs ceased in the 1970s (because of health and environmental concerns), concentrations in Lake Michigan trout declined from about 5–6 ppm (mg per kg) to less than 2 ppm by the late 1980s (Stow et al., 1995). Since the 1980s there has been little change in the fish burden of these persistent chemicals (figure 10.11).

Zooplankton species richness depends on land use (Dodson & Lillie, 2001). Ponds in agricultural watersheds have about half as many microcrustacean species as ponds in least-impact (reference sites). Agricultural ponds with grassy buffer strips have more species than ponds without the buffers. Chemical contamination of these ponds is one possible factor that affects species richness. Laboratory studies of the water flea *Daphnia magna* reveal numerous ecologically important effects of agricultural contamination that may contribute toward decreased species richness in agricultural ponds. These effects occur even at the low concentrations frequently found in aquatic habitats. For example, the common triazine Atrazine® (herbicide) increases male production (changes *Daphnia* reproductive strategy) at a few parts per billion (ppb) (Dodson et al., 1999) and interferes with amphibian sexual development at 0.1 ppb (Hayes et al., 2002). In an extensive survey of natural amphibian populations, Hayes et al. (2003) found an increased abundance of abnormal animals in aquatic sites in agricultural watersheds, compared to relatively undisturbed sites.

Bioaccumulation and Biomagnification

Ecological effects of toxic chemicals occur through either direct exposure (as in the Atrazine® example) or indirectly via food web effects. Many natural and

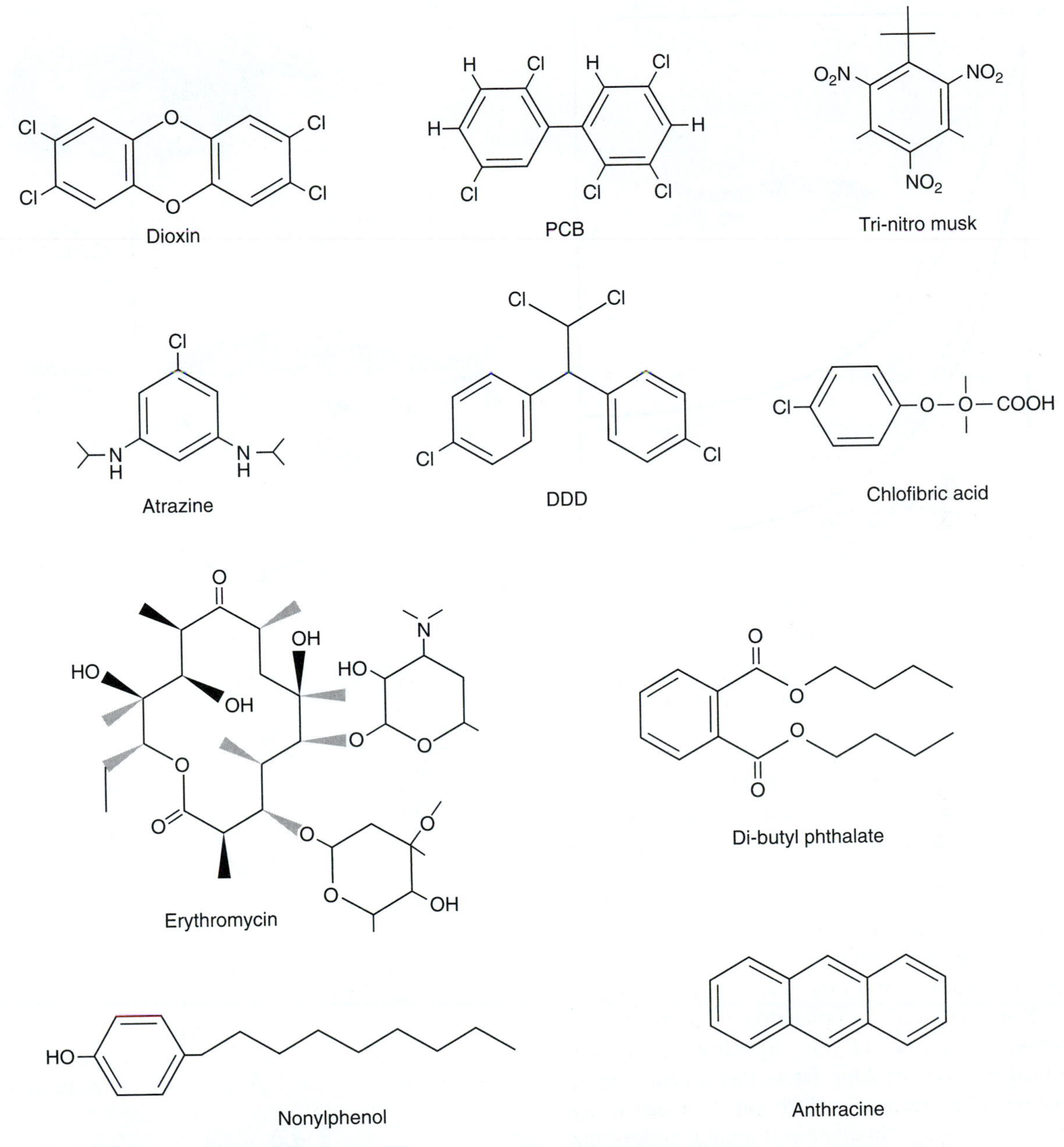

Dioxin
PCB
Tri-nitro musk
Atrazine
DDD
Chlofibric acid
Erythromycin
Di-butyl phthalate
Nonylphenol
Anthracine

FIGURE 10.10 Rogue's gallery of common chemical contaminants in freshwater.

anthropogenic chemicals are more soluble in fats (**lipids**) than in water, and therefore concentrate (**bioaccumulate**) in the organisms. Examples of natural **lipophilic** (fat-soluble) chemicals are the carotenoid pigments. Natural carotenoids color animals but are typically nontoxic. The organic forms of mercury (such as methyl-mercury) are produced both naturally by bacteria under anaerobic conditions (for example, in the bottom of ponds) and by human activity, as when coal is burned for fuel (Ulrich et al., 2001). These organic-mercury compounds are lipophilic and extremely toxic. Examples of lipophilic anthropogenic organic chemicals include many industrial chemicals, pesticides, pharmaceuticals, and personal care products.

Many toxic chemicals can be detected at low (parts per billion or parts per trillion) concentrations

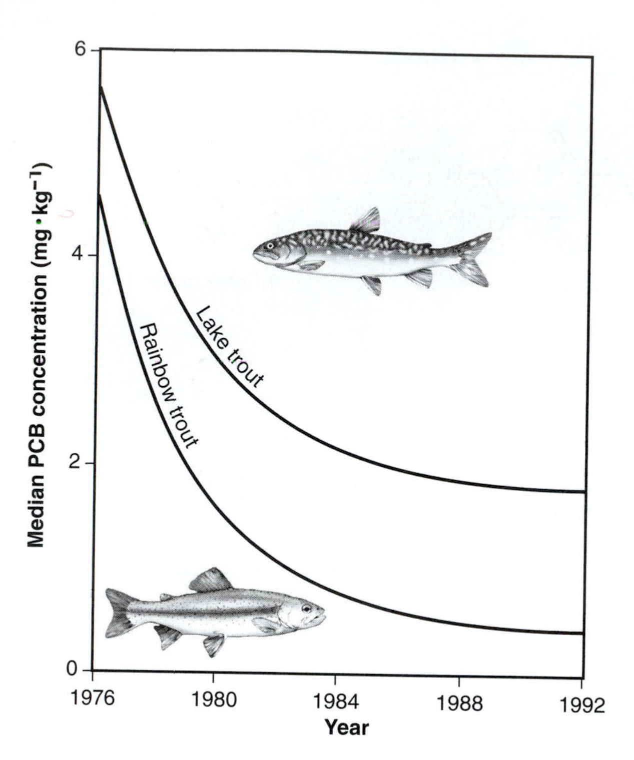

FIGURE 10.11 Average PCB concentration in Lake Michigan lake trout and rainbow trout. **Source:** Modified from Stow et al., 1995.

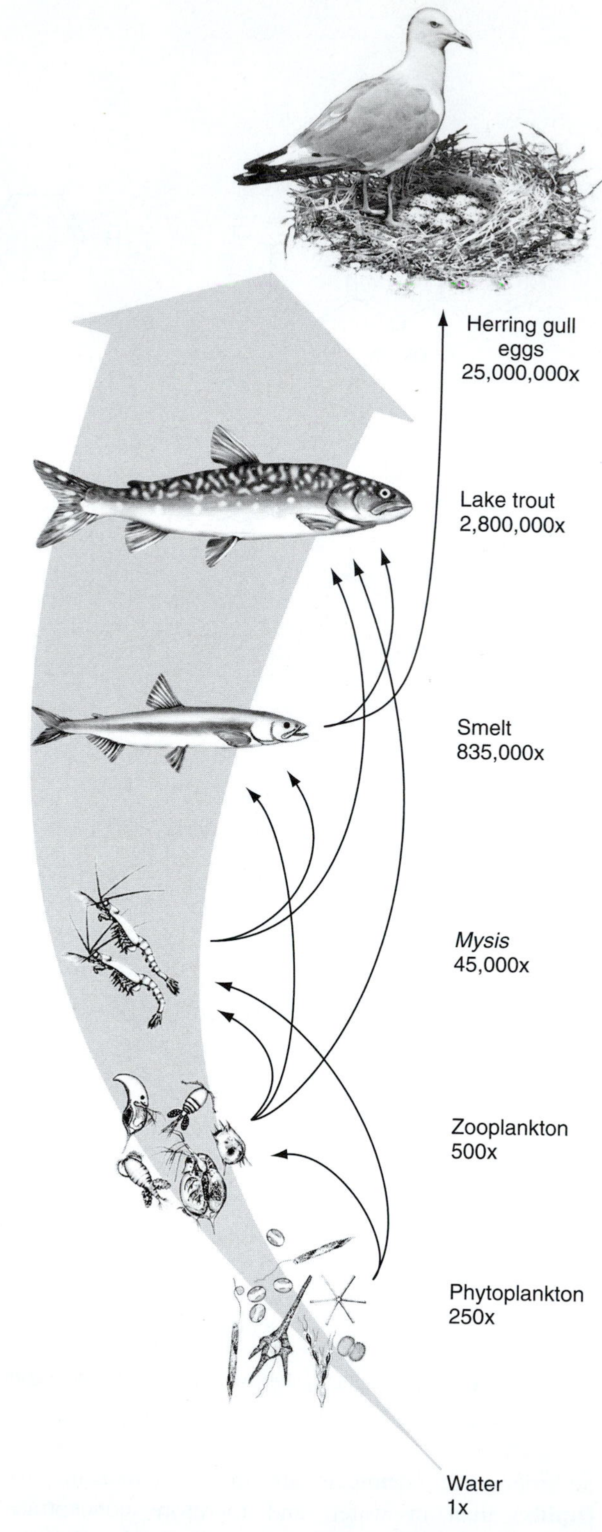

FIGURE 10.12 An example of bioaccumulation. The concentration factors are for PCBs relative to the concentration in the water, which was estimated to have been about 5×10^{-9} g l^{-1} or 5 parts per trillion (Norstrom et al., 1978). The factor of 25 million biomagnification refers to herring gull eggs, which contained about 125 mg PCB kg^{-1} wet weight. See figure 10.11 for the chemical structure of a typical polychlorinated biphenyl.

in water. These levels are not directly dangerous to aquatic life. However, because of bioaccumulation, the chemicals can accumulate to dangerous levels in living organisms.

Biomagnification is the process by which lipophilic chemicals are passed up the food chain, from prey to predator. Phytoplankton bioaccumulate lipophilic chemicals from the water, as do all organisms. However, when zooplankton consume phytoplankton, a portion of the lipophilic chemicals dissolved in phytoplankton fat is then transferred to zooplankton fat. Because each zooplankton eats many times its own weight in algae and retains most of the lipophilic chemicals, zooplankton will have higher concentrations in their fat compared to concentrations in algae. This process continues up the trophic structure. Predators on the herbivores further concentrate the fat-soluble chemicals, and so on. A rule of thumb is that lipophilic organic chemicals are 10 times as concentrated at each successive trophic level.

Top carnivores and scavengers can easily have concentrations a million times higher than what is found in water (figure 10.12). A study of the Lake Ontario food web showed that in the 1970s, gull eggs had a PCB concentration more than 25 million times that dissolved in the lake water (Norstrom et al., 1978). Other top predator and scavenger species, such as fish-eating birds and turtles, have shown similar high concentrations of biomagnified anthropogenic organic chemicals (Colborn et al., 1996).

Metallic ions tend to be only slightly lipophilic. Cadmium and arsenic show some tendency to biomagnify (Chen & Folt, 2000). The organic forms of mercury, such as methyl-mercury, are more strongly lipophilic and show a strong tendency to biomagnify (Cabana et al., 1994; Downs et al., 1998). Wisconsin currently advises its citizens against *ad libidum* consumption of predaceous fishes (such as walleye or pike) from any lake in the state. These fishes contain enough mercury to make their consumption a health concern for the human population.

Clear Lake Example: Bioaccumulation and Evolution of Pesticide Resistance

Clear Lake, California (plate 20) has had a complex history, including a period of recent cultural eutrophication (Suchanek et al., 2001). Clear Lake provides us with one of the first examples of the use of organic pesticides to control a nuisance midge population, with an excellent example of bioaccumulation and evolution of resistance to pesticides.

Clear Lake is a favorite camping site on the coastal plain of northern California. The lake is shallow and productive, with good fishing, beautiful scenery, and hot springs. During the early part of the 20th century, the adult phantom midge *Chaoborus astictopus* (see figure 4.18b for the larva, figure 4.19b for the adult) became abundant enough to be a serious nuisance to campers. There are reports of large drifts of the adults piling up like snowdrifts around nightlights and grills, and of some people having allergic reactions to the midges.

Economic pressures decreed that the midge population had to be reduced (Hunt & Bischoff, 1960). Various *larvicides* and *ovicides* (chemicals that kill larvae or eggs, such as copper and arsenic) were used between 1919 and 1949 as midge poisons, with little success. In 1949, the organic insecticide DDD (related to DDT; see figure 10.10) was added to the entire lake. Six barges towed by tugboats put 14,000 gallons of DDD slurry into the lake, producing a final concentration of 14 ppb (parts per billion). This treatment quickly killed all the midges in the lake. Midges did not reappear until 1951, and they were not at nuisance levels until 1953, when the lake was treated again. Most of the midges were killed, but enough remained to initiate another whole-lake treatment in 1954, at 20 ppb DDD. The biologists were surprised that this higher concentration killed only about 99% of the midges. The lake was treated again with DDD in 1956, and biologists noted that the midges appeared to be becoming resistant to the insecticide.

After the lake was dosed at 20 ppb DDD in 1957, the midge population was only slightly reduced. After 1959, the lake was treated with other pesticides, with generally the same pattern of midge reduction followed by a rapid recovery. The lake was treated with methyl-parathion each year from 1962 to 1975, with a development of larval resistance similar to what was seen with DDD.

Before the 1950s, there were about 1000 nesting pairs of grebes at Clear Lake. Grebes are a kind of slender, small, diving duck that eats fishes (figure 10.13). In 1955, about 100 dead grebes washed up on shore. Biologists were mystified as to the cause—laboratory exposure trials had shown that the concentration of DDD in the water was not toxic to vertebrates, and there was no evidence of infectious disease. In December 1957, 75 dead grebes were collected. They showed no evidence of disease, but an analysis of their tissue revealed 1600 ppm DDD in their body fat. In 1958, further analysis showed:

- 1–10 ppm DDD in flesh of sunfish, bluegill, and frogs
- 10–100 ppm in flesh of bass, bullhead, crappie, hitch, blackfish, and carp
- >100 ppm in flesh of white catfish and grebes

By 1959, only about 15 pairs of grebes were nesting on Clear Lake, and there were no young produced.

FIGURE 10.13 Western Grebe, a piscivorous diving duck living on Clear Lake, California (see plate 20 for a photo of the lake).

Biologists still hadn't figured out that biomagnification was going on, but they were beginning to be suspicious of DDD.

In 1962, Rachel Carson published "Silent Spring," an essay on the ecological effects of organic pesticides. She used data from Clear Lake and other sources to warn us of the dangers of bioaccumulation of chlorinated organic chemicals such as DDD. A story that Carson might have used became clear decades later. Cook et al. (2003) show that the disappearance of lake trout and whitefish from the Laurentian Great Lakes in the 1940s was due in large part to dioxin contamination.

The Clear Lake story makes a fundamental contribution to both ecosystem ecology and ecotoxicology. In 1949, no one envisioned the consequences of large-scale pesticide use, such as loss of nontarget birds and development of resistant midges. Experiences like those at Clear Lake (DDD) and the Great Lakes (PCBs and dioxin) show dramatically that human tinkering with nature changes the chemical milieu of the biosphere in ways that present us with novel and unpredictable scenarios.

> The supreme good is like water, which nourishes all things without trying to.
>
> It is content with the low places that people disdain.
>
> Thus, it is like the Tao.
>
> *Stephen Mitchell, 2000,*
> *Chapter 8,* Tao Te Ching: A new English version.
> *New York: Harper.*

Study Guide

Chapter 10 Aquatic Ecosystems: Chemical Cycles

Questions

1. How can lakes be titrated by acid deposition? Are lakes in your area experiencing titration? Where are they in the process?
2. What was the mechanism that caused Lake Nyos, Cameroon to explode?
3. Why do aquatic animals use gills instead of breathing oxygenated water into lungs?
4. In a temperate-zone stratified lake in the summer, the vertical oxygen profile can be orthograde, clinograde, or heterograde, depending on the level of productivity. Nitrogen gas is also dissolved in lake water. What kind of vertical nitrogen profile would you expect, and is the profile related to productivity?
5. Would Lake Mendota, Wisconsin immediately become oligotrophic if no more phosphorus was imported into the watershed?
6. Explain how nitrogen cycles through the aquatic and terrestrial ecosystems. Does energy cycle?
7. Managers of sewage treatment plants are required to meet standards for nitrogen and phosphorus concentration in the effluent of their plants. It is costly to remove excess nitrate and phosphate from sewage. Why are the processes of denitrification and possibly phosphine generation of interest to managers of sewage treatment plants?
8. In a lake contaminated with pesticides, which one of the following organisms living in that lake will contain the maximum amount of a bioaccumulated pesticide: small plankton-eating fishes, zooplankton, large piscivorous fishes, phytoplankton, or scavenger water birds such as gulls? Define the process of bioaccumulation.
9. Use the Internet or other sources to identify the 10 most important anthropogenic contaminants of aquatic systems in your local area. What do you mean by "important"? Choose from ground water, lakes, or streams.
10. Use the logic of *surface-to-volume ratio* to answer the following questions:
    a. In the fall, which freezes first—a small pond or a large lake?
    b. Is bioaccumulation the same for large and small algae, in the same water (P, N, PCB)?
    c. Is competition for N or P related to algal size? Which size wins?
    d. Do small ponds and large lakes receive the same "dose" of pesticides from their watershed?
    e. How does sinking rate and dissolving rate relate to particle size?
    f. Are fog droplets less acidic than raindrops, in the same air?
    g. Consider two 1-liter jars, each containing lake water. In one jar, suspend 1 gram of coarse sand. In the other jar, suspend 1 gram of clay. Then measure the phosphate concentration in each jar. Which will have the higher phosphate concentration? (Hint: Phosphate tends to stick to the surface of clay particles. Is there a difference in size of sand and clay particles?)
11. What was the role of limnologists in persuading the U.S. Congress to draft the U.S. Clean Water Act of 1972? (Note: If you prefer, you can choose a different governmental agency and a different piece of clean water policy.)

Words Related to Aquatic Ecosystem Chemical Cycles

acute
adverse
anthropogenic
bioaccumulate
biogeochemistry
biomagnification
buffer
carcinogenic
chronic
colloid
detritus
dissociate
dolomite
dy
ecotoxicology
effluent
endocrine disruptor
environmental toxicology
EPA
equilibrium
facultative
gill
gyttja
heterocyst
homeostatic
hydrate
lethal
limestone
lipophilic
loading
marl
methemoglobinemia
mutagenic
pesticide
ppb
ppm
seston
sink
source
stoichiometry
teratogenic
titrate
USGS
whiting
will-o-the-wisp
winter-kill

Major Examples and Species Names to Know

Chemicals

alanine
alum
aluminosilicate
amino acid
ammonia
ammonium hydroxide
ATP
atrazine
bicarbonate
carbonate
carbon dioxide
carbonic acid
DDD
DIC
DIN
dioxin
DIP
DNA
DOC
DOP
ferric
ferrous
humic acid
leach (verb)
lipid
musk
nitrate
nitric acid
nitrite
nitrogen gas
nitrogen oxides
nonylphenol
orthophosphate
oxygen gas
ozone
PAH
PCB
pharmaceutical
phosphate
phosphine
phosphorus
phthalate
PIP
plasticizer
POC
POP
protein
silica
silicate
SRP
sulfuric acid
tectonic
TN
TP
TSP
urea
uric acid

Organisms

Asterionella, Cyclotella, Aphanizomenon, Gleotrichia, Microcystis, Daphnia, diving beetles, back swimmers, water boatmen, *Chaoborus,* spiders, mud minnows, pike, grebe

Sites

Lake 226, ELA, Ontario, and the influence of phosphate on primary productivity

One of the lakes used as examples of cultural eutrophication (Gull Lake, Lake Wingra, Lake Mendota, Lake Thonotosassa, or Clear Lake)

Clear Lake California, the clear lake midge (*Chaoborus tetans*), DDD, grebes, and the concepts of biomagnification, bioaccumulation, and the evolution of pesticide resistance

Lake Nyos, Cameroon, an exploding lake

The concept of cultural eutrophication; examples include those from Lake Wingra, Lake Mendota, Clear Lake, or Gull Lake

What Was a Limnological Contribution of These People?

Art Hasler

Gene Likens

Dave Schindler

Jake Vallentyne

Additional Resources

Further Reading

Hutchinson, G. E. 1957. *A treatise on limnology. Volume I: Geography, physics, and chemistry.* London: John Wiley & Sons. 1015 pp.

Rand, G. M., and S. R. Petrocelli. 1985. *Fundamentals of aquatic toxicology.* New York: Taylor and Francis. 666 pp.

Schlesinger, W. H. 1997. *Biogeochemistry: An analysis of global change.* 2nd ed. San Diego, CA: Academic Press. 588 pp.

Siegel, A., and H. Sigel. eds. 1997. Mercury and its effects on environment and biology. Volume 24 of Series: Metal ions in biological systems. New York: Marcel Dekker.

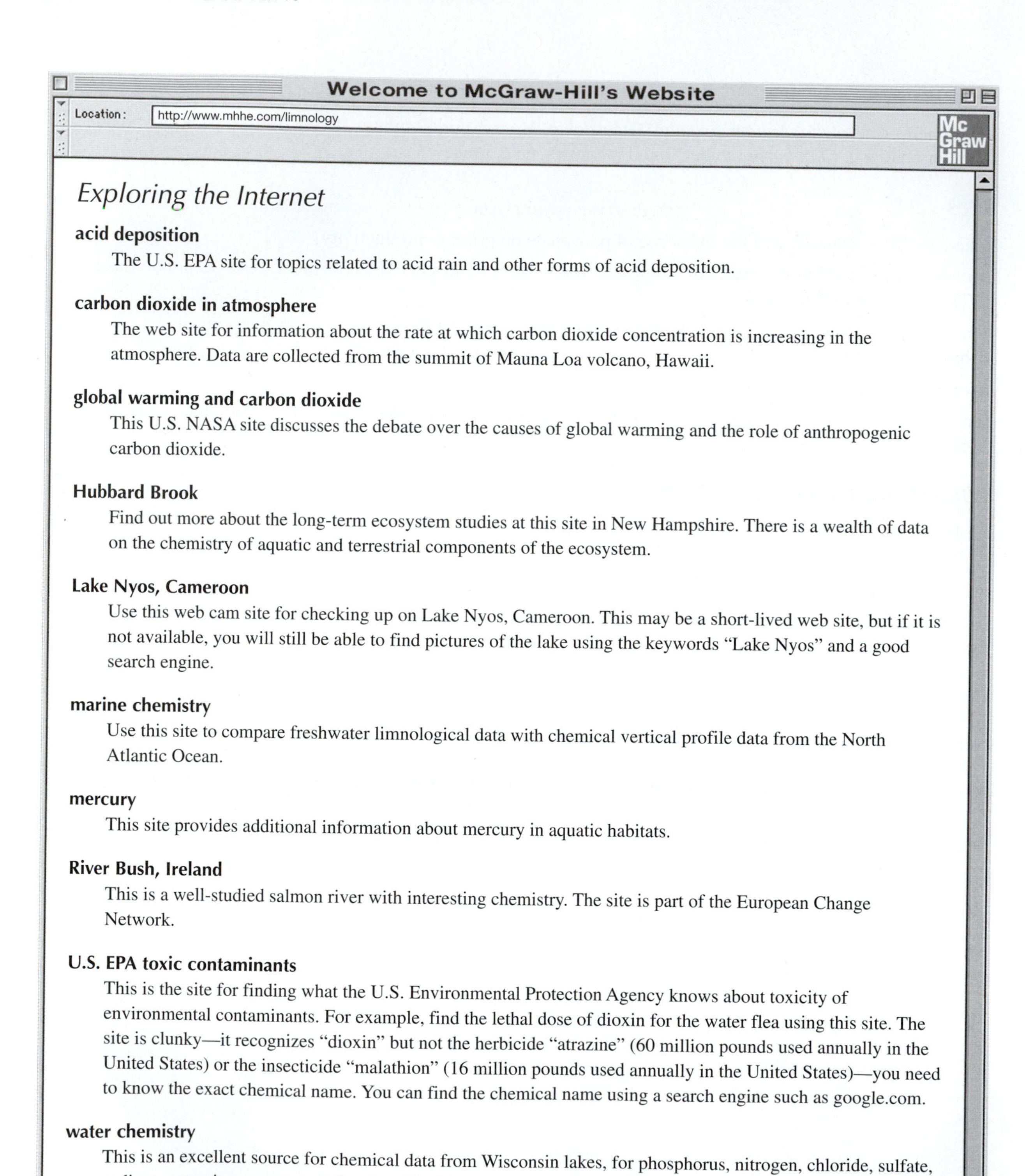

Exploring the Internet

acid deposition

The U.S. EPA site for topics related to acid rain and other forms of acid deposition.

carbon dioxide in atmosphere

The web site for information about the rate at which carbon dioxide concentration is increasing in the atmosphere. Data are collected from the summit of Mauna Loa volcano, Hawaii.

global warming and carbon dioxide

This U.S. NASA site discusses the debate over the causes of global warming and the role of anthropogenic carbon dioxide.

Hubbard Brook

Find out more about the long-term ecosystem studies at this site in New Hampshire. There is a wealth of data on the chemistry of aquatic and terrestrial components of the ecosystem.

Lake Nyos, Cameroon

Use this web cam site for checking up on Lake Nyos, Cameroon. This may be a short-lived web site, but if it is not available, you will still be able to find pictures of the lake using the keywords "Lake Nyos" and a good search engine.

marine chemistry

Use this site to compare freshwater limnological data with chemical vertical profile data from the North Atlantic Ocean.

mercury

This site provides additional information about mercury in aquatic habitats.

River Bush, Ireland

This is a well-studied salmon river with interesting chemistry. The site is part of the European Change Network.

U.S. EPA toxic contaminants

This is the site for finding what the U.S. Environmental Protection Agency knows about toxicity of environmental contaminants. For example, find the lethal dose of dioxin for the water flea using this site. The site is clunky—it recognizes "dioxin" but not the herbicide "atrazine" (60 million pounds used annually in the United States) or the insecticide "malathion" (16 million pounds used annually in the United States)—you need to know the exact chemical name. You can find the chemical name using a search engine such as google.com.

water chemistry

This is an excellent source for chemical data from Wisconsin lakes, for phosphorus, nitrogen, chloride, sulfate, sodium, potassium, oxygen, and carbon dioxide.

Exercise

Water Chemistry

How does the water composition compare for the ocean, Great Salt Lake, Trout Lake, and Lake Mendota? (Note: If you can find the data, you are free to use any other lakes.)
Fill in the following table:

mg/liter	Lake Mendota	Trout Lake	Bog 12–15 (Trout Bog)	Great Salt Lake	Ocean
DIC as bicarbonate					
Cl					
SO_4					
Ca					
Mg					
Na					
K					
Fe					
Mn					
Specific					
conductance					
umohs cm^{-1}					

Hint: search Google.com using "Great Salt Lake Water Composition" for some ions. There is also useful conductivity data at http://wow.nrri.umn.edu/wow/under/parameters/conductivity.html and http://www.ugs.state.ut.us/online/PI-39/PI39PG9.htm.

Look on the NTL-LTER website under data catalog (you need to log on again, as you did for the species richness exercise; the chemical data are only taken a few times a year, so ask for an entire years' worth of data). DIC is listed under "nutrients" data set. The other ions are listed under "major ions" dataset. The NTL-LTER site has years of data—choose any data for this exercise.

Use the Internet to find additional data to complete the table.

Answer the following questions:

1. Is the relative proportion of ions similar among the different kinds of water?
2. What are the major ions in each of these water bodies?

11

"White Pond and Walden are great crystals on the surface of the earth, Lakes of Light."

Henry David Thoreau, 1854, in "Walden"

outline

Water In Landscapes

THE LANDSCAPE CONCEPT

In this chapter, we explore the interaction of water bodies with the land, using concepts and techniques of limnology and *landscape ecology*. The concept of **landscape** can be thought of as being made up of different patches, characterized by different organisms and environments, such as lakes, streams, and wetlands, intertwined in a complex mosaic with terrestrial communities. Limnology examines the ways in which physical and chemical processes produce and cause change in streams and lakes, and the interaction between the pattern of patches and ecological process—that is, the biological causes and consequences of patchy environment (Gergel & Turner, 2002).

Lakes and streams have shape, size, and position in the watershed. These large-scale landscape characteristics change with time—lakes and streams have a history and a future as they change through developmental stages of birth, aging, and eventual extinction. By understanding the relations of water with the spatial landscape and with time, we can determine the human uses that are possible for a given lake or stream.

This chapter first explores the pattern of water in a landscape, as if we were looking closely at the parts of a patchwork quilt. This close look at lakes and streams is followed by a description of the relationship between patches (landscape position and the hydrological cycle), where aquatic features come from, and how they change over time.

SHAPES AND SIZES OF LAKES AND STREAMS

Morphometry is the quantitative description of the **shape** and **size** of lakes or streams.

Shape

Shape is defined by ratios of linear dimensions. Lake shape (seen from above, as represented on a map—see figure 11.1) depends greatly on the surrounding topography. Lakes have a tendency toward roundness because of shoreline erosion, but they often have complex shapes because of the bedrock substrate and patterns of watershed erosion. Reservoirs (plate 12) are particularly prone to having irregular, tree-like **(dendritic)** outlines because they are often dammed stream valleys.

Lakes are very flat, three-dimensional objects. The depth is nearly always much less than the width. For example, Lake Mendota, Wisconsin is about 6500 meters wide and a maximum of 27 meters deep. This shape is flatter than a pancake!

Shore Length

Lake *shore length* is a piece of data that is often used to describe a lake. However, there is something odd about this parameter. The value of parameters such as lake volume and lake area depends on the resolution of the measuring stick (measuring scale). The smaller the resolution, the more accurate the value (the value is closer to the true value). That is, if you use contour intervals at every meter, your volume estimate will be more accurate than if you use 2-meter contour intervals. As the measuring stick is made smaller, estimates of lake volume or lake area converge on the true value.

The convergence does not happen for lake shore length. The value for lake shore length depends on the size of the measuring tool and the length does not converge—the smaller the measuring scale used, the longer the length. Shoreline length is *scale-dependent.* This is because using a smaller measuring stick allows you to include smaller zigzags of shoreline, which makes the shoreline longer. However, on average, there is a zig for every zag, so the area does not change with the longer shoreline measurement (figure 11.2).

Shoreline lengths are typically given in meters. However, the lengths are actually most often measured from maps or photographs, or with survey instruments. No one (except an owner of lakeshore property) goes to a lake with a meter stick to measure the shoreline. Thus, the measuring scale actually used is probably in the range of 10 to 100 meters. This standard scale is never stated when lake shore lengths are published, but it would be a good idea.

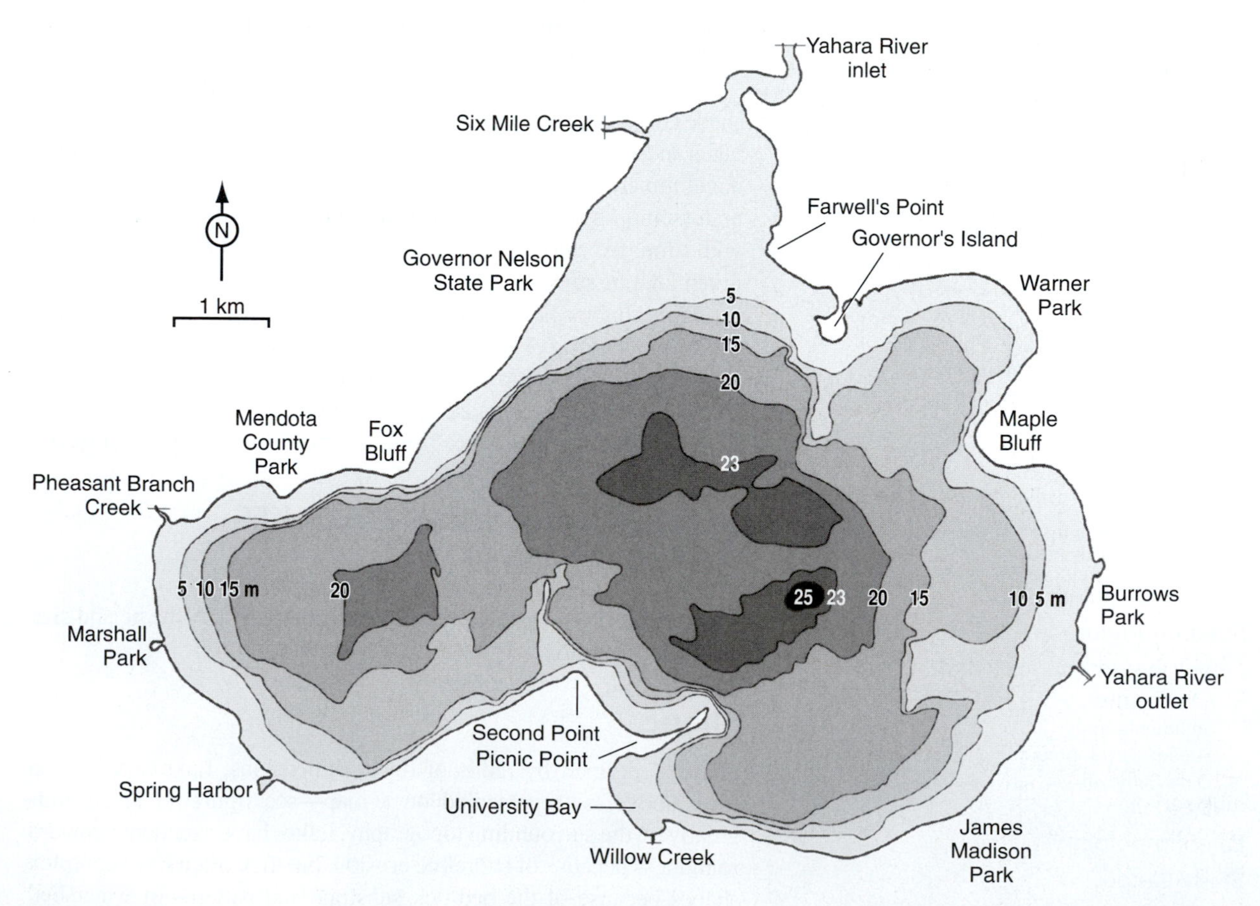

FIGURE 11.1 Depth contours (m) for Lake Mendota, Wisconsin.

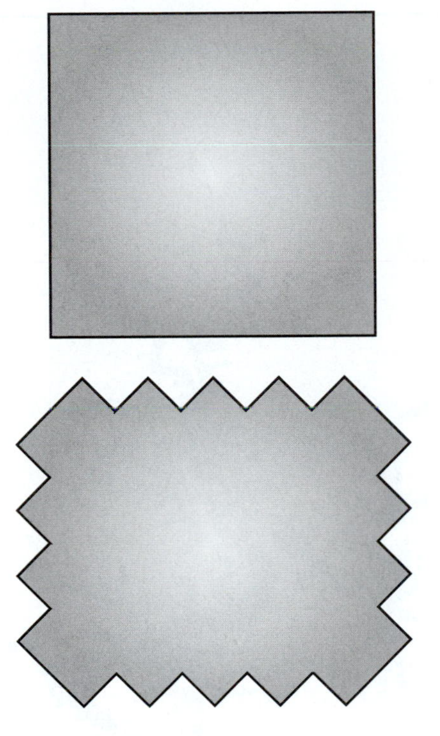

FIGURE 11.2 Illustration of the effect of shoreline roughness. The two surfaces have the same area, but the lower surface has a longer perimeter.

The *Index of Shoreline Development* compares the measured shore length (at the standard scale) to the minimum length possible for a lake of a given size. The minimum circumference of a lake is that of a circle having the area of the lake. Using the Euclidean formula for the area of a circle, $A = 1/2\ \pi\ r^2$, and the relationship between radius and circumference, $C = 2\pi r$, the circumference of a circle of area A is:

$$C = \text{square root } (8\pi A)$$

The index of shoreline development is the measured length divided by the minimum circumference calculated using A. Bull Shoals Reservoir, Michigan, is an example of a lake with an unusually large index of shoreline development (figure 11.3). The index of shoreline development is scale dependent, like shoreline length.

Area

Lake area, one measure of lake size, is calculated by carefully surveying the lake to produce a map. In the past, the surveying was done with optics. Today, global positioning system (GPS) technology (based on 24 satellites circling the Earth) is readily available and is used to survey lakes. Technicians can visit prominent features along the shoreline and record their exact position. Alternatively, an image (map image or aerial or satellite photo) of the lake can be digitized. Excellent topographic maps are available on the Internet (see the USGS site, topozone.com). Software is readily available to then calculate the area, given the position of points on the shoreline or a digitized image. An alternative low-tech method is to weigh a cutout of the lake map and then use a piece of paper of known area (based on the scale of the map) for comparison.

Depth

Lake depth obviously depends on where the depth is measured (see figure 11.1). Traditionally, researchers estimated depth by lowering a weighted and calibrated line. However, modern depth-finding technology allows depth to be measured electronically or optically more quickly and, often, more accurately. The depth contour lines shown in figure 11.1 were calculated based on depth measurements made at many points on the lake. These depth measurements can easily be made with a handheld GPS unit to record location and an electronic depth finder. A lake map with contour lines is called a **bathymetric** map; the lines can also be called **isobaths.**

Volume

Lake volume, a measure of lake size, is calculated by dividing the lake into a number of horizontal strata (layers) and estimating the volume of each stratum. The volume of a layer is the average area of that layer times the depth interval. Figure 11.1 shows depth contour lines for Lake Mendota, Wisconsin, used to calculate the volume. The method for calculating lake volume is given as an exercise at the end of this chapter.

Lake volume is used to estimate *average depth* of a lake (*z*, written with a line over the letter). The average depth is the lake volume divided by the surface area.

Watershed or Catchment Basin

The area of landscape that contributes to the water supply of a lake or stream is called either the **watershed** or the **catchment** basin (figure 11.4). Limnologists who prefer the term *catchment* consider the watershed to be the dividing line between catchment basins. However, in this text, "watershed" means the surface basin that drains into a lake.

Watersheds are defined for surface water. Groundwater flow can follow the pattern of surface water flow, but ground water can also have a subsurface (hyporheic)

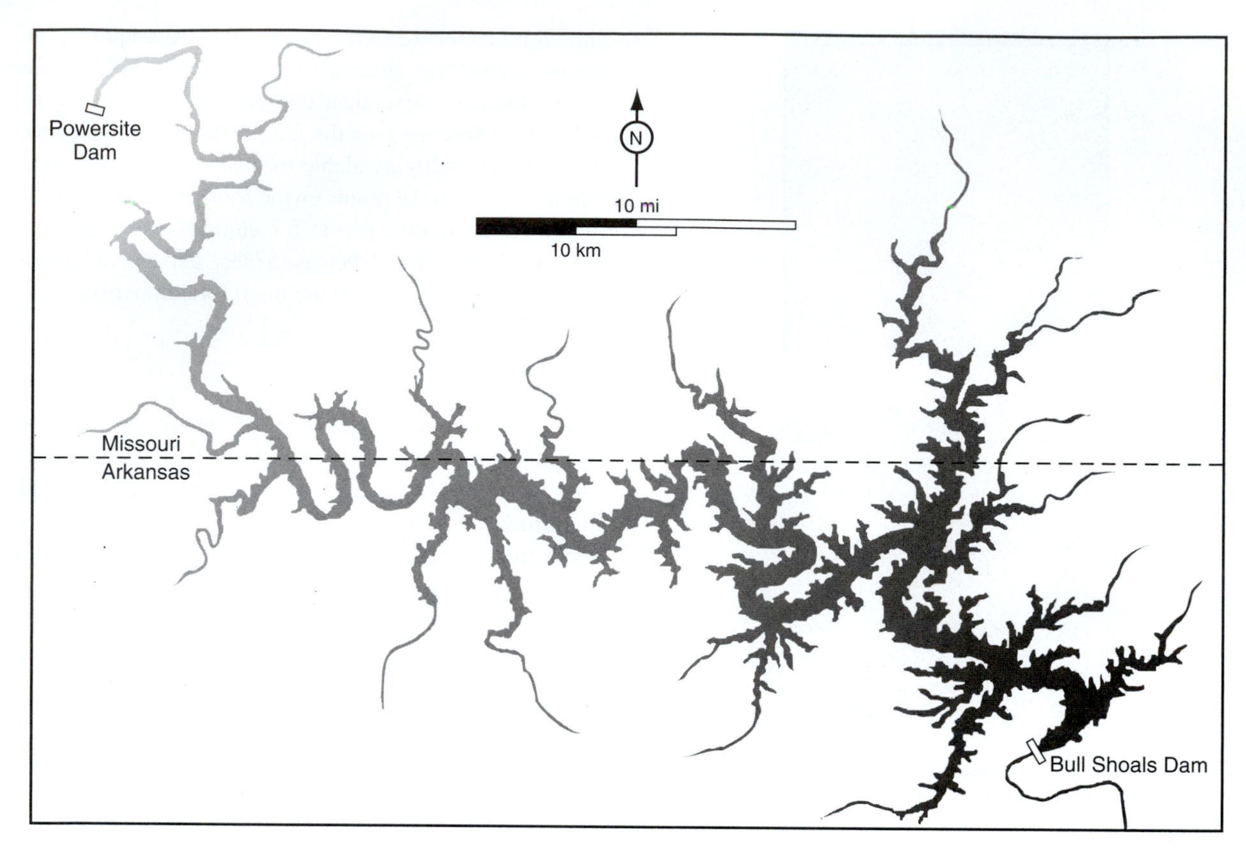

FIGURE 11.3 A map of Bull Shoals Reservoir, Michigan. Notice the highly irregular "dendritic" shoreline, characteristic of a reservoir in a drowned river valley (in this case, the Missouri River). This reservoir has a high index of shoreline development at this scale of measurement (roughly 0.1 miles).

flow pattern that depends on the subterranean geology and is not the same as the surface pattern.

STREAM MORPHOMETRY

Streams have a different shape compared to lakes because water flows in streams, producing an elongated, branched pattern. The branching pattern of streams in a watershed depends on the geography. Streams in mountains tend to be straight, while streams that drain very flat areas will **meander** (move back and forth across the landscape in a serpentine path). Relatively slow-moving streams also often have multiple, or *braided,* channels. For flowing water, we can calculate the velocity, gradient, cross-sectional area, and discharge.

Velocity

Stream velocity (figure 11.5) is measured along the direction of flow. Velocity represents the rate of movement of a particle in the water. The velocity depends on the particle's distance from shore, the water depth, and the shape of the bottom. Because water flow is often turbulent, average velocity can be difficult to measure—some water even flows upstream due to turbulence and eddies!

Gradient

Stream gradient is defined as the drop in elevation over a given stretch of flow. Gradient is typically measured at a large scale, over tens to thousands of meters. Gradient is calculated using data from landscape survey or from **topographic** maps that show elevation contours and stream courses (figure 11.5).

Cross-Sectional Area

This area is estimated from careful measurements of depth along a transect taken at right angles to the flow direction. A rough estimate of cross-sectional area is obtained by multiplying one-half times the greatest

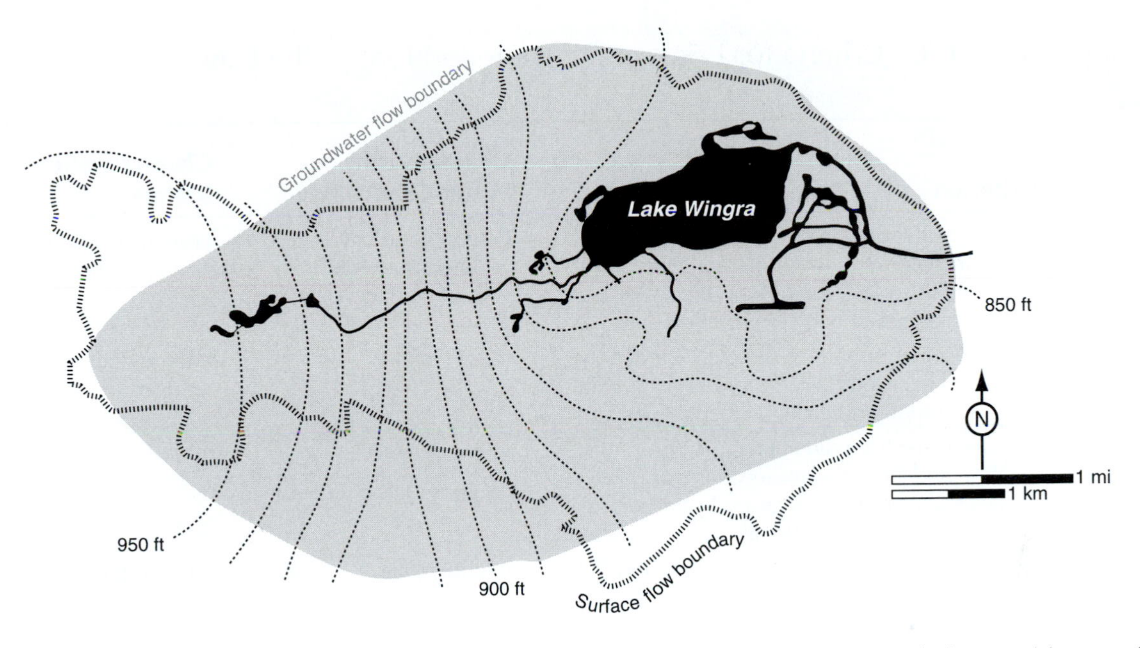

FIGURE 11.4 A topographic map of a stream and its watershed, with Lake Wingra, Wisconsin at the bottom of the watershed. The map has elevation contours, with intervals of 10 feet between contours. The stream gradient is the change in altitude along the length of the stream. Note that the boundary of surface flow is different from the boundary of ground flow.

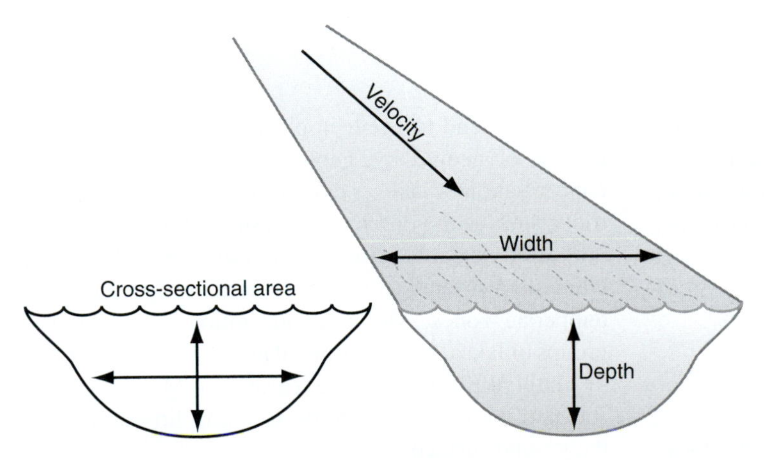

FIGURE 11.5 A model of a stream, illustrating stream velocity, width, depth, and cross-sectional area.

depth times the stream width (distance across the stream, at right angles to flow).

Discharge

Discharge (or flux) is a measure of the amount of water being carried by the stream per unit time. It is estimated by multiplying average velocity times the cross-sectional area. Discharge is typically given in units of cubic feet per second, acre feet per day, or cubic meters per second. Small pulses of high discharge are called **spates,** while major peaks of discharge are called *floods.*

LANDSCAPE POSITION

The relative position of lakes and streams in the watershed has important influences on many aspects of water, including morphology or gradient, water chemistry, primary productivity, and species diversity.

The position of a particular feature in the landscape is available either from topographic maps or from electronic databases called geographic information systems **(GIS)** (an address is given for an informative website at the end of this chapter). GIS systems typically include digital information on location (the data is "spatially

Table 11.1 Criteria for Lakes at Different Positions in the Landscape

Position	Criterion	Characteristic Water Chemistry	Characteristic Biology
−3	Receive water only from direct precipitation on the lake's surface (rain, melting ice, or snow) or surface runoff from the watershed. Water leaves only by evaporation.	Lowest conductivity, lowest nutrient concentrations	Lowest productivity and biodiversity
−2	Receive water only from direct precipitation or surface runoff. Water leaves as evaporation and groundwater seepage.	Intermediate	Intermediate
−1	Receive water from direct precipitation, surface flow, and groundwater seepage from −2 lakes. Water leaves as groundwater seepage or as an intermittent stream.	Intermediate	Intermediate
1	Receive water from rain, surface flow, and groundwater seepage. Water leaves as groundwater seepage and as permanent stream flow.	Intermediate	Intermediate
2–??	Receive water from rain, surface flow, ground seepage, and stream flow from the previous position lakes. Water leaves by evaporation and as groundwater seepage and stream flow.	Highest conductivity and nutrient concentrations	Highest productivity and biodiversity

explicit"), land use and ownership, demographics, and characteristics of the hydrology, soil, and geology. GIS technology allows limnologists to ask questions about relationships among many aspects of the landscape, such as the relationship between lake position in a landscape and the chemistry, physics, and biology of the lake.

Lake Position

The concept of *lake position* (Webster et al., 1996) was developed to explain the pattern of water chemistry seen in lakes of northern Wisconsin. The lakes varied in total salinity (measured as conductivity) and nutrient concentration in a seemingly random pattern over the landscape (plate 25). The randomness disappeared when lake position was taken into account—the conductivity of a lake was strongly correlated with lake position in the landscape.

An index of lake position can be calculated by taking into account several criteria (table 11.1; figure 11.6). The negative categories in table 11.1 are lakes without a permanent surface outlet.

Lakes tend to occur in clusters in suitable combinations of geography, geology, and climate. A large concentration of lakes is called a *lake district.* Lake districts, such as the one in northeastern Wisconsin (plate 26), are associated with recent continental glaciation, which is restricted to the more northern (or southern) temperate zones. Mountain glaciation also produces groups of lakes, even in tropical areas. The lake districts in southern Florida and the Yucatan in Mexico are associated with extensive karst (eroding limestone near the ground surface).

Landscape position is observed at a large scale within a catchment basin. At a smaller scale, details of the shape and slope of the catchment basin can affect within-lake processes. The *geomorphic-trophic* concept describes the dependence of lake community structure and productivity on indirect effects of catchment morphology. For example, the Toolik Lake region of northern Alaska lies in the foothills of a major mountain range. The distribution of fishes among lakes depends on dispersal corridors for the fishes. Lakes upstream of waterfalls have depauperate fish communities, which, in turn, through trophic cascades, affects

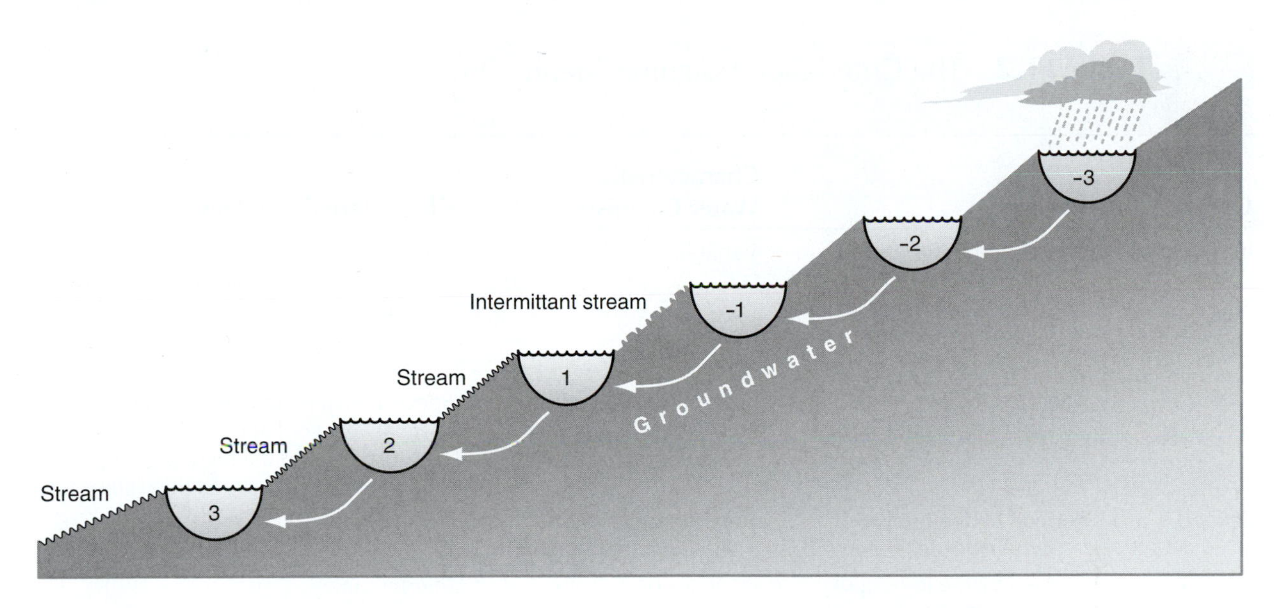

FIGURE 11.6 A model of lake position.

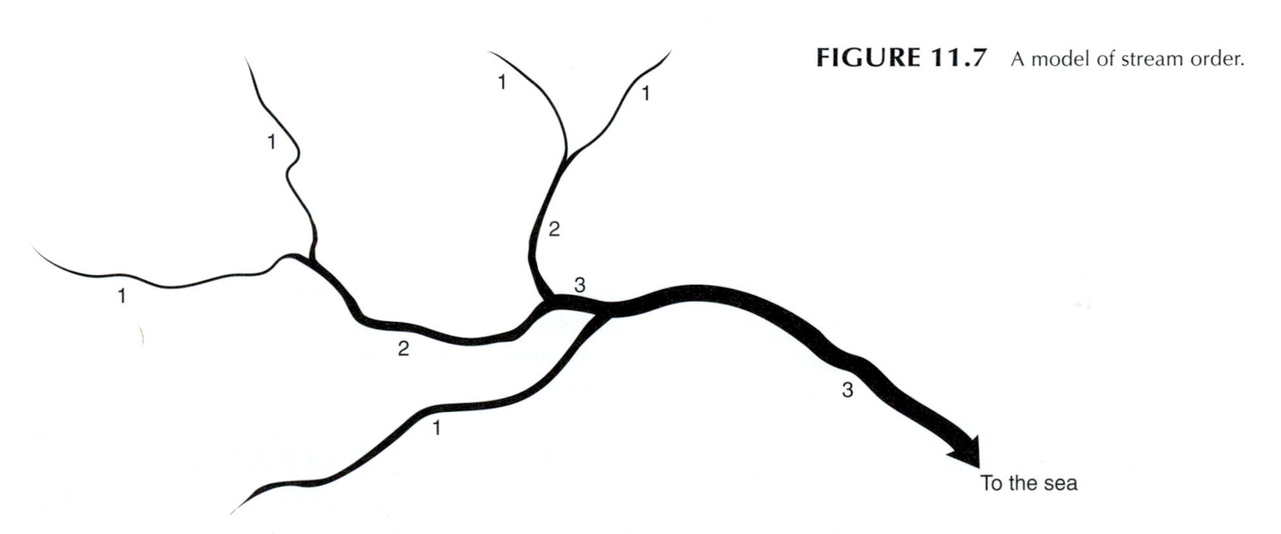

FIGURE 11.7 A model of stream order.

zooplankton abundance and the rate of primary productivity (Hershey et al., 1999).

Stream Order and River Continuum Concept

The concept of *stream order* (figure 11.7) classifies streams according to their relative position in the watershed (Strahler, 1964). The concept of stream order was important in developing the *River Continuum concept* (Vannote et al., 1980), which explored the biological consequences of a hypothesized gradient of allochthonous-to-autochthonous production from first- to higher-order streams. The River Continuum concept was developed for stream systems in wooded areas of temperate eastern North America to synthesize ideas of productivity and aquatic insect community composition. The basic idea is that first-order streams are shaded by trees, so energy input to the stream is in the form of allochthonous plant material, especially dead leaves. Larger (higher-order) streams are wide enough to allow sunlight to reach the aquatic substrate, so autochthonous productivity is possible by algae and macrophytes. Even larger streams are turbid because of sediment load or algae blooms, and the main energy source for aquatic insects is finely divided organic material washing down

Table 11.2 The Criteria for Assigning Stream Order

Order	Criteria	Characteristic Water Chemistry	Characteristic Biology
Intermittent	The stream flows part of the year, depending on seasonal climate events such as snow melt or rainfall patterns.	Variable	A mixture of aquatic and terrestrial biology, with some species adapted to living in either habitat
First	The stream flows all year and often comes from a spring.	Depends on soil and bedrock in a specific, small area	Species adapted to flowing water, often relatively cool and clear water, coarse substrate, and the highest gradient and stream velocity in the watershed
Second	Formed by the joining of two first-order streams.	Depends on soil and bedrock in watershed	Intermediate
Third, etc.	Formed by the joining of two of the previous-order streams.	Depends on soil and bedrock in watershed	Intermediate
Highest order—7 to 12: large rivers	Formed by the joining of two of the previous-order streams.	Confluence of smaller streams integrates ever-larger watershed area; pollution and turbidity often increase with stream order.	Species often adapted to turbid water, silt substrate, and low flow associated with a low gradient; turbid water is often low productivity because of light limitation.

from upstream. Three major *functional* groups of aquatic insects identified in the concept are:

- *Shredders* such as certain case-building caddisfly larvae, which chew and tear up leaves;
- *Scrapers* such as mayfly nymphs, which remove algae and aufwuchs from surfaces; and
- *Collectors* such as certain caddisfly larvae and fly larvae, which filter fine particles from water.

The River Continuum concept states that while all three groups are present throughout a stream system, shredders tend to dominate low-order streams, scrapers dominate intermediate streams, and collectors dominate the higher-order turbid rivers.

The River Continuum concept inspired a great deal of research, which in turn led to an understanding that there is not typically a clear productivity gradient in streams (see the more detailed discussion in Allan, 1995). For example, the concept is not particularly effective for understanding tropical or boreal streams stained with tannin, or streams in deserts or grasslands, or low-gradient rivers with extensive wetlands. The River Continuum concept is useful because it provides a baseline for making comparisons. Some generalizations about water chemistry and ecology related to stream order are given in table 11.2 and figure 11.7.

INTERACTIONS BETWEEN SURFACE AND GROUND WATERS: THE HYDROLOGICAL CYCLE

Surface water is water on the surface of the Earth, such as streams, lakes, and wetlands. **Ground water** is water in the permanently saturated zone below the surface. Deep reservoirs of water, often in porous rock, are called **aquifers.** Water also occurs in the soil, a dynamic reservoir between surface water and ground water. This soil (or gravel) reservoir is called the **aquitard.**

In his essay on "The Lake as a Microcosm," Stephen Forbes (1887) described the lake as a small world, separate from the surrounding landscape, that has all the parts and processes of the world at large—the macrocosm—but on a smaller scale (quoted in chapter 1). The micro-

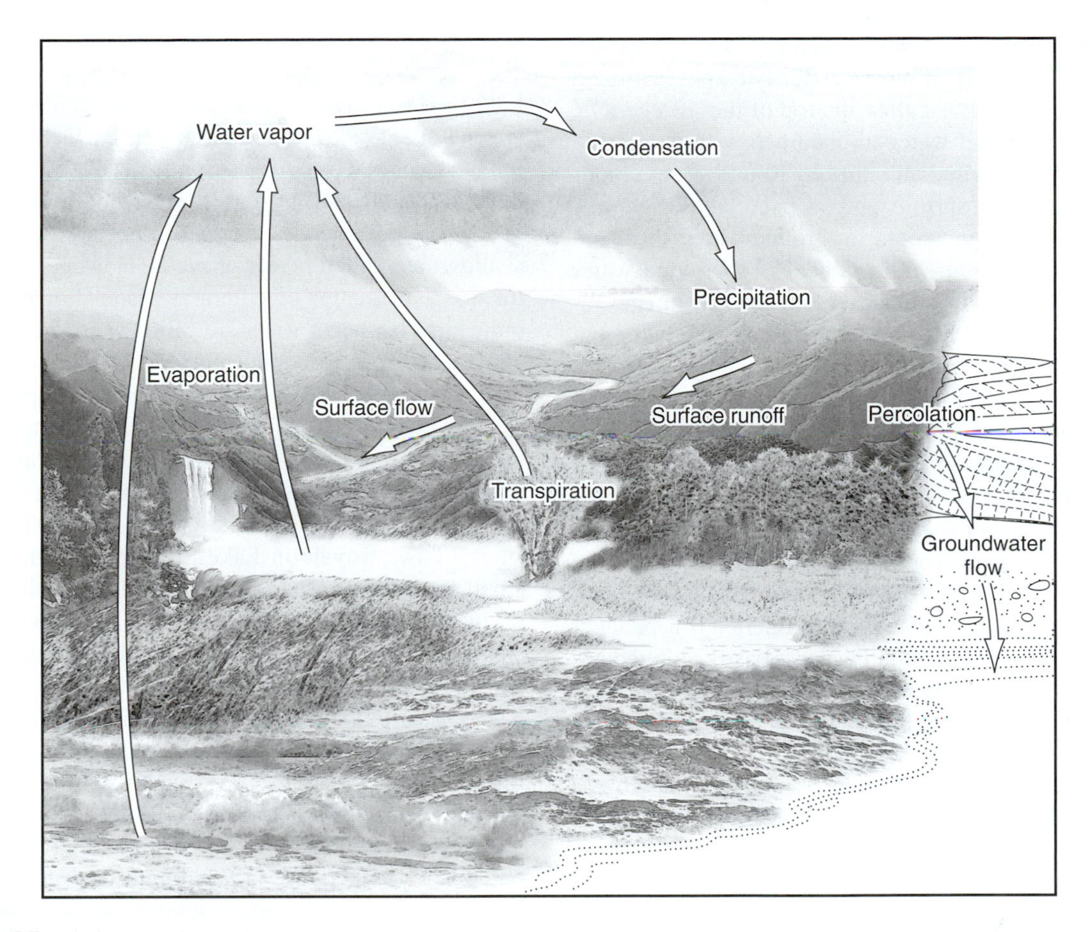

FIGURE 11.8 A model of the hydrological cycle.

cosm concept is useful when it allows us to generalize understanding of a smaller (and presumably simpler) system to reality in general. However, the concept encourages limnologists to focus on lakes and streams and to forget that lakes are a part of a larger system, which includes the atmosphere, and surface and groundwater watersheds draining into the lake or stream.

Jack Vallentyne (1974) documents the importance of the watershed to lake chemistry and biology, with emphasis on phosphate as a limiting factor for blue-green algal growth. His book is important because it emphasizes watershed effects on lake processes. One result of looking at lakes and streams as part of the landscape is the realization that public policy directed toward reducing pollution in fresh water must address the entire watershed, not just the lake or stream.

The larger system, including surface, ground, and atmospheric water, is called the *hydrological cycle* (figure 11.8). At the global scale, the hydrological cycle can be considered a closed system. Although there is a steady input of water to the Earth from outer space, and a steady loss of water (mostly as hydrogen and oxygen) back to space, the amount of water on Earth has remained essentially constant over millions of years.

Water in the atmosphere, water frozen in polar ice and glaciers, water in the soil and aquifers, geological water (water in rocks), and surface water, including streams, wetlands, lakes, estuaries, and the oceans, is all the same water in the hydrological cycle (figure 11.8). Each of these reservoirs of water has a *water budget*—a description of the total amount of water coming in and leaving the reservoir. Freshwater reservoirs include ground water, lakes, the atmosphere, streams, wetlands, and estuaries.

Ground Water

Water is stored in and slowly moves through moderately to highly permeable rock strata such as sandstone or fractured limestone. This ground water is recharged from the surface—rain percolating through the soil. The depth at which water becomes saturated in the soil or rock is called the top of the *water table.* A mass of ground water trapped below impermeable clay or rock is called an *aquifer.* Ground water comes to the surface as springs and seepage lakes. **Seepage** lakes have no inlet except for ground water (these are lakes with position −2 in table 11.1). Ground water flows into lakes via subsurface springs. **Recharge** water flows into the lake bottom, through the lake, and out as **discharge** water.

When flowing in caves or through sand or gravel, ground water contains organic material from the surface: bacteria, and animals (including small crustaceans and insects) that eat the bacteria.

Ground water is a critical source of drinking water, which is obtained from springs or, more often, from wells. About 50% of the world population is dependent upon ground water; 95% of rural America is dependent upon ground water. Ground water is easily polluted because it moves slowly and is out of sight. Because it is relatively inaccessible, ground water is difficult to clean once it is polluted.

Ground water is different from surface water (water in streams, lakes, wetlands, and estuaries) because it contains no gases, is under pressure, and usually moves slowly through rock (although fractures provide channels for high-speed transport). Ground water tends not to mix, so plumes (of saline water or contaminated water) can move great distances. Volatile substances (contaminants) cannot evaporate. Ground water sometimes comes to the surface as a *spring.* The spring can then be the source of a stream.

Fossil water is water in aquifers that are no longer recharging. Water is *mined* when it is taken from an aquifer faster than the aquifer is recharged. Mining an aquifer can:

- Reduce the pressure of the ground water (reduce **artesian** pressure); lower artesian pressure then means springs flow less strongly and pumping of ground water may become necessary;
- Cause land subsidence (as in Mexico City);
- Cause saline water intrusion (from other deeper aquifers or the ocean) and therefore a reduced water quality; and
- Contribute to **desertification** (creation of a desert) as water tables are lowered and spring flow declines.

Lakes

Lakes are bodies of fresh or saline (but not marine) still water. Lakes grade into streams (rivers), which are flowing fresh water. The size criterion for a lake is vague (see table 1.2 in chapter 1). Large water bodies are called *lakes,* and smaller ones (perhaps a hectare or less and shallow) are called *ponds* or *pools.* However, this distinction is a matter of common usage. For example, quite large bodies of water are called *ponds* in Maine.

Lakes typically contain pelagic zooplankton. However, even quite small pools may contain zooplankton. The smallest bodies of water containing pelagic zooplankton are rock pools with a surface area of about 0.1 m^2, reported by Ranta et al. (1987) along the coast of Sweden and by Van Buskirk and Smith (1993) on the shore of Isle Royale in Lake Superior. Besides being small, ponds can often be **ephemeral** (short duration), drying up during dry parts of the year. In the spring, small pools (**vernal** ephemeral pools) develop from snow melt and rain, with the basin sealed by frozen soil. As soon as the soil thaws, many of these ephemeral pools dry up. By way of contrast, the largest water bodies (by surface area) are Lake Superior, at almost 10^{11} m^2, and the Caspian Sea, at 4×10^{11} m^2.

The water budget for a lake is a list of all inputs and outputs of water (figure 11.9a). Inputs of water to lakes include precipitation, surface inlets of streams, surface flow, and ground flow (springs). Water outputs include evaporation and ground flow. *Drainage* lakes, which have a visible surface outlet, provide an additional water output from a lake system.

Once the water budget of a lake is known, it is possible to estimate the amount of time, or residence time, that water spends in the lake. Flushing rate is the reciprocal of residence time:

$$\text{Residence time (years)} = \text{lake volume } (m^3)/\text{discharge rate } (m^3\ yr^{-1})$$

$$\text{Flushing rate } (yr^{-1}) = 1/\text{residence time}$$

The water budget for Lake Mendota, Wisconsin (figure 11.9b) can be used to calculate residence time. Most often, discharge rate is taken to mean the amount of water that flows out of the lake through the outlet. If this value is used for discharge rate, water in Lake Mendota has a residence time of about 8.8 years. If, however, we use the total discharge rate (using both stream flow and evaporation), the residence time of water in the lake is

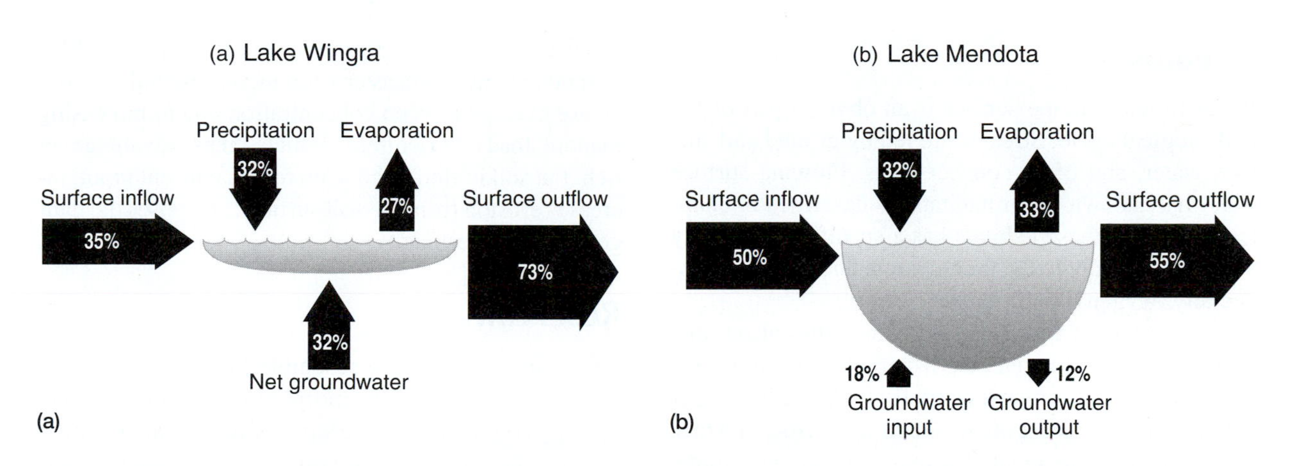

FIGURE 11.9 (a) A model of the annual water budget for Lake Wingra, Wisconsin. (b) A model of the annual water budget for Lake Mendota, Wisconsin. The percent of input and output are on an annual basis. **Source:** Data from IES, 1999 and Brock, 1985.

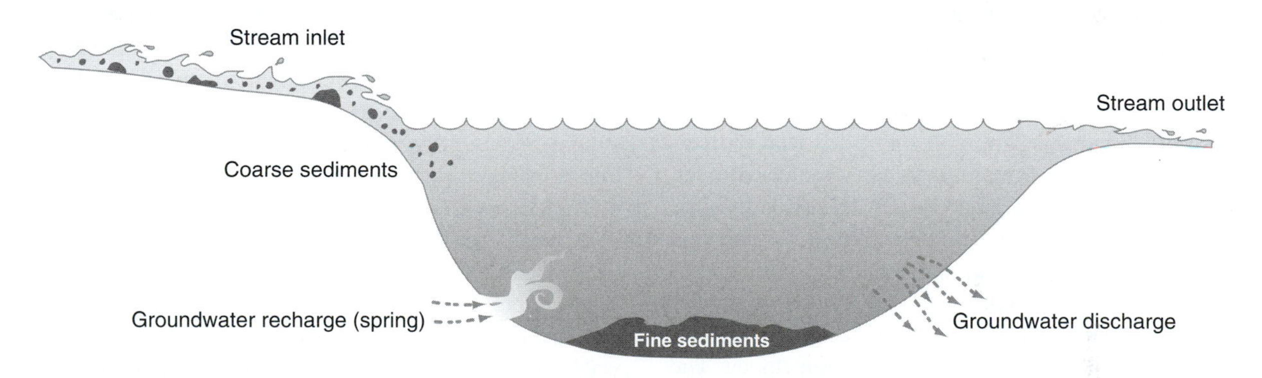

FIGURE 11.10 A model of the benthic landscape.

about 5.5 years. Therefore, it is important to clearly explain discharge rate.

Lakes have complex benthic environments because of variation in water depth and topography, **sedimentation** patterns, and the location of underwater springs (figure 11.10). Sedimentation is the process by which particles drop out of moving water. Large particles (stones and sand) drop out first, while fine silts take the longest to leave suspension. Sedimentation rate is typically highest near stream mouths as they enter into the lake.

The position of lakes in a watershed has an effect on the chemistry and biology of the lake (Webster et al., 1996). Water runs downhill, either on the surface or as ground water. Water moving on the surface evaporates, concentrating any dissolved chemicals. As water moves over and through the soil and substrate, it collects additional chemicals. Thus, water tends to increase in total conductivity, salinity, hardness, and acid-neutralizing capacity (ANC), especially if the water is in contact with limestone or sandstone.

Large lakes can affect climate on a scale of hundreds of kilometers. For example, *lake effect snows* are produced by cold winter air moving across the Laurentian Great Lakes. The prevailing northwest winds pick up moisture as they move across the unfrozen lakes, depositing the moisture as snow along southeastern shores. During the summer, the lakes are cooler than the surrounding land and affect the flow of weather, which, in turn, affects terrestrial vegetation. For example, the permanent low pressure associated with Lake Michigan is a major factor in determining plant species distribution in Wisconsin (Curtis, 1959). The lake distorts the weather pattern, allowing northern plant species to grow unusually far south near the lake's western shore.

Streams

Water flowing on the surface is an obvious part of the hydrological cycle. Streams are fed by ground and surface water, and often connect lakes. Flowing surface water has its own set of habitats, as described in chapter 7. Each of these habitats has a specific community of organisms adapted to the flow rate and chemical (especially oxygen) conditions.

As surface streams transport water, they also transport sediments. In general, faster currents transport larger particles. Conversely, as flow slows, the larger particles fall to the bottom (sediment) first. Surface flow tends to be associated with a much larger volume of groundwater flow. This hyporheic (or interstitial) flow through sediments is the home of a group of obscure and tiny insects and crustaceans.

Streams flood periodically. The *flood pulse concept* is used to further understand the dynamic stream-land interaction (Junk et al., 1989). In the temperate zone, the flooding is often in the spring, when snow melt corresponds with increased rainfall on still-frozen or relatively saturated soil. Streams rise, carry more sediment, and spread out over the adjacent low area **(floodplain).**

The interaction between high water during a flood and the flat floodplain is an important part of the chemistry and biology of streams. Floodplains are typically areas of high terrestrial productivity because of the annual addition of nutrients when the stream floods. The nutrients deposited during flooding contribute to luxurious plant growth (riparian vegetation) when the water recedes. Much of this plant material dies during the winter and is available as food for aquatic organisms during the next flood. Pools left behind by receding floods tend to be highly productive and often fishless, making a perfect habitat for aquatic insects, crustaceans, and amphibians.

Humans also find floodplains to be a valuable habitat. Because floodplains are productive and are located near water, the land and the associated riparian vegetation are used extensively for grazing and agriculture. Both uses can lead to loss of riparian vegetation and to increased erosion, which, in turn, leads to a more turbid stream. Compared to streams with an intact floodplain, streams in abused floodplains show an increased range of discharge, with greater discharge and erosion during flooding and less discharge during dry periods. An overgrazed floodplain has exposed earth, little vegetation, and a shallow stream lying in a wide and often deeply cut stream bed. Overgrazing increases water temperature by removing shade, converts riffles to sand and mud bars as erosion increases, and may decrease average oxygen concentration due to increasing manure load. Agriculture, which takes advantage of rich, flat soil in floodplains, increases nutrient input, increases erosion from the soil surface, and poses the danger of use of pesticides.

Reservoirs

Reservoirs superficially resemble lakes, but there are important differences in morphology, hydrology, and ecology (figure 11.11). Reservoirs lie behind a dam in a river valley, and the reservoir is typically fed by a single major stream or river. If reservoir water level is managed for power production or flood control, there will be little or no littoral zone vegetation.

Many reservoirs release water near the base of the dam. This cold and often hypolimnetic release can have strong effects on organisms downstream of the dam. The water leaving the dam has a fast current and the stream bed is often **canalized,** or artificially formed into a narrow channel. Besides being cold, the hypolimnetic water can be either anoxic (if the water has been stratified long enough) or supersaturated with oxygen and nitrogen (if the water has mixed recently). Water supersaturated in gasses adversely affects fish gills and can result in fish kills. The cold, hypolimnetic water creates a stream habitat different from the relatively warm and often turbid river water flowing into the reservoir. For example, the Colorado River below the Glen Canyon Dam is a cold, hypolimnetic discharge that is unsuitable for native fishes, but is an excellent habitat for exotic trout (Walters et al., 2000).

Wetlands

Wetlands are regions of shallow surface flow with slow to imperceptible currents. Wetlands have deeper water during a flood pulse, but they are also wet even between floods. Water is typically slow-flowing or stagnant in wetlands because of large surface area and presence of rooted and emergent macrophytes (plates 13 and 37). The slow flow in wetlands allows them to act as filters, trapping sediments and associated organic and inorganic nutrients. Water in wetlands, relative to that of adjacent streams and lakes, is often warmer, richer in nutrients, and well supplied with light. These conditions make wetlands more productive than surrounding surface water.

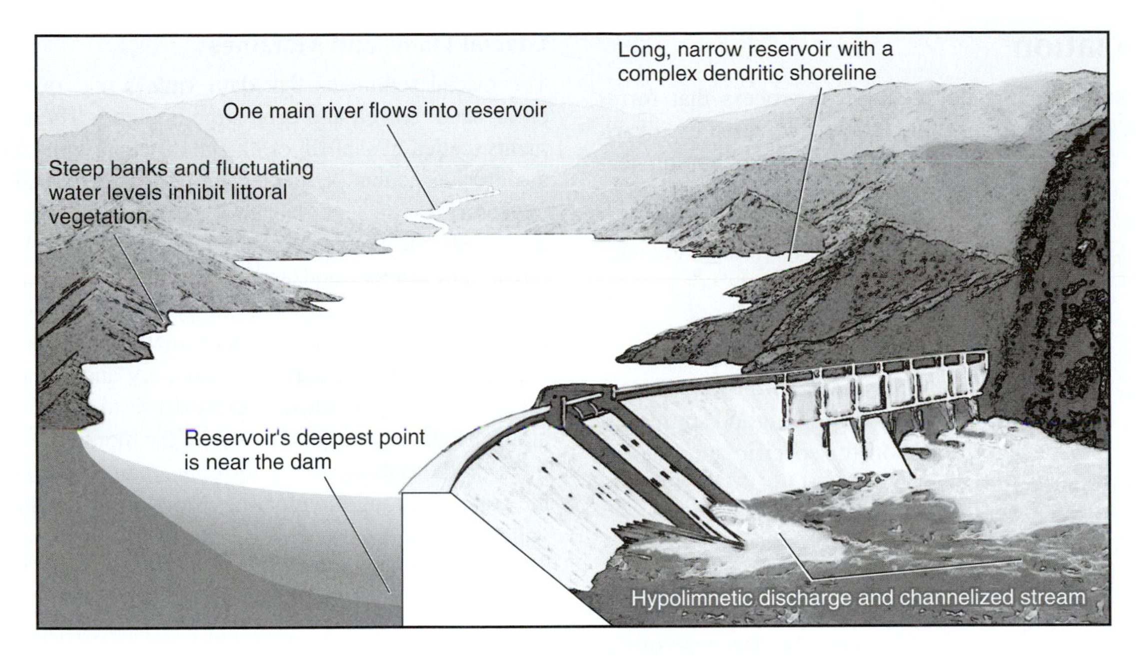

FIGURE 11.11 Diagram of a typical reservoir and dam. The reservoir lies in a steep-sided river valley. This reservoir has a hypolimnetic discharge.

Estuaries

Estuaries are slow-moving streams and associated brackish wetlands that form where streams meet the sea. Like inland wetlands, **estuarine** habitats are often highly productive.

Atmosphere

Water is stored in the Earth's atmosphere as vapor—relative humidity is a measure of air's water content. Relative humidity is the amount of moisture contained in the air, as compared with that of air completely saturated with water at the given temperature. Warm air holds more water than cold air: 1 m^3 of air holds 4.8 grams of water at 0°C, but 30 grams of water at 30°C.

Evaporation is often a significant part of the water leaving a lake. For example, on an annual basis, Lake Wingra, Wisconsin loses as much as 26% of its incoming water through evaporation (see figure 11.9). In extreme dry, sunny climates, the majority of water that leaves a lake is often by evaporation. This invisible flow of water has strong effects on the lake's heat budget, the vertical circulation of water, and the concentration of solutes.

The great Tao flows everywhere.
All things are born from it, yet it doesn't create them.
It pours itself into its work, yet it makes no claim.
It nourishes infinite worlds, yet it doesn't hold on to them.
Since it is merged with all things and hidden in their hearts, it can be called humble.
Since all things vanish into it and it alone endures, it can be called great.
It isn't aware of its greatness; thus it is truly great.

Stephen Mitchell, 2000
Chapter 34, *Tao Te Ching: A New English Version.*
New York: Harper.

LAKE ORIGINS: HOW LAKES ARE MADE

The formation and life history of lakes is an important part—the dynamic dimension—of landscape ecology of lakes. The following discussion will include major ways in which lake basins are formed.

Glaciation

Glaciation: is a major geological process that forms lakes. Nonglacial lakes lie in basins created by a variety of agents, including erosion and sedimentation, tectonic and volcanic activity, and animal behavior.

> "Of all the larger features of the landscape, the lake is the youngest and the most temporary"
>
> E. A. Birge, 1907

Local geographies, such as *Geography of Wisconsin* by R. W. Finley, (1975), are often good sources of the history of lake formation for specific areas. A detailed resource for lakes of the world is G. E. Hutchinson's *Treatise on Limnology,* volume 1 (1957).

Continental glaciers are caused by accumulation of snow during cool climate conditions. Accumulated snow compacts to form ice, and the mass slowly spreads out, like syrup on a pancake. Because of the location of the continents, continental glaciers have been restricted to the northern continents, although each bout of continental glaciation is accompanied by extension of mountain glaciers in the southern continents. Continental glaciers have been occurring for about the last 1.5 million years. Typically, glaciers have been present for about 100,000 years, and then melted for about 20,000 years of better weather during an interglacial period. The last bout of continental glaciation ended about 10,000 years ago. The continental glaciers were layers of ice as deep as 2000 to 3000 meters. (!). These immense sheets of ice were able to rework the Earth's surface on a large scale. Most of the lakes in northern North America and southern South America are of glacial origin. Glaciers create basins for lakes by dropping sediments that dam valleys and rivers, leaving icebergs in sediments, and by gouging, grinding, and depressing the Earth's surface.

Alpine Glaciation

Smaller mountain (alpine) glaciers advanced significantly as recently as during the "little ice age," a period of lower temperatures between 1570 and 1860. These glaciers deepened earlier basins or produced new basins. Alpine glaciers often dig out a basin at the base of a steep, semicircular cliff **(cirque).** If the basin fills with water, it is called a **tarn.** Tarns are often part of extraordinarily beautiful scenery (plates 2 and 30).

Glacial Dams and Moraines

The glacial sediments that dam valleys and rivers are clays, sand, gravel, and even large boulders. These sediments (called glacial **till** or glacial drift) are remnants of the erosive action of glaciers moving across bedrock. **Moraines** are piles of glacial till. Ground moraines were deposited beneath glaciers, lateral moraines formed along the sides of glaciers, and terminal moraines are piles of till pushed ahead of glaciers and dumped as glaciers melt at the edges. When the accumulation/melting ratio is 1, the edge position of the glacier is stationary and the glacier acts like a conveyor belt, delivering debris to a single band along the glacier margin. The size of the moraine depends more on the stability of the terminus than on the size of the glacier making the moraine. Moraines often dam up valleys or streams. For example, the Madison lakes in the vicinity of the University of Wisconsin are lying in basins in the scoured-out bed of the Yahara River behind ground moraines. As is the case with many lakes associated with moraines, the Madison lakes are lying in basins mostly filled with glacial till: Lake Mendota is at most 27 meters deep, but the basin is as much as 100 meters deeper.

Devil's Lake in the Baraboo Hills of Wisconsin is an unusual example of a lake formed by terminal moraines, in the preglacial bed of the Wisconsin River (figure 11.12) (Finley, 1975). The lake water is about 13 meters maximum depth, resting on another 107 meters of glacial till in the basin below the water. The lake is dominated by striking purple cliffs made of Precambrian quartzite. These cliffs tower 150 meters above the lake to the east and west. The Wisconsin River has been flowing through the Baraboo hills, slowly cutting a deep canyon, for hundreds of millions of years. Before the last glacial episode, the canyon cliffs were on either side of the river flowing to the south. As the glacier moved south and west, lobes of thick ice moved into the north and south ends of the canyon, pushing terminal moraines ahead. Just before the two lobes met, the glacier began melting. When the glacier melted, it dumped enough till to effectively seal the north and south ends of the canyon. Devil's Lake currently fills the basin between the cliffs and the two terminal moraines. The Wisconsin River now flows to the south and east of the Baraboo hills, in a broad valley carved out by the glacier.

Kettle Pond

The *kettle pond* is formed among moraines (plate 9). As the glaciers melted, they dumped accumulated till mixed with large pieces of ice. Blocks of ice melted, sometimes

Before glaciation

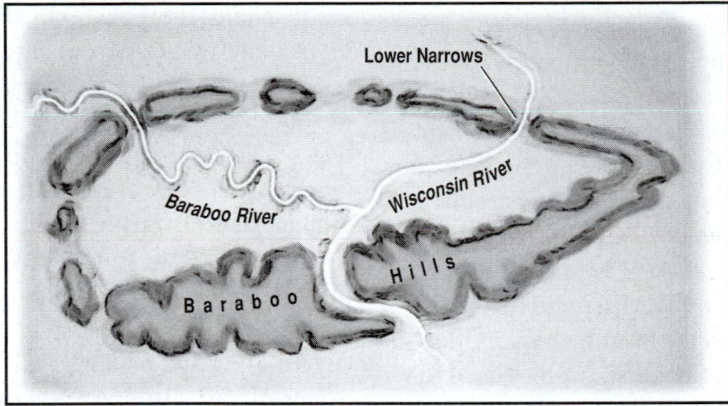

During glaciation

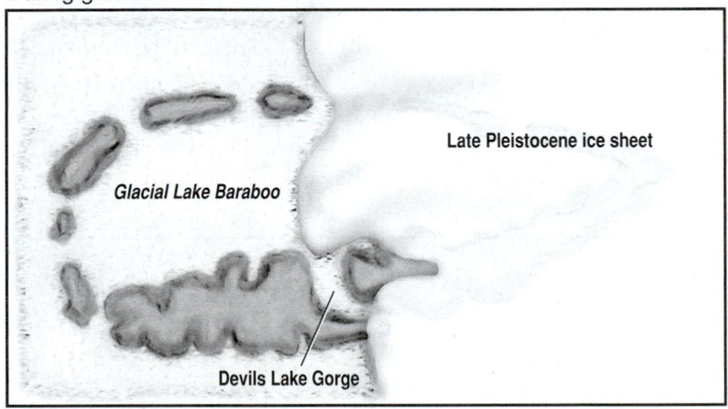

After glaciation

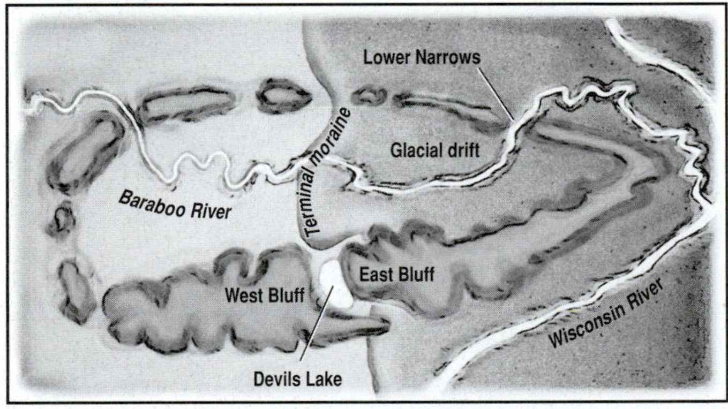

FIGURE 11.12 The interactions among glaciers, topography, the Wisconsin River, and the formation of Devil's Lake, Wisconsin. The Baraboo hills are a range of glacier-resistant quartzite.

producing a basin within the glacial till. These basins typically are round with steep sides, 11 to 50 meters high. The ponds are usually about 100 meters across and only 1 or 2 meters deep. Kettle ponds are very susceptible to erosion, and many are rapidly filling with sediments produced by erosion caused by overgrazing or other shortsighted land uses. The kettle moraine area in southcentral Wisconsin is a region rich in kettle ponds. It marks the remnants of till and blocks of ice piled up between two lobes of the last continental glacier.

Alluvial Dam

Glaciers produce massive amounts of sediments in watersheds. Sediments carried by stream currents are called **alluvium.** Even after the glaciers have melted, rivers in glaciated terrain can be partially dammed by sediments produced by glacial action, forming large, long lakes in the river valley. When two rivers meet, the one that is flowing faster will dump sediments into the other slower stream, forming an alluvial dam. For example, Lake Pepin, Wisconsin is located in the Mississippi River behind a dam formed by sediments from the rapidly flowing Chippewa River. Similarly, Lake St. Croix, Wisconsin is at the end of the slowly moving St. Croix River, where it flows into the more rapidly flowing Mississippi.

Plunge Basin

As the glaciers melted, they released torrents of water. Waterfalls off glaciers produced deep plunge basins as the water fell into lateral or terminal moraines. Plunge basins are associated with gravel and sand bars, unlike kettle ponds. For example, Fish and Sand Lakes in Waushara County, Wisconsin are lakes in plunge basins in lateral moraines that separate the unglaciated portion of southwestern Wisconsin (the *driftless area*) from the rest of the state (figure 11.13). The deep Fayetteville Green Lake, New York is an example of a meromictic plunge basin (Brunskill & Ludlam, 1969).

Blue Lake, along with several other small lakes, lies in a spectacular plunge basin in an old bed of the Columbia River in the Coulee region of eastern Washington (figure 11.14 and plate 40). Blue Lake is downstream of a large glacial lake, Lake Missoula, that formed as many as 18 times in western Montana, when ice dammed a narrow river valley through the Idaho mountains (figure 11.15) (Alt and Hyndman, 1986). Each time the dam formed, the lake grew to about the size of Lake Ontario, and was about 650 meters deep. Because it floats and melts under great pressure, ice makes a poor dam. When

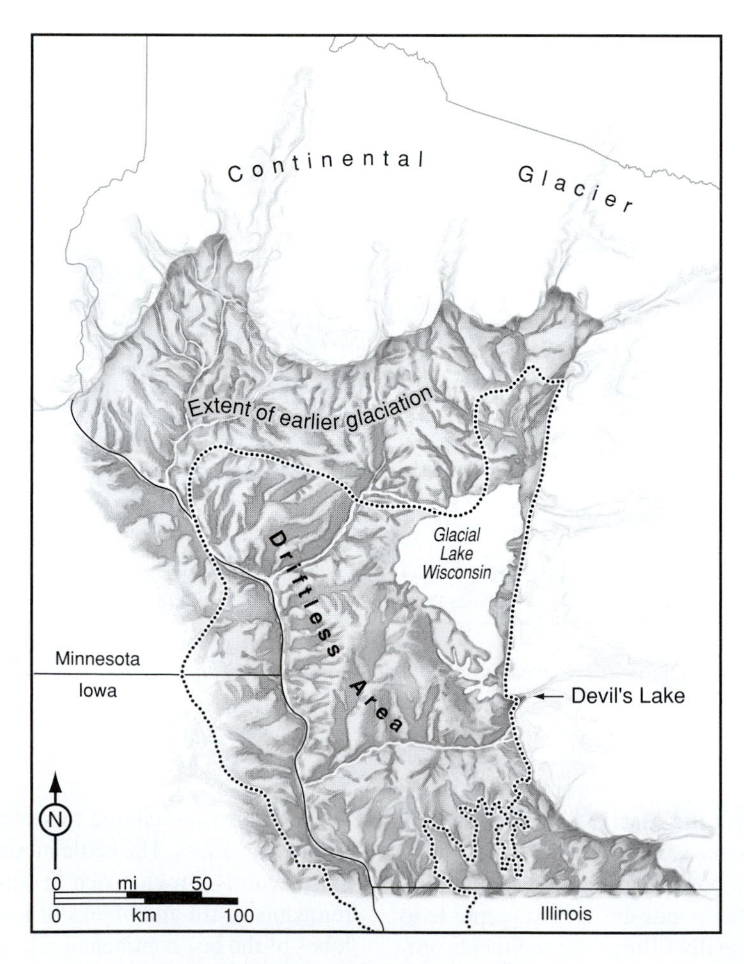

FIGURE 11.13 A map of the driftless (unglaciated) area of Wisconsin.

the huge lake formed behind the ice dam, it was only short lived. When the dam broke, a huge flood of water poured down the riverbed. This torrent flowed across Idaho and into Washington. Part of the flood poured over a cliff of resistant basalt and into a plunge basin (figure 11.14). Each time the dam broke, the plunge basin was deepened, forming the depression that is the current site of Blue Lake and its sister lakes. The evidence for multiple versions of Lake Missoula is recorded as old lake shores high on hillsides (plate 39).

Glacial Scouring

The immense continental glaciers scoured out large basins in bedrock. For example, the Lauretian Great Lake basins were gouged out as the last glacier moved south (uphill) (Powers and Robertson, 1966). These lakes are deep: Lake Superior is 307 meters deep; Lake Michigan is 265 meters deep. The bottom of these lakes is well below sea level (figure 11.16). The lakes are surrounded by old beaches (the beaches of Lake Chicago can still be seen 17, 12, and 7 meters above the present lake, with the highest beach 1 to 3 kilometers back from the present shoreline). These large and deep lakes retain animals such as the crustaceans *Mysis* and *Limnocalanus* and white fish that are typical of Arctic estuarine habitats (Dadswell, 1974). Arctic organisms got to the middle of North America by being pushed! In the central part of North America, glaciers formed in the Arctic and moved south (but slightly uphill). Because they were moving uphill, they pushed before them water filled with Arctic organisms. The waters quickly lost their saline character and became oligotrophic lakes.

FIGURE 11.14 Several lakes in the Grand Coulee region of Washington. From the bottom: Soap Lake, Lake Lenore, and Blue Lake, showing the basalt cliff, old river bed and water fall, and the lakes in a plunge basin. The valley was carved by water rushing out of glacial Lake Missoula, hundreds of km to the east.

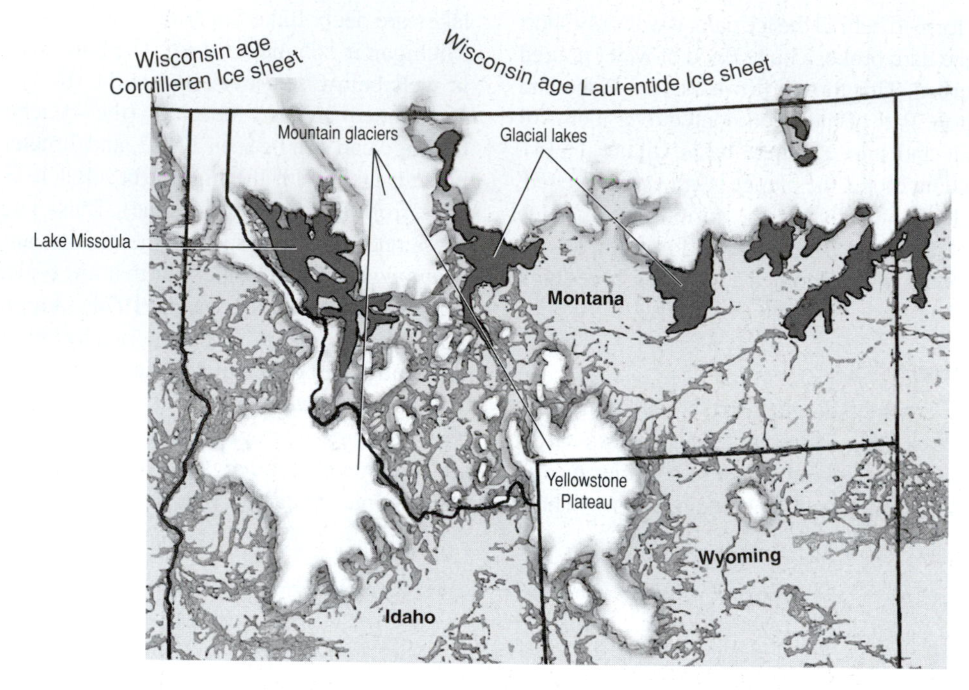

FIGURE 11.15 The extent of continental and mountain glaciation in Montana during the last ice age (Wisconsin age glaciation), showing the position of several proglacial lakes, including Glacial Lake Missoula.

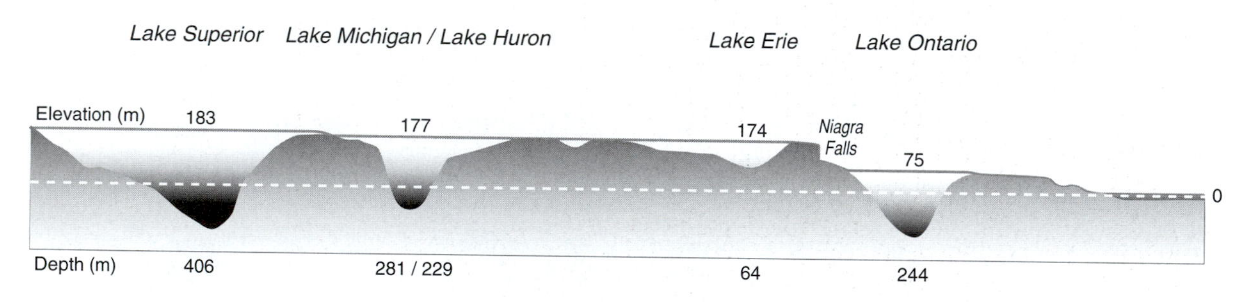

FIGURE 11.16 Cross section of the Laurentian Great Lakes.

Because the lakes formed ahead of the advancing glacier, they are called **proglacial** lakes. The Great Lakes were proglacial lakes, and they still persist in the deep basins dug by the glacier. Many lakes that were dammed by glacial ice, such as Glacial Lake Wisconsin (see figure 11.13) and Glacial Lake Missoula (see figure 11.15), were also proglacial lakes, but have entirely disappeared. We know the lakes did exist because the bottoms remain as sandy plains. Occasional large boulders, glacial **erratics,** are found embedded in the sand, evidence of icebergs that would have rafted the boulders out onto the proglacial lakes.

A special kind of *subglacial* lake forms beneath glaciers (Siegert, 2000) in basins scoured by glacial activity. Sonar exploration of Antarctic ice has revealed several subglacial lakes—the largest is Lake Vostok (78°28′ S, 106°48′ E), lying under glacial ice about 4 kilometers thick! This is a huge lake—the size of Lake Erie—about 14,000 km^2 in area with a maximum water depth of about 500 meters. Sonar data suggest the water is not saline, and probably is the result of geothermal heating at the base of the ice (Jouzel et al., 1999). These lakes are near freezing, pressurized, and are in perpetual darkness. It has been suggested that Lake Vostok is about as suitable for life as possible subsurface water on Mars or subglacial seas on the Jovian moon Europa.

Many of the smaller lakes in northern Wisconsin resemble Canadian shield lakes (e.g., in Ontario) in that they are usually shallow, lie on exposed bed rock, and have very thin soil (plates 26 and 28).

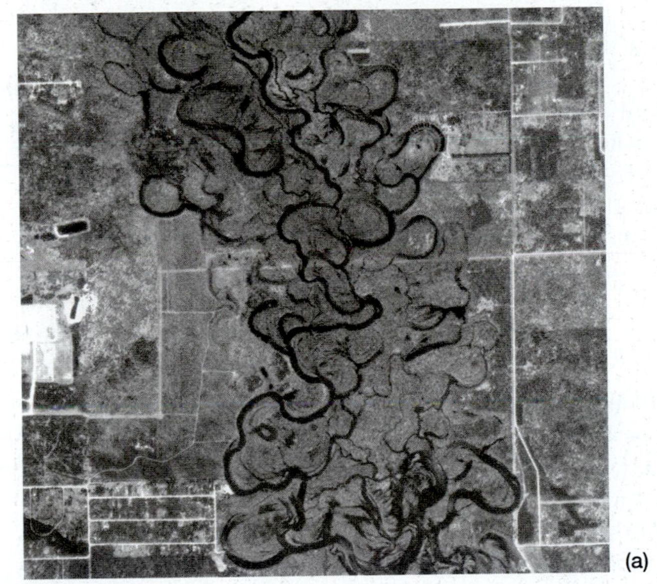

(a)

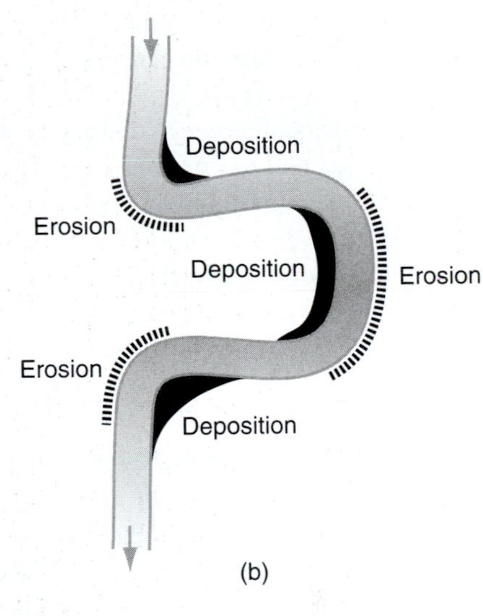

(b)

(c)

FIGURE 11.17 (a) A meandering stream with associated oxbow lakes (USGS photo). (b) A model showing erosion and sediment deposition sites along an oxbow. (c) An oxbow pool in the East River, near the Rocky Mountain Biological Laboratory, Colorado, 1978. The river flows from right to left.

Glaciers are so massive that they depress the continental surface to about one-third the height of the glacier. After the glacier melts, the crust slowly rebounds. Hudson's Bay is approximately the center of the depressed area in North America. The crust is expected to rise about 75 meters more, which will drain Hudson Bay. As the crust slowly rises, the Laurentian Great Lakes will become shallower and many streams will change course.

Glaciers are wonderful lake-making machines. However, lakes do occur in unglaciated areas, the result of a number of processes that create basins that can hold water.

Oxbow Lakes

Rivers can make lakes in low-gradient valleys by meandering. Streams meander (wander back and forth) in wide, flat valleys with small gradients (figure 1.17a and c). As the river channel snakes back and forth, the outer bank erodes (where the current is strongest) and sediment is deposited along the inner bank where the current and erosion is slowest (figure 11.17b). A small curve in a stream can grow into a wide loop (plate 27).

This erosional process continues until loops cut themselves off and the stream flows across the base of the loop and dumps sediments into the old riverbed (as shown in figure 11.17a and 11.17c). The resulting half-moon portion of the riverbed, dammed off from the main channel, is called an **oxbow** *lake.*

Sinkholes in Karst Regions

In unglaciated areas, several processes produce basins for lakes. Sinkholes occur in areas having a layer of partly eroded limestone near the Earth's surface (**karst** regions). Sinkholes are typically shallow and more or less round—for example, the pool in the Owens Valley of eastern California (plate 7). Streams often flow through caves in limestone. Rainwater dissolves limestone, producing caverns. If the cavern is near the surface, and if it collapses so as to seal the basin, a small lake can be created in a steep-sided, rock-walled basin. Much of the Yucatan Peninsula of Mexico and northern Guatemala is a karst region dotted with sinkholes, such as the Laguna de Petenxil (Cowgill et al., 1966), which often contain pools, and sometimes crocodiles, along with human bones and artifacts. The implication is that humans were sacrificed by being thrown into the steep-sided holes.

Rock Pools

Small pools (10 centimeters to a few meters across) are often seen in depressions along the top and edges of bare rock cliffs. Rock pools are found in sandstone, which weathers easily, and harder rock such as granite (see figure 1.1 and plate 14). Rock pools do not occur in rock that fractures easily, such as limestone. Under optimal conditions, rock pools perform self-assembly. If the stone holds water (sandstone is especially sponge-like, without having fractures to drain off water), the rock pool can enlarge itself. When the surface of the stone freezes in winter, water held near the surface of the stone freezes and expands, breaking off bits of the stone and creating a small depression. If the depression holds water, the process accelerates. If the broken chips and sand are removed from the depression by wind or driving rain, a basin is made available for a small pool. Rock pools are usually temporary, filling from rain or snowmelt and evaporating to dryness several times a year. Rock pools are the only kind of lake in most nonglaciated desert areas, especially if there are no streams (for oxbows) or limestone (for sinkholes).

Frost Polygons

In the Arctic, the soil is permanently frozen—the permafrost. A meter or so at the surface of the tundra thaws each summer, but then refreezes each winter. This cycle of freezing and thawing can produce shallow basins called *frost polygons,* with trenches between the polygons. The polygons are about 10 meters across and less than a meter deep. The flat land near the Arctic coast of Alaska is covered with innumerable polygon and trench ponds (figure 11.18).

FIGURE 11.18 An aerial view of tundra lakes and ponds (polygons) on the coastal plain near Point Barrow, Alaska, about 25 June 1976. The lakes are about 100 meters long, and the frost polygons are about 10 meters in diameter. The lakes still have floating ice.

Tectonic Activity

Continent-sized pieces of the Earth's surface (**tectonic** plates) move slowly relative to one another. Where two plates meet, the surface of the Earth will sometimes break (fault or rift), with one plate moving downward relative to the other. The fault lines can be short (1 kilometer or so) or thousands of kilometers long. The African rift lakes, such as Lake Tanganyika, occur along just such a major fault line, which extends north to the Dead Sea in Israel and on to Lake Baikal in Russia. These lakes tend to be millions of years old and hundreds to thousands of meters deep. These are lakes on a grand scale!

The basin created by broken crust is called a **graben** ("a ditch" in German) and can host a lake. Grabens are often filled with a deep lake lying on even deeper sediments. For example, Lake Baikal, Russia is a 1741-meter-deep lake (more than a mile!) on top of more than 3000 meters of lake sediments (figure 11.19). The broken crust often forms cliffs on the sides of a graben lake.

Volcanic Activity

The basin of Lake Superior was formed about 1.2 billion years ago, in the last major **orogeny** (mountain building) episode in northern Wisconsin and southwest Ontario. Volcanoes spewed out as much as 24,000 cubic miles of basalt, an outpouring that was associated with movement of crustal rocks, producing the Lake Superior basin. The basin was then filled with Precambrian and Cambrian sediments. Erosion over the past 550 million years cut into the sediments, forming river valleys. Glaciation hollowed out the basin over the last 1.5 million years.

FIGURE 11.19 Lake Baikal, Russia, a graben lake in a deep rift in the Earth's surface. (a) Drawing of the western lake shore, showing steep cliffs. (b) A model of a rift valley.

Volcanic activity produces craters, also called **calderas,** which, in some cases, hold water. Crater Lake, in Oregon, is a good example of an ultraoligotrophic caldera lake lying in a deep, steep-sided crater (plate 3). The lake basin was formed as magma subsided within the crater. This lake dates to about 6500 years ago, shortly after the mountain (Mt. Mazama) last blew its top. An older and smaller volcanic crater lake is the one that hosts the Lago di Monterosi (figure 11.20) in central Italy (Hutchinson et al., 1970). This lake is about 25,000 years old, in a crater just a few meters deep and mostly filled with wind-blown sediments. Lake Nyos, Cameroon, is a crater lake associated with a still-active volcano (see chapter 10). Guatemala's Lago de Atitlán is an example of a younger caldera lake associated with volcanic activity. This lake lies in a subsidence basin surrounded by still-active volcanoes. The lake itself is probably only a few thousand years old, although the age of the basin may be several times that.

If the volcano is still at all active, the crater lake will probably have an unusual and acidic water chemistry because magma is a source of chemicals such as sulfur (plate 4). Crater lakes have a small watershed (basically, the inside of the crater) and are therefore oligotrophic—for example, Lago di Monterosi and Crater Lake.

Meteorites

Impacts of rocks from the sky also produce deep basins (impact craters) (plate 28). These basins rarely hold water, perhaps because the impact is so strong it fractures the underlying bedrock, making a leaky basin. Impact craters can be recognized by the modified bedrock containing "shocked" minerals, or by the shape of the basin, which is either round or elliptical. The impact of a stony meteorite in northern Argentina created several basins that filled with water and became lakes (Schultz & Lianza, 1992). A much older impact site in northern Québec, Canada has been modified by geological processes into an unusual donut-shaped basin and lake, Lake Manicougan (plate 28).

Biological Activity

Several species of relatively large animals can make lakes. Some animals actually build dams, others wallow, and humans have made innumerable lakes.

Beaver Ponds

Beaver ponds are backed up against a twig and earthen dam built across a stream or river by large rodents (plate 6). These ponds are typically found associated

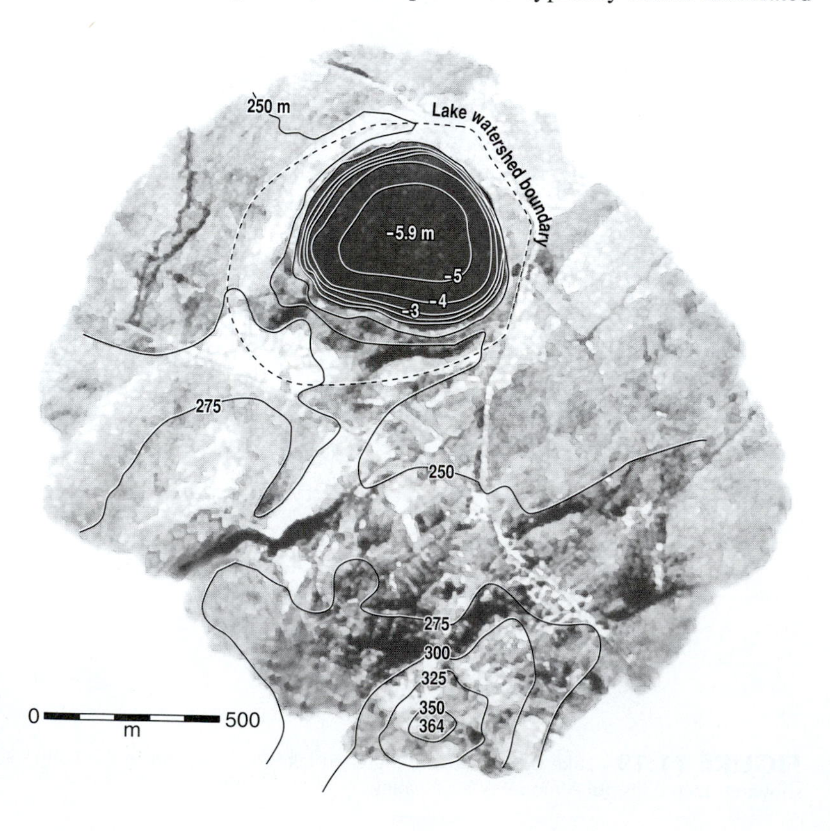

FIGURE 11.20 Lago di Monterosi, Italy. This topographic map shows the small and circular watershed characteristic of a lake in a volcanic crater.

with streams and in areas with deciduous trees, especially aspen or willows, which serve as food for the rodents. Human trapping and hunting have greatly reduced the number of beaver in North America since the European invasion (1492). Before this increase in hunting pressure, there were as many as 400 million beavers widely distributed between the Arctic tundra and northern Mexico (Naiman et al., 1988). This many beavers would have built millions of dams in streams, producing wetlands, ponds, and lakes.

Wallows

Shallow, temporary ponds in prairies are visited by large herbivores, such as cows (and previously bison and even rhinoceros species, before the great Pleistocene extinction). The large herbivores drink, wade, and roll about (**wallow**) in the mud. This discourages plant growth. When the pool dries out, the exposed, disturbed soil can be removed by the wind, deepening the basin. These pools are called by the Spanish name **playa** in the southwestern United States (plate 11).

Human Activity

Human activity has resulted in unbelievable numbers of reservoirs (plates 12 and 28), livestock ponds (plate 10), farm ponds for fishing and swimming, urban retention ponds, and quarry ponds. Even in desert areas, as in northern Mexico, there appears to be at least one artificial pond per square kilometer, and artificial agricultural ponds are even more common in the mid-latitude temperate zone.

Some of the oldest artificial ponds are **salterns,** which are small basins (usually 10 to 100 meters in diameter) built near the sea for the purpose of evaporating salt water. For example, on the coast of the Yucatan Peninsula, the Mayan people built stone-walled pools about half a meter deep (plate 8). Seawater could be let into the pond through a channel at high tide, and the channel was then closed. After a few weeks, only a layer of salt remained on the saltern's bottom.

STREAM ORIGINS

Streams develop spontaneously as the power of flowing water cuts channels into the watershed. **Erosion** is the process of moving material downstream. The stronger the gradient, the higher the velocity of water flow, the greater the power of flowing water, and the larger particles that can be moved. Slow-moving water carries only fine sediments, such as **silt** (fine clay particles), while steep gradients can carry large (meter-scale) stone **boulders.**

A drainage basin is a system of connected streams that drain a watershed. Streams can capture additional drainages (watersheds) by eroding (cutting down) the ridges that separate watersheds.

Stream valleys in glaciated areas show clear signs of having been much larger at an earlier time. It is common to see a small stream in a much larger valley (plate 27), with large cobbles high above the present stream. The large valley indicates a higher discharge, probably related to melting of glacial ice.

> "It [the Big Blackfoot River in western Montana] runs straight and hard—on a map or from an airplane it is almost a straight line running due west from its headwaters at Rogers Pass on the Continental Divide to Bonner, Montana, where it empties into the South Fork of the Clark Fork of the Columbia. It runs hard all the way.
>
> . . . From its headwaters [on the Continental Divide] to its mouth it was manufactured by glaciers. The first sixty-five miles of it are smashed against the southern wall of its valley by glaciers that moved in from the north, scarifying the earth; its lower twenty-five miles were made overnight when the great glacial lake [Glacial Lake Missoula] covering northwestern Montana and northern Idaho broke its ice dam and spread the remains of Montana and Idaho mountains over hundreds of miles of the plains of eastern Washington. It was the biggest flood in the world for which there is geological evidence; it was so vast a geological event that the mind of man could only conceive of it but could not prove it until photographs could be taken from earth satellites."
>
> *Norman Mclean, 1976, "A River Runs Through It"*

DEVELOPMENT OF LAKES AND STREAMS OVER TIME

Glacial lakes have a life span, depending on the depth at which they begin, of only a few thousand or tens of thousands of years. Animals that live in fresh water have existed for about 500 million years. During this long stretch of time there have been few periods of glaciation. There was a major episode of glaciation at the end of the Permian, about 255 million years ago. Another major glaciation episode began about 1.5 million years ago. Thus, freshwater aquatic plants and animals have lived, for most of their history, in lakes and streams of nonglacial origin.

Sedimentation

The present abundance of glacial lakes is a temporary richness in the landscape. These lakes will fill up with sediments within the next few thousand years or be resculpted by the next round of continental glaciation. Lakes become more shallow and eventually become wetlands by one of two paths: sedimentation with mostly inorganic material such as silt, sand, and gravel; or accumulation of organic sediments (**peat** is compressed plant material preserved in aquatic habitats).

Drainage lakes in an extensive watershed tend to fill with inorganic sediments eroded from upstream. Lake Mendota, Wisconsin is an example of this developmental process: The lake is slowly filling in via sedimentation. The sediments are a mixture of organic and inorganic materials from erosion enhanced by current agricultural and construction practices. Although Lake Mendota has a maximum depth of 27 meters of water, the basin is much deeper. About 70 to 100 meters of mostly inorganic sediments (gravel, sand, and mud) have accumulated in 9000 years. An extreme example of slow sedimentation is that of the Italian lake, Lago di Monterosi. Here, the rate of inorganic sedimentation was very slow—about 2.4 meters in 23,000 years. At the other extreme are reservoirs with intensive agriculture or grazing in their watershed. The Lago Loíza was built to provide drinking water for San Juan, Puerto Rico, in 1953 (Webb & Soler-López, 1997). Erosion in the watershed and sedimentation in the basin have been rapid because of removal of vegetation and large floods (these are not independent events!). The reservoir was built with a maximum water depth of about 24 meters but had a maximum depth of 15 meters in 1994. In only 41 years, half the original capacity of the reservoir became filled with sediment.

Instead of sediment, peat accumulates in seepage lakes, especially those with a floating vegetation mat and acidic water. The vegetation mat is composed of species adapted to having their roots in or near the water. It is possible to walk on the floating mat, but it provides unstable footing, moving with a wave-like undulation. Dead vegetation accumulates under the floating mat. Decay of organic material is slower than production of plant material in cool, acidic water, especially if oxygen is absent. Undecayed plant material accumulates on the bottom of the lake. When this material is compressed, it becomes peat. As this vegetable matter partly decays, it produces organic acids that color the water brown. Ponds with floating mats are called *quaking bogs* (plate 15).

Stream Development

Streams also show development over time. Unlike lakes, streams tend to persist. While lakes are made by outside forces (such as glaciers), streams make themselves. Lakes slowly fill in with sediments, while streams remove sediments, as long as they have a gradient. Sources of water, such as springs, begin cutting into the soil. As time goes on, erosion deepens the stream bed. This process continues indefinitely. Streams can persist for tens of millions of years and cut spectacular canyons, such as the Grand Canyon of the Colorado River or the Goose Necks of the San Juan River (plate 27). Eventually, erosion flattens all gradients, or tectonic movements of the Earth's surface change continental gradient patterns.

Streams typically have a higher gradient near their sources. Sediments are carried downstream and, over time, deposited in valleys toward the sea. Young streams tend to be straight, with high current velocities that carry large amounts of sediments in *braided* channels. Braided streams have many connected, relatively straight channels separated by bars of large-sized sediments (gravel cobble). As streams age, meanders develop (if the gradient is not too great). The change in morphology as streams age has been used (Ward et al., 2000) to interpret the fossil record of the great mass extinction at the end of the Permian period (255 million years ago). Sediments laid down at the time of the extinction suggest a major terrestrial catastrophe, such as would be caused by a major asteroid impact (similar to the impact that probably caused dinosaur extinctions 65 million years ago). South African stream beds changed suddenly from braided to straight, indicating a sudden increase in erosion and implying that a natural catastrophe had removed terrestrial vegetation.

Dam Removal

Reservoirs typically do not endure as long as natural lakes because the age of the reservoir is determined by the capacity of the basin and the nature of the dam (Stanley & Doyle, 2003).

As reservoirs age, they collect sediments from the input streams. Reservoir basins typically fill rapidly because they receive water from high-gradient streams carrying heavy sediment loads. Dams are typically built

FIGURE 11.21 Dam removal on Koshkonong Creek, Rockdale Wisconsin in 2000. The remnants of a low dam are visible in the foreground. The stream is cutting through and resuspending the fine mud sediments that had accumulated behind the dam. **Source:** photo Dscn0143 by Martin Doyle.

in steep stream valleys for purposes of flood control, water storage, or for generation of hydroelectric power. The steep valleys necessarily come with fast flow and high sedimentation rate. Thus, reservoirs typically fill up within decades or a few centuries, while natural lakes can persist for millennia.

When a dam is removed, huge amounts of fine sediments are resuspended by the flowing stream (figure 11.21). For example, a major reason for retaining dams on the Colorado River is that there is currently no way to manage the huge amount of sediment that would be mobilized with dam removal (Walters et al., 2000). Sediments accumulated in a reservoir are typically fine-grained silt that is high in nitrate and phosphate, which poses a long-term threat to riparian habitats and lakes downstream, as well as to estuarine and marine habitats (Stanley & Doyle, 2002).

Dam Failure

The materials and design used in dams also can limit the age of a reservoir. The longest-lasting dams are made of cut stone or cement reinforced with metal. Earthen dams are particularly short-lived. An important design principle is to have a way for water to overflow the dam without eroding the dam. A **spillway** is often cut into rock near the dam as an overflow device (see figure 11.11). The spillway allows surface water to escape safely without flowing over the dam itself. Also, safe dams often are designed to have a deep (hypolimnetic) discharge. Pipes can be placed in the bottom of the dam so that water can be released at will, to lower water level in case of flooding and often to generate electricity. If these principles are not followed, the results can be disastrous.

Two famous catastrophic dam failures in North America are the 1928 collapse of St. Francis Dam, California, and especially the 1889 Lake Conemaugh disaster and Johnstown's Flood. The latter provides a poignant example of the extreme potential danger of poorly designed and maintained dams (Cambor, 2000). Starting in 1839, the dam for a reservoir (later called Lake Conemaugh) was constructed on the South Fork tributary of the Little Conemaugh River near South Fork, Pennsylvania, about 10 miles northeast of Johnstown in a steep Allegheny mountain valley. (A woodcut of Lake Conemaugh can be seen at the website listed at the end of this chapter.) The reservoir was constructed as a reservoir for a canal and railroad system. The dam was composed of a stone core covered with clay and earth. Hypolimnetic discharge was possible through four 2-foot diameter cast iron pipes (sluice pipes). A spillway was cut in rock alongside the dam to allow for safe release of surface water when the reservoir was at maximum capacity. The dam was 900 feet across, 72 feet high, and 270 feet thick at the base. It had a steep outer face and a more gently sloping face into the lake. The reservoir was probably never deeper than about 20 meters, with a maximum surface area of about 180 hectares. The basin began to fill in 1853 when the sluice pipes were closed.

In 1857, the Pennsylvania Railroad bought the land and lake basin from the Commonwealth of Pennsylvania. In 1862, the dam developed large leaks during a summer rainstorm, but the lake level was low and the sluice pipes were open in time to avert disaster. After the failure, the lake was drained. In 1875, the railroad sold the property to a local politician, who removed the sluice pipes to sell for scrap iron in 1879 before selling the property to a developer. The developer organized and built the South Fork Fishing and Hunting club, a country club association of steel, banking, and related businessmen from Pittsburgh, Philadelphia, and New York.

The dam was repaired in a slipshod manner in 1879, and the lake was then refilled, stocked with fishes (black bass), and graced with sailboats and iceboats. The spillway was blocked by a fish fence built to keep fishes in the lake. By 1881, the lake, repaired with earth and wood, had no hypolimnetic discharge and a blocked spillway. During an intense rainstorm on 31 May 1889, the reservoir overflowed the dam, which eroded away in minutes. The burst of water killed over 2209 people and injured thousands more in the valley below, in the Johnstown area. The lake water traveled in a wave 11 to 13 meters high and 1 kilometer wide for 22 kilometers down the valley, dropping about 150 meters in elevation in a few minutes. All of the club members survived. The site of the lake is now a U.S. National Park Service *National Memorial,* providing camping and nature walking facilities.

PALEOLIMNOLOGY

Evidence for the formation and development of lakes comes from analysis of lake sediments. The sediments are cored with a pipe and the sediment plug is pushed out of the tube. A core of a bog lake of glacial origin might be 10 or even 20 meters long. The longest cores are from ancient lakes. A 1.4-kilometer core taken from Lake Biwa, Japan, yielded microfossils as old as about 430,000 years (Nishino & Watanabe, 2000).

Lake sediment can be dated and analyzed for microfossils. Dating is usually achieved by measuring concentrations of radioactive elements in the sediments. Small, resistant parts of organisms that fossilize include pollen grains, diatom and dinoflagellate shells, pieces of water fleas, and the hard head capsules of insect larvae.

A natural radioactive isotope, ^{14}C (carbon 14), is used for dating lake sediments. This dating technique depends on the steady rate of radioactive decay of ^{14}C in sediments. The ^{14}C is created in the upper atmosphere: A neutron combines with ^{14}N to produce ^{14}C and a proton. Photosynthesis incorporates both ^{14}C and the more common and stable ^{12}C into organic matter. Once the ^{14}C enters organic matter and is buried in lake sediment, the radioactive isotope slowly decays. The half-life of ^{14}C is about 5760 years, so that carbon between about 500 and 50,000 years old can be dated. The technique assumes the rate of formation of ^{14}C has remained constant, and is the same now as in the past (although production has increased in the last few decades because of atmospheric nuclear tests). In fossil organic material, the ratio between ^{14}C and ^{12}C gives an estimate of the age of the carbon.

Other radioactive elements are also used for dating sediments. Radioactive lead (^{210}Pb) is a decay product of the radioactive gas radon. ^{210}Pb has a half-life of about 22.26 years, so it can be used to date sediments 1 to 150 years old—too young to be dated by the ^{14}C method. For sediments too old for the ^{14}C method, there are slower-decaying radioactive elements, such as the potassium-argon (^{40}K–^{40}Ar) dating technique. Radioactive potassium incorporated in sediments decays slowly into argon, allowing dating from less than 1 million years back to 4 billion years.

In some lakes, especially meromictic lakes, age estimates from radioactive decay can be calibrated with direct counts of **varves** in sediments. Varves are distinct sediment layers, composed of a light and a dark layer. One varve is produced per year. Varves are characteristic of oligotrophic lakes, in which there are few benthic animals to stir up sediments. Varves are especially characteristic of meromictic lakes, in which there is little or no disturbance of the sediments by benthic animals because of the anoxia in the monimolimnion. Varves can also be used to estimate the age of fossil lakes, such as the lakes that produce "oil shale" sediments in western Colorado, or the Early Triassic fossil lakes in New Jersey studied by Amy McCune (1996).

Ash layers from volcanic eruptions and special sediments associated with historical events near a lake can also be used to calibrate dates from radioactive decay. For example, Lago di Monterosi (Hutchinson et al., 1970) has a layer of charcoal dated at between 150 and 285 B.C. There are written records of a road built past the lake in 171 B.C. It is quite possible that the charcoal layer is the result of the road building.

The age range for carbon dating is good for late Pleistocene and recent sediments—remember that most glacial-origin lakes are less than 12,000 years old. Even Lago di Monterosi (also called *Ianula*), which is at least 23,000 years old, can have its entire history dated using the ^{14}C method (Hutchinson et al., 1970).

An excellent example of reading the history of a lake from its sediments comes from Lago di Monterosi (see figure 11.20), a small crater lake in central Italy (Hutchinson, 1970). Professors Edmondson (then at University of Washington) and Bonatti (University of Pisa) took the first two cores in 1959, Hutchinson collected rock and water samples in 1961, and Clyde Goulden collected another core in 1964.

Based on experiences with bog lakes in North America, Edmondson expected the Italian cores to be several meters long. Instead, the first core could be pushed into the sediments only about 250 centimeters. Edmondson thought he had hit a rock or a solid part of the sediments, so he pushed harder and broke the coring device. When the replacement device arrived, Edmondson once again found only about 250 centimeters of sediment. He decided this was better than nothing, so he sent the sediment cores to Hutchinson. Much to everyone's surprise, the short core contained information about the lake history back to its beginning, about 23,000 years ago. Hutchinson saw to it that the chemistry, pollen, and crustacean and insect microfossils were analyzed. In summary, the group found that:

Before 23,000 years before present (B.P.): At the oldest (bottom) part of the core (23,000 years old) *Artemisia* (sage brush-like plants) and water lily pollen indicated a lake 1 to 3 meters deep, surrounded by tundra-like vegetation in a dry, cool landscape. The ratio of inorganic to organic sediment remains suggested moderate erosion and sedimentation rate.

23,000–13,000 B.P.: Pollen indicated that the lake was surrounded by pine trees and *Artemisia* during this time when the continental glaciers covered northern Europe. The lake experienced low sedimentation and low productivity.

13,000–5,000 B.P.: The surrounding vegetation was changing to oak, fir, grasses, and hazelnut as the glaciers melted. Minimum lake depth occurred during this time of cool and dry climate. Analysis of aquatic macrophyte pollen indicated that the lake was inhabited by plants currently found in water less than 1 meter deep. Archaeological data indicated that humans were present, but they left no evidence in the lake sediments.

5,000–1,200 B.P.: Compared to older sediments, the sediments of this age contain more oak, chestnuts, and weedy species such as nettles. Pollen analysis suggested the lake was about 8 meters deep, indicating a wetter climate than had occurred previously. The maximum rate of sedimentation occurred during this time. During this time, pastoralists and farmers moved into the area. As the Roman culture developed, the area around the lake became more heavily populated and cultivated. Romans constructed the road *Via Cassia* through the lake's catchment basin. Burning of vegetation associated with road building probably produced an ash layer in the sediments. Roman records record the date of construction as 171 B.C. The ash was ^{14}C dated as 150–285 B.C.—so in this case, historical records nicely corroborate the carbon dating technique.

1,200–present: Pollen in superficial sediments indicated a continuing increase in oak and other deciduous trees in the surrounding forests. Increasing agriculture is indicated by the increase in cereal and weed pollen.

Thus, the story of Lago di Monterosi is the story of the great sweep of European history, seen from a limnological perspective. The book is a little grubby, but it can be read by those who know the secrets of paleolimnology.

Paleolimnologists have been analyzing microfossils, nutrients, metals, and pigments in lake sediments for half a century, but it is only recently that technological advances have opened two additional doors on the past: recovery of ancient DNA and resurrection ecology (Jeppesen et al., 2001). Using polymerase chain reaction **(PCR)** technology, trace amounts of DNA from organic material in lake sediments can be amplified to amounts large enough to sequence. This allows for species-specific identification of buried material and for identification of (some) inhabitants of the lake over time. *Resurrection ecology* refers to studies of animals (copepods and branchiopods) hatched from decades-old resting eggs (Kerfoot et al., 1999). For example, Hairston et al. (1999) hatched *Daphnia* eggs from sediments of Lake Constance, Germany to reveal an extraordinary example of rapid microevolution. The originally oligotrophic lake experienced cultural eutrophication with the attendant cyanobacteria blooms, peaking in about 1980. The lake has recently been restored to a more oligotrophic condition, and the cyanobacteria are less abundant. Hairston et al. (1999) hatched out resting eggs from sediment deposited over the period of the last 40 years. They fed the resurrected *Daphnia* toxic cyanobacteria and measured *Daphnia* growth. *Daphnia* from before and after the eutrophic period were sickened by the toxic algae and grew less than *Daphnia* from the period of cyanobacteria peak abundance. The differences in growth relative to diet are associated with genetic differences. Thus, in this example of resurrection ecology, the *Daphnia* population adapted (change genetically) at a rapid rate (on the time scale of a few years) to a changing food supply.

SUMMARY

This chapter focused on the landscape-scale processes that create and destroy lakes and streams, on the connection of aquatic habitats to other features of the landscape. Today, the most significant feature of landscapes is the human influence (Vitousek et al., 1997). For example, as of 1997, humans had transformed about one-half of the Earths' surface (watershed) and were using about one-half of all the Earth's fresh surface water, mostly for agriculture. Water is limited in many parts of the world and is likely to soon be of great political importance. Chapter 12 explores the human connection to aquatic habitats, via their modes of appreciation and their attempts at management.

Study Guide

Chapter 11 Water in Landscapes

Questions

1. How do you measure the volume of a lake? Practice being a physical limnologist in a lake near you by measuring the lake's volume using information in this chapter (see the following exercise).
2. How do you measure discharge of a stream? Practice being a physical limnologist by measuring a nearby stream's volume using information in this chapter.
3. Should the 1977 water input balance the water output for Lake Mendota (see figure 11.11b)? What could explain the difference? Notice that the value for groundwater is a net value. What happens to the estimate of water residence time if there is a high rate of recharge and discharge through the sediments?
4. What processes are causing lakes to fill up in your local area?
5. For streams in your local area, how well does stream primary productivity and insect community composition fit the predictions of the River Continuum concept?
6. How does the hydrological cycle for a specific place depend on local conditions of geography, geology, and climate? To explore this question, draw the hydrological cycle for your local area.
7. Many processes in limnology are scale-dependent. Debate the assertion that "landscape ecology can be done at any scale."
8. Can you categorize your local lakes and streams according to lake position in the landscape and stream order?
9. How does the flood pulse concept integrate the limnology of streams, floodplains, and ponds?

Words Related to Water in Landscapes

Use Figure 11.22 to practice using limnological words related to concepts about water in the landscape.

alluvium
aquifer
aquitard
artesian
bathymetric
boulder
caldera
canalization
catchment
cirque
dendritic
desertification
discharge
ephemeral
erosion
erratics
estuary
floodplain
GIS
GPS
graben
ground water
isobath
karst
landscape
meander
moraine
morphometry
orogeny
oxbow
PCR
peat
playa
proglacial
recharge
saltern
sedimentation
seepage
shape
silt
size
spate
spillway
tarn
tectonic
till
topographic
varve
vernal
wallow
watershed

Across

2. Means the water is stained brown with DOC, in bog lakes
6. Another word for the catchment basin of a lake or stream
7. The small lake in a cirque
9. A kind of lake that lacks a surface outlet—no stream flows out of the lake
10. It holds water underground

Down

1. The measurement of the size and shape of a lake
3. This lake type is found in volcanic regions, a result of subsidence
4. To move back and forth in an aimless curved pattern
5. Salt is precipitated in this kind of shallow pond
8. A layer of lake sediment representing one year of accumulation

FIGURE 11.22

Major Examples and Species Names to Know

Mirror Lake and Hubbard Brook, New Hampshire, studied by Gene Likes, are examples of a well-studied watershed containing a stream and a lake.

Devil's Lake, Wisconsin, is an unusual example of a lake formed behind a dam created by a terminal moraine.

Laurentian Great Lakes, in northcentral North America, are examples of the proglacial lakes and of lakes formed by glacial scouring.

Lake Pepin, Wisconsin, a wide place in the Mississippi River; a lake behind a dam caused by sediments dumped in the slow-moving Mississippi by the fast-moving Chippewa River

Lake Mendota, Wisconsin, as an example of a lake formed by glacial scouring and a morainal dam

Lago de Monterosi, an important paleolimnological site in Italy, studied by Hutchinson and his colleagues

Lake Conemaugh, Pennsylvania, a reservoir that became deadly when its earthen dam failed

What Was a Limnological Contribution of These People?

G. Evelyn Hutchinson

Norman Mclean

Exercises

Lake Morphology

Practice calculating the volume of a lake, using a map with depth contours. Find a map for a lake near you. For example, look at the map for Lake Waubesa, Dane County, Wisconsin, on the Center for Limnology website (see the list at the end of the chapter) (choose "Lake Information"). Note: You can use any bathymetric lake (check the URLs at the end of this chapter) that has clear contour lines.

Print the map. You will need to estimate the area at the depths corresponding to the depth contours. For this exercise, use the surface area given on the web site. There are three ways to estimate the area at the contour interval depths. Choose one method:

1. Use a digitizing pad.
2. From a printed copy, cut around the outline of the entire lake and weigh the cut-out. Use this weight to convert mg of paper to m^3 of lake area using the surface area value from the web site. Then cut around the first contour interval (for example, 2 meters). Weigh the remaining cut-out, which represents the area at a depth of 2 meters. Use the conversion factor to convert the weight of this smaller area to m^3. Repeat for each depth contour.
3. Print or trace the lake onto graph paper. Count the number of squares covered by the map of the entire lake. Use this number of squares to convert from squares to m^3 of lake area, using the surface area value from the web site. Then count the number of squares inside the first contour interval (for example, 2 meters). Use the conversion factor to convert the weight of this smaller area to m^3. Repeat for each depth contour.

A stratum is a layer of the lake between two contour intervals. To calculate the volume of a stratum, use the equation from Hutchinson (1957). The equation is complicated because the area of the top of the layer (A_m) is always larger than the area of the bottom of the layer (A_n), so we need to use an average area. The thickness of the layer is (m–n) meters:

$$V = (1/3) * (A_m + A_n + \mathrm{SQRT}(A_m * A_n)) * (m - n)$$

Note: The area of the deepest spot is zero.
Fill in the following table for your lake. Be sure to calculate the area of each layer. You can get the area at the surface (depth = zero meters) from the web site. The area of the lake at the next depth contour (for example, 2-meter contour) is the entire area at that depth—not just the ring between two contour lines.

Lake name: ________________ Web site for lake data: ____________________________

Lake location (use Topozone.com to find the decimal coordinates)

Latitude: ____________________ Longitude: ___________________

Fill in the following table:

Depth (meters)	Area of Stratum (m^2)	Thickness of Stratum (meters)	Volume of Stratum (m^3)
Surface (0)			

Total volume of the lake = ________________ units __________

Total volume of the lake given on the web site: _____________ units ______________

Explain the difference in values.

Landscape Ecology

The purpose of this exercise is to explore using a satellite image that has several different layers available, including land use. The exercise also provides practice in using a USGS topographic map. These images are for Wisconsin sites, but can be easily modified for other places.

Directions:

Open your Internet browser.

Open Google.com.

Search for "Wiscland land cover."

Click on "View a Landcover Image."

Zoom in on Lake Mendota in Dane County.

Select the "DNR Watersheds" level. Most of the levels are selected by default, except for "DNR Regions" and "Geo Management Units."

Adjust the amount of zoom so that the two catchment basins of Lake Mendota are entirely visible. The watershed lines (boundaries of the catchment basins) are the wiggly, light-blue lines. One catchment basin includes the lake and extends to the northwest. The other basin extends to the north of the lake. These two basins include all the surface area that potentially drains into Lake Mendota.
Click on the "Landcover Key" to view the color codes for the various land covers.
Fill in the following table by estimating percent land cover.

Landcover Type	Percent of the Two Watersheds
Urban	
Agriculture	
Grassland	
Forest	
Open water	
Other	
Total	

Zoom in more to see the individual pixels. Using what you know about distances in Madison, or using additional information from Topozone.com, estimate the size of each pixel (the length of the side) in meters.

Each pixel is _________ m on a side.
Go to topozone.com and find:

The elevation of Lake Mendota: ___________________ units = _____________

Click on the name of the lake in the table of information (under "Place") to see the topographic map. Click on Picnic Point. In what direction does Picnic Point point? ____________

Click on the base of Picnic Point. Choose the 1:25,000 scale. Find the elevation of the highest hill in the Picnic Point—Second Point area on the shore of Lake Mendota (the contour lines are 10 feet apart—look for the 900 ft. line, and follow it): _________ ft
What are the coordinates of the hill in decimal degrees? (Click on the hill, choose DD.DDD at the bottom of the map.) The coordinates of the cursor (on top the hill) are at the top of the map:

Latitude: _________________

Longitude: _________________

Additional Resources

Further Reading

Cambor, K. 2000. *In sunlight, in a beautiful garden.* New York: Perennial Press.

DeMers, M. N. 2000. *Fundamentals of geographic information systems.* 2nd ed. New York: John Wiley & Sons.

Finley, R. W. 1958. *Geography of Wisconsin.* Madison: University of Wisconsin Extension.

Hutchinson, G. E. 1957. *A treatise on limnology. Volume I: Geography, physics, and chemistry.* London: John Wiley & Sons.

Mitsch, W. J., and J. G. Gosselink. 1986. *Wetlands.* New York: Van Nostrand Reinhold.

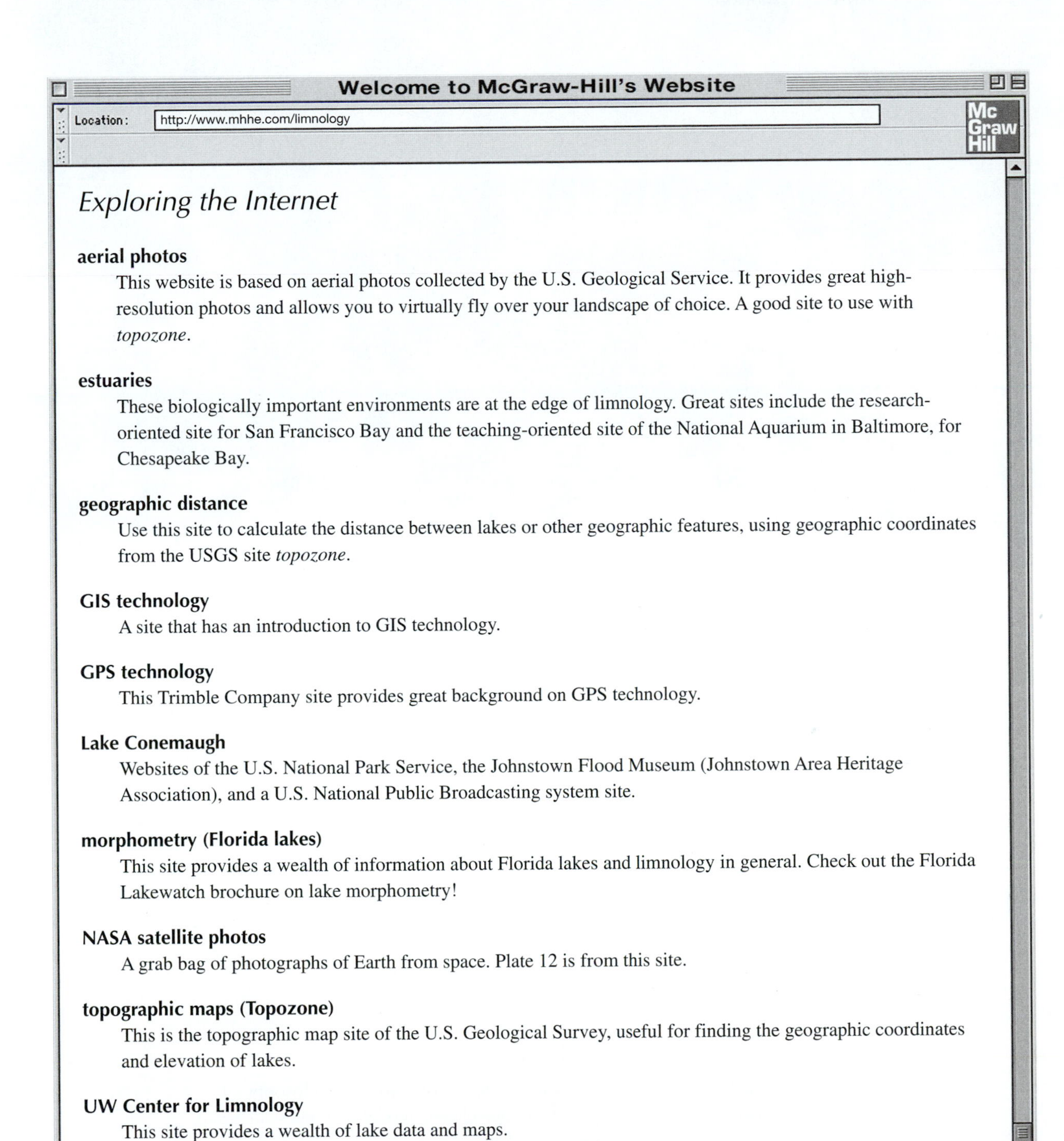

Welcome to McGraw-Hill's Website

Location: http://www.mhhe.com/limnology

Exploring the Internet

aerial photos

This website is based on aerial photos collected by the U.S. Geological Service. It provides great high-resolution photos and allows you to virtually fly over your landscape of choice. A good site to use with *topozone*.

estuaries

These biologically important environments are at the edge of limnology. Great sites include the research-oriented site for San Francisco Bay and the teaching-oriented site of the National Aquarium in Baltimore, for Chesapeake Bay.

geographic distance

Use this site to calculate the distance between lakes or other geographic features, using geographic coordinates from the USGS site *topozone*.

GIS technology

A site that has an introduction to GIS technology.

GPS technology

This Trimble Company site provides great background on GPS technology.

Lake Conemaugh

Websites of the U.S. National Park Service, the Johnstown Flood Museum (Johnstown Area Heritage Association), and a U.S. National Public Broadcasting system site.

morphometry (Florida lakes)

This site provides a wealth of information about Florida lakes and limnology in general. Check out the Florida Lakewatch brochure on lake morphometry!

NASA satellite photos

A grab bag of photographs of Earth from space. Plate 12 is from this site.

topographic maps (Topozone)

This is the topographic map site of the U.S. Geological Survey, useful for finding the geographic coordinates and elevation of lakes.

UW Center for Limnology

This site provides a wealth of lake data and maps.

12

"... the University of Wisconsin has been noted not only for the philosophy of service to the state ... but for the practical courses which deal with every factor of life in the state."

C. McCarthy, 1912, "The Wisconsin Idea"

outline

The Citizen Limnologist

LIMNOLOGY AND SOCIETY

Limnology is more than chemistry, physics, and ecology. It also has a place in society, outside the ivory tower of academia. Politics, local knowledge, human exploitation, and academic science have been a part of limnology since the very beginning of the science (Schneider, 2000). This chapter focuses on the contributions that limnologists can make to society.

This chapter is dedicated to Arthur Hasler, who taught his students the concept of the citizen limnologist. During his career, Professor Hasler taught a limnology course to thousands of undergraduates at the University of Wisconsin–Madison. He called his course "Conservation of Aquatic Resources" because he felt it was important that every educated Wisconsin voter knew the basics of limnology and was prepared to apply them to societal issues. He was in this way adapting "The Wisconsin Idea"—the boundaries of the university are the boundaries of the state.

As stated in the Preface, water is a critically important human resource—a theme that is woven into every aspect of our culture. Humans have a long history of having either not enough water during droughts or in deserts, or too much water in floods. If you don't know how to swim, water can be terrifying and deadly. Water can be sinister or beautiful, dangerous or life-saving, valuable or costly, private property or a common resource. Humans have always been obsessed with water. Cave art was mentioned in the Preface—Bayly (1999) believes that ancient Australian rock art was used to map locations of camps and rock pools. Modern society, in any part of the world, is still occupied with the use, apportionment, distribution, and purification of water.

Despite the value of fresh water, this resource is being "intensely modified and degraded" by human activity (Johnson et al., 2001). Water is fast becoming a limited resource for the human species—more than half of all accessible surface fresh water is used directly or indirectly by humanity (Vitousek et al., 1997). In addition to use, human activity in watersheds (such as agriculture and urbanization) changes land cover, lake stratification, and streamflow characteristics. Global warming, whether caused by people or not, is happening, and will also affect all aspects of the hydrological cycle (Meyer et al.,

1999). Water quality is often important, but in many parts of the world, availability is more important than quality (figure 1.1).

Many components of society have a strong interest in applying limnological knowledge to important and complicated social problems. In this chapter, we will explore the mechanisms of observation, management and assessment techniques, and a model for restoration of aquatic habitats.

> A limnologist is a zoologist who, during the summertime, studies chemical and botanical aspects of geological problems in readily accessible lakes, 15m deep, located in the vicinity of universities.[1]
>
> *J. R. Vallentyne, 1969*

[1]Referee's comment to Vallentyne's definition: The head of the world's leading eutrophication project [Vallentyne] shows understandable nostalgia for those 15-m days. University lakes of his student days have now shoaled to less than 14-m depth. A modern limnologist is best defined as a biogeochemist and self-taught systems analyst, whose favorite systems are imbedded in an exponentially increasing matrix of septic tanks.—E. S. Deevey

Observation

Human priorities for water management are based on how we view water. Where there are multiple groups with an interest in water management, there are multiple ways of simply seeing water, freshwater habitats, and aquatic resources (Cairns et al., 1993; Lee, 1993). To understand how limnological knowledge can be applied to social or political situations, it is useful to understand how we observe the world around us.

An Observation Model

Imagine you are standing in front of a lake or stream. What you see is the result of a complex process of your interpretation of sensory information. Access to "physical reality" is indirect (figure 12.1)—the remnants of sensory information after it passes through an individual's mental models (perspective). These models are the result of our past experiences, knowledge,

FIGURE 12.1 The relationship between interpretations of nature and the physical reality. Human senses are required to experience physical reality. The sensory information is filtered by physiological capabilities of the human system, and then organized and interpreted (given meaning) by mental models or filters. **Source:** The lake photo is by Bill Feeny.

state of physical comfort, expectations, and intentions—the mental models used to interpret the "meaning" of the sensory input. For example, look at the individuals in plate 27. All four people are looking at the same river and canyon view, but they have very different interpretations. The person on the left (your author) was thinking, "there are mayflies down there in the riffles, but it's too warm for trout." The next person had just said, "thank God we all moved west, so we can enjoy this scenery." The next person had just said, "I could cross the river, climb the gooseneck, and make camp before dark." And the person on the right was thinking, "we are too close to the edge."

Citizen Perspectives

The first impression one has of water in nature often develops through sight. Aquatic habitats are typically interpreted as "scenery"—a desirable part of a landscape. Property value is enhanced by being on a shoreline. Why is this?

A single glance at a water body can support a wealth of interpretation of lake or stream conditions, whether the viewer is a citizen or a citizen limnologist. In addition to providing the viewer with limnological information, the view will automatically produce a variety of feelings. For example, when scenery causes a viewer to feel an emotion, the scene is said to be *romantic*. As you read the poems and quotations in this chapter, pay attention to the images and emotions that move through your mind. The following poem is an example of the romantic style used in a limnological context.

Four limpid lakes, —four Naiades
Or sylvan deities are these,
In flowing robes of azure dressed;
Four lovely handmaids, that upheld
Their shining mirrors, rimmed with gold,
To the fair city in the West.
By day the coursers of the sun
Drink of these waters as they run
Their swift diurnal round on high,
By night the constellations glow
Far down the hollow deeps below,
And glimmer in another sky.
Fair lakes, serene and full of light,
Fair town, arrayed in robes of white,
How visionary ye appear!
All like a floating landscape seems
In cloud-land or the land of dreams,
Bathed in a golden atmosphere!

H. D. Longfellow, January 15, 1870, "The Four Lakes of Madison" (see plate 1)

The poem by H. D. Longfellow is an extreme example of the romantic style of description. Using a different palette of colors to induce different emotions, Thoreau (1845) creates a transcendental atmosphere in his description of the color of the clear water of Walden Pond:

"I have discerned a matchless and indescribable light blue, such as watered or changeable silks and sword blades suggest, more cerulean than the sky itself, alternating with the original dark green on the opposite sides of the waves, which last appeared but muddy in comparison. It is a vitreous greenish blue, as I remember it, like those patches of the winter sky seen through cloud vistas in the west before sundown. Yet a single glass of its water held up to the light is as colorless as an equal quantity of air."

In a similar romantic vein, Aggasiz (1850) expressed his impression of Lake Superior:

"The huge basin is filled with clear, icy water, of a greenish cast."

Walka's poem about the San Juan River uses very different conventions than those used by the Victorians, but achieves visual and emotional fireworks, along with expressing an intimate personal connection to a particular river:

The River

Every morning at dawn
the blond river
lifts herself like a mare
from the stony terraces
shakes off the dark
and gathers into a slow canter
through oxblood hills

She comes down sleek and heavy
with mud tendrils of steam curling
from her flanks
Light rides on her back
and in the evening

> scatters from her mane
> in flakes of copper
> The river clatters off between umber hills
> and into the appaloosa night
> She goes on down
>
> *Ann Weiler Walka, 1993, "The River" (see plate 27) Waterlines: Journeys of a Desert River. Flagstaff, AZ: Red Lake Books.*

Sense of Place

Globally, there is a growing need to manage our valuable aquatic resources. The first step along the road of successful management is appreciation. Appreciation of a natural resource in its landscape context is what Western nature writers call a *sense of place*. The poems by Longfellow and Walka express sense of place. A sense of place makes the place special, and special things are sacred, with special rules for their use. The "sacred" mental model is a pre-scientific way of appreciating (or managing or proscribing) water resources. Modern limnologists tend not to think of lakes and streams as being sacred, but it is important to know that for millennia, humans have considered some lakes and streams special, requiring special behavior. For example, the lakes of southern Wisconsin are ringed with animal effigy mounds, which were probably related to the special importance of lakes in the culture of ancient peoples (Birmingham & Eisenberg, 2000). Lake Mendota, which Arthur Hasler liked to call the "most-studied lake in the world," is overlooked by scores of effigy mounds (figure 12.2). This ancient human tradition of making lakes special places is probably related to our current interest in conservation of natural areas.

> There was a time when meadow, grove and stream,
> The earth, and every common sight,
> To me did seem
> Appareled in celestial light,
> The glory and freshness of a dream
>
> *W. Wordsworth, 1908, Intimations of Immortality*

Water in the landscape is an essential part of human culture. Lakes and streams have always provided us with food and essential drinking and bathing water. However, in addition to exploitation, people sometimes set a lake or stream aside from common use to be a special place for recreation, contemplation, and even worship. For individuals who do not swim, lakes can also be dangerous places to be avoided.

One way to make water sacred is to include it in religious rituals, wherein water is a favorite religious metaphor. Many religions teach that spiritual beings live in water, in special lakes or streams. While Western religions avoid deification of individual lakes or

FIGURE 12.2 Location of animal effigy mounds (each about 10 meters long, and usually on an overlook about 50 meters above the lake) near Lake Mendota, Wisconsin. Common shapes are bears and other mammals, swallows and other birds, and long lines that may represent "water spirits." **Source:** From Birmingham and Eisenberg, 2000.

streams, they do teach that the idea of water is central to interpreting the human experience. Water is seen as a precious part of creation, a gift given to people, but also as a powerful part of nature that can be used to correct people and dispense justice, as in Noah's flood. Stories about paradise are heartbreakingly beautiful and aquatic, featuring lush landscapes with pools, fountains, and streams. Most religions use water to cleanse both the body and the spirit, and often use water, especially river water, for special or even daily ritual washing.

Sacred lakes were a key component of ancient culture in Europe. These lakes were used as meeting places for momentous events. Sacred lakes are often dangerous places to visit. Klesius (2000) gives a partly fanciful description of events that might have led to the occurrence of a smashed-up ship, horse skeletons, and personal artifacts in a small coastal bog (now called Nydam Mose bog in Denmark). He suggests that a raiding Viking ship attacked a coastal village but lost the fight, and that the victorious villagers threw the ship, horses, and belongings into the bog in celebration. This seems like a waste of horses and a good ship—perhaps there is another explanation . . .

The heroic poem *Beowulf* describes the dark side of a Danish lake from a manuscript probably written about 300 A.D. (A **fen** is a wetland, a low-lying, marshy area.) In this poem, Hrothgar, an ancient king of the Danes, says to the hero Beowulf:

> "I have heard some of my subjects . . . speak of having seen two such enormous monsters haunting the fenland. One of them, so far as they could tell, looked like a woman; but the other, misshapen brute, trod the wilderness in the form of a man, though his size was greater than a human being's. The country people used to call him Grendel. But they know nothing about his progenitor, or whether any other mysterious beings were begotten before him. They live in an unvisited land among wolf-haunted hills, windswept crags, and perilous fen-tracks, where mountain waterfalls disappear into mist and are lost underground. The lake which they inhabit lies not many miles from here, overhung with groves of rime-crusted trees whose thick roots darken the water. Every night you can see the terrible spectacle of fire on the lake. No one knows how deep it is. Although the antlered hart will sometimes take refuge in that forest after a long chase by the hounds, it will sooner give up its life at the lake's edge than try to escape by plunging in. It is no inviting spot. Frothing waves rise blackly to the clouds when the wind provokes terrifying storms, until skies weep rain in thickening air . . ."
>
> *Beowulf 20, 1957* translation, David Wright.

Some cultures still take their sacred lakes seriously. For example, the first edition of a recent limnology text contains a picture of a lake sacred to a group of North American people. The people objected to having the image of the lake published, and threatened a lawsuit if the lake was shown in the next edition. The picture was withdrawn.

In other cultures, it is apparently acceptable to publish a picture of a sacred lake, as long as the rules are followed. For example:

> "The famous 'Wish fulfilling Lake' Khecheopalri is considered by the Sikkimese people as most sacred (figure 12.3). Many folklores and legends are associated with its formation and shape. The lake water is used for rites and rituals only. Fishing and boating are strictly prohibited. It is situated in the midst of a pristine forest at an altitude of 1700 m above sea level in the western part of Sikkim state, India" (Jain et al., 1999).

Most modern societies no longer set aside sacred lakes. Instead, we make laws that regulate water and land use. The U.S. Clean Water Act of 1972 provided rules and guidelines for cleaning up our water surface and ground waters and established the Environmental Protection Agency (**EPA**) to implement the act. Challenges to the health of our aquatic resources can be *point-source* (coming from an easily identifiable pipe or other source)

FIGURE 12.3 The sacred Lake Khecheopalri in northern India. **Courtesy of Dr. Jain.**

or *nonpoint,* in which contamination is spread out over part of the watershed (as by agriculture), deposited from the atmosphere, or contributed by ground water.

Many communities have solved major point-source problems of water pollution by focusing on placement of discharge pipes or smokestacks. For example, Lake Washington was contaminated for decades with phosphorus-rich sewage from Seattle and dozens of smaller towns. When the point-sources (sewage effluent pipes) were directed toward Puget Sound (part of the Pacific Ocean) rather than into the lake, lake water clarity improved dramatically (Edmondson, 1991).

Nonpoint pollution is often difficult to **remediate** (improve the situation) because it often occurs at large spatial and temporal scales and is produced by many sources. For example, thousands of North American lakes have been acidified by acid deposition (Keller et al., 2002). The acid (mostly sulfuric and nitric) was produced in coal-fired power plants and smelters in the industrial midwestern United States, and then deposited as rain, fog, snow, and dust in New England and eastern Canada. Acidification was particularly intense downwind from Ontario's Sudbury nickel smelter. An international effort of governments, environmentalists, anglers, and power companies has produced policy leading to a reduction in emissions and a decrease in acid deposition in these lakes (figure 12.4). Many of today's challenges to water quality are related to the ever-expanding human population: habitat destruction and an increase in pollutant-producing land use such as agriculture, urban development, and industrial activity.

The critical importance of lakes and streams is shown worldwide by efforts to protect and manage fresh water. As appreciation for the value and beauty of aquatic resources grows, we can look forward to a landscape with ever more special freshwater places.

Contrasting Cultural Perspectives

Societal response to fresh water is *complex*—there are many conflicting and strongly held opinions about how to use and manage fresh water. Two contrasting ways of appreciating lakes and streams can be illustrated by considering American Lake Mendota, Wisconsin and the European (German-Swiss) Lake Constance. Both lakes are rather large, temperate, and more-or-less eutrophic. They both rest in a densely populated watershed with major agricultural and urban land use. Both have a well-known limnological research station on their shore. Both lakes provide many important major

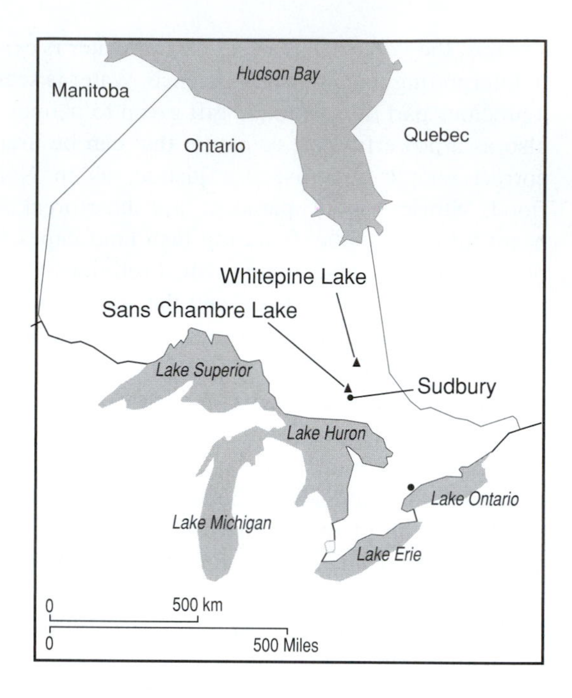

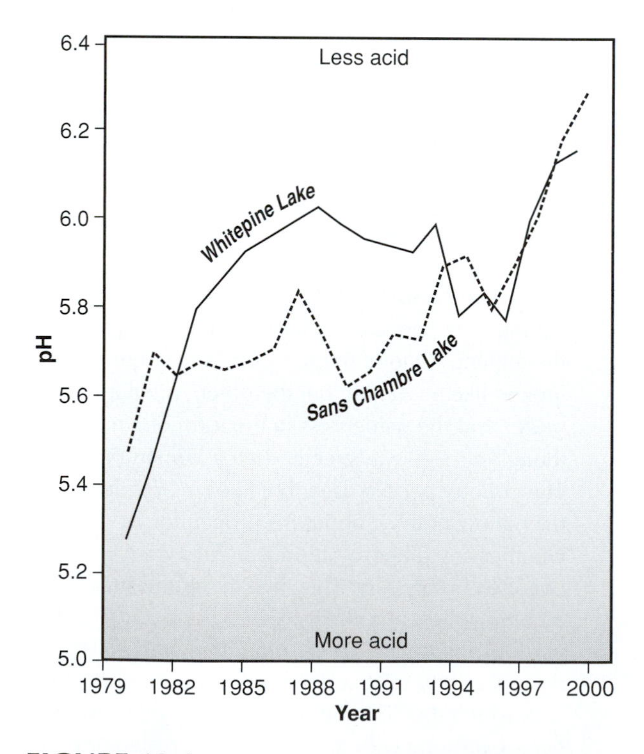

FIGURE 12.4 A success story for limnology and environmental management of a nonpoint source pollutant. The long-term data in the two graphs show the increase in pH (decrease in acidity) for two lakes downwind from (north of) U.S. power plants and the Sudbury smelters. The two lakes that are recovering from acid deposition are Whitepine Lake and Sans Chambre Lake. **Source:** Data from Keller et al., 2002.

ecosystem services to the surrounding human populations and both lakes are favorite places enjoyed by large numbers of people. However, the visitor's experience of the two lakes is quite different. The two experiences might be called Quiet Time and Party Time.

Quiet time At Lake Constance, one sees fleets of sailboats. Powerboats glide slowly and quietly in and out of the marinas, in most cases with the assistance of small and silent electric motors. Along the bank are paved walks occupied by tourists, people out for a walk or a cuddle on the benches, anglers, and grandparents and children feeding ducks and swans. Somewhere out in the lake, water skiing occurs, but it is not evident from shore. There is a smell of water, algae, perhaps a touch of goose droppings. The atmosphere is conducive to strolling, talking, and even dreaming. People making noise receive a frown in return.

Party time On Lake Mendota, there are sailboats, but also fleets of powerboats driven by huge gasoline motors. These boats thunder about the lake, creating wakes that crash into the shore. Boat drivers are exhilarated by the speed and sound of their boats and are proud to own such a powerful machine. The average boat on Lake Mendota now has a motor rated at more than 85 horsepower (plate 1). Water skiers and anglers are crowded (plate 31), while special, small but powerful boats—"jet skis"—produce huge rooster-tail wakes as they zoom across the lake. People also walk and fish along the shore, and the terrace of the University of Wisconsin Memorial Union is a favorite meeting place. The atmosphere is often permeated with the roar of the boats and the air sometimes smells faintly of gasoline along with the algae and bird droppings. Swimmers or canoeists are in some danger of being run over. Boaters anchored near shore often broadcast loud music. There are signs that discourage feeding ducks, but some people, compelled perhaps by their European ancestry, can't avoid the temptation. The overall atmosphere is the result of a muscular, extroverted aesthetic that values speed and excitement. It's party time!

Changing times The way society relates to fresh water can change over time. As the resource becomes more limited and therefore more valuable, cultural perspectives and management styles are forced to change. For example, as the human population of the Lake Mendota watershed grows, spring floods are becoming more frequent. This is a direct consequence of changing land use, especially the increase of impervious surfaces (roofs, roads, parking lots) that shunt water rapidly to the lake. When the lake water is high, boaters need to be particularly careful about their wakes. During a high-water period in the summer of 2000, boaters had their speed constrained by a "no wake" rule for several weeks. Some were greatly annoyed at this restriction, but many people found they could enjoy the lake in ways that did not create wakes and loud noise. The local citizens are now considering alternate ways of using the lake. This process of redesigning a management policy is not easy. One way of achieving a policy optimal for all interested parties is to include authorities, such as limnologists, in a process called *adaptive management.* Before exploring this special management style, consider what limnologists have to offer to society.

The Limnological Perspective

In addition to the cultural perspectives of citizens in general, limnologists have additional, special perspectives. When a limnologist looks at an aquatic habitat, he or she immediately has interpretations about the ecology, chemistry, and physics of the water.

For an example of a limnological interpretation, consider the color of pure, deep water, which is perceived to be blue or light-green. The limnologist understands that this color is the result of the relationship between light's wavelength and its tendency to scatter (see chapter 2).

To a modern limnologist, the water color gives important clues to the chemistry and biology of the water. Table 12.1 presents scientific, limnological interpretations of water color. Notice that table 12.1 gives "scientific" interpretations of sensory data only; there is no mention of aesthetics or other mental models for interpreting water color.

In this section, we have looked at some of the perspectives (mental models) humans use to interpret sensory information about lakes. In the next section, Freshwater Management, we focus on mental models that make lakes special. In earlier times, these models were religious. Today, they are more likely to be related to dire necessity or to a naturalism based on environmentalism and conservation of natural resources.

FRESHWATER MANAGEMENT

There are many opportunities for managing aquatic habitats, and limnologists are anxious to contribute to solutions. Limnological history is full of attempts to address large-scale, complex societal issues related to fresh water. Past areas of contribution include issues of

Table 12.1 Some Scientific Limnological Generalizations About the Color of Water

Water Color	Limnological Interpretation
Clear	Shallow, pure water (plate 14), saline water (plate 8) or toxic water. A handful of water from a blue oligotrophic lake will be clear.
Blue	Deep and relatively clear water (plates 2 and 3)
Green-white	*Glacial milk,* caused by finely ground rock (plate 30). A milky opaqueness can also be caused by suspended sulfur in volcanic lakes (plate 4).
Green	Green algae or floating macrophytes, usually in summer (plates 20 and 31)
Yellow-brown	A diatom bloom, in spring or early summer
Blue-green	Blue-green algae, often as floating scum in summer (plate 19)
Brown to black	Humic acids (DOC) such as tannin (plate 15)
Red	Clays associated with erosion (plate 12), or red-colored algae blooms
Gray	Overcast sky, reflected clouds (plate 1)
Glory or halo	A phenomenon seen in lake water. If, on a clear day with bright sunlight, you look directly at your shadow on water, the head of the shadow is surrounded by a glowing halo. If the surface is wavy, the glory is made up of alternate bright and dark streaks.

cultural eutrophication, acid deposition, toxicology, and fisheries management. How best can the citizen limnologist continue to contribute to society?

Biological Integrity

Biological integrity, sometimes called environmental health, is a concept used in evaluation and management of freshwater habitats. Integrity implies that all parts of a system are present and working together in a natural way. One definition of integrity is that all the aquatic species are present and can interact as they would have before the stream experienced human impact in its watershed. Another definition is that the freshwater habitat is able to provide the goods or services desired by society (Vitousek et al., 1997). *Ecosystem services* in freshwater systems include:

- Maintenance of biological diversity (including desirable fishes);
- High rate of production of desirable species;
- Being a source of safe drinking water for people, plants (agriculture), and animals;
- Being a useful water for industry and agriculture;
- Purification of sewage; and
- Providing a pleasing recreation experience.

The freshwater systems that provide these services are complex, in terms of species, chemistry, physics, and cultural perspectives. For simple situations, such as eutrophication of a single lake, it may be possible for a single authority (a limnologist!) to provide the data and policy necessary to optimize management. For example, the management and restoration of Lake Washington was achieved based on the advice of one or a few scientists and engineers (Edmondson, 1991). The single-authority management style is most effective if the economic and environmental stakes are low—if little money is involved and if threats to the ecological community and environment are minimal (figure 12.5). With higher stakes, trained managers or engineers can often solve problems and achieve successful management of the resource. However, in the case of complex (often large-scale) systems with high stakes involved, simple management approaches often fail. What is needed is a management style that recognizes and uses the complexity of the system.

Adaptive Management

The adaptive management model is useful in a complex situation that does not have a simple limnological solution (figure 12.5). Examples of complex limnological systems involving social and political issues include fisheries (salmon) management, cultural eutrophication, acid deposition, groundwater management, and invasive species.

Most limnological research is *basic* research, characterized by intense interest by the scientists, but of low direct value for solving complex important questions of immediate interest to society. Thomas Kuhn (1962) called most basic research "normal science." Although he was more interested in the evolution of scientific ideas, he also pointed out that normal science is best used for learning more and more about less and less, rather than solving society's big problems.

Adaptive management has three main steps (Lee, 1993).

Step 1—Assembly

Bring together all the interest groups who need a solution to a complex problem. Relying solely on scientific authority to solve a complex social problem is a mistake and has led to much mismanagement (Ludwig, 2001). In complex systems, there are people (**stakeholders**) with different perspectives and different vested interests. Examples of stakeholders include federal and state managers, scientists, conservationists, farmers, fishermen, anglers, and businesses that make a living exploiting a freshwater resource.

Many of these stakeholders will resist meeting together and, at first, they will not even speak the same language. The process of identifying the problems and modeling the system works to generate a common stakeholder language. When assembled, the stakeholders can develop a conceptual model of the system in question. This model identifies primary drivers of goods and services, mechanisms of interactions, and historical constraints. This model is often complex, reflecting system complexity. This modeling step is an excellent way to ensure that future management policy meets the needs of society, especially when there are difficult issues and many stakeholders.

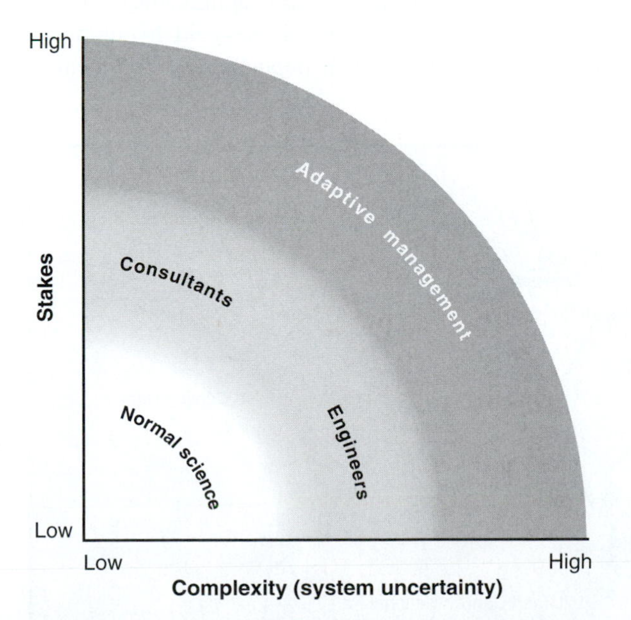

FIGURE 12.5 Adaptive management techniques are most useful for complex systems with high stakes and a high level of uncertainty. **Source:** Based on Funtowicz & Ravetz, 1994.

This stage of assembly is the stage of negotiation—stakeholders attempt to optimize management for their own interests. An adaptive management approach requires two additional steps: a monitoring program and an adaptive stage in which policy is reevaluated.

Step 2—Monitoring

The stakeholders design a monitoring program to gather data needed to optimize management of the problem. Monitoring often makes use of appropriate *indicator* species—species that are relatively abundant, easy to observe, sensitive to environmental stress, and can provide an early warning of environmental deterioration (Cairns et al., 1993). For example, common species of stream organisms can be classified according to their tolerance to low oxygen levels (Hilsenhoff, 1987), an indicator of organic pollution (see figure 13.2). In addition to indicator species, chemical and physical data can produce useful and cost-effective information.

To optimize monitoring success of a complex freshwater habitat, it is wise to use many different measurements. For example, an Index of Biological Integrity (IBI) developed by Karr et al. (1986) integrates several measures of biological integrity in three broad categories: fish species composition, trophic composition, and fish abundance and condition. The goal of the IBI is to integrate information from individual, population, community, zoogeographic, and ecosystem levels into a single, ecologically based index of the quality of a water resource. IBIs are currently widely used in water management and in policy making.

In a complex system, it is often impossible to rely on normal science to answer critical questions—there simply isn't time or resources for carefully designed laboratory or field experiments. Instead, since a management policy is needed anyway, the best approach is to design the *management policy* so that even if the initial management policy is less than successful, learning can occur based on monitoring. Adaptive management experiments can help resolve scientific uncertainty about how the system works and provide a basis for best management practices. To take advantage of the knowledge of the entire scientific community, monitoring data must be published and discussed in peer-reviewed papers and at national meetings.

Step 3—Adaptation

After a year or two, stakeholders use what they have learned from the initial monitoring to *redesign* (adapt) policy. This step is difficult for managers because it may put them in the light of having used bad policy—they sometimes prefer to continue to use an old policy rather than to change. This process of designing new policy, followed by monitoring and learning, can be *repeated.* Long-term data are often particularly valuable, because what we learn over a couple of years may be misleading. Long-term monitoring data are often particularly valuable in showing both short-term and long-term patterns of change (Magnuson, 1990).

Lake Powell Example

The Lake Powell–Glen Canyon Dam–Grand Canyon–-Mead Lake system of the Colorado River (figure 12.6) provides an excellent example of an adaptive management program that is working (Walters et al., 2000). This system is large and complex, with economic stakes in the billions of dollars and with several ecological communities at risk. The Glen Canyon Dam was completed in 1963. Water flow was initially managed to provide flood control and water for agriculture and urban uses. By the early 1980s, the Bureau of Reclamation was monitoring limnological and ecological aspects of the river and surrounding ecosystem, and in the late 1990s, adaptive management techniques were adopted to help design future management policy.

The basic goods and services in this system include regulating water for hydroelectric power, storing water for agriculture, providing water and sandbars for recreational use, maintaining a trout fishery and stocks of native fishes, restoring beaches and sand bars for camping, and maintaining riparian plant communities in the face of aggressive invader species. Multiple vocal and often powerful stakeholders represent each of these issues. Major stockholders include the U.S. National Park Service, U.S. Bureau of Reclamation, U.S. Fish and Wildlife Service, the Arizona Fish and Game Department, the Colorado Basin States, limnologists, archaeologists, hydroelectric companies, recreational groups focused on the reservoirs and the river, representatives of agriculture, and environmentalist nongovernmental organizations (**NGO**s).

Figure 12.7 shows a conceptual simulation model, generated by stakeholders and used to explore best management policy. This model gives a nice picture of the complexities of the system!

Stakeholders made an experimental policy decision to allow an artificial flood—a greater-than-normal release of water—in 1996. From this, limnologists and ecologists learned how flooding affected fish stocks, beaches, and riparian vegetation. For example, floods are apparently necessary to maintain sand bars that are prized camping locations for whitewater rafters. Policy can now be redrafted to allow for the possibility of artificial floods, when conditions (especially time of year and reservoir volume) permit.

A possible future, large-scale management experiment includes the release of warm, epilimnetic water (the Glen Canyon Dam currently releases cold, hypolimnetic water) to favor native fish populations. The extent to

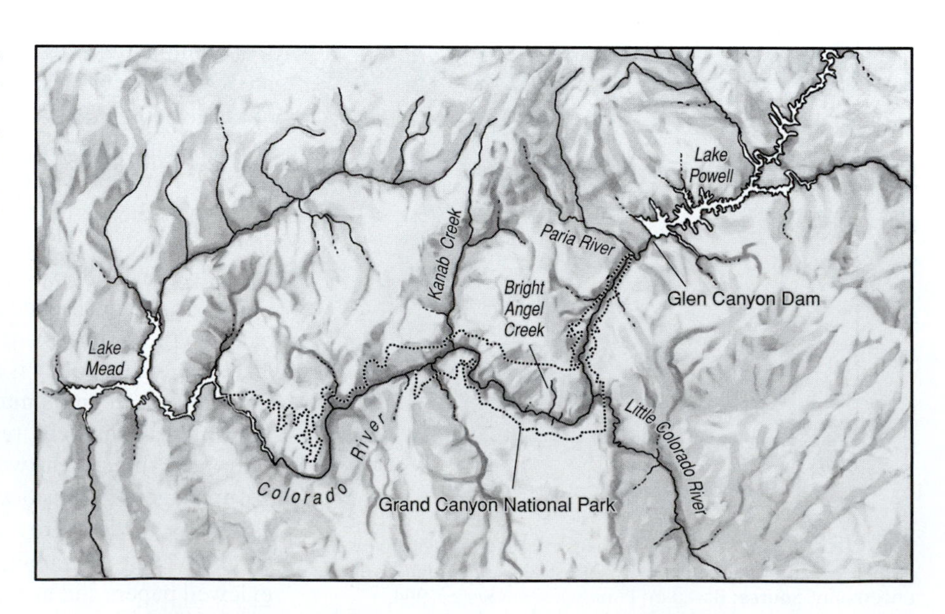

FIGURE 12.6 Map of the Colorado River system undergoing an adaptive management program. Water flows from Lake Powell, through the Glen Canyon Dam, through the Grand Canyon, and into Lake Mead. **Source:** Map based on Walters, 2000.

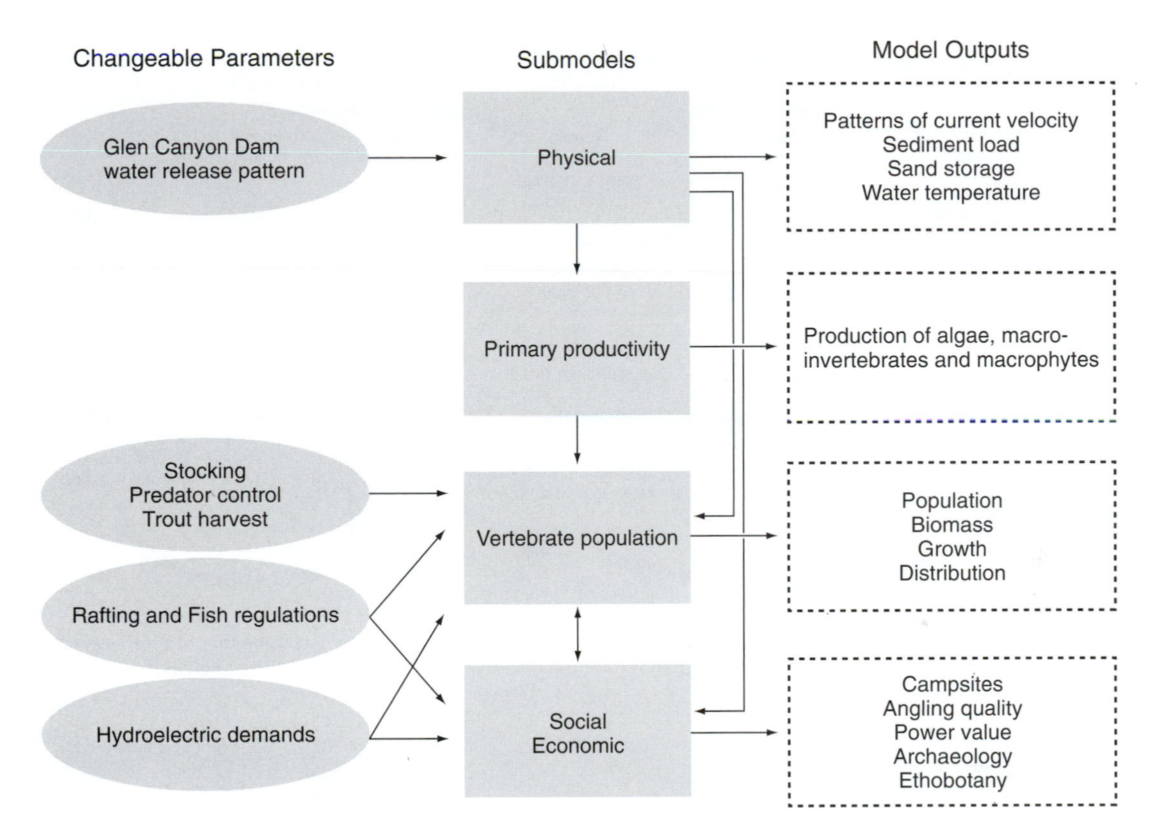

FIGURE 12.7 A conceptual model for management of the Colorado River from Glen Canyon Dam to Lake Mead. Parameters listed in the ovals are inputs to the models. These inputs can be manipulated to see how changes in the parameters affect outputs of the submodels (listed in the boxes). Outputs for each submodel are in the dotted boxes.

which warm water is released will be constrained by fears that warm water will also favor expansion of populations of exotic and aggressive fish species such as carp.

Several limnologists have developed graphical *ecological management models* that map out the steps of adaptive management (Cairns et al., 1993; Norman Yan, pers. com.). These models provide systematic, step-wise procedures for identifying, assessing, and developing policy for complex freshwater systems (figure 12.8). These models are useful for managing complex systems at all levels, from local to international, because the models provide a road map for navigating complex issues.

Restoration of Aquatic Communities

People often look at the current condition of a lake, stream, or wetland and envision a restoration of the landscape feature to "what it once was." The goal is often to regain lost services (such as potable water, a wildlife refuge, or a valuable fishery) or to improve aesthetics (a "natural" appearance or habitat for aquatic organisms)—to **restore** a past condition (figure 12.9). Restoration is a complicated process that requires more time than is sometimes available to individual researchers or managers (Carpenter & Lathrop, 1999).

There are significant costs associated with reversing cultural eutrophication, so lake managers need to know how much to reduce phosphate inputs into a lake. In a brilliant analysis, Lathrop et al. (1998) related phosphate input to the probability that a nuisance blue-green algal bloom will occur. They adopted this approach because they felt the public may be more likely to notice extreme events, such as an algal bloom, than they are to notice a change in, for example, the average algal abundance. They used a long-term data set for watershed-scale phosphate loading to develop a predictive model, taking into account losses of phosphate to the sediments and through the outlet of Lake Mendota, Wisconsin. Currently, there is a 60% probability that a blue-green bloom will occur on a given summer day. The model predicts that reducing phosphate input by half will result in a 20% probability of a bloom on a given summer day. This kind of prediction can then be used to develop management policy for the

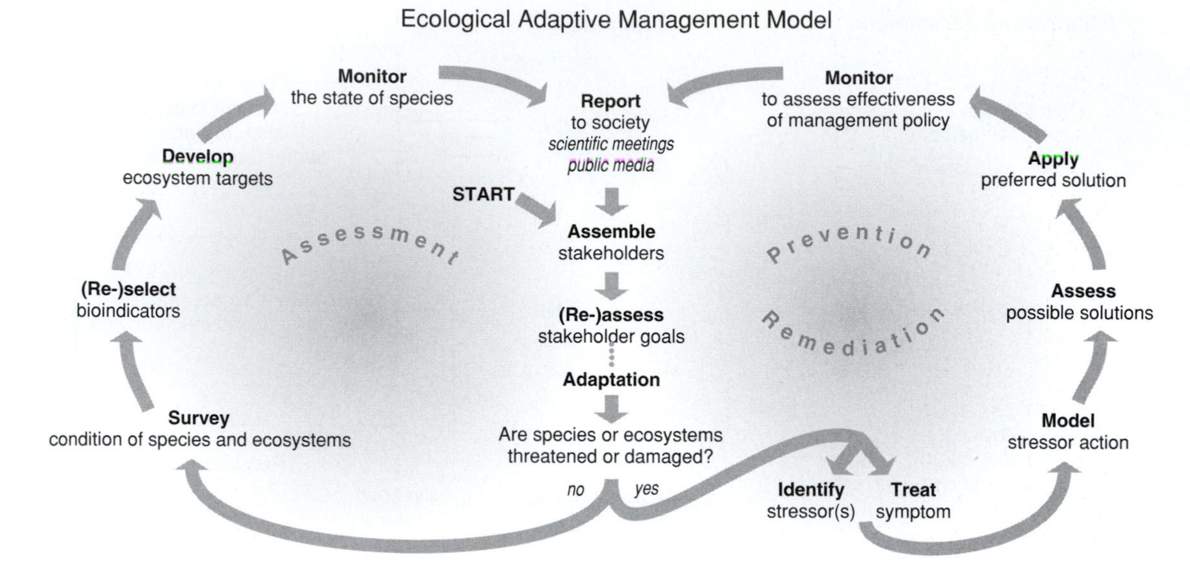

FIGURE 12.8 A step-wise model, using principles of adaptive management, that can be used for either routine assessment or remediation of aquatic resources. **Source:** Thanks to Norman Yan for the graphical model, which has been slightly modified.

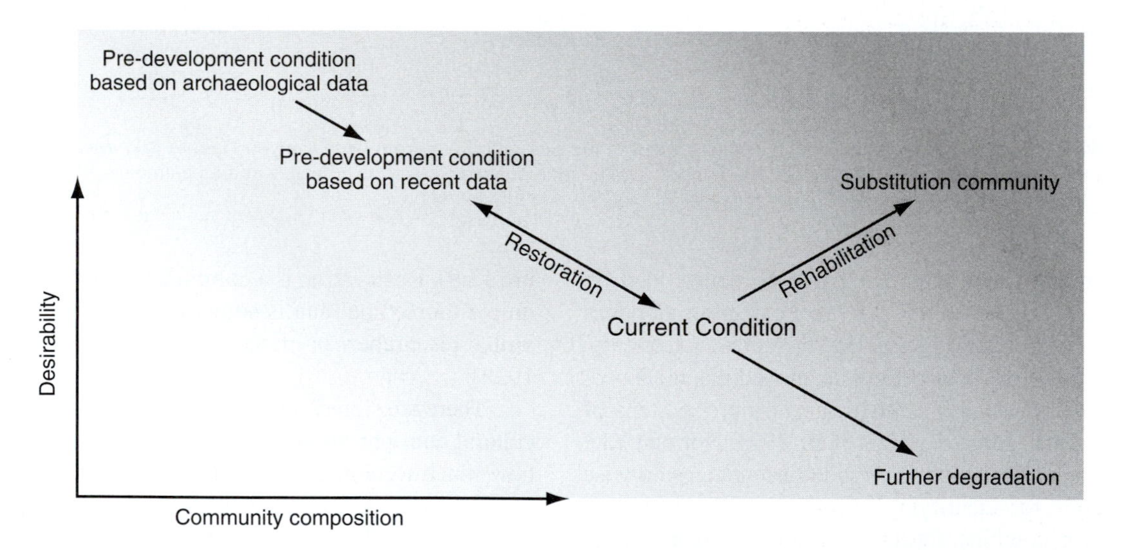

FIGURE 12.9 A model showing the distinction between restoration and rehabilitation, and the difference between historical and fossil evidence for past community structure and ecosystem functioning. **Source:** Based on a model developed by John Magnuson.

lake—policy that takes into account the cost of phosphate reduction and the public's perception of water quality.

While the local population may retain a memory of the lake in the old days, there may be little scientific information on the lake's past community structure or ecosystem functioning. Both historical and paleolimnological studies can provide useful, base-line data for defining the target of restoration of a degraded aquatic community (figure 12.9). For example, hypereutrophic Lake Thonotosassa, Florida has been heavily polluted with agricultural and sewage waste, and there is strong interest in restoring the lake to its prepollution condition. Paleolimnological studies suggest that Thonotosassa was a naturally eutrophic lake before pollution, and that the restoration goal for the lake might therefore be a eutrophic rather than an oligotrophic lake (Brenner et al., 1996).

Although less direct than written records, fossil evidence often suggests that communities were impacted significantly by early human exploitation. For the U.S.

Atlantic coastal fisheries, Jackson et al. (2001) report that historical records of fish stocks indicate that current fish populations are degraded compared to the earliest reports of the fishery, and that archaeological evidence suggests that the fish stocks were much higher than the historical records indicate, before European settlement. This means that historical records may mislead restoration ecologists as to what is possible in a specific location.

Writing about wetland restoration, Zedler (2000) emphasizes the complexity of the process: "It takes more than water to restore a wetland." She reminds us that the restoration process is constrained by landscape setting, habitat type, hydrological regime, soil properties, topography, nutrient supplies, disturbance regimes, invasive species, seed banks, and the global decline in biodiversity. She also finds that because of the complexity of the restoration process, in light of the current low level of ecological understanding of ecological interactions, it is difficult to predict how a restoration project will turn out.

European aquatic environmentalists are excited about restoration of native macrophyte vegetation, especially in shallow lakes (Jeppesen et al., 1999; Scheffer, 1998). The reeds, lily pads, and related emergent and submergent aquatic plant species have been extirpated in many lakes over the last century or so. This vegetation is being restored using a combination of watershed management of nutrients, management of fish populations, and, in some cases, active replanting.

Sometimes, the desire is to **rehabilitate** a community, by improving the desirability of the species present, but to produce a different community than was present in the past (figure 12.9). The massive salmonid stocking effort for the Laurentian Great Lakes is a good example of rehabilitation in response to the loss of a desirable lake trout fishery. This rehabilitation project requires balancing the effects of natural predation with sport and commercial fishing pressure, in a context of variable climate and food supply (Jones et al., 1993). In the Great Lakes, several species of non-native salmon have been stocked successfully, while restoration of the native lake trout has met with limited success.

When lakes are drawn down for drinking or irrigation purposes, the fine sediments of the dry lake bed can cause significant air quality problems. When water was diverted from Mono Lake, California, dust became such a nuisance that the US Clean Air Act was used to require reduced water diversion (Aron and Patz, 2001). An even worse problem developed around Lake Texcoco in Mexico City, Mexico (Lindig-Cisneros and Zedler, 2000). People began living near large (about 30,000 ha) and shallow Lake Texcoco as much as 10,000 years ago. Agriculture has been practiced along the lakeshore for millennia. Water diversion and land use changes by modern agriculture and urban development caused the lake to slowly disappear. By 1960, the lake had become a derelict saline wasteland, which caused major dust storms that were shutting down the airport and producing dangerous air pollution for the city. The dry lake bed has been partly re-flooded using storm sewer water and sewage effluent. Because extensive urban land use surrounding the lake, exotic wetland species, disappearance of native organisms, extreme (cultural) eutrophication, and salinization, it is unlikely that managers will be able to re-establish the rich wetland and pelagic communities used by inhabitants of the Aztec empire of 600 years ago. However, the lake is now home for a number of fish, bird and macrophyte species, and is providing some degree of water purification, aquifer recharge, flood control and climate modification. Migrating water fowl use the surrounding vegetation as a resting place. The lake holds the possibility of providing a bit of (modified) nature in the middle of an urban desert. The Lake Texcoco Project success has encouraged limnologists and managers to seek additional funds to reclaim and rehabilitate more of the former lake basin.

Successful restoration of aquatic resources often requires attention to the watershed as well as the freshwater site. Landscape ecology models can be used successfully to explore consequences of alternate management programs (Turner et al., 2001). Just as with adaptive management, many possible restoration scenarios can be tried in virtual reality, before actually putting policy into practice. For example, managers can look at the effects of alternate land use and riparian development on the restoration of wetlands or the sediments revealed when dams are removed. Software for simulation models is available from Gergel and Turner (2002).

SUMMARY

Lakes and streams provide many human services. We use the water for drinking, agriculture, power generation, fishing, recreation, for conservation of biological diversity, and for aesthetic and spiritual experiences. As we use water, we mold the shape, appearance, and biology of lakes and streams. Optimizing the many disparate uses and services of lakes and streams will require all of our limnological knowledge, political skill, cultural wisdom, and appreciation of aquatic life.

Study Guide

Chapter 12 Human Interaction with Lakes and Streams

Questions

1. Do Longfellow, Thoreau, and Walka use the same literary style to describe nature? Which person expresses the strongest sense of place?
2. What senses besides sight are important in the relationship between the observer and freshwater systems? (Hint: Check out the sense of smell, in reference to salmon, and Arthur Hasler—see Beckel, 1987.)
3. Can you use your personal experience to give specific examples of lakes with different colors? What do these colors indicate?
4. In the quotation from Beowulf, how would the chemical limnologist interpret the phrase, "Every night you can see the terrible spectacle of fire on the lake"? (Hint: Refer back to chemistry chapter 10.)
5. What was your first significant experience with lake or stream water?
6. Do you know a story about a "sacred" or special lake? Write a reflective essay showing your own sense of place. Interview someone who has a particularly strong sense of place.
7. In what way is one aesthetic view of a lake better than another?
8. What practices are you familiar with that contribute toward aquatic resource integrity?
9. At the end of chapter 1, you were asked, "What does it mean to think like a limnologist?" Now that you have completed this text, how do you answer this question? In what ways is your understanding of limnology richer and more sophisticated?

Words Related to the Citizen Limnologist

EPA	rehabilitate	stakeholder
NGO	restore	

Crossword Puzzles

As you have read through this text, you have come across a number of words used either exclusively in limnology, or used in a particular way in limnology. Use the crossword puzzle in figure 12.10 to review your grasp of limnological jargon. These words have been used somewhere in the text and are standard limnological words not used in general conversation. Figure 12.11 is crossword puzzle that focuses on important limnological acronyms from the entire book.

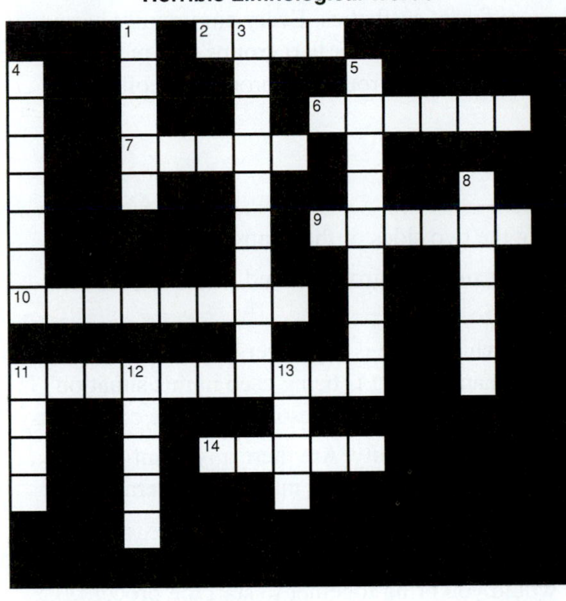

FIGURE 12.10

Across

2. Also called a gyre; a circular current characteristic of turbulent flow
6. Organic sediment on the lake bottom
7. A small or sudden flood
9. The basin formed by a large-scale fault or rift of the Earth's surface
10. The community of microorganisms living on submerged surfaces
11. The lake does not completely mix, even though the surface waters mix
14. A distinct layer of lake sediment representing one year's accumulation

Down

1. Horizontal limestone near the Earth's surface, often eroded into sinkholes
3. A productivity category for tea-colored lakes of low productivity, often bogs
4. A basin formed in or near a volcano, caused by subsidence
5. Underground stream flow, within the streambed sediments
8. A periodic wave movement at the scale of the entire lake
11. Calcium carbonate precipitated during times of intense photosynthesis
12. A kind of shallow lake associated with meandering streams
13. The lake at the base of a cirque

FIGURE 12.11

Across

1. A U.S. federal agency charged with protecting water quality
3. This tells you where you are located in the landscape or lakescape
5. A program of the National Science Foundation that supports research over long time periods
7. Refers to a gas dissolved in water; the gas is produced by photosynthesis
9. A high-energy phosphorus-containing molecule essential in metabolic reactions

Down

2. The largest limnological society in North America
4. Light energy used in photosynthesis
6. A lake district in western Ontario, site of whole-lake experiments
7. Zooplankton do this to escape fish predation during the day
8. The total amount of energy fixed during photosynthesis

Exercise

Exploring Complexity

1. Describe your sense of place in an essay, poem, drawing, or photograph. Share your description with another student or with the class.
2. With your whole class, do a role-playing exercise to explore different aesthetics. Every community has a controversial aquatic issue. Divide the class into three groups. Each of the first two groups will present its strongest possible case for its position on the controversy. Then the third group will produce a policy that satisfies as many people as possible. The groups are:
   a. People on one side of the controversy
   b. People on the other side of the controversy
   c. The "County Board" that will develop policy after hearing the two sides of the controversy

   Before the County Board declares the final policy, try switching sides for groups A and B, and repeat the presentations. This allows the two groups to argue for both sides.
3. Identify a complex limnological system in your community for which there is controversy about how the system should be managed. Is there any evidence that adaptive management is being used in this situation? Is there an opportunity for adaptive management to be useful in finding an optimal resolution to the controversy?
4. Identify a major freshwater system that is being managed in your community. Are there any point-sources or nonpoint sources of contamination that affect this system adversely? Identify and interview several stakeholders to find out if everyone sees these sources of contamination as a problem.
5. Is there a freshwater resource in your community that could be managed using the ecological management model scheme shown in figure 12.8? Which stakeholders would you bring together to start the process?

Additional Resources

Further Reading

Cech, T. V. 2003. *Principles of water resources: History, development, management, and policy.* New York: Wiley. 446 pp.

Lee, K. N. 1993. *Compass and gyroscope: Integrating science and politics for the environment.* Washington, DC: Island Press. 243 pp.

Leopold, A. 1968. *A Sand County almanac and sketches here and there.* London: Oxford University Press.

Miller, C. 2002. *After the deluge: Integrated watershed management in the Red River Valley.* Chapter in "The Farm as Natural Habitat," by D. L. Jackson and L. L. Jackson (eds.) Washington, DC: Island Press. This chapter reports on the success and failures of an adaptive management approach to solving the flooding problem in Minnesota. After the severe Red River floods of 1997, the Minnesota government funded a professionally mediated and adaptive process to resolve conflict among farmers, urban land owners, environmentalists, and other interested stakeholders.

Mitchell, M. K., and W. B. Stapp. 1997. *Field manual for water quality monitoring: An environmental education program for schools.* Dubuque, IA: Kendall/Hunt Publishing. 277 pp.

USDA, Natural Resources Conservation Service. 1996. *National handbook of water quality monitoring.* 450-vi-NHWQM. (Available online.)

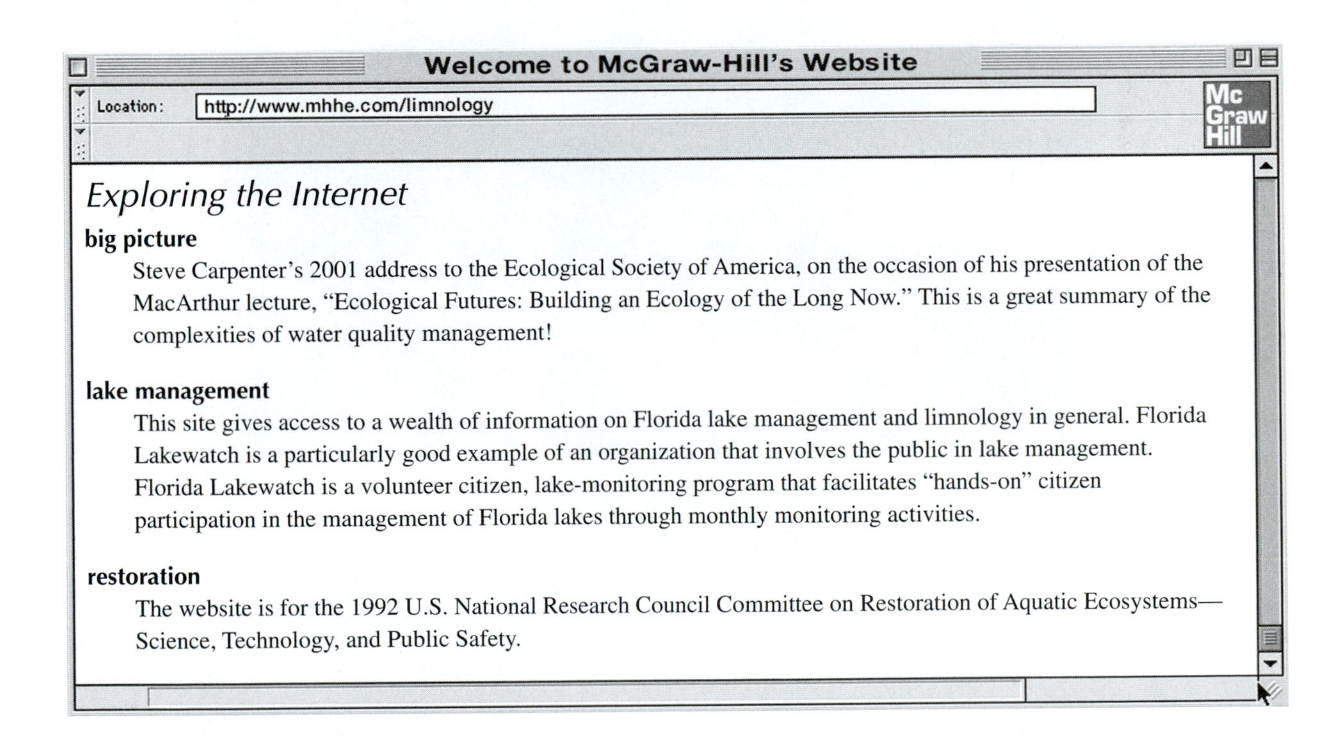

Exploring the Internet

big picture

Steve Carpenter's 2001 address to the Ecological Society of America, on the occasion of his presentation of the MacArthur lecture, "Ecological Futures: Building an Ecology of the Long Now." This is a great summary of the complexities of water quality management!

lake management

This site gives access to a wealth of information on Florida lake management and limnology in general. Florida Lakewatch is a particularly good example of an organization that involves the public in lake management. Florida Lakewatch is a volunteer citizen, lake-monitoring program that facilitates "hands-on" citizen participation in the management of Florida lakes through monthly monitoring activities.

restoration

The website is for the 1992 U.S. National Research Council Committee on Restoration of Aquatic Ecosystems—Science, Technology, and Public Safety.

13

"When in doubt, measure."

Unknown

outline

Field and Laboratory Exercises

BACKGROUND

The 10 field and laboratory exercises in this chapter have been field tested for several years. Each of the field exercises is designed to take an entire 8-hour day. They work well with a class of about 20 students, divided up into groups of 2 to 4 students. If this time schedule is not appropriate, most of these exercises can be easily converted into two or three 4-hour lab periods. For example, field collections can be made during one lab, and the analysis and write-up can be done the following week. Similarly, the laboratory exercises are designed for 8-hour lab periods, but can easily be broken into shorter time segments.

These exercises are designed using a model pioneered by the field courses taught by the Organization for Tropical Studies. This style includes constructing a hypothesis and developing an experimental design (often while in the field), sampling or making observations to test hypotheses, using statistical tests when appropriate, and writing short reports in a notebook, due at the end of the day. The only homework is reading material in preparation for the next exercise.

STATISTICAL ANALYSIS SUPPLEMENT

Statistical thinking is a critical part of limnology. These exercises provide opportunities to use statistical tests. Material to assist in statistical analyses is given at the end of this chapter as appendices. The Spearman rank correlation and chi-square tests are referred to in several of the 10 exercises. These tests were chosen to be simple, quick, and appropriate for the exercises.

The two statistical appendices are:

1. *The Spearman Rank Correlation Test*
2. *The χ^2 Test of Independence, with emphasis on the χ^2 Median Test*

SUPPLEMENTAL READING

Two books are recommended for use with the exercises:

Pond life: A guide to common plants and animals of North American ponds and lakes by G. K. Reid. 1987. New York: Golden

Press. This picture book and field guide is a valuable resource for identifying aquatic organisms. There may be other illustrated guides written specifically for your local area.

Ecology and classification of North American freshwater invertebrates, 2nd ed. by J. H. Thorp and A. P. Covich. 2001. San Diego, CA: Academic Press. This is a great resource for the lab exercises. It includes dichotomous identification keys and illustrations of diversity and anatomy.

FIELD EXERCISE PROTOCOL

It takes planning to ensure that students have the best possible field experience! Before going into the field, it is an excellent idea that everyone has a written schedule of the events of the day. Staying on schedule makes for efficient use of time and avoids frustration at the end of the day. If all participants have this schedule, everyone can help the class stick to the schedule. For example, the schedule for a day might look like the one shown in table 13.1.

Field work is accompanied by a number of safety concerns and permission requirements. For example, students who cannot swim or have medical issues relating to water or being out of doors must inform the instructor before the class goes into the field or on water. Students who cannot swim must wear personal floatation devices at all times when on the water. Any time vertebrate animals are caught or handled, there must be up-to-date and proper approvals and permissions from the appropriate fish and game organizations. Your class and activities must be covered by an approved animal care protocol. Vehicles must be safe and drivers must be properly trained and have the appropriate permissions. Bring along a first-aid kit for each vehicle (and have one available in the laboratory).

For these laboratory exercises that take several hours, it is a good idea to organize a rotation among the students and staff to bring treats and beverages for combating the potential afternoon attach of hypoglycemia. A little food in the mid-afternoon can make a field or laboratory exercise a lot more enjoyable!

Table 13.1 A Schedule for 19 June Limnology Lab (Vertical Patterns in a Lake)

Activity	Time
Start lab: Meet in classroom, take quiz on reading materials for this lab	8:30 A.M.
Introductions of staff and students.	8:45
Discussion of Spearman rank correlation. Start write-up in notebook: Null hypothesis, alternate hypothesis, justification, data table; let the instructor check your hypotheses	9:00
Leave the classroom.	10:15
Arrive at the boat dock, find your group, don life jackets, load limnological and personal gear onto the boats.	10:30
Leave the dock on boats.	11:00
Sample plankton on the way to the middle of the lake. Pause at selected sites along a transect from shore to the middle, and take your samples. At the center of the lake, drop anchor, do temperature and oxygen vertical profiles, and take Secchi disk readings. When the data have been collected, eat lunch. Test water from deep in the lake with your senses, explore your amphibious roots (go swimming).	
Start back from center of lake.	2:15
Arrive at shore, return life jackets, carry all other equipment to classroom.	2:45
Be back in classroom, start the rest of the write-up. Note that your time is limited and that the report is due at the end of class.	3:00
Lab report is due at end of class.	4:30

Vertical Patterns in a Lake

LAB 1

In this field trip, students collect limnological data from a boat. Students who have difficulty swimming should inform the instructor.

Wear

Prepare for total immersion. You might go swimming or you might fall in. Be prepared for the predicted weather (bring sun screen or rain gear). Bring a lunch and all the required materials (notebook, hand lens, *Pond Life* book).

Goals

- Learn to measure temperature and oxygen along a vertical depth gradient (vertical profiles).
- Measure light penetration using a Secchi disk.
- Collect zooplankton using a plankton net.
- Collect plankton, observe the zooplankton and algae with the unaided eye, and then return the sample back to the lake.
- Test a hypothesis about plankton abundance and distance from shore. Use the Spearman rank correlation test.
- Practice writing up a lab report.
- Have fun boating and plumbing the depths of your lake!

Introduction

Review the material in chapter 2 and read the material on the Spearman rank correlation test (Appendix 1) and this information sheet.

In class, before going to the field, discuss the Spearman rank correlation test. The instructor will lead a discussion of:

Correlation
Hypothesis testing
Null hypothesis
Alternate hypothesis
Limnological justification of the alternate hypothesis
Rejecting versus failing to reject the null hypothesis

Lab Activity

There will be a demonstration of calibration of the oxygen meter and use of a Secchi disk. The instructor will demonstrate techniques for measuring vertical profiles of oxygen and temperature, measuring depth of light penetration, and catching zooplankton.

The lab report in your notebook is due at the end of the lab period.

Form small groups, three or four people per group, for efficient use of the equipment. When you take samples, make sure that each member of the group has a chance to both use equipment and record data.

Before leaving the classroom, discuss the experimental design and hypothesis testing contained in this exercise.

Experimental Design—Hypothesis Testing

With your group and the instructor, develop a testable hypothesis that will allow you to use the Spearman rank correlation test. Your hypothesis will be focused on a relationship between phytoplankton, zooplankton, and distance from shore. As you go out in the boats, each group will take a series of vertical tows from 3 meters to the surface (this samples most of the epilimnion). Start sampling near shore, where the water is 4 meters deep, and also sample at 100-meter intervals, toward the middle of the lake. At each location, take three replicate vertical tows and combine them into one jar. The samples are collected (filtered) by the plankton net, and each sample is put into a small jar.

Write the number of the site on a tape label on the bottom of the jar. The first site, nearest shore, is site 1, the next site is site 2, and so on.

You may notice that the jars hold different amounts of sample, so when you get to the center of the lake, make the volume of the jars the same by adding water filtered through the plankton net. Assume that lake depth increased as you went along the transect from near shore to the lake middle. Thus, the numbers on the bottoms of the jars represent both the rank of the distance from shore (1 = nearest shore) and the rank of depth (1 = most shallow).

Without looking at the depth labels, rank the jars according to amount (density) of phytoplankton. For example, if you sampled at six locations, you have six jars, and you rank the jars from 1 (most plankton) to 6 (least plankton). Record the ranks of plankton density and depth in your notebook; for example, as in table 13.2.

As you collect data at the middle of the lake (Secchi disk depth; temperature and oxygen data), record the information in your notebook.

Secchi disk depth is the depth, to the nearest centimeter, at which the disk just disappears from view. The depth is the average between the depth at which the disk disappears and the depth at which it reappears. When the disk is lowered to the Secchi disk depth, mark the place where the line enters the water, and use a meter stick to measure from the number of centimeters above nearest meter marking that is in the water. Record the Secchi disk depth to the nearest centimeter.

For the oxygen and temperature instruments, follow the instructions for use and calibration supplied by the manufacturer.

Notebook

After lunch, or during travel back to the classroom, you can begin to write up the notebook report. At the top of the first page of each report, write the exercise and class name, your name, the date, the site name, and location. For example:

Table 13.2 An Example of How Data Are Ranked for the Spearman Test

Variable	Ranks					
Plankton abundance	2	3	1	5	4	6
Depth (or distance from shore)	1	2	3	4	5	6

Vertical Patterns in a Lake

Zoology 316: Limnology

Jane Doe

19 June 2001

Lake Mendota

43° 4′37″N, 89° 24′28″ W

Limit the report on this exercise to *three pages* plus one page of raw data:

Title information

Plankton data (for your group)

Test of the plankton hypothesis

Answers to questions about the plankton test

Vertical data (for your group)

The vertical profile graph

Answers to the questions about the vertical data

Answer the question, "Is phytoplankton abundance correlated with depth (or distance from shore)?" Use the Spearman rank correlation test. What is the null hypothesis, the alternate hypothesis (one-tailed or two-tailed?), and what is the justification for the alternate hypothesis? Write this up in your notebook.

Draw a single graph for the Secchi disk depth and temperature and oxygen profiles. Use figures 2.8 and 2.13 as your models.

Answer the following questions:

1. What is the vertical pattern of temperature in the lake? On the graph, indicate the thermocline and explain how you determined its location. Label the hypolimnion and epilimnion.
2. What is the vertical pattern of oxygen concentration in the lake? Is this a eutrophic lake? Why?
3. What is the Secchi disk depth and what is the euphotic zone depth? Indicate the euphotic zone on the graph.
4. What is the pH of surface and deep hypolimnetic water? Is there a significant difference? To answer this question, pairs of students measure the pH of surface water and deep water. Each pair makes one measurement of each kind of water. Because this is a learning situation, pairs of students need to show their data to instructors while it is still possible to take another measurement. After having the data approved by an instructor, students record the data in the notebook. The class then pools the data using a table such as that shown in table 13.3

Table 13.3 An Example of a Data Table Used to Convert Observations of Pairs of Students into Pooled Class Data

The overall median for this example set of data is 7.85. Use these pooled observations and directions in Appendix 2 for the Chi-square median test.

Habitat	Pair 1	Pair 2	Pair 3	Pair 4	Pair 5	Pair 6	Pair 7	Pair 8
Surface water	8.1	8.3	8.0	8.1	8.1	8.4	8.2	7.9
Deep water	7.5	7.4	7.5	7.6	7.3	7.2	7.8	7.5

Equipment

Suitable boats and life jackets

Per group: bucket, six 125-ml jars and tops, label tape, sharpie, meter stick

Oxygen and temperature meter with long probe cord. The cords on the oxygen and temperature probes need to be long enough to reach to the bottom of the lake.

A pH meter. This meter can have a short cord, if there are Van Dorn bottles and line for collecting water at various depths. The Van Dorn line must be marked in meters and must be longer than the depth of the lake.

Plankton nets and line. The line needs to be at least 4 meters long, and marked in meters.

Secchi disk and line. For the Secchi disk, there needs to be at least one length of line marked in meters for each group. This line needs to be longer than the expected Secchi disk depth.

For long labs (all day or all afternoon), it is a good idea to organize afternoon treats. Plan to have a break in the middle of the afternoon to eat cookies or health food or even ice cream.

LAB 2

Zooplankton in a Lake

This field trip introduces techniques for catching zooplankton and for analyzing the samples back in the lab.

Wear

Wear swimming gear under long pants and shirt. Bring a lunch, plenty of water, and all the required materials (notebook, *Pond Life* book). Do you need to bring sun screen? Think about dark glasses (polarizing glasses are especially amusing for looking into the water).

Goals

- Learn to take quantitative samples of zooplankton and identify the kinds of zooplankton living in a specific lake.
- Have fun working on boats, see a whole new world of zooplankton, and swim!

Introduction

Before coming to class, review the material in chapter 4 and the Rotifer, Arthropod, and Crustacean sections in *Pond Life* (pages 80 and 85–91).

Lab Activity:

Meet in the classroom.

Discuss the concept of "one-tailed" and "two-tailed" statistical tests, in the context of the Spearman rank correlation test (Appendix 1).

The field part of the lab will be on a suitable lake. Each group will take a series of vertical zooplankton samples from boats. We will return to the lab and analyze the samples the same afternoon. While on the lake, collect data for vertical profiles of temperature and oxygen, and make a Secchi disk depth determination.

Both in the classroom and at the dock, the instructor will demonstrate how to collect zooplankton samples in lakes using a Schindler-Patalas trap. This trap is a clear plastic box with a hinged top and bottom. As the trap is lowered into the water (at about 1 m sec^{-1}), the top and bottom are vertical, allowing water to move through the trap—no plankton are captured. When the descent stops, the top and bottom swing shut, trapping a volume of water. When the trap is brought to the surface, the water can be drained through the net attached to the side of the trap, and the zooplankton originally in the trapped volume of water are concentrated in the net's bucket. The plankton are washed out of the bucket using 95% ethanol and a wash bottle, into a labeled sample jar. For the label, use a pencil to write on a piece of paper and put the label inside the jar. On the label, write the name of the lake, date, group name, and sample depth.

Traditionally, plankton samples were preserved using formaldehyde. However, serious health concerns suggest that this chemical should only be used with extreme safety precautions, including protective clothing. If you can smell formaldehyde, you have probably been exposed to an unsafe level (Black & Dodson, 2003). Alcohol is a much safer alternative and an excellent preservative.

Larger zooplankton traps (30 liters) seem to work better than the smallest size available (12 liters). The smaller trap appears to collect plankton as it is lowered.

Experimental Design—Hypothesis Testing

With your group and the instructor, discuss a testable hypothesis that will involve use of the Spearman rank correlation test. A good question to use as the beginning point is, "Is zooplankton abundance correlated with oxygen concentration?" What is the null hypothesis? What is the alternate hypothesis? What is the ecological justification for the alternate hypothesis? Use the Spearman rank correlation test (Appendix 1) to reject or fail to reject the null hypothesis. Is this a one- or two-tailed test?

Before you go out to the lake, spend time as a class designing your data collection protocol. Each group will be collecting zooplankton, but after the samples are counted by each group, pool the data for the entire class. Each group needs to collect and analyze the samples in the same way. Agree on:

The depths that will be sampled (do you start at the surface or at 1 meter?)

The number of samples to be taken at each depth

In the field, work with the others in your group. Record in your notebook data and a graph for the Secchi disk depth and temperature and oxygen profiles. Sample zooplankton at each 2 meters of depth, starting at the surface. Use a Schindler-Patalas trap to collect zooplankton.

When you get back to the lab, analyze the zooplankton samples. *Read this entire section before you do anything*

Look at the first sample (see the following for details of the analysis), determine what to record, and draw the organism (or organisms) on a blackboard or large piece of paper. Also draw a class data table. To do this, you will need to agree on the number of significant figures to be retained. The table might look like the one shown in table 13.4.

First Sample The class chooses a depth, and each group prepares the sample from that depth for analysis. To analyze each sample, decant alcohol off a sample jar into another jar. Pour the zooplankton in the sample into a petri dish that has been marked off into "lanes" that are just wide enough to cover the entire field of view at a medium magnification of your dissecting microscope. Use a pipette to draw off additional alcohol, without removing zooplankton. Scan the dish to decide which is the most common kind of zooplankton. Confer with the rest of the

Table 13.4 An Example of a Data Table Designed for Pooling Class Data

The Schindler-Patalas trap volume for this example is 20 liters.

Depth (m)	Group 1	Group 2	Group 3	Total Number per Jar	Average Number per Jar	Variance in Number per Jar	Average Number per Liter
0	0	23	7	30	10	139	0.5
2	6	15	37	52	19.3	254	1.0
4	123	44	35	etc.	etc.	etc.	etc.
8	14	96	88				
10	81	55	62				
12	12	60	3				

class to choose a species for counting. Then, scan the dish, using the lanes, to count the number of organisms of the chosen species. Record the abundance and depth first on the blackboard and then in your notebook.

Succeeding Samples Arrange the analysis protocol within your group so that each person has the opportunity to count at least two samples. Count your first sample, then give it to your neighbor who will also count the sample. Compare your results and recount if necessary. If your results are within a few percent of each other, it isn't necessary to continue comparing results. Enter the number into the class data table.

Notebook

Start a new page and record the title information for the exercise, as described in the previous exercise (Lab 1).

After you have collected all your data, draw the following graph and answer the following questions in your notebook. You are limited to *three pages* plus one page of raw data.

Answer the following questions:

1. Using your group data, draw a vertical profile graph with oxygen and temperature.

   Write a paragraph comparing this vertical profile to any that were done earlier in the same class. What are the differences between the graphs? How do these differences relate to weather that has occurred between the two sample dates? Does this second set of data reflect your growing experience with limnological procedures?
2. Draw a graph showing the vertical distribution of zooplankton in the lake. Graph class-average numbers with error bars of 1 standard error using class data. The standard error (SE) of the mean (average) for data at a specific depth is:

$$\text{SE} = (\text{variance/number of replicates})^{1/2}$$

$$\text{Variance} = (\text{sum of } (\text{mean} - \text{each replicate})^2\,) / (\text{number of replicates} - 1)$$

3. Is there a correlation between average zooplankton abundance (of a common species) and oxygen concentration? Use the Spearman rank correlation test on the data in table 13.4. Start by ranking the depths and the average number of zooplankton at each depth. Discuss and then write the null hypothesis and the alternate hypothesis. Write down the limnological justification of your alternate hypothesis. State your conclusion, based on the results of the statistical test. In light of the results of the statistical test, explain how you would redesign your alternate hypothesis.
4. What are the common kinds of zooplankton? List and draw them and be prepared to identify these kinds as part of preparation for the lab final exam.

Equipment

Boats and life jackets

Per group: Schindler trap (make sure it works before the lab). Make sure the bucket is OK and that it comes off. Line is marked off in meters.

Fourteen 2-oz (60-ml) jars with caps, stiff paper labels, pencils, bucket for holding jars and other small gear, 1 liter of 95% alcohol, wash bottle filled with alcohol

Temperature and oxygen meter and cable

Secchi disk and line

In the lab, for each student: dissecting microscope, petri dish, 120-ml jar, dropper, dissecting needle-like poker, and an alcohol wash bottle and a large jar for waste alcohol

Remember the afternoon treats.

LAB 3

Becoming Familiar with Rotifers

This inside lab allows students to practice handling rotifers, to observe rotifer behavior, and to become familiar with rotifer morphology.

Wear

This is an inside lab.

Goals

The purpose of this exercise is to become familiar with recognizing and handling rotifers and understanding how they move. The main goals of this lab are to:

- Become familiar with what free-swimming rotifers look like under low magnification.
- See and draw several different kinds of rotifers. The purpose of drawing organisms is to train the eye to see them more clearly in greater detail. Make the drawings large—only four to six per page.
- Practice using dichotomous keys to identify rotifers to the genus level.
- Estimate swimming speed of several rotifers, using millimeter paper and a dissecting microscope.
- Describe the swimming behavior of one or more rotifers.
- Learn to pick up rotifers with an Edmondson pipette.
- Make a wet mount microscope slide.
- Use the eyepiece micrometer to measure the body length of rotifers.
- Become familiar with and draw the internal anatomy of at least one living rotifer.

Introduction

To prepare for this lab, review the Rotifer section in chapter 4 and the rotifer material in *Pond Life.*

Lab Activity:

Before the lab begins, the instructor collects rotifers from lakes and ponds. It is often necessary to visit a number of sites before one is discovered that is rich in species and has a wide range of sizes. Use a fine-mesh plankton net (e.g., 60-μm mesh) to collect and concentrate the rotifers. These samples can be put into liter jars. Pour the sample through a 500-μm mesh net to remove larger organisms.

Get Sample Each student uses a wide-bore pipette to subsample a jar containing rotifers. A wide-bore pipette is also called a *turkey baster,* and is used in cooking. Place enough of the rotifer sample into a petri dish to cover the bottom of the dish. Probably the best place to get rotifers is from near the bottom of the sample jar.

Notebook

Start a new page and record the title information for the exercise, as described in the directions for Lab 1. Then, complete the following modules, writing in the notebook as directed.

Recognize Rotifers Scan the sample at moderate magnification (about 12×) using your dissecting scope. If your samples are from nature, you will have a variety of different kinds of organisms. Which of these are rotifers? Rotifers are smaller than you might expect and mostly transparent, so they are difficult to see (many are about 0.1 mm long), even with a dissecting microscope. They move smoothly through the water, but can give small hops if disturbed.

They may have spines and a shell (lorica) and a flexible foot, but no evidence of legs or other jointed appendages. Many rotifers appear to be tiny, clear plastic bags filled with nothing much. At first, it is easy to confuse rotifers with protists and nauplii. Look at figures 4.1 and 4.2, and compare the images of rotifers to those of protists in figure 3.8 and nauplii in figure 4.13 and plate 23. Ask your instructor for assistance!

Draw the outlines of three to five rotifers in your notebook. Beware of spending too much time drawing, and make your drawings large. Allot only a few minutes to each outline. Even though these are microscopic animals, draw them large enough to cover a quarter of a page.

Practice drawing what you see, instead of drawing your idea of what you are supposed to be seeing. Avoid drawing "stick figures" and avoid copying drawings from resource materials. The best approach is to relax and let your hand move in response to what you see. Make each outline large enough to see and not cramped—about six per page. Concerns about "I can't draw" are irrelevant to this exercise. If you draw what you see, you will be amazed by the results.

Identify the Rotifers Using your drawings and a standard key, such as can be found in the rotifer chapter in Thorp and Covich (2001), identify the specimens to the genus level.

Estimate Swimming Speed

First: Students in the class need to do this part of the exercise together, at the same time. Working with another person, each student estimates the swimming speed of the most common kind of rotifer. To measure the distance traveled, slide a piece of millimeter graph paper under the petri dish. One person keeps track of time, and one person watches the rotifer, estimating how far it swims in a given amount of time (try 10–30 seconds).

You will be pooling class data, so it is important that swimming speed data are collected in a consistent manner. After you and your partner have estimated the first swimming speed, participate in a class discussion. Compare your data with those of others, and describe your method. As a class, agree on details of the measuring procedure. At this time, also agree on which species of rotifer to measure—pick two to four additional rotifer species, covering a wide range of body size.

Then: With another person, measure the swimming speed of the additional rotifer species. Record the data in your notebook. Make three replicate measurements of speed for each kind of rotifer.

Make Wet Mount of Rotifer Pick up one or more rotifers of the same kind with an Edmondson pipette. Place the rotifer and the water associated with it (less than one drop) onto a microscope slide. Place a millimeter-sized shred of lens paper next to the drop, and lower a cover slip onto the rotifer, water, and paper. The purpose of the paper is to give the rotifers a little room so they are not squashed, but not enough room to allow them to swim freely. Scan the slide at low power (about 25–40×) using the compound microscope. It is best if the microscope is fitted with an eyepiece micrometer. You will eventually find a rotifer either swimming or somewhat trapped between the coverslip and the slide.

Measure Rotifer First, find out how many eyepiece micrometer units are in one millimeter. Do this by looking at a piece of filter paper through the compound microscope at the different magnifications. Make a table in your notebook. Include columns for the magnification, the number of units per millimeter, and the length of a unit in micrometers. If your microscope does not have an eyepiece micrometer, you can still measure the diameter of the field of view (using the millimeter paper), and use this diameter (at different magnifications) to estimate the length of specimens such as rotifers.

Make Graph In your notebook, make a graph of the relationship between rotifer body size and the median swimming speed. On a blackboard or other group facility, make a table of class data, similar to table 13.5.

In your notebook, graph Average Speed as a function of Average Length, using the pooled class data. You will have as many points as there are rotifer species in your study.

See Internal Anatomy Make a wet mount of a large rotifer. In your notebook, draw the internal anatomy. Label your drawing as completely as possible. Review figure 4.1, or identify parts from memory, and ask your instructor how you have done.

Equipment

- Sample of mixed species of live rotifers
- Notebook and pencil
- Dissecting microscope with illumination from below
- Compound microscope fitted with an eyepiece micrometer
- Lens paper, dissecting needle (to lower cover slips), microscope slides, coverslips, millimeter graph paper, stop watch
- Petri dish
- Large-bore pipette with rubber bulb ("turkey-baster")
- Edmondson pipette, which is made from a short piece of narrow glass tubing that has been heated and drawn out into a fine tip. A short piece of rubber tubing is attached to the blunt end of the glass tubing. The other end of the tubing is closed with a bit of glass rod. The pipette fits easily in one hand, and water can be drawn into or expelled from the fine pipette by squeezing on the rubber tubing.
- Drawings of rotifer morphology
- Remember to arrange for afternoon treats.

Table 13.5 A Table for Pooling Class Data that Can Be Used to Compare Rotifer Body Size and Swimming Speed

Rotifer Species	Group 1 Length	Group 2 Length	Group 3 Length	Average Length	Group 1 Speed	Group 2 Speed	Group 3 Length	Average Speed
A								
B								
etc.								

LAB 4

Becoming Familiar with Cladocerans

This inside lab allows students to practice handling cladocerans, to observe cladoceran behavior, and to become familiar with cladoceran morphology.

Wear

This is an inside lab.

Goals

The main goals of this lab are to:

- Become familiar with what cladocerans look like under low magnification.
- See and draw a representative cladoceran, first with the unaided eye (for a large *Daphnia*) and then with a dissecting microscope.
- Learn how to immobilize a large *Daphnia.*
- Observe the feeding mechanism of *Daphnia.*
- Become familiar with and draw the internal anatomy of a large, immobilized *Daphnia.*
- Estimate swimming speed of large and small cladocerans, using millimeter paper and a dissecting microscope.
- Make a wet mount microscope slide of one or more cladocerans.
- Use the eyepiece micrometer to measure the body length of cladocerans.
- Graph cladoceran swimming speed versus body length.

Introduction

To prepare for this lab, review the Branchiopod section in chapter 4, plate 24, and the cladocera material in *Pond Life.*

Lab Activity

The instructor will have a culture of *Daphnia magna* on hand. These animals can be purchased from a biological supply house or cultured using methods described in the Branchiopoda chapter in Thorp and Covich (2001). However, it is more convenient to locate someone in the community who cultures *Daphnia*—check local biology departments and state and private environmental labs. Have several *Daphnia* per student, but keep the culture at no more than about 25 animals per liter.

Also, the instructor can collect a variety of cladocerans from local lakes and ponds before class begins. Use a medium-fine net (about 120-μm mesh). Make several tows with the net and put each concentrated sample into a liter jar. Then pour this combined sample through a coarse filter (about 5-mm mesh) to remove insects and plant material. If there appear to be more than about 25 animals per liter, add some filtered pond water to reduce the density. If the samples are to be kept for a day or more, dilute them even more. (Chilling a summer sample in a refrigerator is not a good idea—at least some of the different forms will quickly die from the cold.)

Get Sample Start with the *Daphnia magna* culture. Using a wide-bore pipette (in this case, use an eye-dropper type of pipette, with an opening of about 2–3 mm), place a large *Daphnia* into a 100-ml jar filled with filtered lake water. Watch the animal swim to get an idea of how it moves through the water.

Draw the Swimming *Daphnia* First, draw the *Daphnia* with your unaided eyes. Observe the swimming animal for at least 5 minutes before you begin drawing. The purpose of this exercise is to have the opportunity to see as much detail as possible, without using a microscope. As you do this exercise, you will be training yourself to see zooplankton.

Immobilize the *Daphnia* Put a pin-head-sized speck of grease (Vaseline) in the middle of a small, dry petri dish. Use the wide-bore pipette to capture a large *Daphnia.* Put the water containing the *Daphnia* on top of the speck of grease. Under the dissecting scope, use a needle to position the *Daphnia* onto the speck. The *Daphnia* will become stuck in the grease. Avoid using so much grease that the *Daphnia* is enveloped, but if the *Daphnia* won't stick, start over with a slightly larger speck. Gently fill the petri dish to half-full with filtered lake water.

Feed the *Daphnia* Fill a small-bore (Pasteur) pipette (or an Edmondson pipette) with the dense algae suspension. Place the tip of the pipette in front of the *Daphnia* and slowly release a small cloud of algae. Notice how the algae is captured by the *Daphnia,* and how the algae is processed by the *Daphnia.* You will see the thoracic legs capturing the algae, the green mass moving toward the mouth, where the algae is chewed with the mandibles, and then swallowed. Follow the movement of the algae into and through the gut.

Time the movement of algae through the gut. Start timing when you introduce the algae. In your notebook, record the time it takes for food to move through the gut.

Draw the *Daphnia* at Low Magnification In your notebook, draw the immobilized *Daphnia,* showing the following internal and external parts. Review figure 4.8a and plate 24, and work with a partner. Draw and label:

compound eye
simple eye
first antenna
rostrum
second antenna
dorsal branch of the second antenna
muscles of the second antenna
mandible
mouth
gut
anus
thoracic legs
postabdomen
postabdominal claw
carapace
tail spine
heart
brood chamber
ovaries
eggs or embryos (if present)

When you are finished, return the *Daphnia* back into a jar with filtered lake water.

See Cladoceran Diversity Place several squirts (using the large-bore pipette) of the cladoceran sample (from a local lake) into your petri dish. Scan the sample at moderate magnification using your dissecting scope. Cladocerans are diverse in form, but will resemble the *Daphnia* you just saw more than they will resemble copepods or rotifers. Look for animals with a carapace that covers the thorax and abdomen, and that swim with branched second antennae (although these may be difficult to see in the smaller forms).

Draw the outlines of three to five different cladocerans in your notebook.

Make Wet Mounts (Microscope Slides) of Cladocerans Pick up one or more of the smaller *Daphnia* in a small amount of water, using a large-bore pipette. Put the animal onto a microscope slide and remove most of the water with a screen-tipped pipette, except for about one large drop. Place a millimeter-sized shred of regular paper next to the drop, and lower a cover slip onto the specimen, water, and paper. The purpose of the paper is to give the cladoceran(s) a little room. Scan the slide at low power using the compound microscope fitted with an eyepiece micrometer. You will eventually find the cladoceran trapped between the coverslip and the slide.

Identify the Cladocerans Using your drawings, wet mounts, and a standard key, such as can be found in Dodson and Fry (2001), identify the specimens to the genus level. The drawings in figures 4.7, 4.8, and plate 24 may also be helpful.

Measure a Cladoceran Measure the body length of a wet-mounted cladoceran using the eyepiece micrometer and the conversion table in your notebook. (Use the calibration table for the micrometer with your millimeter paper, from Lab 3.)

Estimate Swimming Speed This section requires the pooling of class data, so it is necessary for class members to do this section together.

First: Working with another person, estimate the swimming speed of a large *Daphnia magna.* Obtain at least five separate measurements for the animal. To measure the distance traveled, use a ruler next to a jar, or beneath a petri dish. One person keeps track of time, and one person watches the cladoceran (with the unaided eye, or at low magnification), estimating how far it swims in a given amount of time (try 10 seconds).

Make a wet mount of the *Daphnia,* and measure its length.

Come together as a class, and compare results and standardize your protocol:

Details of measuring swimming distance (two-dimensional or three?)

Number of replicate estimates

Length of time to observe each swimming bout

Which species to observe

Then: Repeat the swimming speed and length measurements on an intermediate-sized *Daphnia,* a small *Daphnia,* and other (probably small) cladocerans if they are available and as identified by the class members. Record the data in your notebook and on the blackboard (table 13.6).

Make a Graph In your notebook, make a graph of the relationship between cladoceran body size (X-axis) and the median swimming speed. Label the axes and include the units. Use the class data. Write a paragraph about the relationship you see:

Is there a relationship?

What generalization can you make about the relationship between body size and swimming speed? Is the same generalization true for dogs, people, and horses?

Equipment

Notebook, pencil,

Daphnia magna culture (see p. 329 for details about the culture)

Mixed cladocera samples from local lakes and ponds

Dense algae culture (can be purchased from biological supply houses)

Filtered lake water (pour clean lake water through a fine-mesh net, less than 100-μm mesh)

Dissecting microscope with illumination from below

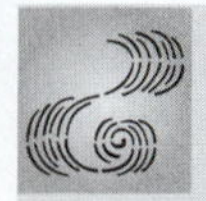

Table 13.6 An Example of a Data Table for Pooled Class Data to Compare Body Length to Average Swimming Speed

Use a table similar to table 13.5 to pool data from individual groups.

Name	Species	Length (mm)	Average Swimming Speed (mm sec^{-1})
Smith and Doe	*Daphnia magna*	3.05	7.5
	Daphnia magna	2.17	6.6
	Daphnia magna	1.20	3.2
	Bosmina	0.75	3.6
	Ceriodaphnia	0.88	2.5
Kashian and Jung	*Daphnia magna*	3.10	8.5
etc.	*Daphnia magna*	2.15	7.6
	Daphnia magna	1.22	3.3
	Bosmina	0.72	3.2
	Ceriodaphnia	0.90	2.1

Compound microscope fitted with an eyepiece micrometer, small petri dish

Large-bore pipette of the eye-dropper variety (opening 2–3 mm)

Turkey baster

Small-bore pipettes with rubber bulb (Pasteur or Edmondson)

Two dissecting needle per student, microscope slides, coverslips, mm graph paper, stop watch, lake water, 100-ml jars

Vaseline

Becoming Familiar with Copepods

LAB 5

This inside lab allows students to practice handling rotifers, to observe rotifer behavior, and to become familiar with rotifer morphology. The instructor may wish to turn this into two labs, focusing on calanoid and cyclopoid copepods separately.

Wear

This is an inside lab.

Goals

The main goals of this lab are to:

- Become familiar with what copepods look like under low magnification.
- See and draw two to three different copepods, including your largest and smallest species. Use a dissecting microscope to draw a male and female body of each species.
- Use a dichotomous key to identify one or more species of copepods.
- Estimate swimming speed of a large and small species, using millimeter paper and a dissecting microscope.
- Dissect copepods for identification.
- Use the eyepiece micrometer to measure body length.
- Graph copepod swimming speed versus body length.

Introduction

To prepare for this lab, review the copepod section in chapter 4 and the copepod material in *Pond Life.*

Lab Activity

The instructor can collect a variety of copepods from lakes and ponds before the class begins. Use a moderately coarse mesh net (about 120-μm mesh). Make several tows with the net, and put each concentrated sample into a liter jar. Then pour this combined sample through a coarse filter (about 5-mm mesh) to remove insects and plant material. If there appear to be more than about 25 animals per liter, add some filtered pond water to reduce the density. If the samples are to be kept for a day or more, dilute them even more. (Chilling a summer sample in a refrigerator is not a good idea—at least some of the different forms will quickly die from the cold.)

It is also useful to have an alcohol-preserved sample of a large copepod on hand. The instructor can pick it out of a sample and preserve it in 70% alcohol before class.

Get Sample of Copepods Collected from Nature For drawings and dissections, use a wide-bore pipette to pick up a few copepods from the live samples. Place the copepods into a petri dish with enough lake water to cover the bottom of the dish.

Even with the unaided eye, it is possible to distinguish between cyclopoids and calanoids. Use table 13.7 to identify your copepods. You will also be using the information in table 13.7 when you use a microscope to look more closely at copepods.

When you estimate the swimming speed, use a wide-bore pipette to capture live copepods (of the large and small species). For the live observations, fill the petri dish nearly to the brim with filtered lake water and put the copepods into the petri dish.

Table 13.7 A Guide to the Identification of Copepods at the Ordinal Level

Use this information, along with figure 4.14 and plate 23 to identify copepods.

Calanoida	Cyclopoida	Harpacticoid
Anterior part of the body (the metasome, also called the prosome) is much broader than the posterior part of the body (urosome) and there is an abrupt change in width between the two parts.	Metasome width diminishes gradually to the less narrow urosome.	Metasome and urosome about the same width.
First antennae with 23–25 segments. First antennae reach to near the end of the cephalosome, to near the end of the body.	First antennae with 6–17 segments. First antennae shorter than the metasome, often about as long as the first segment of the metasome.	First antennae with 5–9 segments. First antennae about half as long as the first metasome segment.
Fifth legs in the male are asymmetrical. Fifth legs in both the males and females are dissected off the body for observation for identification of the specimen to species.	Fifth legs in male and female symmetrical, with no more than three segments. To see the fifth legs, dissect the copepod to separate the urosome, and turn it ventral side up. The fifth legs will be attached to the anterior margin of the ventral side of the first segment of the urosome.	Fifth pair of legs in male and female similar and symmetrical. Fifth legs are dissected off the body for identification to species.
Adult females carry eggs in a single mass fastened to the ventral surface of the urosome, near the joint with the metasome. Some release eggs into the water.	Adult females carry egg masses in two sacs, attached on either side of the urosome and attached to the first urosome segment.	Adult females carry eggs in a single mass attached to the ventral surface of the first segment of the urosome.
Mostly planktonic species, living their entire life in open water.	Different species live in different habitats, from open water, to surfaces of plants, to the interstitial spaces in coarse sediments. Some species have an ectoparasite habit.	These are benthic species (meiofauna 5 living in or on sediments or vegetation, or psammon 5 living among sand grains). A few species venture into the open water.

This capture process is an important part of your experience with how copepods swim. Compare the experience with the previous experience of capturing a live *Daphnia.* Write a paragraph in your notebook comparing the swimming behavior of *Daphnia* and copepods.

Draw Free-Swimming Copepods Scan the sample at moderate magnification using your dissecting scope. In your samples from nature, you will have a variety of different kinds of organisms. Which of these are copepods?

Look for the largest and smallest adult copepods in your sample, and draw their outlines in your notebook. If there are additional copepods, draw another kind.

Recognize the Nauplius Stage The nauplius is a larval stage that hatches out of the egg. The nauplius has a roughly triangular body and three pairs of appendages (made up of just a few segments each): the first and second antennae and the mandible (see figure 4.12 and plate 23).

Each time the copepod molts, the body adds segments and appendages and existing appendages become more complex (with more segments and spines).

Draw a nauplius along with the two or three outlines of adult copepods that you have already drawn in your notebook.

Recognize Adult Copepods

Calanoids: The females carry eggs in two bunches, beneath the abdomen. The males have a modified (bent and flattened) right first antenna, and the left antenna is identical to the adult female left and right antennae.

Cyclopoids: The females carry eggs in two bunches on either side of the abdomen. Males, which may be only half as long as the females, have both right and left first antennae bent. Also, in adults, the last segment of the abdomen (just anterior to the fork) is shorter than the next-to-last segment. In juveniles, the last segment is longer than the next anterior segment (the long terminal segment divides when the last juvenile copepodid stage molts into the adult).

You have drawn adult copepods in your notebook. Now label these as either cyclopoid or calanoid copepods and as adults or juveniles.

Draw the Morphology of a Large Copepod Using a dissecting microscope, draw the adult male and female of the large copepod species, with more detail than you used in the first drawing. If the instructor has provided preserved animals, use these. Otherwise, you can pick out the largest animals, drop them into a petri dish with 70% ethanol, wait 10 minutes, and then begin the drawing. Draw the whole animal, with its appendages. Label the following parts. If you don't know where a part is, check figure 4.12 and plate 23, or ask the instructor.

- simple eye
- first antenna
- metasome
- urosome
- thoracic legs
- caudal rami (the two branches at the end of the urosome)
- caudal setae (at the posterior end of the copepod)
- eggs or embryos (if present)
- enlarged and jointed first antenna (if present).

Dissect the Copepod Start with several preserved *adult* copepods, both males and females, if possible. If only live animals are available, kill a few by dropping them into 70% ethanol.

Dissect the large copepod in a small drop of thin Hoyer's mounting medium on a microscope slide (Dodson and Fry, 2001, give the recipe for Hoyer's medium). Transfer at least one male and one female, or up to two of each, to the small drop of Hoyer's, using your dissecting needles, being careful to transfer as little alcohol as possible. Use your needles to position a specimen on its back. Hold it down with one needle. Use the other needle to dissect off parts of the copepod.

Calanoid Copepods It is best to use males for the identification. While holding the body down with one needle, use the other as a knife to slice off the fifth pair of thoracic legs. The fifth legs are the most posterior thoracic legs and are about half as long as the fourth pair. Then slice off the pair of first antennae from the anterior end of the head. Place the fifth legs and the antennae near the rest of the body, and go on to the next specimen. It may be necessary to place the parts near the edge of the Hoyer's medium so the parts stay in place with a good orientation. Make sure the first antennae are visible and stretched out. Work fast—the thin Hoyer's evaporates and quickly becomes too thick for dissection.

Cyclopoid Copepods It is best to use females for identification. Cut between the posterior base of the fourth thoracic legs and the genital segment. The tiny fifth legs are on the narrow genital segment at the anterior end of the urosome. Make sure the posterior surface of the urosome stays facing up. Slice off the fourth pair of legs and place them, posterior (concave) face upwards, near the urosome. Repeat the dissection with another male (if available), and then dissect two females. Make sure the first antennae are visible and stretched out.

When you are done with the dissection, label the slide and let the preparation dry on a slide warmer until the Hoyer's is thick. This assures that the copepod parts will remain where you put them, when you add the coverslip. When the Hoyer's is thick (about 5 minutes) add another large drop of Hoyer's and gently lower on the cover glass. Let the slide rest on the slide warmer for 15 minutes, if possible, before looking at it. This allows the clearing process to begin and dries the Hoyer's a little around the edges of the cover glass, so the cover glass will not easily slide off.

Draw Key Characters Draw the fifth legs of the male and female of the large species, and draw a male right and left first antenna. Include as much detail as you can see at high magnification. Leave enough time for the speed estimate (about an hour).

Identify the Copepods Using your drawings and a standard key, such as can be found in Thorp and Covich (2001), identify the specimens to the genus level.

Estimate Swimming Speed This section requires the pooling of class data, so it is necessary for class members to do this section together.

First Working with another person, estimate the swimming speed of a large copepod. Obtain at least five separate measurements for the animal. To measure the distance traveled, use a ruler next to a jar or beneath a petri dish. One person keeps track of time, and one person watches the copepod (with the unaided eye, or at low magnification), estimating how far it swims in a given amount of time (try 10 seconds).

Make a wet mount of the copepod and measure its length.

Come together as a class and compare results and standardize your protocol:

Details of measuring swimming distance (two-dimensional or three?)

Petri dish or jar?

Number of replicate estimates

Length of time to observe each swimming bout

Which species to observe

The swimming behavior of copepods is a combination of slow gliding and fast jumps. Agree on what you are going to measure.

Then Repeat the swimming speed and length measurements on an intermediate-sized copepod and a small copepod. Record the data in your notebook and on the blackboard, using a table such as that shown in table 13.8.

Table 13.8 An Example Data Table Useful for Presenting Pooled Class Data to Compare Copepod Body Length with Swimming Speed

Name	Species	Length (mm)	Average Swimming Speed (mm sec^{-1})
Smith and Doe	*Mesocyclops*	1.7	
	Diacyclops	1.5	
	Eucyclops	1.3	
	Diaptomus minutus	1.1	
	Diaptomus clavipes	2.5	
Li and Joseph	*Mesocyclops*	1.80	
etc.	*Diacyclops*	1.65	
	Eucyclops	1.25	
	Diaptomus minutus	1.05	
	Diaptomus clavipes	2.55	

Make a Graph In your notebook, make a graph of the relationship between cladoceran body size (X-axis) and the median swimming speed. Label the axes and include the units. Use the class data. Write a paragraph explaining which swimming behavior component you measured. Also, compare the copepod relationship between size and speed to the cladoceran relationship. What generalization can you make?

Equipment

Notebook and pencil

Live samples of a large and a small species

Preserved copepod sample of a large species, with enough specimens so each student has six to eight available for drawing and dissection. After the students have selected their copepods, they can trade with a neighbor, so each student has several males and females.

Dissecting microscope with illumination from below; compound microscope fitted with an eyepiece micrometer

70% ethanol, large petri dish, large-bore pipettes with rubber bulb

Thin Hoyer's mounting medium, as described in the Branchiopod chapter of Thorp and Covich (2001)

Two dissecting needles, microscope slides, coverslips, millimeter graph paper, stop watch, filtered lake water

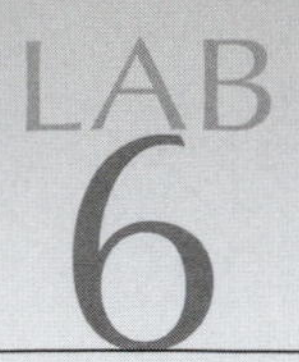

Aquatic Insects in a Stream

In this field trip, students experience the stream habitat firsthand, become familiar with aquatic macroinvertebrates, and practice doing science in the field. This field trip is most successful in a stream with both slow- and fast-moving water, and with both muddy pools and rocky reaches. The smaller first- or second-order streams are best for this exercise.

Wear

Wear swimming gear under long pants and a shirt. Wear outer clothes that can get wet and very muddy. There will be mosquitoes (wear a long-sleeve shirt) and poisonous plants along the stream bank, such as wild parsnip, poison ivy, and nettles (wear long pants). You will be wading in a stream, so wear footgear that can get wet and will protect you from broken glass hidden in deep mud. Tevas or other sandals that strap onto the feet are best. You will be wading in deep mud, and probably falling into the stream, so come prepared. Everyone will be expected to get wet. Bring a lunch, plenty of water, and the required materials (notebook and *Pond Life* book).

The instructor will bring a first-aid kit. If the water is likely to be polluted, bring antibacterial hand wipes, and water to wash off your hands after sampling.

Goals

- Learn about aquatic insects in a stream setting.
- Experience stream morphology and stream microhabitats.
- Have fun looking at bugs and wading in mud and moving water.
- Practice using the chi-square test.

Introduction

Meet in the classroom. Prepare by reading the material on the chi-square test, this sheet, and the insect and mollusk material in chapters 4 and 5, and in *Pond Life.* Be able to identify the orders of insects and distinguish a snail from a clam. Focus on the big picture in the chi-square handout.

In the classroom, discuss how to sample for aquatic insects in streams. Your instructor will demonstrate the use of an aquatic dip net and a picking pan.

Your instructor will also discuss stream microhabitats. Riffles are fast-flowing stretches that will have relatively coarse substrate, and will be easy to sample with the kick sample technique (see Kick Sample Directions). Slow-moving water in pools will have fine mud or sand substrate, and will be more of a challenge (will take longer) to sample using kick samples.

Kick Sample Directions

- Stand in the stream—do not wade deeper than about 0.5 meter into the stream. Place the aquatic dip net at arms' length downstream from you, with the net frame placed on the stream bottom. If you are working in pairs, one person holds the net, standing downstream from the net, and the other person stands upstream of the net.
- With the toe and heel, the "kicker" digs, scrapes, and churns the surface of the sediments to dislodge macroinvertebrates. Dig at least 10 centimeters into the sediments.

- Make sure the plume of silt flows into the net. Avoid kicking sand, gravel, or rocks into the net.
- After 2 to 3 minutes of kicking, stop and swish the net through the water to wash silt out of the net. If there is mud, slosh the net up and down in the stream in such a way as to wash mud, but not invertebrates, out of the sample. Remove any large debris by hand—rocks, leaves, sticks, the occasional fish, etc. Place the sample in a white pan (picking pan) that has 1 centimeter of clean water in it. Scan the net for organisms clinging to the mesh, and put them into the pan.

Recognize Aquatic Macroinvertebrates Spend about an hour becoming familiar with common aquatic macroinvertebrates. These are the organisms you see when you dump the contents of an aquatic dip net into a picking pan. It is best to start becoming familiar with these animals by kick-sampling a rocky reach of a stream.

Draw the outlines of each kind of macroinvertebrate in your notebook. After collecting organisms, create a temporary zoo by combining examples of all the different kinds of organisms the class finds. Put each kind into its own jar, and work as a class to label each kind of macroinvertebrate. Label your drawings.

Experimental Design—Hypothesis Testing

The class and the instructor design a hypothesis using concepts of stream biology. The hypothesis must be testable using the chi-square test.

Constraints to consider as a group are the:

- Total time available
- Time it takes to do a single kick sample and count the organisms
- Minimum number of samples you need to take to perform a chi-square test
- Advantages of taking more than the minimum number of samples
- Number of people in the class
- Time required for doing the chi-square test

Depending on the constraints specific to your class, you can test one or more hypotheses.

Answer the following questions:

1. Are there more species of aquatic insects in fast water compared to slow water? Take several kick samples in fast and slow water. Record species per kick sample in repeated samples of the two habitats (table 13.9). Then use a chi-square median test to analyze counts of species (see Appendix 2).
2. Are there more individual aquatic insects (per kick sample) in fast water compared to slow water? As for the previous question, take multiple samples and use the chi-square median test to answer the question (see Appendix 2).
3. Are there more air-breathing macroinvertebrates in slow-water substrate, compared to fast-water substrate? As for the previous question, take multiple samples and use the chi-square median test to answer the question (see Appendix 2).

Notebook

In your notebook, record standard name, date, and location data (see the directions for Lab 1).

State the question or questions that you are studying. For each question, give the null hypothesis, the alternate hypothesis, the limnological justification of the alternate hypothesis, the data table, the contingency table, the results of the statistical test, and your conclusion (to reject, or fail to reject, the null hypothesis, which is the answer to the question). You are limited to writing three pages for this lab.

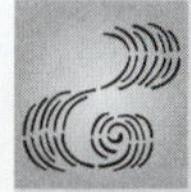

Table 13.9 An Example of Data You Might Collect from Slow and Fast Parts of the Streams

The numbers are number of species per kick sample, for 10 samples in each habitat. The overall median here is 6. Using this information, you can fill in a contingency table (such as table 13.14)—for example, there are 10 kick samples from pools that had fewer than the median number of species. After you complete the data table, perform the chi-square median test (see Appendix 2).

Pool (slow water)	3	4	5	7	5	8	5	6	1	4	3	2
Riffle (fast water)	5	8	7	8	9	6	3	7	9	12	10	8

Equipment

Vehicles and up-to-date driver permissions

Enough aquatic D-nets and picking pans so there is one net and one pan for each pair of students

Per group: Bucket with six medium-sized jars with tops, meter stick, white tape, marking pen

For the entire group: 26 finger bowls (or the large plastic jars that are finger-bowl shaped) to be used for the practice ID, before beginning the data collection

Aquatic Macrophytes and Multivariate Techniques

LAB 7

This field trip can take place either along a lake or a pool of a river. Students gain experience with aquatic macrophytes and with applications of concepts of community ecology and functional morphology.

Wear

Prepare for total immersion, swimming, lots of mud, strong sunlight, and rain. Wear something on your feet for wading in deep, sticky mud: sandals or old shoes that won't come off in deep mud. Bring a lunch and all the required materials (notebook, hand lens, *Pond Life* book).

Goals

- Learn to identify common aquatic macrophytes.
- Use point sampling.
- Create and use a scatter diagram (an introduction to multivariate techniques).
- Have fun wading in mud and swimming.

Location

Choose two sites with abundant aquatic macrophytes, including rooted, nonemergent species. One of the sites should be a long shoreline with a gradient of exposure to waves—from protected to directly exposed. The other site can be a single location in a different lake or stream.

Introduction

Meet in the classroom. Review material on this sheet and in the aquatic macrophyte pages in chapter 5 and in *Pond Life*. A useful resource to have during the day is *Through the Looking Glass: A Field Guide to Aquatic Plants* by Borman et al. (1997). Field work will be done along the shore of a lake or quiet water of a large river.

Experimental Design—Hypothesis Testing

Question: Aquatic macrophytes are diverse and ecologically important members of lake communities. The various macrophyte species have different forms, such as emergent, rooted, floating leaves, or free-floating. *Are these different species found in different habitats?* Approach this question by collecting lots of macrophytes and looking at the effects of light penetration (measured by water depth) and disturbance (measured by potential exposure to waves in a particular habitat). Communities, being made up of numbers of species, require special (multivariate) techniques for their description.

One such multivariate technique is the *scatter diagram.* To make a scatter diagram, label the X- and Y-axes of a graph using two (more-or-less independent) environmental variables, such as water depth and degree of exposure. Then, for each sampling site (combination of water depth and wave height), find the location on the graph and write the names of the species you find at those conditions. Repeat this process several times, and you have a scatter diagram—a map of the range of environmental conditions for which you find each species.

(To have enough data, sample at least three depths at four different points along a shoreline, along a gradient of wave action.) Draw a contour line around each species. The contour is a boundary that includes all the observations for a single species. When you look at the diagram, you can ask: Do all species occur at all combinations of conditions? Which species overlap? Figure 13.1 is an example of a scatter diagram.

The main goal of the lab is to create a scatter diagram for the common macrophytes found along the margin of a lake or river. At four places along the shore, you will take point samples perpendicular to the shore. Sample at four depths along an imaginary line running straight out into the lake. The depths are ankle deep, knee deep, waist deep, and 2 meters (the 2-meter depth will require total immersion—bend down and gather as many stems as you can). If you have willing swimmers in the group, go for 3 meters.

A point sample is taken by grasping two handfuls of aquatic macrophytes at the base (and including any associated, nonrooted plants). Take two point samples at each depth.

Collect macrophytes at four sites along the shore of your lake or pool. Choose these sites, if possible, so they have a range of exposure to waves.

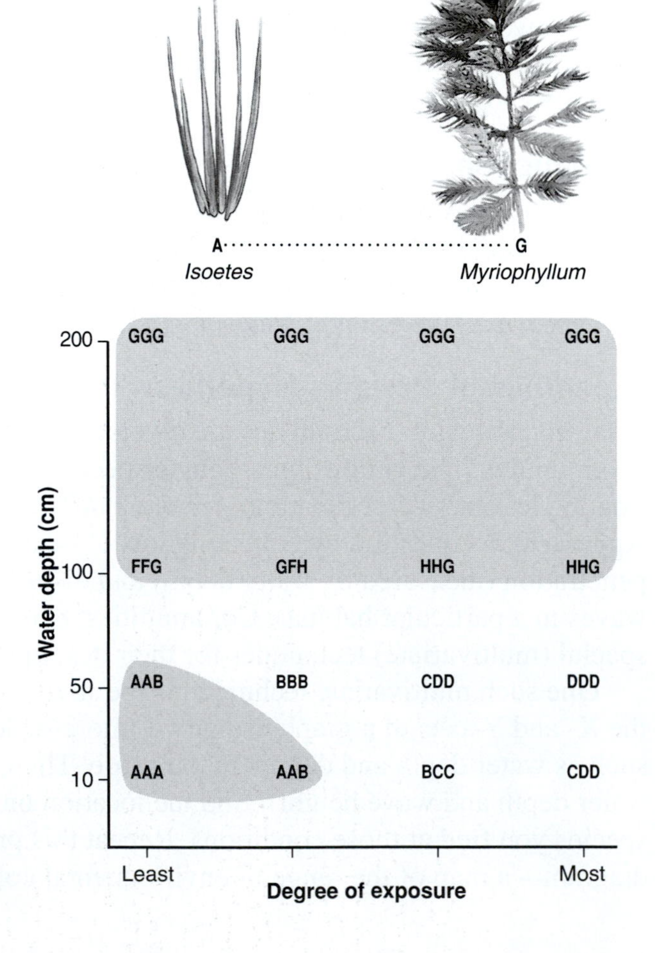

FIGURE 13.1 An example of a scatter diagram. The contours representing suitable habitat have been drawn for two macrophyte species: *Isoetes* and *Myriophyllum*.

Notebook

In your notebook, record standard name, date, and location data (see the directions for Lab 1).

Record in your notebook the location, depth of the three samples at each of four sites, and species you find at each depth. Also estimate exposure (to wave action) at each site, by consensus—agree on a ranking for the four sites. Write these four sites (1–4) along the X-axis of your scatter diagram. Put depth along the Y-axis (with ankle depth nearest the origin)—see the example graph in figure 13.1. Then record your data on this graph. Choose a different symbol for each kind of macrophyte. For all the different kinds of macrophytes, plot the exposure and depth data. Draw a boundary around each set of symbols of the same kind. These polygons can be used to predict what species will be found at a specific depth and disturbance level. If possible, take the graph to a different site to see if the predictions work.

Draw the outline of each kind of macrophyte in your notebook.

Answer the following questions:

- When you look at the scatter diagram, do all species occur at all combinations of conditions? Do all the species polygons overlap? What is the appropriate multivariate null hypothesis?
- What is the relationship between macrophyte morphology and disturbance?
- If you took the scatter diagram to a different site, describe how well its predictions were realized.
- Design a simple hypothesis and experimental design to investigate effects of other environmental factors on the macrophyte community.

Equipment

Vehicles

Meter sticks

Identification keys to aquatic plants

One large poster board, magic markers

Buckets for bringing macrophyte specimens to the group. Use the specimens for drawing. With the entire group, agree on what you call each kind of aquatic macrophyte.

LAB 8 Fishes in Lakes and Streams

In this field exercise, you seine fishes from two different lakes or streams. You will use the data you collect on fishes to test a hypothesis. Answer questions in your notebook. Because this lab focuses on catching, handling, and identifying fishes (vertebrate animals!), you must be sure that you have the proper approvals and permissions from the appropriate fish and game organizations and that your class and activities are covered by an approved animal care protocol.

Wear

Prepare for total immersion. You will be using a seine to catch fishes, so you will be wading in deep water (up to your neck). Wear appropriate foot gear. There will also be an opportunity to swim. Bring a lunch, your notebook, and the *Pond Life* book. If you are collecting in moving water, be aware of the danger of being swept away.

Goals

- Learn to seine fishes and identify common fishes from two sites.
- Practice using the chi-square (especially the chi-square median) test and the Spearman rank correlation test.
- Have fun wrestling with the seine, catching fishes, and being in lakes.

Introduction

Meet in the classroom. Prepare for the lab by studying the fish section in chapter 5 and in *Pond Life,* pages 120–128. Also, review the use of the chi-square test.

The instructor will discuss methods for sampling and identifying fishes and will provide you with identification keys, such as the keys to genera found in George C. Becker's (1983) *Fishes of Wisconsin* (available on the Internet).

Your instructor needs to be sure that the class has the necessary permits for collecting fishes.

Field work will be done from the shore of two different aquatic sites. This lab works best if the two sites are known sources of fishes that differ significantly in limnological characteristics, such as water clarity, productivity, or land use.

Experimental Design—Hypothesis Testing

With your group and the instructor, design a null and alternate hypothesis about the length of fishes, which you will collect at two sites. Before you sample, write down your null hypothesis, your alternate hypothesis, and the ecological justification for the alternate hypothesis. (For example, consider effects of productivity, lake or stream size, community structure, and angling.)

To answer your null hypothesis, measure 50 fishes at each of the two sites. Find the overall median fish size, and then use a chi-square median test to test the null hypothesis.

Notebook

In your notebook, record standard name, date, and location data (see the directions for Lab 1).

Answer the following questions:

1. What are the fish species lists for the two lakes?
2. Are the same species dominant in the two lakes?

3. Are the fishes you caught at the first site the same size as those caught at the second site? Write data and do the statistical analyses in your notebook, then write out the answers to the questions. Discuss with the class: What is the appropriate statistical test for answering this question?

Equipment

Transportation

Permits

Large beach seines

Buckets

Picking pans

Forceps

Meter sticks or fish measuring sticks

LAB 9

Chemistry and Productivity of a Sewage Pond

This is a field trip to an extreme environment. Chemical concentrations and environmental conditions change over a wide range daily. Processes at sewage treatment plants provide an opportunity for limnologists to see ecology and water chemistry in action in an applied setting. A leading ecologist, E. P. Odum (1971, p. 437), wrote "A study of a secondary treatment plant makes an excellent (and highly relevant) exercise for an ecology class."

Wear

Rubber boots are a good idea because sewage ponds are often contaminated with bacteria and dangerous chemicals such as PCBs or heavy metals. If rubber boots are not available, use foot gear that can get wet and muddy. It is not necessary to wade into the water, but the edges of these ponds are often unstable and very muddy. There is no shade, so be prepared for a lot of bright sun. Bring all the required materials (notebook, hand lens, *Pond Life* book). Bring binoculars and a bird ID book if you desire them. Bring a lunch, but only eat if you have washed your hands and/or used antibacterial hand wipes. The instructor will bring a first-aid kit.

Goals

- Learn about changes in pH, temperature, and oxygen concentration in a hypereutrophic pond, as photosynthesis starts up in the morning.
- Estimate the rate of primary production in the pond.
- Have a guided tour of the sewage treatment facilities, if possible.
- Enjoy watching a great sunrise and hearing the birds wake up.

Introduction

To prepare for this exercise, read this sheet and the material about oxygen in chapters 2 and 8.

Meet at the classroom at an agreed-upon time in the dark of the morning, just before dawn. The reason for the early hour is to get to the pond at about sunrise. We will work at the pond for about 4 hours.

We will measure a portion of the diel chemical and physical changes in a hypereutrophic sewage lagoon. Take measurements each hour, starting at (or just before) the sun rises. Measure temperature pH and oxygen concentration of the water. Dangle the probes in the water. If the water is shallow, tie the probes to poles so they can be put into the water and not touch bottom. Repeat these measurements each hour for the next 4 hours.

At sunrise, there will probably be no or very little oxygen in the water. Three hours after sunrise, there should be more than 4 milligrams of oxygen in the water. At this time, fill the dark bottles with pond water. Place the bottles in the water, but near shore in very shallow water, so they are just submerged. Wait for 1 hour. Do the regular measurements on the pond water, and also record oxygen concentrations in the light and dark bottles.

Changes in pH, temperature, and oxygen concentration are related to primary productivity (mg C fixed per liter per day, and mg C fixed per m^2 per day).

Estimate the rate of gross photosynthesis (gross primary productivity) using changes in oxygen concentration in the pond and in "dark bottles." To do this, measure the change in oxygen concentration in the lake between the third and fourth hour (when the dark bottles are in the water). This observed increase in oxygen concentration is called *net primary productivity,* and is due to two processes: *photosynthesis* and *respiration.* Respiration (which takes oxygen out of the water) is measured by the decrease in oxygen in the dark bottles.

Gross primary productivity:	The total amount of oxygen produced per liter per hour, by photosynthesis.
Respiration:	The amount of oxygen used per liter per hour. (You measure this with the dark bottles.) Respiration is a positive number.
Net primary productivity:	The increase in oxygen produced per hour that is observed in the pond. (You measure this by sampling the pond.)

Net primary productivity = Gross primary productivity *minus* respiration

Gross primary productivity = Net primary productivity *plus* respiration

The respiration rate is estimated by the rate of disappearance of oxygen from dark bottles. Oxygen meters typically tell you the concentration of oxygen, in milligrams of oxygen per liter. Subtract the average $[O_2]$ in the dark bottles at the end of the hour from the $[O_2]$ you started with. Divide by 1 hour. This is a rate, and a positive number.

Notebook

In your notebook, record standard name, date, and location data (see the directions for Lab 1).

In your notebook, answer the question: "What is the rate of gross and net primary productivity in a hypereutrophic pond (measured on a per hour basis)? Include a graph of the hourly change in temperature, oxygen, and pH in your report. What organisms are living in the pond? The Teaching Assistants will check off your less-than-one-page answer and graph.

Equipment

Vehicles

Temperature-oxygen meters

pH meters

Secchi disks

Aquatic D-nets

Long poles for dangling Secchi disks and probes into the water

For each group: bucket, four dark bottles, one meter stick, 250-ml beaker (plastic), four 60-ml jars

Aquatic Resource Integrity

In this field exercise, students look at streams or lakes from the perspective of evaluating aquatic resource integrity.

Wear

Wear outdoor clothes—prepare for the possibility of stepping in water or mud.

Goals

Learn to use two tools for evaluation of the health (integrity) of aquatic resources.

- Calculate habitat quality using an evaluation sheet.
- Calculate the Stream Invertebrate Community Index using kick sample data.
- Become familiar with parameters of aquatic habitat assessment.
- Review knowledge of aquatic macroinvertebrates.
- Practice stream kick-sampling.
- Measure dissolved oxygen (DO) in a stream.
- Review the concept of aquatic resource integrity.

Introduction

Read the sections in chapter 11 about aesthetics and about aquatic resource integrity.

Break the class into groups of two people each, and have each pair fill out evaluation sheets. Visit two stream sites and two lake sites. Either the instructor or the class can choose sites beforehand based on knowledge of the local area—seek sites that are likely to differ in aquatic resource integrity.

Aquatic Resource Assessment At each site, each pair of students performs the two evaluations. Also, each pair can measure DO, using an oxygen meter.

The Habitat Quality Evaluation Sheet (table 13.10) provides a step-by-step protocol for evaluating habitat of an aquatic community. This evaluation can be applied to a stream, a lake, or a wetland.

The Stream Invertebrate Community Index (SICI) is based on data from a kick sample. Take the sample in a riffle, and sample until you have 50 to 200 macroinvertebrates. The SICI is then calculated using the graphical data sheet (figure 13.2).

The SICI is a reflection of the kind of physical and biological conditions prevalent in a stream. The highest index values will occur in streams with high dissolved oxygen and low biological oxygen demand. The lowest values will occur in streams inhabited by animals tolerant of low oxygen.

Notebook

In your notebook, record standard name, date, and location data (see the directions for Lab 1). Give the location for each site you visit. In addition, record the time of day, percent cloud cover, and any precipitation that has occurred during the last 3 days.

Table 13.10 The Integrated Habitat Quality Evaluation Sheet

Date: ______________

Location: ______________ County: ______________ State: ______________

Stream or lake name: ______________

Name of evaluator: ______________

	Write Score Here	Value 12	9	6	3
1. Substrate condition—In riffle areas, how much of the gravel, cobble, or boulder is covered with mud or sand?		0–25% covered	25–50% covered	50–75% covered	75–100 % covered
2. Water level—Annual variation in depth, measured from annual low water mark.		Less than 50 cm	0.5–1 m	1–2 m	More than 2 m
3. Coarse woody debris, undercut banks, and large boulders—As % of stream or littoral zone.		More than 10%	5–10%	2–5%	Less than 2%
4. Pools and littoral zone vegetation—Area covered by root mats and submerged vegetation		100–50%	25–50%	10–25%	0–10%
5. Bank stability—Amount of the bank that is stabilized by vegetation or impervious material, such as boulders, rip-rap, or lunker structure. The alternative is bare unstable and eroding soil.		100–75%	50–75%	25–50%	0–25%
6. Riparian vegetation zone—Average width (include grasses, lawns or pasture).		More than 6 cm	6–4 m	4–2 m	Less than 2 m
7. Land use in adjacent watershed (within 50 m)—Percent of the watershed area in undisturbed land uses (e.g., forest, grassland).		100–90%	90–70%	70–40%	Less than 40%
8. Point source pollution along bank—Pipes, ditches carrying pollution per 1 km.		None	One	One–five	More than five
9. Nonpoint source pollution along bank—Feedlots, confined animals, houses, industrial plants, roads crossing stream, in the km above the site.		None	One	One–five	More than five
10. Aesthetics		Exceptional	High	Low	Offensive
Total Score					

Aquatic Resource Integrity Total Score	
120–91	Highest habitat quality, the safest and most valuable resource
90–61	Moderate habitat quality, but improvement is possible
60–31	Be seriously concerned, restoration program is indicated
Less than 31	Take immediate action—the water is a danger to health of humans and other organisms.

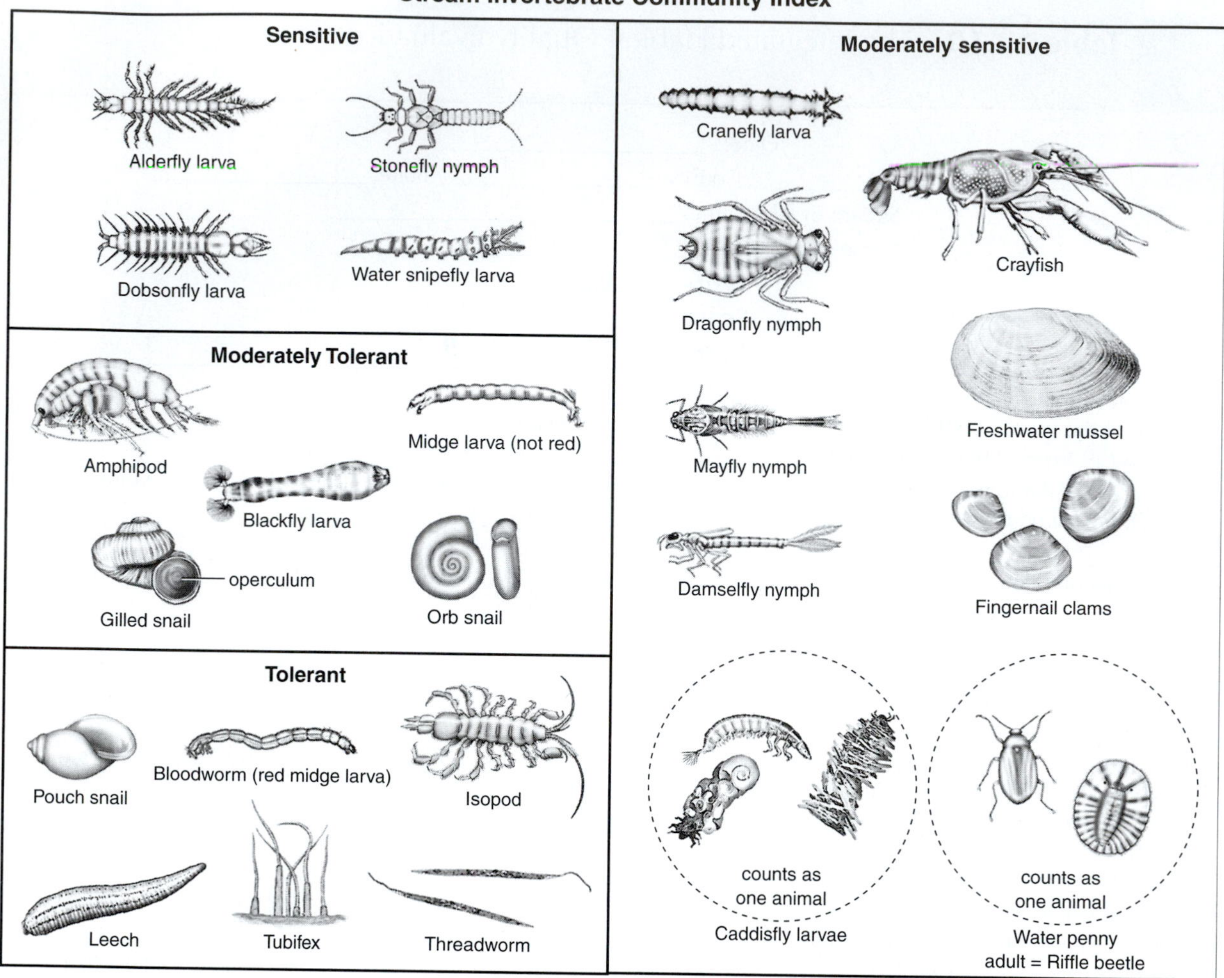

1. Circle the animals found in your sample of aquatic insects

---

2. Tally the number of animals circled in each category, and multiply the number of animals by the number given at right.

Sensitive ________ X 4 = ________

Moderately sensitive ________ X 3 = ________

Moderately tolerant ________ X 2 = ________

Tolerant ________ X 1 = ________

Total score = ________

---

3. Use the total score (Stream Invertebrate Community Index) to assess water quality:

SICI Score	Quality
> 25	Excellent
16–25	Good
8–15	Fair
< 8	Poor

FIGURE 13.2 The Stream Invertebrate Community Index (SICI)—an assessment tool for evaluating stream water quality.

At each site, perform the evaluations and measure DO and record the data.

Answer the following questions:

1. How is the habitat quality index related to the stream invertebrate community index? In your class, you had several pairs of students estimating the same indices. Combine all the indices for each site into a common data table (such as table 13.11), and calculate the variance for the two kinds of evaluations.
2. Which index has the higher coefficient of variation? What does this difference mean in terms of usefulness of the indices? What are possible reasons for differences seen by different pairs of observers (e.g., why would two pairs produce different estimates of habitat quality)?
3. What is the relationship between sediment size (from fine particles of mud, to sand, gravel, and cobble) and species richness of aquatic macroinvertebrates?
4. For the sites with the lowest indices of habitat quality and of stream invertebrate community, what practices could increase the habitat quality or improve the stream invertebrate community?
5. What is the relationship among habitat quality, the invertebrate index, and the measured dissolved oxygen (DO)?

Equipment

Vehicles

Habitat Quality Evaluation Sheets, Stream Invertebrate Community Index Sheets

Aquatic dip nets for kick sampling, picking pans, DO meters

Pencils, notebook, 30-m tape

The Habitat Quality Evaluation Sheet is modified from material produced by the River Alliance.

The Stream Invertebrate Community Index sheet is based on the version created by Michael A. Miller, Wisconsin Department of Natural Resources.

Table 13.11 An Example of Data Related to Stream Assessment

The individual scores are averaged and a variance is calculated for each score. The coefficient of variation (CV), a useful way to compare relative amount of variability among different parameters, is the standard deviation (the square root of the variance) divided by the average. The CV is often written as a percent, by multiplying the ratio times 100%.

Names	Habitat Quality Index	Stream Invertebrate Community Index
Zao and Flaherty	107	3.2
Allen and Santos	80	3.5
Schell and Haubenschildt	95	3.7
Mean	94	3.07
Variance	183	0.06
Coefficient of variation	14.4%	7.3%

appendix 1

Spearman Rank Correlation Test

In research, we frequently wish to know how two sets of data are related: Do the values of one variable tend to increase as the values of the other variable increase? If the two sets of data tend to increase together, they are directly (or positively) correlated; if one increases as the other decreases, there is an inverse (or negative) correlation.

For example, the data might be two measurements taken on a series of fishes, such as body length and swimming speed. If faster swimming is associated with larger fish, we would say there is a positive correlation.

The Spearman rank correlation test uses the *ranks* (rather than the actual values) of the two sets of variables. For example, a string of pH values 6.3, 6.0, 6.4, and 6.8 can be ranked from lowest to highest, keeping the sequence of the data, as 2, 1, 3, 4. These ranks are used to calculate a statistic, the correlation coefficient called r_s.

In a correlation test, the null hypothesis is that the two sets of data appear to be correlated only due to accidents of sampling, and that the calculated value of r_s is due purely to chance. The test allows us to calculate the chance of finding the observed value of r_s if the null hypothesis is true. Large values of r_s are very unlikely if the null hypothesis is true. When the r_s value is large enough (improbable enough), we are allowed to reject the null hypothesis. The null hypothesis is rejected if the chance of getting the calculated value of r_s is less than 0.05 (5%). This probability depends on sample size. Even small values of r_s can be improbable if the sample size is large enough (see table 13.12).

To calculate r_s:

1. Write the two sets of data in two columns (one column for one variable, the other column for the other variable), with each related pair of values on the same line.
2. Rank (from 1 to *n,* where *n* is the number of pairs of data) the numbers in each column. If there are ties within a column (several measurements with the same value), then assign all the measurements that tie the same median rank. Note: Ties reduce the power of the test, and ties can be avoided by measuring with as fine a scale as possible. Design your experimental method to avoid ties in collecting data!

   If you think there is a positive correlation between the two sets of data, then number both columns from the lowest to the highest value. If you think there is a negative correlation, number one column from lowest to highest value and the other column from highest to lowest. If you are not sure, try both ways of ranking, and use the larger of the two r_s values you calculate.
3. Subtract the ranks of each pair of data.
4. Square each difference.
5. Sum the squared differences.
6. Multiply the sum of squared differences by 6: this is 6*S*.
7. $r_s = 1 - \{6S/(n^3 - n)\}$
8. Look up the probability of your r_s in table 13.12. Use the appropriate value of *N* (the number of pairs of data).
9. If your calculated r_s value is larger than the critical value given in the table for your *N,* you may reject the null hypothesis at the level indicated (0.05 level or 0.01 level).

The calculation of r_s is a little complicated. Table 13.13 gives an example of a test of correlation using some made-up data.

In table 13.13, the sum of the squares is 8.5.

Six times the sum of squares is 51.

Therefore, the $r_s = 1 - \{6S/(n^3 - n)\} = 1 - \{51/(1000 - 10)\} = 1 - (51/990) = 1 - 0.052 = 0.948$

The r_s value of 0.948 (with $N = 10$) is larger than both the 0.05 and 0.01 critical values in table 13.12, so we can reject the null hypothesis with some confidence.

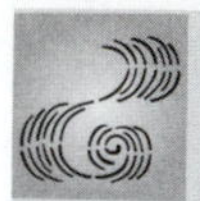

Table 13.12 Values of r_s at Two Different Rejection Levels

To reject the null hypothesis, the r_s value must be greater than the critical value for $p = 0.05$. For example, if you have eight pairs of data, and your r_s value is 0.70, you may reject your null hypothesis at the 0.05 level but not at the 0.01 level.

N The Sample Size	r_s Value $p = 0.05$	r_s Value $p = 0.01$
4	1.000	
5	0.900	1.000
6	0.829	0.943
7	0.714	0.893
8	0.643	0.833
9	0.600	0.783
10	0.564	0.746
12	0.506	0.712
14	0.456	0.645
16	0.425	0.601
18	0.399	0.564
20	0.377	0.534
22	0.359	0.508
24	0.343	0.485
26	0.329	0.465
28	0.317	0.448
30	0.306	0.432

Table 13.13 Examples of Data that are Analyzed According to the Spearman Rank Correlation Method

This is an example (with one tie). The *X* and *Y* values are both ranked from smallest to largest value, so this is a test for a positive correlation.

X Values	Rank of *X*	*Y* Values	Rank of *Y*	Difference Between Ranks	Difference Squared
5	**4**	40	**3.5**	0.5	0.25
7.5	**9**	55	**7**	2	4
4	**3**	40	**3.5**	−0.5	0.25
6	**6**	50	**6**	0	0
2	**1**	15	**2**	−1	1
9	**10**	80	**10**	0	0
7	**8**	65	**9**	−1	1
5.5	**5**	45	**5**	0	0
6.5	**7**	60	**8**	−1	1
3	**2**	10	**1**	**1**	**1**
				Sum of squared values =	**8.5**

appendix 2

The χ^2 Test of Independence, with Emphasis on the χ^2 Median Test

When your data consist of counts that fall into distinct categories, you may use a χ^2 (chi-square) test to determine the significance of differences between two (or more) independent groups of observations.

Groups: Sets of observations. The observations need to be assigned to categories.

Categories: Some aspect of the individuals (e.g., size class, gender, above or below the overall median, etc.)

This statistical tool allows comparison of two groups to see if the distribution of the number of observations is different in the various categories.

The null hypothesis is that the difference between two distributions of observations is due only to sampling error.

For example, if you have collected fishes from two lakes, you might have the alternate hypothesis that the number of male and female fishes are different in the two (that the sex ratio is significantly different). You would test the null hypothesis by first making a contingency table, with the categories (genders) as columns and the groups (lakes) as rows. Then you record the number of fishes you collected (table 13.14).

Table 13.14 has only two categories (gender) and two groups (lakes). However, in many situations, there can be more than two categories and/or groups. For example, your alternate (two-tailed) hypothesis might be that small, medium, and large fishes are distributed differently in the two lakes (table 13.15).

CHI-SQUARE MEDIAN TEST

A common question is, "Is there a difference between the two independent groups in relationship to the overall median?" Another way of saying this is, "Does one group of observations (measurements) tend to be above the overall

Table 13.14 An Example of a Contingency Table Used In the Chi-Square Test

The numbers in each cell are the number of fishes collected for that category. Only whole numbers can be in the cells for the observed data.

	Number of Males	Number of Females	Total Number of Fishes
Lake 1	15	36	51 in Lake 1
Lake 2	25	18	43 in Lake 2
Total per gender	40 Males	54 Females	94 grand total

Table 13.15 An Example of a Test of Null Hypothesis About Size Distributions

In this contingency table, the categories (columns) are size classes and the groups (lakes) are rows.

	Small Fishes	Medium Fishes	Large Fishes	Total Fishes
Lake 1	10	21	20	51
Lake 2	8	10	25	43
Total per size class	18	31	45	94 grand total

median, and one group below the median?" This kind of question can be tested with the χ^2 median test.

For example, say you had the following measurements on fishes from two lakes:

Lake 1: 17, 18, 19, 23, 23, 23, 24, 24, 25, 25, 26, 28, 31, 32, 33, 39

Lake 2: 6, 20, 24, 24, 25, 25, 25, 25, 25, 26, 28, 29, 35, 38, 43, 45, 45, 45, 46

Find the *overall* median. The median is the middle value of all the measurements (all groups taken together). The overall median is the median of all your measurements in all categories and groups. If there is an odd number of measurements, the median is the middle number. If there is an even number of measurements, the median is the average of the two middle numbers. In this example, there are 35 measurements, so the overall median is value of the 18th largest measurement, or 25.

To test the null hypothesis of no difference in central tendencies (except for sampling error), construct a contingency table with two categories (above or below the overall median) as columns and the groups (lakes) as rows. There can be more than two groups (e.g., lakes).

Do not use the measurements that are exactly equal to the median. Since there are seven values in this example that are equal to the median, the total number of observations will be 35 − 7 = 28. The contingency table is shown in table 13.16.

CHI-SQUARE CALCULATION

To calculate the χ^2 statistic, you want to find the average distribution of observations for the two (or more) groups. The χ^2 statistic will be larger the more the two observed distributions deviate from the average distribution. The average distribution has an expected value (E) in each cell of the contingency table, just as there is an observed number (O) in each cell. The calculation of chi-square proceeds as follows:

1. Prepare the contingency table, with an observed value (O) in each cell.
2. Calculate the expected (E) values for each cell. Choose a cell and multiply the row total times the column total and divide by the grand total of observed values. Write the expected value in each cell, next to the observed value. For the chi-square test to work properly, the grand total should be at least 20 and the *expected* value in any cell should be at least 5. While observed values are whole numbers, expected values can be decimal fractions. An example of this calculation of expected values is given in table 13.17.
3. After you calculate the observed and expected values, subtract the expected value from the observed value in each cell: (0 − E). Don't worry about the sign of the difference.
4. Square the differences for each cell $(O - E)^2$
5. Divide the squared value for each cell by the expected value of each cell: $(O - E)^2/E$
6. Add up all the results of the division for each cell. This is the χ^2 statistic.

   $$\chi^2 = \text{Sum}\ [\ (O - E)^2/E\ ]$$

   For the data in table 13.17, the chi-square value is 2.33.

7. Satisfy yourself that the χ^2 statistic is larger the more different the observed values are from the expected (average) values for each cell.
8. The number of degrees of freedom is the number of rows minus one times the number of columns

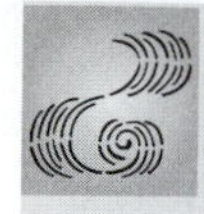

Table 13.16 A Contingency Table Demonstrating a Chi-square Median Test

The total number of observations has been classified relative to the overall median, using the data given in the example in the text.

	Number Below the Median	Number Above the Median	Total Number of Fishes
Lake 1	8	6	14
Lake 2	4	10	14
Total per category	12	16	28 grand total

minus one. For the example in table 13.17, there are two rows and two columns, so there is only 1 degree of freedom. The test using table 13.15 has 2 degrees of freedom.

9. Use the information in table 13.18 to calculate the probability of finding your calculated χ^2 statistic. Note that this is a two-tailed probability. If you have a one-tailed test, divide the probability in the table by 2.

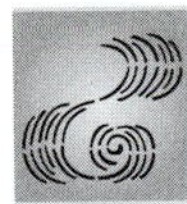

Table 13.17 An Example of Observed and Expected Values in a Chi-square Contingency Table

The data are from table 13.16.

	Below the Median	Above the Median	Total Number of fishes
Lake 1	8 = Observed	6 = Observed	14
	6 = Expected	8 = Expected	
Lake 2	4 = Observed	10 = Observed	14
	6 = Expected	8 = Expected	
Total per category	12	16	28 grand total

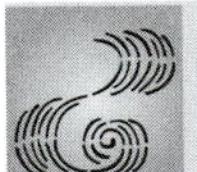

Table 13.18 Chi-square Table

The numbers in the table are the critical values for the chi-square statistic, which depends on the probability level and the number of degrees of freedom. To reject the null hypothesis, the χ^2 value must be greater than the critical value for $p = 0.05$. For example, if you have 1 degree of freedom, and your χ^2 value is 7.5, you may reject your null hypothesis at the 0.05 level and at the 0.01 level, but not at the 0.001 level. Values are from Rohlf & Sokal, 1981.

Degrees of Freedom	I = 0.05	I = 0.01	I = 0.001
1	3.841	6.635	10.828
2	5.991	9.210	13.816
3	7.815	11.345	16.2662
4	9.488	13.277	18.467
5	11.070	15.086	20.515

references

Literature Cited

Ackerman, J. D., B. Sim, S. J. Nichols, and R. Claudi. 1994. Review of the early life history of zebra mussels (*Dreissena polymorpha*)—Comparisons with marine bivalves. *Canadian Journal of Zoology* 72: 1169–79.

Alexander, R. M. 1979. *The invertebrates.* Cambridge, UK: Cambridge University Press.

Allan, J. D. 1987. Macroinvertebrate drift in a Rocky Mountain stream. *Hydrobiologia* 144: 261–68.

Allan, J. D. 1995. *Stream ecology: Structure and function of running waters.* New York: Chapman & Hall.

Allen, T. F. H. 1998. Community ecology. In *Ecology,* ed. S. I. Dodson et al. New York: Oxford University Press.

Allen, T. F. H., and T. W. Hoekstra. 1992. *Toward a unified ecology.* New York: Columbia University Press.

Allen, T. F. H., J. A. Tainter, and T. W. Hoekstra. 1999. Supply-side sustainability. *Systems Research and Behavioral Science* 16:403–27.

Alt, D., and D. W. Hyndman. 1986. *Roadside geology of Montana.* Missoula, MT: Montana Press.

Anbar, A. D., and A. H. Knoll. 2002. Proterozoic ocean chemistry and evolution: A bioinorganic bridge? *Science* 297: 1137–42.

Anderson, K. L., and D. J. Leopold. 2002. The role of canopy gaps in maintaining vascular plant diversity at a forested wetland in New York State. *Journal of the Torrey Botanical Society* 129: 238–50.

Andrews, J. H., editor 1986. *Nuisance vegetation in the Madison Lakes: Current status and options for control.* A committee report. Madison, WI: University of Wisconsin-Madison Press.

Angert, E. R., K. D. Clements, and N. R. Pace. 1993. The largest bacterium. *Nature* 362:239.

APHA. 1998. *Standard methods for the examination of water and wastewater.* 20th ed. Prepared and published jointly by American Public Health Association, American Water Works Association, Water Pollution Control Federation. American Public Health Association, Washington, DC.

Arengo, F., and G. A. Baldassarre. 1999. Resource variability and conservation of American flamingos in coastal wetlands of Yucatan, Mexico. *Journal of Wildlife Management* 63: 1201–12.

Arnott, S. E., N. D. Yan, J. J. Magnuson, and T. M. Frost. 1999. Interannual variability and species turnover of crustacean zooplankton in Shield Lakes. *Canadian Journal of Fisheries and Aquatic Sciences* 56: 162–72.

Aron, J. L. and J. A. Patz (editors) 2001. *Ecosystem change and public health: A global perspective.* Baltimore, MD: The Johns Hopkins University Press.

Baldauf, S. L. 2003. The deep roots of the eukaryotes. Science 300: 1703–1705

Baird, W. 1850. *The natural history of the British Entomostraca.* London: Ray Society.

Baker, L. A., A. T. Herlihy, P. R. Kaufmann, and J. M. Eilers. 1991. Acidic lakes and streams in the United States: The role of acidic deposition. *Science* 252: 1151–54.

Barnes, R. D. 1980. *Invertebrate zoology.* 3rd ed. Philadelphia: Saunders.

Barry, M. J. 1994. The costs of crest induction for *Daphnia carinata. Oecologia* 97: 278–88.

Bayly, I. A. E. 1986. Aspects of diel vertical migration in zooplankton, and its enigma variations. pp.349-368 In *Limnology in Australia,* eds. P. De Deckker and W. D. Williams, CSIRO, Melborne. Dordrecht, Netherlands: Junk.

Bayly, I. A. E. 1999. Review of how indigenous people managed for water in desert regions of Australia. *Journal of the Royal Society of Western Australia* 82:17–25.

Beaton, M. J., and P. D. N. Hebert. 1997. The cellular basis of divergent head morphologies in *Daphnia. Limnology and Oceanography* 42: 346–56.

Beckel, A. L. 1987. Breaking new waters: A century of limnology at the University of Wisconsin. *Transactions of the Wisconsin Academy of Sciences, Arts and Letters.* Special Issue.

Becker, G. C. 1983. *Fishes of Wisconsin.* Madison: University of Wisconsin Press. Available online at *http://www.seagrant.wisc.edu/greatlakesfish/becker.html.*

Bell, G. 2001. Neutral macroecology. *Science* 293: 2413–18.

Bennett, E. M., T. Reed-Andersen, J. N. Houser, J. R. Gabriel, and S. R. Carpenter. 1999. A phosphorus budget for the Lake Mendota watershed. *Ecosystems* 2:69–75.

Berendonk, T. U., and M. B. Bonsall. 2002. The phantom midge and a comparison of metapopulation structures. *Ecology* 83: 116–28.

Berzins, B. 1958. Ein planktologisches Querprofile. *Report. Institute for Freshwater Research. Drottningholm* 39: 5–22.

Birge, E. A. 1897. Plankton studies on Lake Mendota. II. The crustacea from the plankton from July 1894, to December, 1896. *Transactions of the Wisconsin Academy of Sciences, Arts, and Letters* 11:274–448.

Birge, E. A. 1906. The oxygen dissolved in the waters of Wisconsin Lakes. *Transactions of the American Fisheries Society* 35:142–163.

Birge, E. A. 1907. The respiration of an inland lake. *Transactions of the American Fisheries Society* 36: 223–241.

Birmingham, R. A., and L. E. Eisenberg. 2000. *Indian Mounds of Wisconsin.* Madison: University of Wisconsin Press.

Black, A. R., 1993. Predator-induced phenotypic plasticity in *Daphnia pulex:* Life history and morphological responses to *Notonecta* and *Chaoborus. Limnology and Oceanography* 38: 986–96.

Black, A. R., and S. I. Dodson. 2003. Ethanol: A better short-term preservation technique for freshwater Branchiopoda. *Limnology and Oceanography: Methods volume 1.*

Boersma, M., L. De Meester, and P. Spaak. 1999. Environmental stress and local adaptation in *Daphnia magna. Limnology and Oceanography* 44: 393–402.

Bohle-Carbonell, M. 1986. Currents in Lake Geneva. *Limnology and Oceanography* 31: 1255–66.

Bollens, S. M., B. W. Frost, and J. R. Cordell. 1994. Chemical, mechanical and visual cues in the vertical migration behavior of the marine copepod *Acartia hudsonica. Journal of Plankton Research* 16: 555–64.

Borman, S., R. Korth, and J. Temte. 1997. *Through the looking glass: A field guide to aquatic plants.* Wisconsin Department of Natural Resources Publication Number FH-207-97. Merrill, WI: Reindl Printing.

Bormann, F. H., and G. E. Likens. 1970. The nutrient cycles of an ecosystem. *Scientific American* 223: 92–101.

Boronat, L., M. R. Miracle, and X. Armengol. 2001. Cladoceran assemblages in a mineralization gradient. *Hydrobiologia* 442: 75–88.

Borror, D. J., and D. M. DeLong. 1964. *An introduction to the study of insects.* 2nd ed. New York: Holt Reinhart Winston.

Brenner, M., T. J. Whitmore, and C. L. Schelske. 1996. Paleolimnological evaluation of historical trophic state conditions in hypereutrophic Lake Thonotosassa, Florida, USA. *Hydrobiologia* 331: 143–52.

Brewer, M. C. 1998. Mating behaviours of *Daphnia pulicaria,* a cyclic parthenogen: Comparisons with copepods. *Philosophical Transactions of the Royal Society of London B.* 353: 805–15.

Brinkhurst, R. O. 1974. *The benthos of Lakes.* With the collaboration of R. E. Boltt, M. G. Johnson, S. Mozley, and A. V. Tyler. New York: St. Martin's Press.

Brinkhurst, R. O., and S. R. Gelder. 2001. Annelida: Oligochaeta, including Branchiobdellidae. In *Ecology and classification of North American freshwater invertebrates.* 2nd ed., eds. J. H. Thorp and A. P. Covich, 431–64. San Diego: Academic Press.

Brittain, S. M., J. Wang, L. Babcock-Jackson, W. W. Carmichael, K. L. Rinehart, and D. A. Culver. 2000. Isolation and characterization of microcystins, cyclic heptapeptide heptotoxins from a Lake Erie strain of *Microcystis aeruginosa. Journal of Great Lakes Research* 26: 241–49.

Brock, T. D. 1985. *A eutrophic lake: Lake Mendota, Wisconsin.* New York: Springer-Verlag.

Brönmark, C. and L._A. Hansson. 1998. The biology of lakes and ponds. Oxford. Oxford University Press. 216 pp.

Brooks, J. L., and S. I. Dodson. 1965. Predation, body size, and composition of the plankton. *Science* 150: 28–35.

Brunskill, G. J., and S. D. Ludlam. 1969. Fayetteville Green Lake, New York. I. Physical and chemical limnology. *Limnology and Oceanography* 14: 817–29.

Buchsbaum, R. 1976. *Animals without backbones.* Chicago: University of Chicago Press.

Burgis, M. J., and P. Morris. 1987. *The natural history of lakes.* Cambridge, UK: Cambridge University Press.

Burkalova, L. E., A. Y. Karatayev, and D. K. Padilla. 2000. The impact of *Dreissena polymorpha* (Pallas) on unionid bivalves. *International Review of Hydrobiology* 85: 529–41.

Butzel, H. M., and J. Fischer. 1983. The effects of purine and pyrimidines upon transformation in *Tetrahymena vorax,* Strain V_2S^1. *Journal of Protozology* 30: 247–50.

Byron, C. J., and K. A. Wilson. 2001. Rusty crayfish (*Orconectes rusticus*) movement within and between habitats in Trout Lake, Vilas County, Wisconsin. *Journal of the North American Benthological Society* 20: 606–14.

Cabana, G., A. Tremblay, J. Kalff, and J. B. Rasmussen. 1994. Pelagic food chain structure in Ontario lakes: A determinant of mercury levels in lake trout (*Salvelinus namaycush*). *Canadian Journal of Fisheries and Aquatic Sciences* 51: 381–89.

Cáceres, C. E., and M. S. Schwalbach. 2001. How well do laboratory experiments explain field patterns of zooplankton emergence? *Freshwater Biology* 46: 1179–89.

Cairns, J. Jr., P. V. McCormick, and B. R. Niederlehner. 1993. A proposed framework for developing indicators of ecosystem health. *Hydrobiologia* 263: 1–14.

Camazine, S. 2003. Patterns in nature. *Natural History* 112: 34–41.

Cambor, K. 2000. *In sunlight, in a beautiful garden.* New York: Perennial Press.

Canter, H. M., and J. W. G. Lund. 1951. Studies on plankton parasites. III. Examples of the interaction between parasitism and other factors determining the growth of diatoms. *Annals of Botany* 15: 359–71.

Capriulo, G. M., K. Gold, and A. Okubo. 1982. Evolution of the lorica in tintinnids: A possible selective advantage. *Annales de l'Institut Oceanographique* (Supplement) 58: 319–23.

Cardinale, B. J., and M. A. Palmer. 2002. Disturbance moderates biodiversity-ecosystem function relationships: Experimental evidence from caddisflies in stream mesocosms. *Ecology* 83: 1915–27.

Carpenter, S. R., J. F. Kitchell, and J. R. Hodgson. 1985. Cascading trophic interactions and lake exosystem productivity. *BioScience* 35: 159–72.

Carpenter, S. R., and R. C. Lathrop. 1999. Lake restoration: Capabilities and needs. *Hydrobiologia* 396: 19–28.

Carson, R. 1962. *Silent spring.* Greenwich, CT: Fawcett.

Caswell, H. 1972. On instantaneous and finite birth rates. *Limnology and Oceanography* 17: 787–91.

CavalierSmith, T. 1993. Kingdom Protozoa and its 18 phyla. *Microbiological Reviews* 57: 953–94.

Cech, T. V. 2003. *Principles of water resources: History, development, management, and policy.* New York: Wiley.

Chapman, P. M. 2001. Utility and relevance of aquatic oligochaetes in Ecological Risk Assessment. *Hydrobiologia* 463: 149–69.

Cheer, A. Y. L., and M. A. R. Koehl. 1987. Paddles and rakes: Fluid flow through bristled appendages of small organisms. *Journal of Theoretical Biology* 129: 17–39.

Chen, C. Y., and C. L. Folt. 2000. Bioaccumulation and diminution of arsenic and lead in a freshwater food web. *Environmental Science and Technology* 34: 3878–84.

Chiavelli, D. A., J. W. Marsh, and R. K. Taylor. 2001. The mannose-sensitive hemagglutinin of *Vibrio cholerae* promotes adherence to zooplankton. *Applied and Environmental Microbiology* 67: 3220–25.

Colborn, T. 1985. The use of the stonefly, *Pteronarcys californica* Newport, as a measure of biologically available cadmium in a high-altitude river system, Gunnison Co., Colorado. PhD. thesis. University of Wisconsin, Madison.

Colborn, T., D. Dumanoski, and J. P. Myers. 1996. *Our stolen future.* New York: Dutton Press.

Colborn, T., and M. Smolen. 2002. Cetaceans and contaminants. In *Toxicology of Marine Mammals,* eds. J. G. Voss, G. D. Bossart, M. Fournier, and T. O'Shea. London: Taylor and Francis Books Ltd. UK.

Cole, G. A. 1961. Some calanoid copepods from Arizona with notes on congeneric occurrences of *Diaptomus* species. *Limnology and Oceanography* 6: 432–42.

Cole, G. A. 1994. *Textbook of limnology.* 4th ed. Prospect Heights, IL: Waveland Press.

Cole, L. C. 1954. The population consequences of life history phenomena. Quarterly phenomena. Quarterly Review of Biology 29: 103–137.

Confer, J. L, and J. M. Cooley. 1977. Copepod instar survival and predation by zooplankton. *Journal of the Fisheries Research Board of Canada* 34: 703–706.

Connell, J. H. 1978. Diversity in tropical rainforests and coral reefs. *Science* 199: 1302–10.

Connell, J. H. 1961. The influence of interspecific competition and other factors on the distribution of the barnacle *Chthamalus stellatus. Ecology* 42: 710–723.

Cook, R. E. 1977. Raymond Lindeman and the trophic-dynamic concept in ecology. *Science* 198: 22–26.

Cook, P. M., Robbins, J., Endicott, D. D., Lodge, K., Guiney, P. D., Walker, M. K., Zabel, E. W., and Peterson, R. E. 2003. Effects of Ah receptor-mediated early life stage toxicity on lake trout populations in Lake Ontario during the 20th century. *Environmental Science and Technology. in press.*

Cooper, S. D., L. Barmuta, O. Sarnelle, K. Kratz, and S. Diehl. 1997. Quantifying spatial heterogeneity in streams. *Journal of the North American Benthological Society* 16: 174–88.

Cooper, S. D., S. Diehl, K. Kratz, and O. Sarnelle. 1998. Implications of scale for patterns and processes in stream ecology. *Australian Journal of Ecology* 23: 27–40.

Cooper, S. D., and C. R. Goldman. 1980. Opossum shrimp (*Mysis relicta*) predation on zooplankton. *Canadian Journal of Fisheries and Aquatic Sciences* 37: 909–19.

Cooper, S. D., and D. W. Smith. 1982. Competition, predation, and the relative abundances of two species of *Daphnia. Journal of Plankton Research* 4: 859–79.

Cottingham, K. L. 2002. Tackling biocomplexity: The role of people, tools, and scale. *Bioscience* 52: 793–99.

Cottingham, K. L., S. R. Carpenter, and A. L. St. Amand. 1998. Responses of epilimnetic phytoplankton to experimental nutrient enrichment in three small seepage lakes. *Journal of Plankton Research* 20: 1889–1914.

Covich, A. P., T. A. Crowl, J. E. Alexander, and C. C. Vaughn. 1994. Predator-avoidance responses in freshwater decapod-gastropod interactions mediated by chemical stimuli. *Journal of the North American Benthological Society.* 13: 285–90.

Covich, A. P., M. A. Palmer, and T. A. Crowl. 1999. The role of invertebrate species in freshwater ecosystems: Zoobenthic species influence energy flows and nutrient cycling. *BioScience* 49: 119–27.

Cowell, B. C., C. W. Dye, and R. C. Adams. 1975. A synoptic study of the limnology of Lake Thonotosassa, Florida. Part I. Effects of primary treated sewage and citrus wastes. *Hydrobiologia* 46: 301–45.

Cowgill, U. M., G. E. Hutchinson, A. A. Racek, C. E. Goulden, Ruth Patrick, and M. Tsukada. 1966. The history of Laguna de Petenxil. *Memoirs of the Connecticut Academy of Arts and Sciences* 17: 1–126.

Crease, T. J., and D. J. Taylor. 1998. The origin and evolution of variable-region helices in V4 and V7 of the small-subunit ribosomal RNA of branchiopod crustaceans. *Molecular Biology and Evolution* 15: 1430–46.

Crowley, T. J. 2000. Causes of climate change over the past 1000 years. *Science* 289: 270–77.

Culver, D. A. 1992. Zooplankton grazing and phytoplankton abundance: An assessment before and after invasion of *Dreissena polymorpha. Journal of Great Lakes Research* 17: 425–36.

Culver, D. A., and G. J. Brunskill. 1969. Fayetteville Green Lake, New York. V. Studies of primary production and zooplankton in a meromictic marl lake. *Limnology and Oceanography* 14: 862–73.

Curtis, J. T. 1959. *The vegetation of Wisconsin.* Madison: University of Wisconsin Press.

Dadswell, M. J. 1974. Distribution, ecology, and postglacial dispersal of certain crustaceans and fishes in eastern North America. National Museum of Natural Sciences. Ottawa. *Publications in Zoology* 11: 1–110.

Damkaer, D. M. 2002. *The copepodologist's cabinet: A biographical and bibliographical history.* Philadelphia, PA: American Philosophical Society.

Davies, J. 2001. In a map for human life, count the microbes, too. *Science* 291: 2316.

Davies, R. W., and F. W. Govedich. 2001. In *Ecology and classification of North American freshwater invertebrates.* 2nd ed., eds. J. H. Thorp, and A. P. Covich, 465–504. San Diego: CA: Academic Press.

Decho, A. W., K. A. Browne, and R. K. Zimmer-Faust. 1998. Chemical cues: Why basic peptides are signal molecules in marine environments. *Limnology and Oceanography* 43: 1410–17.

de Fischer-Foster, C., and C. Brunner. 1849. Recherches sur la température du lac de Thoune. *Mém. Soc. Phys Genève* 12: 255–76. (Reviewed at the beginning of Chapter 7 in Hutchinson, 1957.)

Delebecque, A. 1898. *Les Lacs Français.* Paris. Chamerot et Renouard. France. 436 p.

Delsemme, A. H. 2001. An argument for the cometary origin of the biosphere. *American Scientist* 89: 432–42.

DeMeester, L. 1996. Local genetic differentiation and adaptation in freshwater zooplankton populations: Patterns and processes. *Ecoscience* 3: 385–99.

DeMott, W. R., R. D. Guldati, and E. Van Donk. 2001. Effects of dietary phosphorus deficiency on the abundance, phosphorus balance, and growth of *Daphnia cucullata* in three hypereutrophic Dutch lakes. *Limnology and Oceanography* 46: 1871–80.

Dent, B. D. 1999. *Cartography: Thematic map design.* 5th ed. Boston: McGraw-Hill.

de Saussure, H. B. 1779. *Voyages dans les Alpes, precede d'un essai sur l'histoire naturalle des environs de Genève.* Tome 1, Chap. 1, Le Lac de Genève; Chap 2, de la profondeur de la temperature des eaux du lac. Neuchatel, S. Fauche (quarto; octavo reprint 1803).

Doall, M. H., J. R. Strickler, D. M. Fields, and J. Yen. 2002. Mapping the free-swimming attack volume of a planktonic copepod, *Euchaeta rimana. Marine Biology* 140: 871–79.

Dodds, W. K. 1989. Photosynthesis of two morphologies of Nostoc parmelioides (Cyanobacteria) as related to current velocities and diffusion patterns. *Journal of Phycology* 25: 258–62.

Dodson, S. I. 1979. Body size patterns in arctic and temperate zooplankton. *Limnology and Oceanography* 24: 940–49.

Dodson, S. I. 1984. Predation of *Heterocope septentrionalis* on two species of *Daphnia:* Morphological defenses and their cost. *Ecology* 65: 1249–57.

Dodson, S. I. 1987. Animal assemblages in temporary desert rock pools: Aspects of the ecology of *Dasyhelea sublettei* (Diptera: Ceratopogonidae). *Journal of the North American Benthological Society* 6: 65–71.

Dodson, S. I. 1988. The ecological role of chemical stimuli for the zooplankton: Predator-induced morphology in *Daphnia. Oecologia* 78: 361–67.

Dodson, S. I. 1989. Predator-induced reaction norms. *Bioscience* 39: 447–52.

Dodson, S. I. 1990. Predicting diel vertical migration of zooplankton. *Limnology and Oceanography* 35: 1195–2000.

Dodson, S. I. 1992. Predicting crustacean zooplankton species richness. *Limnology and Oceanography* 37: 848–56.

Dodson, S. I., S. A. Arnott, and K. L. Cottingham. 2000. The relationship in lake communities between primary productivity and species richness. *Ecology* 81: 2662–79.

Dodson, S. I., T. A. Crowl, B. L. Peckarsky, L. B. Kats, A. P. Covich, and J. M. Culp. 1994. Nonvisual communication in freshwater benthos. *Journal of the North American Benthological Society* 13: 268–82.

Dodson, S. I., and V. E. Dodson. 1971. The diet of *Ambystoma tigrinum* larvae from western Colorado. *Copeia* 4: 614–624.

Dodson, S. I., and D. L. Egger. 1980. Selective feeding of red phalaropes on zooplankton in Arctic ponds. *Ecology* 61(4): 755–63.

Dodson, S. I., and D. G. Frey. 2001. Cladocera and other Branchiopods. In *Ecology and systematics of North American freshwater invertebrates.* 2nd edition, eds. A. Covich and J. Thorp, 849–913. San Diego, CA: Academic Press.

Dodson, S. I., A. K. Grishanin, K. Gross, and G. A. Wyngaard. 2003. Morphological analysis of cryptic species in the *Acanthocyclops vernalis* species complex. *Hydrobiologia* 500: 131–43.

Dodson, S. I., and R. A. Lillie. 2001. Zooplankton communities of restored depressional wetlands in Wisconsin. *Wetlands* 21: 292–300.

Dodson, S. I., C. M. Merritt, J.-P. Shannahan, and C. M. Schults. 1999. Low doses of Atrazine increase male production in *Daphnia pulicaria. Environmental Toxicology and Chemistry* 18: 1568–73.

Dodson, S. I., and C. Ramcharan. 1991. Size-specific swimming behavior of *Daphnia pulex. Journal of Plankton Research* 13: 1367–79.

Donald, D. B., N. P. Gurprasad, L. Quinnett-Abbott, and K. Cash. 2001. Diffuse geographic distribution of herbicides in northern prairie wetlands. *Environmental Toxicology and Chemistry* 20: 273–79.

Downs, S. G., C. L. Macleod, and J. N. Lester. 1998. Mercury in precipitation and its relation to bioaccumulation in fish: A literature review. *Water Air and Soil Pollution* 108: 149–87.

Drenner, R. W., J. R. Strickler, and W. J. O'Brien. 1978. Capture probability: The role of zooplankter escape in the selective feeding of planktivorous fish. *Journal of the Fisheries Research Board of Canada* 35: 1370–73.

Dumont, H. J., and S. V. Negrea. 2002. Introduction to the Class Branchiopoda. Guides to the identification of the microinvertebrates of the continental waters of the world. No. 19. Leiden, The Netherlands: Backhuys Publishers.

Edmondson, W. T. 1960. Reproductive rates of rotifers in natural populations. *Memoire dell'Instituto Italiano di Idrobiologia* 36: 439–55.

Edmondson, W. T. 1965. Reproductive rate of planktonic rotifers as related to food and temperature in nature. *Ecological Monographs* 35: 61–111.

Edmondson, W. T. 1991. *The uses of ecology: Lake Washington and beyond.* Seattle: University of Washington Press.

Edmondson, W. T. 1993. Experiments and quasi-experiments in limnology. *Bulletin of Marine Science* 53: 65–83.

Edmondson, W. T. 1994. Sixty years of Lake Washington: A curriculum vitae. *Lakes and Reservoir Management* 10: 75–84.

Edmondson, W. T., and A. H. Litt. 1982. *Daphnia* in Lake Washington. *Limnology and Oceanography* 27: 272–93.

Elliott, J. M. 1984. Numerical changes and population regulation in young migratory trout *Salmo trutta* in a Lake District Stream, 1966–1983. *Journal of Animal Ecology* 53: 327–50.

Elser, J. J., T. H. Chrzanowski, R. W. Sterner, and K. H. Mills. 1998. Stoichiometric constraints on food-web dynamics: A whole-lake experiment on the Canadian Shield. *Ecosystems* 1: 120–36.

Elser, J. J., W. R. Fagan, R. F. Denno, D. R. Dobberfuhl, A. Folarin, A. Huberty, S. Interlandi, S. S. Kilham, E. McCauley, K. L. Schulz, E. H. Siemann, and R. W. Sterner. 2000a. Nutritional constraints in terrestrial and freshwater food webs. *Nature* 408 (6812): 578–80.

Elster, H. J. 1974. History of Limnology. *Mitteilungen. Internationale Vereinigung Limnologie* 20: 7–13.

Elton, C. S. 1927. *Animal ecology.* New York: Macmillan.

Fedorov, A. V., and S. G. Philander. 2000. Is El Nino changing? *Science* 288: 1997–2002.

Fee, E. J. 1976. The vertical and seasonal distribution of chlorophyll in lakes of the Experimental Lakes Area, northwestern Ontario: Implications for primary production. *Limnology and Oceanography* 21: 767–83.

Fee, E. J., R. E. Hecky, S. E. M. Kasian, and D. R. Cruikshank. 1996. Effects of lake size, water clarity, and climatic variability on mixing depths in Canadian Shield Lakes. *Limnology and Oceanography* 41: 912–20.

Findlay, D. L., K. Sem, L. L. Hendzel, G. W. Regehr, E. U. Schindler, and J. A. Shearer. 1994. Biomanipulation of Lake-221 in the Experimental Lakes Area (ELA)—Effects on phytoplankton and nutrients. *Canadian Journal of Fisheries and Aquatic Sciences* 51: 2794–2807.

Finlay, B. J. 2002. Global dispersal of free-living microbial eukaryote species. *Science* 296: 1061–63.

Finley, R. W. 1975. *Geography of Wisconsin.* Madison: University of Wisconsin Press.

Fischer, J. M., J. L. Klug, A. R. Ives, and T. M. Frost. 2001. Ecological history affects zooplankton community responses to acidification. *Ecology* 82: 2984–3000.

Fisher, M. M., and E. W. Triplett. 1999. Automated approach for ribosomal intergenic spacer analysis of microbial diversity and its application to freshwater bacterial communities. *Applied and Environmental Microbiology* 65: 4630–36.

Folt, C., and C. W. Burns. 1999. Biological drivers of zooplankton patchiness. *Trends in Ecology and Evolution* 14: 300–305.

Folt, C., P. C. Schulze, and K. Baumgartner. 1993. Characterizing a zooplankton neighborhood: Small-scale patterns of association and abundance. *Freshwater Ecology* 30: 289–300.

Forbes, S. A. 1925. *The lake as a microcosm.* Peoria, IL: Scientific Association.

Forel, F. A. 1892. *Le Léman, monographie limnologique.* Lausanne: Librairie de l'Université. Switzerland.

Foy, R. H., and R. Rosell. 1991. Fractionation of phosphorus and nitrogen loadings from a northern-Ireland fish farm. *Aquaculture* 96: 31–42.

Frey, D. G. 1982. Questions concerning cosmopolitanism in Cladocera. *Archiv für Hydrobiologie* 93: 484–502.

Fryar S. C., K. M. Yuen, K. D. Hyde, and I. J. Hodgkiss. 2001. The influence of competition between tropical fungi in streams. *Microbial Ecology* 41: 245–251.

Fryer, G. 1968. Evolution and adaptive radiation in the Chydoridae (Crustacea: Cladocera): A study in comparative functional morphology and ecology. *Philosophical Transactions of the Royal Society of London B.* 254 (795): 221–385.

Fryer, G. 1987. The feeding mechanisms of the Daphniidae (Crustacea:Cladocera): Recent suggestions and neglected considerations. *Journal of Plankton Research* 9: 419–32.

Funtowicz, S. O., and J. R. Ravetz. 1994. Uncertainty, complexity, and post-normal science. *Environmental Toxicology and Chemistry* 13: 1881–85.

Garstang, W. 1966. *Larval forms, with other zoological verses.* Oxford, UK: Blackwell.

Gergel, S. E., and M. G. Turner. 2002. Learning landscape ecology: A practical guide to concepts and techniques. New York: Springer.

Gerritsen, J., K. G. Porter, and J. R. Strickler. 1988. Not by sieving alone: Suspension feeding in *Daphnia. Bulletin of Marine Science* 43: 366–76.

Gilbert, J. J. 1966. Rotifer ecology and embryological induction. *Science* 151: 1234–37.

Gilbert, J. J. 1980. Developmental polymorphism in the rotifer *Asplanchna sieboldi. American Scientist* 68: 636–46.

Gilbert, J. J., and R. S. Stemberger. 1985. Control of *Keratella* populations by interference competition from *Daphnia. Limnology and Oceanography* 30: 180–88.

Gillooly, J. F., J. H. Brown, G. B. West, V. M. Savage, and E. L. Charnov. 2001. Effects of size and temperature on metabolic rate. *Science* 293: 2248–51.

Gillooly, J. F., and S. I. Dodson.1999. The relationship of neonate mass and incubation temperature to embryonic development time in a range of animal taxa. *Journal of Zoology* 251: 369–75.

Givnish, T. J. 1995. Plant stems: Biomechanical adaptation for energy capture and influence on species distributions.

Plant stems: Physiology and functional morphology, In ed. B. L. Gartner, 3–49. New York: Chapman & Hall.

Gliwicz, Z. M., A. Ghilarov, and J. Pijanowska. 1981. Food and predation as major factors limiting two natural populations of *Daphnia cucullata* Sars. *Hydrobiologia* 80: 205–18.

Gliwicz, Z. M., and M. G. Rowan. 1984. Survival of *Cyclops abyssorum tatricus* (Copepoda, Crustacea) in alpine lakes stocked with planktivorous fish. *Limnology and Oceanography* 29: 1290–99.

Golterman, H. L. (ed.) 1969. Methods for Chemical Analysis of Fresh Waters. IBP Handbook No. 8. Oxford. England. Blackwell Scientific Publications.

Gomez, A., M. Serra, G. R. Carvalho, and D. H. Lunt. 2002. Speciation in ancient cryptic species complexes: Evidence from the molecular phylogeny of *Branchionus plicatilis* (Rotifera). *Evolution* 56: 1431–44.

Gotelli, N. J. 2000. Null model analysis of species co-occurrence patterns. *Ecology* 81: 2606–21.

Gould, S. J. 1990. *Wonderful life: The Burgess Shale and the nature of history.* New York: Norton. Brock/Springer Series in Contemporary Bioscience. Springer-Verlag. Madison, WI: Science Tech. Publishers.

Graf, J. 1999. Symbiosis of *Aeromonas veronii Biovar sobria* and *Hirudo medicinalis,* the medicinal leech: A novel model for digestive tract associations. *Infection and Immunity* 67: 1–7.

Graham, L. E., and L. W. Wilcox. 2000. *Algae.* Upper Saddle River, NJ: Prentice Hall.

Graymore, M., F. Stagnitti, and G. Allinson. 2001. Impacts of atrazine in aquatic ecosystems. *Environment International* 26: 483–95.

Grossnickle, N. E. 1982. Feeding habits of *Mysis relicta:* An overview. *Hydrobiologia* 93: 101–107.

Hairston, N. G., W. Lampert, C. E. Cáceres, C. L. Holtmeir, L. J. Weider, U. Gaedke, J. M. Fischer, J. A. Fox, and D. M. Post. 1999. Lake ecosystems: Rapid evolution revealed by dormant eggs. *Nature* 401: 446–446.

Hairston, N. G., Jr., and E. J. Olds. 1984. Oecologia 61: Population differences in the timing of diapause: Adaptation in a spatially heterogeneous environment. *Oceologia* 61: 42–48.

Halbfass, W. 1903. Über Seespiegelschwankungen in Madüsee. *Zeitschrift für Gewässerkunde* 6: 65.

Hall, D. J. 1964. An experimental approach to the dynamics of a natural population of *Daphnia galeata mendotae.* Ecology 54: 94–112.

Hammer, U. T. 1995. Saline lake limnology—Saskatchewan style. *Blue Jay* 53: 215–26.

Hampton, S. E., and J. J. Gilbert. 2001. Observations of insect predation on rotifers. *Hydrobiologia* 446: 115–21.

Hanazato, T. 2001. Pesticide effects on freshwater zooplankton: An ecological perspective. *Environmental Pollution* 112: 1–10.

Hanski, I., and M. E. Gilpin, eds. 1997. *Metapopulation biology: Ecology, genetics, and evolution.* San Diego, CA: Academic Press.

Hanson, P. C., T. B. Johnson, D. E. Schindler, and J. F. Kitchell. 1997. *Fish Bioenergetics* 3.0. Madison: Wisconsin Sea Grant Institute. WISCU-T-97-001.

Hasler, A. D. 1947. Eutrophication of lakes by domestic drainage. *Ecology* 28: 383–95.

Hasler, A. D., O. M. Brynildson, and W. T. Helm. 1951. Improving conditions for fish in brown-water lakes by alkalization. *Journal of Wildlife Management* 15: 347–52.

Havel, J. E. 1987. Predator-induced defenses: A review. In *Predation: Direct and indirect impacts on aquatic communities,* eds. W. C. Kerfoot and A. Sih, 263–278. Hanover, NH: New England Press.

Havel, J. E., J. K. Colbourne, and P. D. N. Hebert. 2000. Reconstructing the history of intercontinental dispersal in *Daphnia lumholtzi* by use of genetic markers. *Limnology and Oceanography* 45: 1414–19.

Hawksworth, D. L., P. M. Kirk, B. C. Sutton, and D. N. Pegler. 1995. *Ainsworth & Bisby's dictionary of the fungi.* 8th ed. Cambridge, UK: International Mycological Institute. CAB International. University of Cambridge Press.

Hayes, T. B., A. Collins, M. Lee, M. Mendoza, N. Noriega, A. A. Stuart, and A. Vonk. 2002. Hermaphroditic, demasculinized frogs after exposure to the herbicide, atrazine, at low ecologically relevant doses. *Proceedings of the National Academy of Science of the United States of America* 99: 5476–80.

Hayes, T., K. Haston, M. Tsui, A. Hoang, C. Haeffele, and A. Vonk. 2003. Atrazine-induced hermaphroditism at 0.1 ppb in American leopard frogs (*Rana pipiens*): Laboratory and field evidence. *Environmental Health Perspectives* 111: 568–575.

Hedin, L. O., L. Granat, G. E. Likens, T. A. Buishand, J. N. Galloway, T. J. Butler, and H. Rodhe. 1994. Steep declines in atmospheric base cations in regions of Europe and North America. *Nature* 367 (6461): 351–54.

Helfield, J. M., and R. J. Naiman. 2001. Effects of salmon-derived nitrogen on riparian forest growth and implications for stream productivity. *Ecology* 89: 2403–409.

Hemmingsen, A. M. 1960. Energy metabolism as related to body size and respiratory surfaces, and its evolution. *Reports of the Steno Memorial Hospital and Nordinsk Insulinlaboratorium* 9(2): 6–110.

Hensen, V. 1887. Über die Bestimmung des Planktons oder des in Meere treibenden Materials an Pflanzen und Thieren. *Bericht der Kommission zur Unterssuchung der deutschem Meere* 5: 1–109.

Hepburn, J. S., F. M. Jones, and E. Q. St. John. 1927. A bacteriological study of the pitcher plant liquor of the Sarraceniaceae. Transactions. *Wagner Free Institute of Science of Philadelphia* 11: 1–50.

Hershey, A. E., G. M. Gettel, M. E. McDonald, M. C. Miller, H. Mooers, W. J. Obrien, J. Pastor, C. Richards, and J. A. Schuldt. 1999. A geomorphic-trophic model for landscape control of Arctic lake food webs. *BioScience* 49: 887–97.

Hewitt, M., and M. Servos. 2001. An overview of substances present in Canadian aquatic environments associated with endocrine disruption. *Water Quality Research Journal of Canada* 36: 191–213.

Hilsenhoff, W. L. 1987. An improved biotic index of organic stream pollution. *Great Lakes Entomologist* 20: 31–39.

Hobaek A., and P. Larsson. 1990. Sex determination in *Daphnia magna. Ecology* 71: 2255–68.

Hobbie, J. E., R. J. Dayly, and S. Jasper. 1977. Use of nuclepore filters for counting bacteria by fluorescence microscopy. *Applied and Environmental Microbiology* 33: 1225–28.

Holling, C. S. 1973. Resilience and stability of ecological systems. *Annual Review of Ecology and Systematics* 4: 1–23.

Horne, A. J., and C. R. Goldman. 1994. *Limnology.* New York: McGraw-Hill.

Hrbáček, J. 1962. Species composition and the amount of zooplankton in relation to the fish stock. *Rozpr. Cesk. Akad. Ved Rada Mat. Prir. Ved* 72 (10): 1–116.

Hubbell, S. P. 2001. *The unified neutral theory of biodiversity and biogeography. Monographs in population biology,* No. 32. Princeton, NJ: Princeton University Press.

Hunt, E. G., and A. I. Bischoff. 1960. Inimical effects on wildlife of periodic DDD applications to Clear Lake. *California Fish & Game* 46: 91–112.

Hutchinson, G. E. 1944. Nitrogen in the biogeochemistry of the atmosphere. *American Scientist* 32: 178–95.

Hutchinson, G. E. 1952. The biogeochemistry of phosphorus. In *The biology of phosphorus,* ed. L. F. Wolterink, 1–35. East Lansing: Michigan State College Press.

Hutchinson, G. E. 1957. *A Treatise on limnology. Volume I: Geography, physics, and chemistry.* London: John Wiley & Sons.

Hutchinson, G. E. 1961. The paradox of the plankton. *The American Naturalist* 95: 137–46.

Hutchinson, G. E. 1965a. *The lacustrine microcosm reconsidered. The ecological theater and the evolutionary play.* New Haven, CT: Yale University Press.

Hutchinson, G. E. 1965b. *The ecological theater and the evolutionary play.* New York: John Wiley & Sons.

Hutchinson, G. E. 1967. *A Treatise on limnology. Volume II: Introduction to lake biology and the limnoplankton.* New York: John Wiley & Sons.

Hutchinson, G. E. 1975. *A treatise on limnology. Volume III: Limnological botany.* New York: John Wiley & Sons.

Hutchinson, G. E. (editor Y. H. Edmondson) 1993. *A treatise on limnology. Volume IV: The zoobenthos.* New York: John Wiley & Sons.

Hutchinson, G. E. 1978. *An introduction to population ecology.* New Haven, CT: Yale University Press.

Hutchinson, G. E. 1979. *The kindly fruits of the earth.* New Haven, CT: Yale University Press.

Hutchinson, G. E., E. Bonatti, U. M. Cowgill, C. E. Goulden, E. A. Leventhal, M. E. Mallett, F. Margaritora, Ruth Patrick, A. Racek, S. A. Roback, E. Stella, J. B. Ward-Perkins, and T. R. Wellman. 1970. Ianula: An account of the history and development of the Lago di Monterosi, Latium, Italy. *Transactions of the American Philosophical Society.* New Series 60 (4).

IES. 1999. *Lake Wingra watershed.* Institute for Environmental Studies. Water Resources Management Workshop. Madison: University of Wisconsin-Madison.

Ilmaverta, V. 1975. Diel periodicity in the phytoplankton community of the oligotrophic lake Pääjä, southern Finland. *Annales Botanici Fennici* 12: 37–44.

Ingersoll, C. G. 1999. Laboratory toxicity tests for evaluating potential effects of endocrine-disrupting compounds. In *Endocrine disruption in invertebrates: Endocrinology, testing, and assessment,* eds. P. L. DeFur, M. Crane, C. Ingersoll, and L. Tattersfield, 107–98. SETAC Technical Publication. Pensacola, FL: SETAC Press.

Jack, J. D., and J. J. Gilbert. 1994. Effects of *Daphnia* on microzooplankton communities. *Journal of Plankton Research* 16: 1499–1512.

Jackson, C. R., P. F. Churchill, and E. E. Roden. 2001. Successional changes in bacterial assemblage structure during epilithic biofilm development. *Ecology* 82: 555–66.

Jackson, J. B. C. et al. 2001. Historical overfishing and the recent collapse of coastal ecosystems. *Science* 293: 629–38.

Jain, A., S. C. Rai, J. Pal, and E. Sharma. 1999. Hydrology and nutrient dynamics of a sacred lake in Sikkim Himalaya. *Hydrobiologia* 416: 13–22.

Janssen, J., W. R. Jones, A. Wang, and P. E. Oshel. 1995. Use of the lateral-line in particulate feeding in the dark by juvenile alewife (*Alosa pseudoharengus*). *Canadian Journal of Fisheries and Aquatic Sciences* 52: 358–63.

Jassby, A. D., C. R. Goldman, and J. E. Reuter. 1995. Long-term change in Lake Tahoe (California-Nevada, USA) and its relation to atmospheric deposition of algal nutrients. *Archiv für Hydrobiologie* 135: 1–21.

Jenkins, R. O., T. A. Morris, P. J. Craid, A. W. Ritchie, and N. Ostah. 2000. Phosphine generation by mixed-and monoseptic-cultures of anaerobic bacteria. *Science of the Total Environment* 250 (1–3): 73–81.

Jeppesen, E., P. Leavitt, L. De Meester, and J. P. Jensen. 2001. Functional ecology and palaeolimnology: Using cladoceran remains to reconstruct anthropogenic impact. *Trends in Ecology and Evolution* 16: 191–98.

Jeppesen, E., M. Sondergaard, B. Kronvang, J. P. Jensen, L. M. Svendsen and T. L. Lauridsen. 1999. Lake and catchment management in Denmark. *Hydrobiologia* 396: 419–32.

Jobling, M. 1994. *Fish bioenergetics.* London: Chapman and Hall.

Jones, E. B. G., ed. 1976. *Recent advances in aquatic mycology.* London: ELEK Science.

Jones, M. L., J. F. Koonce, and R. Ogorman. 1993. Sustainability of hatchery-dependent salmonine fisheries in Lake Ontario—The conflict between predator demand and prey supply. *Transactions of the American Fisheries Society* 122: 1002–18.

Johnson, N., C. Revenga, and J. Echeverria. 2001. Managing water for people and nature. *Science* 292: 1071.

Johnson, W. E., and A. D. Hasler. 1954. Rainbow trout production in dystrophic lakes. *Journal of Wildlife Management* 18: 113–34.

Jouzel, J., J. R. Petit, R. Souchez, N. I. Barkov, V. Y. Lipenkov, D. Raynaud, M. Stievenard, N. I. Vassiliev, V. Verbeke, and F. Vimeux. 1999. More than 200 meters of lake ice above subglacial lake Vostok, Antarctica. *Science* 286: 2138–41.

Junk, W. J., P. B. Bayly, and R. E. Sparks. 1989. The flood pulse concept in river-floodplain systems. In *Proceedings of the International Large Rivers Symposium,* ed. D. P. Dodge, *Canadian Special Publication of Fisheries and Aquatic Sciences* 106: 110–27.

Jurine, L. 1820. *Histoire des Monocles, qui se trouvent aux environs de Geneve.* Paris: J. J. Paschoud.

Kalff, J. 2002. *Limnology.* Upper Saddle River, NJ: Prentice Hall.

Kalff, J., and H. E. Welch. 1974. Phytoplankton production in Char Lake, a natural polar lake, and Meretta Lake, a polluted polar lake, Cornwallis Island, Northwest Territories. *Journal of the Fisheries Research Board of Canada* 31: 621–36.

Karatayev, A. Y., L. E. Burlakova, and D. K. Padilla. 1998. Physical factors that limit the distribution and abundance of *Dreissena polymorpha* (Pall.). *Journal of Shellfish Research* 17: 1219–35.

Karatayev A. Y., L. E. Burlakova, and D. K. Padilla. 2002. Impacts of zebra mussels on aquatic communities and their role as ecosystem engineers. In *Invasive aquatic species of Europe—Distribution, impacts and management,* eds. E. Leppäkoski, S. Gollasch, and S. Olenin, 433–46. Dordrecht, The Netherlands: Kluwer Academic Publishers.

Karr, J. R., K. D. Fausch, P. L. Angermeier, P. R. Yant, and I. J. Schlosser. 1986, September. Assessing biological integrity in running waters: A method and its rationale. *Illinois Natural History Survey Special Publication* 5.

Kashian, D. R., and S. I. Dodson. 2003. *in press.*

Keller, W., N. D. Yan, K. M. Somers, and J. H. Heneberry. 2002. Crustacean zooplankton communities in lakes recovering from acidification. *Canadian Journal of Fisheries and Aquatic Sciences* 59: 726–35.

Kendrick, B. 2000. *The fifth kingdom.* 3rd ed. Newburyport, MA: Focus Information Group.

Kerfoot, W. C. 1978. Combat between predatory copepods and their prey: *Cyclops, Epischura,* and *Bosmina. Limnology and Oceanography* 23: 1089–1102.

Kerfoot, W. C., D. L.Kellogg, and J. R. Strickler. 1980. Visual observations of live zooplankters: Evasion, escape, and chemical defenses. In *Evolution and ecology of zooplankton communities,* ed. W. C. Kerfoot, 10–27. Hanover, NH: University Press of New England.

Kerfoot, W. C., J. A. Robbins, and L. J. Weider. 1999. A new approach to historical reconstruction: Combining descriptive and experimental paleolimnology. *Limnology and Oceanography* 44: 1232–47.

Kerfoot, W. C., and A. Sih. 1987. *Predation: Direct and indirect impacts on aquatic communities.* Hanover, NH: New England Press.

Kerr, R. A. 2001. Climate change—It's official: Humans are behind most of global warming. *Science* 291: 566–566.

Kiehl, J. and K. Trenberth. 1997. Earth's annual global mean energy budget. *Bulletin. American Meteorological Society.* 78: 197–208.

Kiesecker, J. M., A. R. Blaustein, and C. L. Miller. 2001, August. Transfer of a pathogen from fish to amphibians. *Conservation Biology* 15 (4): 1064–70.

Kirk, J. T. O. 1994. *Light and photosynthesis in aquatic ecosystems.* 2nd ed. Cambridge: Cambridge University Press.

Kitchell, J. F. ed. 1992. *Food web management: A case study of Lake Mendota.* New York: Springer-Verlag.

Kitchell, J. F. 1998. Physiological ecology. In *Ecology.* eds. Dodson et al., 163–98. New York: Oxford University Press.

Kitchell, J. F., and L. B. Crowder. 1986. Predator-prey interactions in Lake Michigan: Model predictions and recent dynamics. *Environmental Biology of Fishes* 16: 208–11.

Kitchell, J. F., R. V. O'Neill, D. Webb, G. W. Gallep, S. M. Bartell, J. F. Koonce, and B. S. Ausmus. 1979. Consumer regulation of nutrient cycling. *BioScience* 29: 28–34.

Kitchell, J. F., D. J. Stewart, and D. Weininger. 1977. Applications of a bioenergetics model to yellow perch (*Perca flavescens*) and walleye (*Stizostedion vitreum vitreum*). *Journal of the Fisheries Research Board of Canada* 34: 1922–35.

Kitchell, J. F., D. E. Schindler, R. Ogutu-Ohwayo, and P. N. Reinthal. 1997. The Nile perch in Lake Victoria: Interactions between predation and fisheries. *Ecological Applications* 7: 653–64.

Kleiven, O. T., P. Larsson, and A. Hobaek. 1992. Sexual reproduction in *Daphnia magna* requires three stimuli. *Oikos* 65: 197–206.

Klesius, M. 2000. Mystery ships from a Danish bog. *National Geographic* 197(5): 28–35.

Klinger, S. A., J. J. Magnuson, and G. W. Gallepp. 1982. Survival mechanisms of the central mudminnow (*Umbra limi*), fathead minnow (*Pimephales promelas*) and brook stickleback (*Culaea inconstans*) for low oxygen in winter. *Environmental Biology of Fish* 7: 113–20.

Klug, J. L., and K. L. Cottingham. 2001. Interactions among environmental drivers: Community responses to changing nutrients and dissolved organic carbon. *Ecology* 82: 3390–3403.

Knutson, M. G., J. R. Sauer, D. A. Olsen, M. J. Mossman, L. M. Hemesath, and M. J. Lannoo. 1999. Effects of landscape composition and wetland fragmentation on frog and toad abundance and species richness in Iowa and Wisconsin, USA. *Conservation Biology* 13: 1437–46.

Koehl, M. A. R. 2001. Transitions in function at low Reynolds number: Hair-bearing animal appendages. *Mathematical Methods in Applied Sciences* 24: 1523–32.

Koehl, M. A. R., and R. J. Strickler. 1981. Copepod feeding currents: Food capture at low Reynolds number. *Limnology and Oceanography* 26: 1062–73.

Kolar, C. S., and D. M. Lodge. 2001. Progress in invasion biology: Predicting invaders. *Trends in Ecology and Evolution* 16: 199–204.

Koller, G. 1958. *Das Leben des Biologen Johannes Müller,* 1801–1858. Grosse Naturforscher; Bd. 23. Stuttgart, Germany: Wissenschaftliche Verlagsgesellschaft.

Korgen, B. J. 1995. Seiches. *American Scientist* 83: 330–41.

Kozhov, M. M. 1963. Lake Baikal and its life. *Monographiae Biologicae* 11. The Hague, The Netherlands: Junk.

Kraft, C. E., P. J. Sullivan, A. Y. Karatayev, Y. E. Burlakova, J. C. Nekola, L. E. Johnson, and D. K. Padilla. 2001. Landscape patterns of an aquatic invader: Assessing dispersal extent from spatial distributions. *Ecological Applications* 12: 749–59.

Krueger, D. A., and S. I. Dodson. 1981. Embryological induction and predation ecology in *Daphnia pulex. Limnology and Oceanography* 26: 219–23.

Kudo, R. R. 1966. *Protozology.* 5th ed. Springfield, IL: Charles C. Thomas.

Kuhn, T. S. 1962. *The structure of scientific revolutions.* Chicago: University of Chicago Press.

Kumar, R., and T. R. Rao. 2001. Effect of the cyclopoid copepod *Mesocyclops thermocyclopoides* on the interactions between the predatory rotifer *Asplanchna intermedia* and its prey *Brachionus calyciflorus* and *B. angularis.* Hydrobiologia 453: 261–68.

Krueger, D. A., and S. I. Dodson. 1981. Embryological induction and predation ecology in *Daphnia pulex. Limnology and Oceanography* 26(2): 212–23.

Kuzmin M. I. et al. 2000. The first discovery of the gas hydrates in the sediments of the Lake Baikal. Gas hydrates: Challenges for the future. *Annals of the New York Academy of Sciences* 912: 112–15.

LaFee, S. 2000. Meet me at the goo. *New Scientist Magazine* 168 (2266): 44–47.

Lagergren, R., J. E. Svensson, and N. Lundqvist. 2002. Clutch size variation and morphology in a cyclomorphic *Bosmina* population. *Journal of Plankton Research* 24: 653–59.

Lagler, K. F., J. E. Bardach, and R. R. Miller. 1962. *Ichthyology.* Illustrated by William L. Brudon. New York: Wiley.

Lampert, W. 1993a. Ultimate causes of diel vertical migration of zooplankton: New evidence for the predator-avoidance hypothesis. *Archiv für Hydrobiologie. Beih. Ergebn. Limnol.* 39: 79–88.

Lampert, W. 1993b. Phenotypic plasticity of the size at 1st reproduction in *Daphnia*—The importance of maternal size. *Ecology* 74: 1455–66.

Lampert, W. 1994. Phenotypic plasticity of the filter screens in *Daphnia:* Adaptation to a low-food environment. *Limnology and Oceanography* 39: 997–1006.

Lampert, W. 1984. The measurement of respiration. In *Secondary production in fresh waters,* eds. J. A. Downing and F. H. Rigler, 413–68. Oxford, UK: Blackwell Press.

Langeron, M. 1945. *Précis de mycologie.* Paris: Masson.

Langmuir, I. 1938. Surface motion of water induced by wind. *Science* 87: 119–23.

Larsson, P., and S. I. Dodson. 1993. Invited review: Chemical communication in planktonic animals. *Archiv Für Hydrobiologie* 129: 129–55.

Lass, S., M. Boersma, K. H. Wiltshire, P. Spaak, and H. Boriss. 2001. Does trimethylamine induce life-history reactions in *Daphnia? Hydrobiologia* 442: 199–206.

Lathrop, F. C. 1988. Evaluation of whole-lake nitrogen fertilization for controlling blue-green algal blooms in a hypereutrophic lake. *Canadian Journal of Fisheries and Aquatic Sciences* 45: 2061–75.

Lathrop, R. C. 1992. Lake Mendota and the Yahara River chain. In *Food web management: A case study of Lake Mendota,* ed. J. F. Kitchell, 17–29. New York: Springer-Verlag.

Lathrop, R. C., S. R. Carpenter, C. A. Stow, P. A. Soranno, and J. C. Panuska. 1998. Phosphorus loading reductions needed to control blue-green algae in Lake Mendota. *Canadian Journal of Fisheries and Aquatic Sciences* 55: 1169–78.

Lathrop, R. C., B. M. Johnson, T. B. Johnson, M. T. Vogelsang, S. R. Carpenter, T. R. Hrabik, J. F. Kitchell, J. J. Magnuson, L. G. Rudstam, and R. S. Stewart. 2002. Stocking piscivores to improve fishing and water clarity: A synthesis of the Lake Mendota biomanipulation project. *Freshwater Biology* 47: 2410–24.

Learner, M. A., G. Lochhead, and B. D. Hughes. 1978. A review of the biology of British Naididae (Oligochaeta) with emphasis on the lotic environment. *Freshwater Biology* 8: 357–75.

Lee, C. E., and B. W. Frost. 2002. Morphological stasis in the *Eurytemora affinis* species complex (Copepoda: Temoridae). *Hydrobiologia* 480: 111–128.

Lee, G. F. and R. A. Jones. 1991. Effects of eutrophication on fisheries. *Reviews in Aquatic Sciences* 5: 287–305.

Lee, K. N. 1993. Compass and gyroscope: Integrating science and politics for the environment. Washington, DC: Island Press.

Leibold, M. A. 1999. Biodiversity and nutrient enrichment in pond plankton communities. *Evolutionary Ecology Research* 1: 73–95.

Leigh E. G., Jr., R. T. Paine, J. F. Quinn, and T. H. Suchanek. 1987. Wave energy and intertidal productivity. *Proceedings of the National Academy of Sciences of the United States of America* 84: 1314–18.

Lenz, P. H., and D. K. Hartline. 1999. Reaction times and force production during escape behavior of a calanoid copepod, *Undinula vulgaris. Marine Biology* 133: 249–58.

Leslie, M. 2001. The bugs that run the world. *Science* 293: 399–399.

Levins, R. 1968. Evolution in changing environments. *Monographs in Population Biology* 2. Princeton, NJ: Princeton University Press.

Levy, D. A. 1990. Sensory mechanism and selective advantage for diel vertical migration in juvenile sockeye salmon, *Oncorhynchus nerka. Canadian Journal of Fisheries and Aquatic Science* 47: 1796–1802.

Lieberman, O. S., M. Shilo, and J. Varijn. 1994. The physiological ecology of a fresh-water dinoflagellate bloom population: Vertical migration, nitrogen limitation, and nutrient uptake kinetics. *Journal of Phycology* 30: 964–71.

Lilljeborg, W. 1982. *Cladocera suecia*. Facsimile re-issue of the original edition with a prologue, eds. W. Rodhe and D. G. Frey. Stockholm, Sweden: Almquist & Wiksell International.

Lindeman, R. L. 1942. The trophic-dynamic aspect of ecology. *Ecology* 23: 399–418.

Lindig-Cisneros, R., and J. B. Zedler. 2000. Restoring urban habitats: a comparative study. *Ecological Restoration* 18: 184–192.

Litt, A. H. 1988. *Floscularia* drawing on cover. *Limnology and Oceanography* 33(6).

Lodge, D. M., K. M. Brown, S. P. Klosiewski, R. A. Stein, A. P. Covich, B. K. Leathers, and C. Bronmark. 1987. Distribution of freshwater snails: Spatial scale and the relative importance of physicochemical and biotic factors. *American Malacological Bulletin* 5: 73–84.

Lohner, L. M., N. G. Hairston, W. R. Schaffner. 1990. A method for distinguishing subitaneous and diapausing eggs in preserved samples of the calanoid copepod genus *Diaptomus. Limnology and Oceanography* 35: 763–66.

Long, R. A., and F. Azam. 2001. Microscale patchiness of bacterioplankton assemblage richness in seawater. *Aquatic Microbial Ecology* 26: 103–13.

Loose, C. J. 1993. Lack of endogenous rhythmicity in *Daphnia* diel vertical migration. Limnology and Oceanography 38: 1837–41.

Ludwig, D. 2001. The era of management is over. *Ecosystems* 4: 758–64.

Ludyanskiy, M. L., D. McDonald, and D. MacNeill. 1993. Impact of the zebra mussel, a bivalve invader: *Dreissena polymorpha* is rapidly colonizing hard surfaces throughout waterways of the United States and Canada. *Bioscience* 43: 533–44.

Luecke, C., C. C. Lunte, R. A. Wright, D. Robertson, and A. S. McLain. 1992a. Impacts of variation in planktivorous fish on abundance of daphnids: A simulation model of the Lake Mendota food web. In *Food web management: A case study of Lake Mendota,* ed. J. F. Kitchell, 407–26. New York: Springer-Verlag.

Luecke, C., and W. J. O'Brien. 1983. Photoprotective pigments in a pond morph of *Daphnia middendorffiana. Arctic* 36: 365.

Luecke, C., L. G. Rudstam, and Y. Allen. 1992b. Interannual patterns of planktivory 1987–1989: An analysis of vertebrate and invertebrate planktivores. In *Food web management,* ed. J. F. Kitchell, 275–301. New York: Springer-Verlag.

Lund, B. M., T. C. Baird-Parker, and G. W. Gould. 2000. *The microbiological safety of food.* Gaithersburg, MD: Aspen Publishers, Inc.

MacArthur, R. H., and E. O. Wilson. 1963. An equilibrium theory of insular zoogeography. *Evolution* 17: 373–87.

MacIntyre, S. 1998. Turbulent mixing and resource supply to phytoplankton. Physical processes in lakes and oceans. *Coastal and Estuarine Studies,* Volume 54: 561–90.

MacIntyre, S. 1999. Boundary mixing and nutrient fluxes in Mono Lake, California. *Limnology and Oceanography* 44: 512–29.

MacIntyre, S., A. L. Alldredge, and C. C. Gotschalk. 1995. Accumulation of marine snow at density discontinuities in the water column. *Limnology and Oceanography* 40: 449–68.

MacIntyre, S., and R. Jellison. 2001. Nutrient fluxes from upwelling and enhanced turbulence at the top of the pycnocline in Mono Lake, California. *Hydrobiologia* 466: 13–29.

MacIsaac, H. J., H. A. M. Ketelaars, I. A. Grigorovich, C. W. Ramcharan, and N. D. Yan. 2000. Modeling *Bythotrephes longimanus* invasions in the Great Lakes basin based on its European distribution. *Archiv für Hydrobiologie* 149: 1–21.

Madigan, M. T., J. M. Martinko, and J. Parker. 2000. *Brock biology of microorganisms.* 9th ed. Upper Saddle River, NJ: Prentice Hall.

Madill, J., K. A. Coates, M. J. Wetzel, and S. R. Gelder. 1992. Common and scientific names of aphanoneuran and clitellate annelids of the United States of America and Canada. *Soil Biology and Biochemistry* 24: 1259–62.

Magnien, R. E., and J. J. Gilbert. 1983. Diel cycles of reproduction and vertical migration in the rotifer *Keratella crassa* and their influence on the estimation of population dynamics. *Limnology and Oceanography* 28: 957–69.

Magnuson, J. J. 1990. Long-term ecological research and the invisible present. *BioScience* 40: 495–501.

Magnuson, J. J., and R. C. Lathrop. 1992. Historical changes in the fish community. In *Food web management: A case study of Lake Mendota,* ed. J. F. Kitchell, 193–233. New York: Springer-Verlag.

Magnuson, J. J., J. W. Keller, A. L. Beckel, and G. W. Gallepp. 1983. Breathing gas-mixtures different from air: An adaptation for survival under the ice of a facultative air-breathing fish. *Science* 220: 312–14.

Magnuson, J. J., D. M. Robertson, B. J. Benson, R. H. Wynne, D. M. Livingstone, T. Arai, R. A. Assel, R. G. Barry, V. Card, E. Kuusisto, N. G. Granin, T. D. Prowse, K. M. Stewart, and V. S. Vuglinski. 2000. Historical trends in lake and river ice cover in the Northern Hemisphere. *Science* 289 (5485): 1743–46.

Maier, G. 1992. *Metacyclops minutus* (Claus, 1863)—Population dynamics and life history characteristics of a rapidly developing copepod. *Int. Revue ges. Hydrobiol.* 77: 455–66.

Malone, B. J., and D. J. McQueen. 1983. Horizontal patchiness in zooplankton populations in two Ontario kettle lakes. *Hydrobiologia* 99: 101–24.

Mangel, M. and C. W. Clark. 1988. *Dynamic modeling in behavioral ecology.* Princeton, NJ: Princeton University Press.

Margalef, R. 1983. *Limnología.* Barcelona, Spain: Ediciones Omega.

Margulis, L. 1971. Symbiosis and evolution. *Scientific American* 225: 48–57.

Marshall, S. M., and A. P. Orr. 1972. *The biology of a marine copepod.* Edinburgh, Scotland: Oliver & Boyd.

Martin, W. W. 1991. Egg parasitism by zoosporic fungi in a littoral chiromid community. *Journal of the North American Benthological Society* 10(4): 455–62.

Masteller E. C., and E. Obert. 2000. Excitement along the shores of Lake Erie—*Hexagenia*—Echoes from the past. *Great Lakes Research Review* 5(1): 25–36.

Mayr, E. 1963. *Animal species and evolution.* Cambridge, MA: Harvard University Press.

Mazumder, A., W. D. Taylor, D. J. McQueen, and D. R. S. Lean. 1990. Effects of fish and plankton on lake temperature and mixing depth. *Science* 247: 312–15.

McCarthy, C. 1912. *The Wisconsin idea.* New York: MacMillan.

McCune, Amy. 1996. Biogeographic and stratigraphic evidence for rapid speciation in semionotid fishes. *Paleobiology* 22: 34–48.

McIntosh, A. R., B. L. Peckarsky, and B. W. Taylor. 2002. The influence of predatory fish on mayfly drift: Extrapolating from experiments to nature. *Freshwater Biology* 47: 1497–1513.

McNaught, A. S., D. W. Schindler, B. R. Parker, A. J. Paul, R. S. Anderson, D. B. Donald, and M. Agbeti. 1999. Restoration of the food web of an alpine lake following fish stocking. *Limnology and Oceanography* 44: 127–36.

McNeil Alexander, R. 1979. *The invertebrates.* London: Cambridge University Press.

Mellors, W. K. 1975. Selective predation of ephippial *Daphnia* and the resistance of ephippial eggs to digestion. *Ecology* 56: 974–80.

Merritt, R. W., and K. W. Cummins. 1986. An introduction to the aquatic insects of North America. 2nd ed. Dubuque, IA: Kendall/Hunt.

Meyer, J. L. 1994. The microbial loop in flowing waters. *Microbial Ecology* 28: 195–99.

Meyer, J. L., M. J. Sale, P. J. Mulholland, and N. L. Poff. 1999. Impacts of climate change on aquatic ecosystem functioning and health. *Journal of the American Water Resources Association* 35: 1373–86.

Meyer, J. L., J. B. Wallace, and S. L. Eggert. 1998. Leaf litter as a source of dissolved organic carbon in streams. *Ecosystems* 1: 240–49.

Mills, E. L., J. H. Leach and J. T. Carlton. 1993. Exotic species in the Great Lakes. *Journal of Great Lakes Research* 19: 1–54.

Mitchell, M. K., and W. B. Stapp. 1997. *Field manual for water quality monitoring: An environmental education program for schools.* Dubuque, IA: Kendall/Hunt Publishing.

Mitchell, S. 2000. *Tao Te Ching: A new English version.* New York: HarperCollins. NY. USA. 113 pp.

Mittelbach, G. G., C. F. Steiner, S. M. Scheiner, K. L. Gross, H. L. Reynolds, R. B. Waide, M. R. Willigh, S. I. Dodson, and L. Gough. 2001. What is the observed relationship between species richness and productivity? *Ecology* 82: 2381–96.

Monastersky, R. 1995. Iron surprise: Algae absorb carbon dioxide. *Science News* 148: 53.

Morel, A., and D. Antoine. 2002. Small critters—Big effects. *Science* 296: 1980–82.

Moss, B. 1980. *Ecology of fresh waters.* London: Blackwell Scientific Publications.

Moss, B., R. Wetzel, and G. H. Lauf. 1980. Annual productivity and phytoplankton changes between 1969 and 1974 in Gull Lake, Michigan. *Freshwater Biology* 10: 113–21.

Naiman, R. J., C. A. Johnson, and J. Kelley. 1988. Alteration of North American streams by beaver. *Bioscience* 38: 753–61.

Naumann, E. 1917. Beiträge zurKenntnis des Teichmammoplanktons. II. Über das Neuston des Süsswassers. *Biologisches Zentralblatt* 37: 98–106.

Needham, J. G., and G. T. Lloyd. 1916. *The life of inland waters: An elementary text book of freshwater biology for American students.* Ithaca, NY: Comstock.

Newbold, J. D., R. V. O'Neill, J. W. Elwood, and W. Van Winkle. 1982. Nutrient spiralling in streams: Implications for nutrient limitation and invertebrate activity. *American Naturalist* 120: 628–52.

Nilssen, J. P. 1980. When and how to reproduce: A dilemma for limnetic cyclopoid copepods. In *Evolution and ecology of zooplankton communities,* ed. W. C. Kerfoot, 418–26. Hanover, NH: University Press of New England.

Nishino, M., and N. C. Watanabe. 2000. Evolution and endemism in Lake Biwa, with special reference to its gastropod fauna. *Advances in Ecological Research* 31: 151–80.

Norstrom, R., D. Hallett, and R. Sonstegard. 1978. Coho salmon (*Oncorhynchus kisutch*) and herring gulls (*Larus arentatus*) as indicators of organochlorine contamination in Lake Ontario. *Journal of the Fisheries Research Board of Canada* 35: 1401–1409.

O'Brien, W. J. 1979. The predator-prey interaction of planktivorous fish and zooplankton. *American Scientist* 67: 572–81.

O'Brien, W. J., H. I. Browman, and B. I. Evans. 1990. Search strategies of foraging animals. *American Scientist* 78: 152–60.

O'Keefe, T. C., and R. T. Edwards. 2002. The influence of sockeye salmon and alder on nutrient chemistry of streams in Southwest Alaska. *Ecosystems in review.*

Odum, E. P. 1953. *Fundamentals of ecology.* Philadelphia: Saunders.

Odum, E. P. 1971. *Fundamentals of ecology.* 3rd ed. Philadelphia: Saunders.

Odum, H. T. 1956. Primary production in flowing waters. *Limnology and Oceanography* 1: 102–17.

Oliver, M. 1995. *Blue Pastures.* New York: Harcourt Brace.

Olson, R. J., and C. H. Boggs. 1986. Apex predation by yellowfin tuna (*Thunnus albacares*): Independent estimates from gastric evacuation and stomach contents, bioenergetics, and cesium concentrations. *Canadian Journal of Fisheries and Aquatic Sciences* 43: 1760–75.

Osenberg, C. W., E. E. Werner, G. G. Mittelbach, and D. J. Hall. 1988. Growth patterns in bluegill (*Lepomis macrochirus*) and pumpkinseed (*Lepomis gibbosus*) sunfish: Environmental variation and the importance of ontogenetic niche shifts. *Canadian Journal of Fisheries and Aquatic Sciences* 45: 17–26.

Overbeck, J., and R. J. Chróst. 1994. *Microbial ecology of Lake Plusssee.* Volume 105 of *Ecological Studies.* New York: Springer-Verlag.

Pace, M. L., and D. Vaque. 1994. The importance of *Daphnia* in determining mortality rates of protozoans and rotifers in lakes. *Limnology and Oceanography* 39: 984–96.

Pace, M. L., and J. D. Orcutt Jr. 1981. The relative importance of protozoans, rotifers, and crustaceans in a freshwater zooplankton community. *Limnology and Oceanography* 26: 822–30.

Pace, N. R. 1997. A molecular view of microbial diversity and the biosphere. *Science* 276: 734–40.

Paine, R. T. 1969. A note on trophic complexity and community structure. *The American Naturalist* 103: 91–93.

Parejko, K., and S. Dodson. 1990. Progress towards characterization of a predator-prey kairomone: *Daphnia pulex* and *Chaoborus americanus. Hydrobiologia* 198: 51–59.

Pastor, J., R. J. Naiman, B. Dewey, and P. McInnes. 1988. Moose, microbes, and the boreal forest. *BioScience* 38: 770–77.

Paul, M. J., and J. L. Meyer. 1996. Fungal biomass of 3 leaf litter species during decay in an Appalachian stream. *Journal of the North American Benthological Society* 15: 421–32.

Pauly, D., and V. Christensen. 1995. Primary production required to sustain global fisheries. *Nature* 374: 255–57.

Peckarsky, B. L., P. R. Fraissinet, and M. A. Penton. 1990. *Freshwater macroinvertebrates of northeastern North America.* Ithaca, NY: Comstock.

Peckarsky, B. L., A. R. McIntosh, B. W. Taylor, and J. Dahl. 2002. Predator chemicals induce changes in mayfly life history traits: A whole-stream manipulation. *Ecology* 83: 612–18.

Peckarsky, B. L., and M. A. Penton. 1988. Why do *Ephemerella* nymphs scorpion posture—a ghost of predation past. *Oikos* 53: 185–93.

Pennak, R. W. 1957. Species composition of limnetic zooplankton communities. *Limnology and Oceanography* 2: 222.

Pennak, R. W. 1978a. Anomalous primary production in some Colorado alpine lakes, USA. Proceedings. *International Association of Theoretical and Applied Limnology* 20: 434–37.

Pennak, R. W. 1978b. *Freshwater invertebrates of the United States.* 2nd ed. New York: Wiley-Interscience.

Pennak, R. W. 1989. *Freshwater invertebrates of the United States: Protozoa to mollusca.* 3rd ed. New York: Wiley.

Peters, R. H. 1983. *The ecological implications of body size.* Cambridge, UK: Cambridge University Press.

Peterson, B. J., W. M. Wollheim, P. J. Mulholland, J. R. Webster, J. L. Meyer, Jennifer L. Tank, Eugènia Martí, W. B. Bowden, H. M. Valett, A. E. Hershey, W. H. McDowell, W. K. Dodds, S. K. Hamilton, S. Gregory, and D. D. Morrall. 2001. Control of nitrogen export from watersheds by headwater streams. *Science* 292: 86–90.

Peters-Regehr, T., J. Jusch, and K. Heckmann. 1997. Primary structure and origin of a predator released protein that induces defensive morphological changes in *Euplotes. European Journal of Protistology* 33: 389–95.

Petroski, H. 2003. St. Francis dam. *American Scientist* 91: 114–118.

Phillips, W. 1884. The breaking of the Shropshire Meres. *Shropshire Archaelogical Transactions* 7: 277–300.

Pick, F. R., D. R. S. Lean, and C. Nalewajko. 1984. Nutrient status of metalimnetic phytoplankton peaks. *Limnology and Oceanography* 29: 960–71.

Pickett, S. T. A., and P. S. White. eds. 1985. *The ecology of natural disturbance and patch dynamics.* Orlando, FL: Academic Press.

Porter, K. G. 1977. The plant-animal interface in freshwater ecosystems. *American Scientist* 65: 159–70.

Porter, K. G. 1988. Phagotrophic phytoflagellates in microbial food webs. *Hydrobiologia* 159: 89–97.

Porter, K. G., and Y. S. Feig. 1980. The use of DAPI for identifying and counting aquatic microflora. *Limnology and Oceanography* 25: 943–48.

Porter, K. G., J. Gerritsen, and J. D. Orcott, Jr. 1982. The effect of food concentration on swimming patterns, feeding behavior, ingestion, assimilation, and respiration by *Daphnia. Limnology and Oceanography* 27: 935–49.

Post, G. 1983. *Textbook of fish health.* Neptune City, NJ: TFH Publications, Inc.

Powers, C. F., and A. Robertson. 1966. The aging Great Lakes. *Scientific American* 215: 94–104.

Prepas, E. E. 1983. Total dissolved solids as predictor of lake biomass and productivity. *Canadian Journal of Fisheries and Aquatic Sciences* 40: 92–95.

Purcell, E. M. 1977. Life at low Reynolds number. *American Journal of Physics* 45: 3–11.

Raffetto, N. S., J. R. Baylis, and S. L. Searns. 1990. Complete estimates of reproductive success in a closed population of smallmouth bass. *Ecology* 71: 1523–35.

Ragotzkie, R. A., and R. A. Bryson. 1953. Correlation of currents with the distribution of adult *Daphnia* in Lake Mendota. *Journal of Marine Research* 12: 157–72.

Rand, G. M., and S. R. Petrocelli. 1985. *Fundamentals of aquatic toxicology.* New York: Taylor and Francis.

Ranta, E., S. Hällfors, V. Nuutinen, G. Hällfors, and Kai Kivi. 1987. A field manipulation of trophic interactions in rock-pool plankton. *Oikos* 50: 336–46.

Ravera, O. 1953. Gli stadi di sviluppo dei copepodi pelagici del Lago Maggiore. *Mem. Ist. Ital. Idrobiol.* 7: 129–51.

Redfield, A. C. 1934. On the proportions of organic derivatives in sea water and their relation to the composition of plankton. In *James Johnstone Memorial Volume,* pp. 176–192. Liverpool: University of Liverpool Press.

Redfield, A. C. 1958. The biological control of chemical factors in the environment. *American Scientist* 46: 205–21.

Redfield, A. C., B. H. Ketchum, and F. A. Richards. 1963. The influence of organisms on the composition of seawater. In *Comparative and descriptive oceanography,* ed. M. N. Hill, 26–77. New York: Wiley.

Reed-Andersen, T., S. R. Carpenter, D. K. Padilla, and R. C. Lathrop. 2000. Predicted impact of zebra mussel (*Dreissena polymorpha*) invasion on water clarity in Lake Mendota. *Canadian Journal of Fisheries and Aquatic Sciences* 57: 1617–26.

Reid, G. K. 1987. Pond Life: a guide to common plants and animals of North American ponds and lakes. New York: Golden Press.

Reid J. 2001. A human challenge: Discovering and understanding continental copepod habitats. *Hydrobiologia* 453: 201–26.

Reynolds, C. S. 1989. Physical determinants of seasonal change in the species composition of phytoplankton. In *Succession in phytoplankton,* ed. U. Sommer, 9–56. Berlin, Germany: Springer.

Richman, S. E., and S. I. Dodson. 1983. The effect of food quality on feeding and respiration by *Daphnia* and *Diaptomus. Limnology and Oceanography* 28(5): 948–56.

Riessen, H. P. 1999. Predator-induced life history shifts in *Daphnia:* A synthesis of studies using meta-analysis. *Canadian Journal of Fisheries and Aquatic Sciences* 56: 2487–94.

Robertson, D. M., G. L. Goddard, D. R. Helsel, and K. L. MacKinnon. 2000. Rehabilitation of Delevan Lake, Wisconsin. *Lake and Reservoir Management* 16: 155–76.

Rohlf, F. J., and R. R. Sokal. 1981. *Statistical tables.* 2nd ed. New York: W. H. Freeman Press.

Ruttner, F. 1940. *Grundriβ der Limnologie.* Berlin, Germany: W. de Gruvter & Co.

Ruttner, F. 1963. *Fundamentals of limnology.* 3rd ed. Translated by D. G. Frey and F. E. J. Fry. Toronto, University of Toronto Press.

Ryder, R. A., S. R. Kerr, K. H. Loftus, and H. A. Regier. 1973. The Morphoedaphic Index, a fish yield estimator—Review and evaluation. *Journal of the Fisheries Research Board of Canada* 31: 663–88.

Sakai, A. K., F. W. Allendorf, J. S. Holt, D. M. Lodge, J. Molofsky, K. A. With, S. Baughman, R. J. Cabin, J. E. Cohen, N. C. Ellestrand, D. E. O'Neil, I. M Parker, J. N. Thompson, and S. G. Weller. 2001. The population biology of invasive species. *Annual Review of Ecology and Systematics* 32: 305–32.

Sanderson, B. L, T. R. Hrabik, J. J. Magnuson, and D. M. Post. 1999. Cyclic dynamics of a yellow perch (*Perca flavescens*) population in an oligotrophic lake: Evidence for the role of intraspecific interactions. *Canadian Journal of Fisheries and Aquatic Sciences* 56: 1534–42.

Sanderson, J. G. 2000. Testing ecological patterns. *American Scientist* 88: 332–39.

Santer, B., E. Blohm-Sievers, C. E. Caceres, and N. G. Hairston. 2000. Life-history variation in the coexisting freshwater copepods *Eudiaptomus gracilis* and *Eudiaptomus graciloides I. Archiv fur Hydrobiologie* 149: 441–58.

Santos Flores, C. 2001. Taxonomy and distribution of the freshwater microcrustaceans and green algae of Puerto Rico, three contributions to American cladocerology, and a bibliography on West Indian limnology. Ph.D. Dissertation. Department of Zoology, University of Wisconsin–Madison.

Santos-Medrano, G. E., R. Rico-Martinez, and C. A. Velazquez. 2001. Swimming speed and Reynolds numbers of eleven freshwater rotifer species. *Hydrobiologia* 446: 35–38.

Schaper, S. 1999. Evaluation of Costa Rican copepods (Crustacea: Eudecapoda) for larval *Aedes aegypti* control with special reference to *Mesocyclops thermocyclopoides. Journal of the American Mosquito Control Association* 15: 510–19.

Scheffer, M. 1998. *Ecology of shallow lakes.* London: Chapman & Hall.

Scheffer, M., S. Carpenter, J. A. Foley, C. Folke, and B. Walker. 2001. Catastrophic shifts in ecosystems. *Nature* 413: 591–96.

Schindler, D. E., and M. D. Scheurell. 2002. Habitat coupling in lake ecosystems. *Oikos* 98: 177–89.

Schindler, D. E., M. D. Scheuerell, J. W. Moore, S. M. Gende, T. B. Francis, and W. J. Palen. 2003. Pacific salmon and the ecology of coastal ecosystems. *Frontiers in Ecology and the Environment* 1: 31–37.

Schindler, D. W. 1974. Eutrophication and recovery in Experimental Lakes: Implications for lake management. *Science* 184: 897–99. (Fig. 1, p. 897)

Schindler, D. W. 1977. Evolution of phosphorus limitation in lakes: Natural mechanisms compensate for deficiencies of nitrogen and carbon in eutrophied lakes. *Science* 195: 260–62.

Schindler, D. W., S. E. Bayly, B. R. Parker, K. G. Beaty, D. R. Cruikshank, E. J. Fee, E. U. Schindler, and M. P. Stainton. 1996. The effects of climatic warming on the properties of boreal lakes and streams at the Experimental Lakes Area, northwestern Ontario. *Limnology and Oceanography* 41: 1004–17.

Schlesinger, W. H. 1997. Biogeochemistry: An analysis of global change. 2nd ed. San Diego, CA: Academic Press.

Schmidt-Nielsen, K. 1984. *Scaling: Why is animal size so important?* Cambridge, UK: Cambridge University Press.

Schneider, D. W. 2000. Local knowledge, environmental politics, and the founding of ecology in the United States. *Isis* 91: 681–705.

Schultz, P. H. and R. E. Lianza. 1992. Recent grazing impacts on the earth recorded in the Rio-Cuarto crater field, Argentina. *Nature* 355: 234–37.

Schulz, K. L., and R. W. Sterner. 1999. Phytoplankton phosphorus limitation and food quality for *Bosmina. Limnology and Oceanography* 44: 1549–56.

Serena, M., M. Worley, M. Swinnerton, and G. A. Williams. 2001. Effect of food availability and habitat on the distribution of platypus (*Ornithorhynchus anatinus*) foraging activity. *Australian Journal of Zoology* 49: 263–77.

Shannon, C. E., and W. Weaver. 1949. The mathematical theory of communication. Chicago: University of Illinois Press.

Shapiro, J., V. Lamarra, and M. Lynch. 1975. Biomanipulation: An ecosystem approach to lake restoration. In *Water quality management through biological control,* eds. P. L. Brezonik and J. L. Fox, 13–27. Gainesville: University of Florida Press.

Sheldon, S. P., and R. P. Creed. 1995. Use of a native insect as a biological-control for an introduced weed. *Ecological Applications* 5: 1122–32.

Shurin, J. B. 2000. Dispersal limitation, invasion resistance, and the structure of pond zooplankton communities. *Ecology* 81: 3077–86.

Shurin, J. B., J. E. Havel, M. A. Leibold, and B. Pinel-Alloul. 2000. Local and regional zooplankton species richness: A scale-independent test for saturation. *Ecology* 81 (11): 3062–73.

Siebeck, O. 1978. Ultraviolet tolerance of planktonic crustaceans. *Verh Internat Verein Limnol* 20: 2469–73.

Siegel, S. 1956. *Nonparametric statistics for the behavioral sciences.* New York: McGraw-Hill.

Siegert, M. J. 2000. Antarctic subglacial lakes. *Earth-Science Reviews* 50: 29–50.

Sigel, A., and H. Sigel. eds. 1997. Mercury and its effects on environment and biology. Volume 24 of Series: *Metal ions in biological systems.* New York: Marcel Dekker.

Sigurdsson, H. 1987. A dead chief's revenge? *Natural History* 96(8): 44–49.

Smayda, T. J., and C. S. Reynolds. 2001. Community assembly in marine phytoplankton: Application of recent models to harmful dinoflagellate blooms. *Journal of Plankton Research* 23: 447–61.

Smith, V. H. 1983. Low nitrogen to phosphorus ratios favor dominance by blue-green algae in lake phytoplankton. *Science* 221: 669–71.

Smith, V. H., G. D. Tilman, and J. C. Nekola. 1999. Eutrophication: Impacts of excess nutrient inputs on freshwater, marine, and terrestrial ecosystems. *Environmental Pollution* 100: 179–96.

Snell, T. W. 1998. Chemical ecology of rotifers. *Hydrobiologia* 388: 267–76.

Snoeyink , V. L., and D. Jenkins. 1980. *Water chemistry.* New York: Wiley.

Sommer, U., Z. Maciej Gliwicz, W. Lampert, and A. Duncan. 1986. The PEG-model of seasonal succession of planktonic events in fresh waters. *Archiv für Hydrobiologie* 106: 433–71.

Sommer, U., F. Somer, B. Santer, C. Jamieson, M. Boersma, C. Becker, and T. Hansen. 2001. Complementary impact of copepods and cladocerans on phytoplankton. *Ecology Letters* 4: 545–50.

Spaak, P., J. Vanoverbeke, and M. Boersma. 2000. Predator-induced life-history changes and the coexistence of five taxa in a *Daphnia* species complex. *Oikos* 89: 164–74.

Sparrow, F. K. 1959. Fungi. In *Freshwater biology.* 2nd ed. H. B. Ward and G. C. Whipple, 47–94. (W. T. Edmondson, ed.). New York: John Wiley & Sons.

Spencer, C. N., D. S. Potter, R. T. Bukantis, and J. A. Stanford. 1999. Impact of predation by *Mysis relicta* on zooplankton in Flathead Lake, Montana, USA. *Journal of Plankton Research* 21: 51–64.

Sprules, W. G., S. B. Brandt, D. J. Stewart, M. Munawar, E. H. Jin, and J. Love. 1991. Biomass size spectrum of the Lake Michigan pelagic food web. *Canadian Journal of Fisheries and Aquatic Sciences* 48: 105–15.

Stanley, E. H., and M. W. Doyle. 2002. A geomorphic perspective on nutrient retention following dam removal. *BioScience* 52: 693–701.

Stanley, E. H., and M. W. Doyle. 2003. Trading off: The ecological effects of dam removal. *Frontiers in Ecology and the Environment* 1: 15–22.

Stearns, S. C. 1992. *The evolution of life histories.* New York: Oxford University Press.

Stemberger, R. S., and J. J. Gilbert. 1987. Defenses of planktonic rotifers against predators. In *Predation: Direct and indirect impacts on aquatic communities,* ed. W. C. Kerfoot and A. Sih, 2227–229. Hanover, NH: New England Press.

Stemberger, R. S., D. P. Larsen, and T. M. Kincaid. 2001. Sensitivity of zooplankton for regional lake monitoring. *Canadian Journal of Fisheries and Aquatic Sciences* 58: 2222–32.

Sterner, R. W., T. H. Chrznowski, J. J. Elser, and N. B. Gegore. 1995. Sources of nitrogen and phosphorus supporting the growth of bacterioplankton and phytoplankton in an oligotrophic Canadian shield lake. *Limnology and Oceanography* 40: 242–49.

Sterner, R. W., and J. J. Elser. 2002. Ecological stoichiometry: The biology of elements from molecules to the biosphere. Princeton, NJ: Princeton University Press.

Sterner, R. W., J. J. Elser, E. J. Fee, S. J. Guilford, and T. H. Chrzanowski. 1997. The light: nutrient ratio in lakes: The balance of energy and materials affects ecosystem structure and process. *American Naturalist* 150: 663–84.

Stibor, H. 2002. The role of yolk protein dynamics and predator kairomones for the life history of *Daphnia magna. Ecology* 83: 362–69.

Stich, H. B., and W. Lampert. 1984. Growth and reproduction of migrating and nonmigrating *Daphnia* species under simulated food and temperature conditions of diurnal vertical migration. *Oecologia* 61: 192–96.

Stockner, J. G., and E. A. Macisaac. 1996. British Columbia lake enrichment programme: Two decades of habitat enhancement for sockeye salmon. *Regulated Rivers—Research & Management* 12: 547–61.

Stockner J. G., and K. G. Porter. 1988. Microbial food webs in freshwater planktonic ecosystems. In *Complex interactions in lake communities,* ed. S. R. Carpenter, 68–83. New York: Springer.

Stow, C. A., S. R. Carpenter, L. A. Eby, J. F. Amrhein, and R. J. Hesselberg. 1995. Evidence that PCBs are approaching stable concentrations in Lake Michigan fishes. *Ecological Applications* 5: 248–60.

Strahler, A. N. 1964. Quantitative geomorphology of drainage basins and channel networks, section 4-2 in Ven te Chow (ed.) In *Handbook of applied hydrology,* ed. Ven te Chow, section 4-2. New York: McGraw-Hill.

Strickler, J. R. 1984. Sticky water: A selective force in copepod evolution. In *Trophic interactions within aquatic ecosystems,* ed. D. G. Meyers and J. R. Strickler, 187–239. AAA Selected Symposium 85. Boulder, CO: Westview Press.

Strickler, J. R., and A. K. Bal. 1973. Setae of the first antennae of the copepod *Cyclops scutifer* (Sars): Their structure and importance. *Proceedings of the National Academy of Science USA* 70: 2656–59.

Strickler, J. R. and S. Twombly. 1975. Reynolds number, diapause, and predatory copepods. Verh. International Association of Theoretical and Applied Limnology. Verhandlungen. 19: 2943–50.

Strong, A. E., and B. J. Edie. 1978. Satellite observations of calcium carbonate precipitations in the Great Lakes. *Limnology and Oceanography* 23: 877–87.

Summerhayes, V. S. and C. S. Elton. 1923. Contributions to the ecology of Spitsbergen and Bear Island. Journal of Ecology 11: 214–86.

Suschenya, L. M., M. S. Dolbik, V. I. Parfenau, et al. 1981. *The red book of Belarussian SSR: Endangered and protected species of animals and plants.* Minsk, Former Soviet Union: Belarussian Soviet Encyclopedia.

Swammerdam, J. 1669. *Historia insectorum generalis.* Utrecht, Netherlands: Van Drevnen.

Tansley, A. G. 1935. The use and abuse of vegetational concepts and terms. *Ecology* 16: 284–307.

Telschow, A., P. Hammerstein, and J. H. Werren. 2002. The effect of *Wolbachia* on genetic divergence between populations: Models with two-way migration. *American Naturalist* 160: S54–S66 Supplement.

Tessier, A. J., E. V. Bizina, and C. K. Geedey. 2001. Grazer-resource interactions in the plankton: Are all daphniids alike? *Limnology and Oceanography* 46: 1585–95.

Tessier, A. J., and P. Woodruff. 2002. Cryptic trophic cascade along a gradient of lake size. *Ecology* 83: 1263–70.

Thompson, D. W. 1917/1968. *On growth and form* (two volumes). Cambridge, UK: Cambridge University Press.

Thorp, J. H., A. R. Black, K. H. Haag, and J. D. Wehr. 1994. Zooplankton assemblages in the Ohio River: Seasonal, tributary, and navigational dam effects. *Canadian Journal of Fisheries and Aquatic Sciences* 51: 1634–43.

Thorp, J. H., and A. P. Covich. eds. 2001. *Ecology and classification of North American freshwater invertebrates.* 2nd ed. San Diego, CA: Academic Press.

Thienemann, A. 1905. *Biologie der Trichopteren-Puppe.* Jena, Switzerland: G. Fischer.

Thienemann, A. 1925. *Die Binnengewässer: Einzeldarstellungen aus der Limnologie und ihren Nachbargebieten.* Stuttgart, Germany: E. Schweizerbart. This is a series of publications established by Thienemann, covering many aspects of limnology from 1925 to 1992.

Threlkeld, S. T. 1986. Differential temperature sensitivity of two cladoceran species to resource variation during a blue-green algal bloom. *Canadian Journal of Zoology* 64: 1738–44.

Tilman, D. 1977. Resource competition between planktonic algae: An experimental and theoretical approach. *Ecology* 58: 338–48.

Tollrian, R., and C. D. Harvell. 1999. *The ecology and evolution of inducible defenses.* Princeton, NJ: Princeton University Press.

Trautmann, N. M. 2001. *Assessing toxic risk National Science Teachers Association.* Student Edition. Arlington, VA: Cornell Scientific Inquiry Series.

Tuljapurkar, S., and H. Caswell. eds. 1997. *Structured-population models in marine, terrestrial, and freshwater systems.* New York: Chapman & Hall.

Turner, M. G., R. H. Gardner, and R. V. O'Neill. 2001. *Landscape ecology in theory and practice: pattern and process.* New York: Springer.

Tyler, T., W. J. Liss, L. M. Ganio, G. L. Larson, R. Hoffman, E. Deimling, and G. Lomnicky. 1998. Interaction between introduced trout and larval salamanders (*Ambystoma macrodactylum*) in high-elevation lakes. *Conservation Biology* 12: 94–105.

Ullrich, S. M., T. W. Tanton, and S. A. Abdrashitova. 2001. Mercury in the aquatic environment: A review of factors affecting methylation. *Critical Reviews in Environmental Science and Technology* 31: 241–93.

USDA, Natural Resources Conservation Service. 1996. *National handbook of water quality monitoring.* 450-vi-NHWQM. (Available online.)

Usinger, R. L. 1956. *Aquatic insects of California.* Berkeley: University of California Press.

Vadeboncoeur, Y., M. J. Vander Zanden, and D. M. Lodge. 2002. Putting the lake back together: Reintegrating benthic pathways into lake food web models. *Bioscience* 52: 44–54.

Vadeboncoeur, Y., E. Jeppensen, M. J. Vander Zanden, H.-H. Schierup, K. Christoffersen, and D. M. Lodge. 2003. From Greenland to green lakes: Cultural eutrophication and the loss of benthic pathways in lakes. *Limnology and Oceanography in press.*

Vallentyne, J. R. 1969. Definition of a limnologist. *Limnology and Oceanography* 14: 815.

Vallentyne, J. R. 1970, May–June. Phosphorus and the control of eutrophication. *Canadian Research and Development:* 39–43.

Vallentyne, J. R. 1974. The algal bowl: Lakes and men. *Fisheries Research Board of Canada.* Misc. Spec. Pub. 22. Environment Canada. Ottawa. Canada.

Van Buskirk, J., and D. C. Smith. 1993, April. Between the devil and the deep blue lake. *Natural History:* 39–41.

Vandebund, W. J., C. Davids, and S. J. H. Spaas. 1995. Seasonal dynamics and spatial-distribution of chydorid cladocerans in relation to chironomid larvae in the sandy littoral zone of an oligo-mesotrophic lake. *Hydrobiologia* 299: 125–138.

Vander Zanden, M. J., and J. B. Rassumssen. 1996. A tropic position model of pelagic food webs: Impact on contaminant bioaccumulation in lake trout. *Ecological Monographs* 66: 451–77.

Vander Zanden, M. J., J. M. Casselman, and J. B. Rasmussen. 1999. Stable Isotope evidence for the food web consequences of species invasions in lakes. *Nature* 401: 464–67.

Vanni, M. J., and D. L. Findlay. 1990. Trophic cascades and phytoplankton community structure. *Ecology* 71: 921–37.

Vanni, M. J., C. D. Layne, and S. E. Arnott. 1997. Top-down trophic interactions in lakes: Effects of fish on nutrient dynamics. *Ecology* 78: 1–20.

Vannote, R. L., G. W. Minshall, K. W. Cummings, J. R. Sedell, and C. E. Cushing. 1980. The river continuum concept. *Canadian Journal of Fisheries and Aquatic Sciences* 37: 130–37.

Vinebrooke, R. D., S. S. Dixit, M. D. Graham, J. M. Gunn, Y. W. Chen, and N. Belzile. 2002. Whole-lake algal responses to a century of acidic industrial deposition on the Canadian Shield. *Canadian Journal of Fisheries and Aquatic Sciences* 59: 483–93.

Vitousek, P. M., H. A. Mooney, J. Lubchenco, and J. M. Melillo. 1997. Human domination of Earth's ecosystems. *Science* 277: 494–99.

Vogel, S. 1983. *Life in moving fluids: The physical biology of flow.* Princeton, NJ: Princeton University Press.

Vogel, S. 1988. *Life's devices: The physical world of animals and plants.* Princeton, NJ: Princeton University Press.

Vollenweider, R. A. 1976. Advances in defining critical loading levels for phosphorus in lake eutrophication. *Memoires Ist. ital. Idrobiol.* 33: 53–84.

Walka, A. W. 1993. *Waterlines: Journeys of a desert river.* Flagstaff, AZ: Red Lake Books.

Wallace, R. L., and T. W. Snell. 1991. Rotifera. In *Ecology and systematics of North American freshwater invertebrates,* 2nd edition, eds. A. Covich and J. Thorp, 195–254. San Diego, CA: Academic Press.

Walters, C., J. Korman, L. E. Stevens, and B. Gold. 2000. Ecosystem modeling for evaluation of adaptive management policies in the Grand Canyon. *Conservation Ecology* 4 (2): 1st article. Available only online at *www.consecol.org/Journal/* published electronically by the Resilience Alliance.

Waneczek, H. 1930. Untersungen über einige Arten der Gattung *Asplanchna* Gosse (*A. girodi* de Guerne, *A. brightwellii* Gosse, *A. priodonta* Gosse).Annales Musei Zoologici Polonici 8: 109–322.

Ward, H. B., and G. C. Whipple. 1966. *Fresh-water biology.* 2nd ed., ed. W. T. Edmondson. New York: John Wiley & Sons.

Ward, P. D., D. R. Montgomery, and R. Smith, 2000. Altered river morphology in South Africa related to the Permian-Triassic extinction. *Science* 289: 1740–43.

Webb, R. M. T., and L. R. Soler-López. 1997. Sedimentation history of Lago Loíza, Puerto Rico, 1953–1994. San Juan, Puerto Rico: U.S. Geological Survey. Water-Resources Investigations Report 97-4108.

Webster, K. E., T. K. Kratz, C. J. Bowser, J. J. Magnuson, and W. J. Rose. 1996. The influence of landscape position on lake chemical responses to drought in northern Wisconsin. *Limnology and Oceanography* 41: 9710–984.

Weider, L. J., A. Hobaek, P. D. N. Hebert, and T. J. Crease. 1999. Holoarctic phylogeography of an asexual species comples—II. Allozymic variation and clonal structure in Arctic *Daphnia. Molecular Ecology* 8: 1–13.

Welch, E. B., and G. D. Schrieve. 1994. Alum treatment effectiveness and longevity in shallow lakes. *Hydrobiologia* 275–276: 423–31.

Welch, P. S. 1935. *Limnology.* New York: McGraw-Hill.

Wetzel, R. G. 1973. Productivity investigations of interconnected lakes. I. The eight lakes of the Oliver and Walters chains, northeastern Indiana. *Hydrobiological Studies* 3: 91–143.

Wetzel, R. G. 2001. *Limnology: Lake and river ecosystems.* 3rd ed. San Diego, CA: Academic Press.

Whittaker, R. H. 1975. *Communities and ecosystems.* 2nd ed. New York: Macmillan.

Wiegmann, D. D., J. R. Baylis, and M. H. Hoff. 1997. Male fitness, body size and timing of reproduction in smallmouth bass, *Micropterus dolomieuil. Ecology* 78: 111–28.

Wilkens, L. A., B. Wettring, E. Wagner, W. Wljtenek, and D. Russell. 2001 Prey detection in selective plankton feeding by the paddlefish: Is the electric sense sufficient? *Journal of Experimental Biology* 204: 1381–89.

Williamson, C. E., and J. W. Reid. 2001. Copepoda. In *Ecology and systematics of North American freshwater invertebrates,* 2nd ed. ed. A. Covich and J. Thorp, 915–54. San Diego, CA: Academic Press.

Williamson, C. E., and N. M. Butler. 1986. Predation on rotifers by the suspension-feeding calanoid copepod *Diaptomus pallidus. Limnology and Oceanography* 31: 393.

Williamson, C. E., O. G. Olson, S. E. Lott, N. D. Walker, D. R. Engstrom, and B. R. Hargreaves. 2001. Ultraviolet radiation and zooplankton community structure following deglaciation in Glacier Bay, Alaska. *Ecology* 82: 1748–60.

Williamson, C. E., R. S. Stemberger, D. P. Morris, and T. M. Frost. 1996. Ultraviolet radiation in North American lakes: Attenuation estimates from DOC measurements and implications for plankton communities. *Limnology and Oceanography* 41: 1024–34.

Wiltshire, K. H., and W. Lampert. 1999. Urea excretion by *Daphnia:* A colony-inducing factor in *Scenedesmus? Limnology and Oceanography* 44: 1894–1903.

Winberg, G. G. 1966. Growth and metabolic rates in animals. *Uspekhi Sovremennoy Biologii* 6: 107–126.

Wingstrand, K. G. 1978. Comparative spermatology of the crustacea Entomostraca. 1. Subclass Branchiopoda. *Det Kongelige Danske Videnskabernes Selskab Biologiske Skrifter* 22: 1–67 + 20 plates.

Winkler, L. W. 1889. Die Löslichkeit des Sauerstoffs im Wasser. *Ber. dtsch. Chem. Ges.* 22: 1764–74.

Woltereck, R. 1909. Weitere experimentelle Untersuchungen über Artveränderung, speziell über das Wesen quantitativer Artunterschiede bei Daphniden. *Ver. Deutsch. Gesell.* 1909: 110–72.

Wright, R. T., and J. E. Hobbie. 1966. The use of glucose and acetate by bacteria and algae in aquatic ecosystems. *Ecology* 47: 447–64.

Wurtsbaugh, W. A., and T. S. Berry. 1990. Cascading effects of decreased salinity on the plankton, chemistry, and physics of the Great Salt Lake (Utah). *Canadian Journal of Fisheries and Aquatic Sciences* 47: 100–109.

Wyngaard, G. A. 2000. The contributions of Ulrich K. Einsle to the taxonomy of the copepoda. *Hydrobiologia* 417: 1–10.

Wyngaard, G. A., B. E. Taylor, and D. L. Mahoney. 1991. Emergence and dynamics of cyclopoid copepods in an unpredictable environment. *Freshwater Biology* 25: 219–32.

Yan, N. D., and T. W. Pawson. 1997. Changes in the crustacean zooplankton community of Harp Lake, Canada, following invasion by *Bythotrephes cederstroemi. Freshwater Biology* 37: 409–25.

Yu, N., and D. A. Culver. 1999. Estimating the effective clearance rate and refiltration by zebra mussels, *Dreissena polymorpha,* in a stratified reservoir. *Freshwater Biology* 41: 481–92.

Zaret, T. M. 1980. *Predation and freshwater communities.* New Haven, CT: Yale University Press.

Zaret, T. M., and J. S. Suffern. 1976. Vertical migration in zooplankton as a predator avoidance mechanism. *Limnology and Oceanography* 21: 807–13.

Zedler, J. B. 2000. Progress in wetland restoration ecology. *Trends in Ecology and Evolution* 15: 402–407.

credits

Chapter 1

Chapter Opener 1: © PhotoDisc/Vol. 16

Chapter 2

Chapter Opener 2: © PhotoDisc/Vol. 16

Chapter 3

Chapter Opener 3: © PhotoDisc/Vol. 10; **page 70:** "I would like to point out that we [humans] depend on . . .", From J. Davies, 2001, *Science* 291: 2316. Used with permission of the author.

Chapter 4

Chapter Opener 4: © The McGraw-Hill Companies/Carlyn Iverson/photographer

Chapter 5

Chapter Opener 5: © PhotoDisc/Vol. 9

Chapter 6

Chapter Opener 6: © Steven P. Lynch

Chapter 7

Chapter Opener 7: © PhotoDisc/Vol. 19

Chapter 8

Chapter Opener 8: © The McGraw-Hill Companies/Carlyn Iverson/photographer

Chapter 9

Chapter Opener 9: © Michael S. Yamashita/Corbis.

Chapter 10

Chapter Opener 10: © PhotoDisc/Vol. 60; **page 258:** Stephen Mitchell: "The supreme good is like water . . .", From S. Mitchell, 1988, *Tao Te Ching: A New English Version,* Chapter 8. Used with permission from HarperCollins, New York, and Macmillan, London.

Chapter 11

Chapter Opener 11: © PhotoDisc/Vol. 9; **page 277:** Stephen Mitchell: "The great Tao flows everywhere . . .", From S. Mitchell, 1988, *Tao Te Ching: A New English Version,* Chapter 8. Used with permission from HarperCollins, New York, and Macmillan, London.

Chapter 12

Chapter Opener 12: © Richard Hamilton Smith/Corbis; **page 300:** J.R. Vallentyne: "A limnologist is a zoologist who . . ." And E.S. Deevey: "Referee's comment to Vallentyne's definition: . . .", From J. R. Vallentyne, "Definition of a Limnologist," *Limnology & Oceanography,* 1969, 14(5): 815. Used with permission.; **page 301:** The River, From Ann Weiler Walka, "The River," *Waterlines: Journey on a Desert River,* Red Lake Books, Flagstaff, AZ. Used with kind permission of the author.

Chapter 13

Chapter Opener 13: © PhotoDisc/Vol. 13

index

B

C

D

E

F

G

H

N

O

P

Q

R

S

T

U

V

W

X

Y

Z